Classical and Quantum Thermal Physics

Thermal physics deals with interactions of heat energy and matter. It can be divided into three parts: the kinetic theory, classical thermodynamics, and quantum thermodynamics or quantum statistics.

This book begins by explaining fundamental concepts of kinetic theory of gases, viscosity, conductivity, diffusion and laws of thermodynamics. It then goes on to discuss applications of thermodynamics to problems of physics and engineering. These applications are explained with the help of P-V and P-*s-h* diagrams. A separate section/ chapter on the application of thermodynamics to the operation of engines and to chemical reactions, makes the book especially useful to students from engineering and chemistry streams. An introductory chapter on the thermodynamics of irreversible processes and network thermodynamics provides readers a glimpse into this evolving subject.

Simple language, stepwise derivations, large number of solved and unsolved problems with their answers, graded questions with short and long answers, multiple choice questions with answers, and a summary of each chapter at its end, make this book a valuable asset for students.

R. Prasad was professor at the Physics Department, Aligarh Muslim University, Aligarh, India. For over 43 years he taught courses on nuclear physics, thermal physics, electronics, quantum mechanics and modern physics. He also worked at the Institute for Experimental Physics, Hamburg, Germany; at the Swiss Institute of Nuclear Research, Switzerland; at Atom Institute, Wien (Vienna), Austria; at Abduls Salam International Centre for Theoretical Physics, Italy; and at the Variable Energy Cyclotron Centre (VECC), Calcutta, India. His area of specialization is experimental nuclear physics.

Classical and Quantum
Thermal Physics

R. Prasad

CAMBRIDGE
UNIVERSITY PRESS

CAMBRIDGE
UNIVERSITY PRESS

University Printing House, Cambridge CB2 8BS, United Kingdom

One Liberty Plaza, 20th Floor, New York, NY 10006, USA

477 Williamstown Road, Port Melbourne, vic 3207, Australia

4843/24, 2nd Floor, Ansari Road, Daryaganj, Delhi – 110002, India

79 Anson Road, #06–04/06, Singapore 079906

Cambridge University Press is part of the University of Cambridge.

It furthers the University's mission by disseminating knowledge in the pursuit of education, learning and research at the highest international levels of excellence.

www.cambridge.org
Information on this title: www.cambridge.org/9781107172883

First published 2016

Printed in India by International Print-o-Pac Ltd., Noida, U.P.

A catalogue record for this publication is available from the British Library

Library of Congress Cataloging-in-Publication Data

Names: Prasad, R. (Emeritus Professor of Physics), author.
Title: Classical and quantum thermal physics / R. Prasad.
Description: Daryaganj, Delhi, India : Cambridge University Press, [2016] |
  Includes bibliographical references and index.
Identifiers: LCCN 2016030572| ISBN 9781107172883 (hardback ; alk. paper) |
  ISBN 1107172888 (hardback ; alk. paper)
Subjects: LCSH: Thermodynamics. | Quantum theory. | Kinetic theory of gases.
Classification: LCC QC311 .P78 2016 | DDC 536/.7--dc23 LC record available at
https://lccn.loc.gov/2016030572

ISBN 978-1-107-17288-3 Hardback

Dedicated to my parents

Late Smt. Mithlesh Mathur
&
Late Shri Ishwari Prasad Mathur

Contents

Figures

Tables

Preface

The book is designed to serve as a textbook on thermal physics / thermodynamic that may be prescribed to graduate students of physics, chemistry and engineering branches. The book covers all three components of thermal physics: namely the kinetic theory, classical thermodynamics and quantum thermodynamics (quantum statistical mechanics plus thermodynamics), with their applications. Some topics in the book may also be of interest to post graduate students. Since the focus of the book is on 85–90% average and below average students of the class, it is written in simple English with detailed and stepwise derivations starting from first principles. I hope that teachers of the subject and also readers other than the targeted audience, will also like the presentation.

Kinetic theory and transport properties of gases are covered in first two chapters of the book. Maxwell–Boltzmann velocity distribution for an ideal gas, which is mostly derived using the tools of quantum thermodynamics, is obtained in chapter-1 by the method originally used by Maxwell. The four laws of classical thermodynamics and their applications are discussed in chapters 3–8.

A special feature of the book is a separate chapter on the application of classical thermodynamics to chemical reactions (chapter-9), which is generally not covered in books on the subject. Though there are books on the application of thermodynamics to chemical reactions, unfortunately these books do not explain the underlying principles of physics associated with thermodynamics and are, therefore, incomplete. Since chemists use different notations and signs for thermodynamic parameters than those used by physicists and engineers, a separate chapter is included where these differences are clearly mentioned along with their reasons. Each application is explained through an example of an appropriate chemical reaction where technical terms are explained and mathematical derivations are worked out starting from the first principle.

Similarly, engineering applications of classical thermodynamics are discussed in a separate section. These applications are explained with the help of P–V and P–s–h diagrams wherever necessary and are followed by large number of solved and unsolved problems with answers.

Classical thermodynamics is an empirical science based on the behavior of macroscopic systems. On the other hand, quantum thermodynamics is a microscopic theory that uses laws of quantum statistic and the tools of thermodynamics to describe the behavior of systems made up of a large number of identical particles. Essentials of quantum thermodynamics are developed in chapter-10 and their applications to various physical systems are detailed in chapter-11. How quantum thermodynamic treatment of systems overcome the shortcomings found in their classical treatment, has also been elaborated in this chapter.

Formulations of both classical and quantum thermodynamics are applicable only to systems in equilibrium and to processes that are reversible and/or quasi-static. However, real systems are neither in equilibrium nor are processes taking place in the universe reversible. Hence, it is necessary to develop concepts that may be applied to non-equilibrium systems and to irreversible processes, i.e. thermodynamics applicable to real systems. Efforts in this direction have been made and thermodynamics of irreversible processes based on network theorems has been developed recently. Elements of thermodynamics of linear irreversible processes and of more general network thermodynamics are introduced in chapter-12 of the book.

Another distinctive feature of the book is the inclusion of a large number of worked out examples in each chapter. Further, there are sufficient number of unsolved problems with answers, questions with short and long answers and objective questions with multiple choices. Chapter contents are also followed with a summary for revision by students. It is hoped that these features will help students in preparing for examinations, viva and interviews.

Though considerable efforts have been made to remove all errors, I know it is not possible to achieve it, particularly for a project of this size. I, therefore, request readers to kindly point out the errors they find, so that the same may be corrected. I appreciate receipt of healthy and positive criticism that may further improve the presentation.

Acknowledgments

I owe this book to my students and colleagues who encouraged me to write. Few years after my superannuation I came to know that the class notes of my lectures are still being used by students. I am fortunate to have excellent students who not only appreciated my teaching but also persuaded me to write. In this context I would like to mention one Abbas Raza Alvi, who was in my class some thirty years back, and met me just by chance in Sydney, a couple of years back. Abbas, a multi-dimensional personality: engineer, poet, music composer, story writer etc., who had settled in Australia two decades back, surprised me by showing old class notes of my lectures when I visited him in Sydney. Dedicated students like him provided the necessary impetus required to complete such a huge task. I, therefore, thank all my wonderful colleagues and students including Abbas for their significant but not so visible contributions to this project.

I wish to put on record my sincere thanks to all members of my research group who helped in one way or the other in completing this work. As a matter of fact my strength lies in them. I will specially mention the name of Professor B. P. Singh who helped me at each step and in reading the manuscript, pointing out omissions and suggesting alterations. Thank you very much Professor Singh.

I spent the best part of my life at the Aligarh Muslim University: as a student, lecturer, reader, professor, the Chairman of the Department of Physics, and the Dean, Faculty of Science. It is here that I acquired whatever knowledge I have. I sincerely thank the Aligarh Muslim University for providing me with all the support during my stay.

This book was written in three parts at three different places; first four chapters were completed at Sydney, Australia; next six chapters at Boston, USA; and the remaining part was completed at Aligarh, India. I thank my wife Sushma and my daughters-in-law, Pooja and Chaitra, for being excellent hosts and providing congenial atmosphere and nice food, both of which I feel are essential for any creative work.

Acknowledgements will remain incomplete without a mention of Gauravjeet Singh Reen from Cambridge University Press. This highly sophisticated, polite and prompt young man helped me a lot. A big thank you! Gaurav.

I dedicate this book to my parents—my mother, Late Smt. Mithlesh Mathur and my father, Late Shri Ishwari Prasad Mathur. They encouraged me to undertake higher learning and acquire competence.

The Kinetic Theory of Gases

1.0 Kinetic Theory, Classical and Quantum Thermodynamics

Two important components of the universe are: the matter and the energy. Interplay between them creates a variety of processes and phenomenon. In order to understand and appreciate the vast spectrum of happenings around us, it is required to know more intimately the properties of the different forms of matter and their interactions with energy. This may be approached in two different ways. In the first approach, often called the microscopic approach, some assumptions about the nature of the matter present in the universe is made and then the well-known and well-established laws of interaction are applied between the assumed entities of the matter to explain the observed natural phenomenon. The kinetic theory of matter and the statistical mechanics (or quantum statics) are the examples of the microscopic approach. In kinetic theory of matter it is assumed that matter is made of elements, which in turn are made of molecules that are in motion. Molecules of an element are all alike, while molecules of different elements are different. Molecules are themselves assumed to be made of atoms. Having made assumptions about the constitution of the matter, the kinetic theory applies the laws of Newtonian mechanics, like the law of conservation of energy, law of conservation of liner momentum, the law that states that the rate of change of momentum is equal to force, etc. to the molecules and obtain expressions for average properties of the system, like the pressure exerted by a gas, etc.

In the case of quantum statics or statistical mechanics, it is assumed that matter is made of different kinds of identical particles or entities; the number of each type of entity in a given piece of matter is very large and, therefore, the entities follow the laws of quantum statistics. In quantum statistics , the entities can have different discrete energies and their energy distributions are given by distribution laws. The average properties of a given piece of matter may be obtained by the application of the relevant quantum distribution law.

In the second approach, called the macroscopic approach, some very general laws obeyed by macroscopic systems are derived by observing their behavior over a sufficiently long period of time. These laws, that are not specific to any system, are then used to drive the average properties of specific systems. This macroscopic approach is historically called the thermodynamics. In thermodynamics, no assumption about the microscopic constitution of the matter is made, and, therefore, it is immaterial whether the matter is made of molecules or not. This approach has evolved over a considerable period of time, out of the experiments carried out in entirely different contexts. The implications of the laws obeyed by large macroscopic systems were realized much later. There had been controversies about calling this branch of science as thermodynamics, or thermostatic, or equilibrium thermodynamics, etc. which will be discussed in details later. It may, however, be remarked that thermodynamics is a branch of science based entirely on experiments and hence empirical in nature.

1.1   Kinetic Theory of Gases

Out of the three prominent states of matter, solid, liquid and gas, the kinetic theory of gases is perhaps the most developed and complete. Further, under suitable boundary conditions, the kinetic theory of gases may be applied to the solid and the liquid states as well.

1.1.1   What do we expect from a good theory of gases?

Any good theory of gases must be able to explain satisfactorily all experimentally observed facts about gases, like the Boyle's law, Charles' law, Gay-Lussac's law, Graham's law of effusion, Daltons law of partial pressure, magnitudes of specific thermal capacities and their ratios for different gases, their temperature dependence, thermal conductivity and viscosities of gases, etc. The simple classical theory, called the kinetic theory of gases, detailed below, brings out the main properties of gases remarkably well. It, however, fails to explain some finer points as regards to the temperature dependence of some gas properties. It is also not expected that such a simple theory will explain all details of complex gas systems. Laws of quantum statistics, also called statistical mechanics applied to gaseous systems, explain most of the properties that remained unexplained by the kinetic theory.

The first step toward the development of the kinetic theory of gases is to define an ideal gas. An ideal gas is a hypothetical or imaginary gas, properties of which are defined through the following assumptions of the kinetic theory of gases.

1.1.2   Assumptions or postulates of the kinetic theory

1. Gases are made up of molecules. In any given measurable or macroscopic volume there is large number of molecules of the gas.
2. The molecules of a gas are in state of continuous motion and the relative separation between the molecules is much larger than their own dimensions.
3. Molecules exert no force on one another except when they collide. The molecules, therefore, travel in straight lines between collisions with each other or with the wall of the container.
4. Molecules undergo elastic collisions with each other and with the walls of the container. If it is further assumed that the walls of the container are perfectly smooth then during the collisions with the walls the tangential component of the velocity of the molecule will remain unaltered.
5. The molecules are uniformly distributed over the volume of the container.
6. The molecular speeds, that is the magnitudes of the molecular velocities, have values from zero to infinity.
7. The directions of the molecular velocities are uniformly distributed in space.

1.1.3   Justifications and implications of assumptions

According to Avogadro's law, one mole of every gas occupies 22.4 liter of volume at standard temperature and pressure (STP) and contains 6.03×10^{23} molecules of the gas. Since 1 liter of volume

$= 1 \times 10^{-3}$ m^3, there are $\approx 10^{16}$ molecules per cubic millimeter of the volume of a container at (STP) which is a very large number. This justifies the first assumption.

Assumption that the size of molecules is much smaller than the distance of their separation essentially means that large space around each molecule is empty or there is vacuum around each molecule. Molecules in this force-free space move in straight lines till they encounter a collision either with another molecule or with the wall of the container. Assuming that molecules are uniformly distributed, the volume of space that may be associated with each molecule of the gas at STP $\approx \dfrac{22.4 \times 10^{-3} \text{ m}^3}{6.03 \times 10^{23}} \approx 3 \times 10^{-26}$ m^3. Now the size or the diameter of an average molecule is of the order of 10^{-10} m. Taking molecule to be a sphere of radius 0.5×10^{-10} m, the actual volume of the molecule comes out to be $\dfrac{4}{3} \pi \left(0.5 \times 10^{-10}\right)^3$ m $\approx 5 \times 10^{-31}$ m^3. It may now be observed that out of the volume of space of 3×10^{-26} m^3 that each molecule has around it, the actual volume of the molecule is only 5×10^{-31} m^3. It means that space of volume $\dfrac{3 \times 10^{-26}}{5 \times 10^{-31}} = 6 \times 10^4$ times the actual volume of the molecule is available to each molecule as free space or vacuum where there is no other molecule. This justifies the second assumption.

Assumption 3, that molecules exert no force on each other, is not true in case of a real gas and is a major point of difference between a real gas and the hypothetical ideal gas. It may be treated as the first approximation that may be dropped later. Assumptions that collisions are elastic and the container walls are smooth are required to carry out the calculations of momentum transfer in molecular collisions with the walls.

Suppose the total number of gas molecules in the container of volume V is N. The number density of molecules 'n', which is the number of molecules per unit volume, is then $n = N/V$. Therefore, in any volume element ΔV the number of molecules, $N_{\Delta V} = n \Delta V$. Now it may be argued that the assumption of uniform distribution of molecules may break down if one selects the volume element ΔV to be so small that there is no molecule in this volume. It may, however, be realized that elements of infinitesimal small size are taken to carry out operations of differentiation and integration. The term 'infinitesimal' is a relative term. For example, if there is a gas container of 1cubic meter volume, a volume of 1/1000 cubic millimeter is infinitesimally small in comparison to the size of the container. The number of gas molecules at STP in a volume of 1/1000 cubic millimeter is still of the order of 10^6. It may, thus, be seen that the assumption of uniform distribution of gas molecules is valid and does not break down even for the relatively infinitesimal small volumes.

Assumption that molecular speeds vary from zero to infinity is quite justified as even if one starts from the assumption that initially all molecules have same speed, the intra molecular collisions will soon produce a spectrum of speeds. The upper limit of the speed spectrum may remain a question, since no material particle can attain speeds greater than the speed of light. It will, however, be shown that the actual value of the upper limit whether it is infinity or the speed of light does not matter because the speed distribution curve falls almost exponentially for very high values of speeds. Another assumption that the directions of molecular velocities are uniformly distributed in space ensures their randomness, and as many molecules go away from a particular location in the container, on an average same number of molecules reaches there from some other part of the container keeping the average number density of molecules constant in the container. Figure 1.1 (a) shows the velocity vectors of some gas molecules in the container. Actually there should be N such vectors, one for each of the N gas molecules. These velocity vectors are transported parallel to themselves such that they originate from the common point O, in Fig. 1.1 (b). A sphere of radius r is then drawn taking O as the center. The velocity vectors, either themselves or on extension (shown by dotted lines in Fig. 1.1 (b)), cut the surface of the sphere at points P_1, P_2, P_3, ... P_N. Each of these points of intersection indicates a different direction of the motion of the gas molecule. Now in case the directions of motion of gas molecules are uniformly distributed in space then the points of intersection (P's) should also be uniformly distributed on the surface of the sphere of radius r. The total number of points of intersection is N, one for each molecule and the total surface area of the sphere of radius r is $4\pi r^2$. Therefore, for the uniform distribution of the directions of molecular velocities in space, the number of intersection points per unit surface area should be $\dfrac{N}{4\pi r^2}$. On the other hand if we consider a

surface area, ΔS, then the number of intersection points on this area will be $\dfrac{N}{4\pi r^2}\Delta S$.

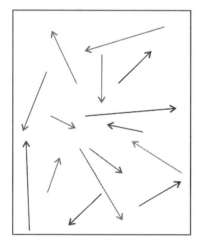

(a) Vectors indicating the velocities of
some gas molecules in the container

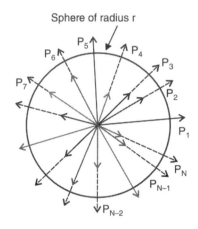

(b) Molecular velocity vectors
transported to the centre O
of a sphere of radius r

Fig. 1.1   (a) shows the velocity vectors for some molecules of the gas in a container. In Fig. 1.1 (b) these velocity vectors are transported parallel to themselves to the origin O, which is the center of a sphere of radius r. The velocity vectors cut the surface of the sphere either themselves or on extension (dotted lines) at points P_1, P_2, P_3,......P_N , each of which gives the direction of motion of a gas molecule. The directions of motion of gas molecules will be uniformly distributed in space if the points of intersection are uniformly distributed on the surface of the sphere.

It is now clear that the number of intersection points on a given area on the spherical surface is equal to the number of gas molecules moving in the direction of that surface area. An element of surface (i.e., a small surface area) on the spherical surface may be conveniently defined in spherical polar coordinates, with the origin of the coordinate system at the center of the sphere as is shown in Fig. 1.2

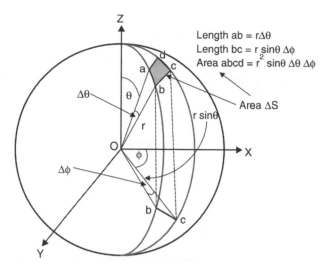

Fig. 1.2   Definition of an element of area in spherical polar coordinates

As is clear from Fig. 1.2, the small element of surface area ΔS is given by the expression,

$$\Delta S = r^2 \sin\theta \, \Delta\theta \, \Delta\varphi \qquad\qquad 1.1$$

The number $N^{(\theta,\varphi)}$ of molecules moving in the direction θ and $(\theta + \Delta\theta)$ and φ and $(\varphi + \Delta\varphi)$ and hitting the shaded area ΔS normally, is equal to the points of intersection on surface, ΔS, and is given by,

$$N^{(\theta,\varphi)} = \frac{N}{4\pi r^2} \Delta S = \frac{N}{4\pi r^2} r^2 \sin\theta \, \Delta\theta \, \Delta\varphi = \frac{N}{4\pi} \sin\theta \, \Delta\theta \, \Delta\varphi \qquad 1.2\ (a)$$

We divide both sides of this equation by the volume V of the container to get,

$$n^{(\theta,\varphi)} = \frac{n}{4\pi} \sin\theta \, \Delta\theta \, \Delta\varphi \qquad\qquad 1.2\ (b)$$

Here n, and $n^{(\theta,\varphi)}$ are, respectively, the number density of gas molecules (N/V) and the number density ($N^{(\theta,\varphi)}/V$) of molecules moving in direction θ & $(\theta + \Delta\theta)$; φ & $(\varphi + \Delta\varphi)$.

The velocity vectors of these $N^{(\theta,\varphi)}$ molecules are in the direction θ within a small spread $\Delta\theta$, and φ within a small spread $\Delta\varphi$, but these molecules may have all possible speeds from zero to infinity. We now select out of $N^{(\theta,\varphi)}$ only those molecules that are moving in the direction θ & $(\theta + \Delta\theta)$ and φ & $(\varphi + \Delta\varphi)$ and have their velocities between v and $(v + \Delta v)$ and denote them by $N^{(\theta,\varphi,v)}$. The velocity vectors of these $N^{(\theta,\varphi,v)}$ molecules will all be confined within two concentric spheres of radii v and $(v + \Delta v)$ as shown in Fig.1.3. The number density of (θ, φ, v) molecules is given by,

$$n^{(\theta,\varphi,v)} = \left(\frac{\Delta n^v}{4\pi} \sin\theta \, \Delta\theta \, \Delta\varphi \right) \qquad\qquad 1.2\ (c)$$

where, Δn^v is the number density of molecules with velocities in the range v to $(v + \Delta v)$.

Since we will consider the molecules that are moving in the direction θ and $(\theta + \Delta\theta)$ and φ and $(\varphi + \Delta\varphi)$, we simply refer them as $n^{(\theta,\varphi)}$, the molecules moving in direction (θ, φ) and will understand that there are spreads of $\Delta\theta$ and $\Delta\varphi$ in the directions without specifying it repetitively. It follows from Eq. 1.2(c) that;

$$n^{(\theta, \varphi)} = \frac{n}{4\pi} \sin\theta\, \Delta\theta\, \Delta\phi \qquad\qquad 1.3$$

Similarly, $N^{(\theta, \varphi, v)}$ will indicate molecules moving in the direction (θ, φ) with velocities between v and $(v + \Delta v)$ and will be called molecules moving in direction θ, φ with velocity v.

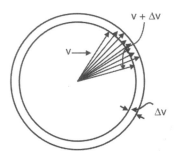

Fig.1.3   Bunch of molecules moving in direction θ and $(\theta + \Delta\theta)$, φ and $(\varphi + \Delta\varphi)$ with velocities between v and $(v + \Delta v)$

1.1.4   Molecular flux

The flux of a system of moving particles is defined as the number of particles crossing an imaginary unit area per unit time. In case of the N ideal gas molecules that are in random motion and are held in a container of volume V, if ΔN molecules cross an imaginary area ΔS in time Δt, then the molecular flux Φ may be given as,

$$\Phi = \frac{\Delta N}{\Delta S . \Delta t} \qquad\qquad 1.3\ (a)$$

In case the imaginary area is taken somewhere within the volume of the container, then there will be two molecular fluxes through the area. One will be the flux of gas molecules crossing the imaginary area from one side of the surface (from right to left in Fig. 1.4 a) to the other and the other of gas molecules crossing the area in the opposite direction (from left to right in Fig. 1.4 a). The two fluxes will also be equal in magnitude because the number density of molecules in the container is constant on an average. If we choose the area on the wall of the container (lower part in Fig. 1.4 a), then also there will be two fluxes: one of gas molecules hitting the area from left to right and the other flux of molecules rebounded by the area on the wall and moving from right to left. Again, the two fluxes will be equal. In case when the area in consideration is at the wall of the container, the flux of molecules reflected from the wall will be related to the flux of incident molecules through the laws of elastic scattering. No such relationship between the two molecular fluxes will exist in case the area is taken somewhere in the volume of the gas.

We now calculate the flux $\Phi^{(\theta, \varphi, v)}$ of (θ, φ, v) molecules through an imaginary area ΔS inside the volume of the ideal gas held in a container of volume V as shown in Fig. 1.4 (b). As shown in Fig. 1.4 (b), the normal to the area ΔS is contained in the shaded plane and makes an angle θ with the direction of motion of (θ, φ, v) molecules. The angle φ is also indicated in the figure. We now construct a cylinder with area ΔS as the base and slant lengths of magnitude $(v\, \Delta t)$ as sides, as shown in the Fig. 1.4 (b). A large number of gas molecules will be contained in this cylindrical volume which may be categorized as

(a) Molecules moving in direction θ & $(\theta + \Delta\theta)$, φ & $(\varphi + \Delta\varphi)$, with their velocities in the range of v & $(v + \Delta v)$. These are the molecules of our interest.

(b) Molecules moving in direction θ & $(\theta + \Delta\theta)$, φ & $(\varphi + \Delta\varphi)$, with their velocities NOT in the range of v & $(v + \Delta v)$. We are not interested in these molecules.

(c) Molecules having their velocities in the range of v & $(v + \Delta v)$ but NOT moving in the direction θ & $(\theta + \Delta\theta)$, φ & $(\varphi + \Delta\varphi)$. These molecules are also not of interest to us.

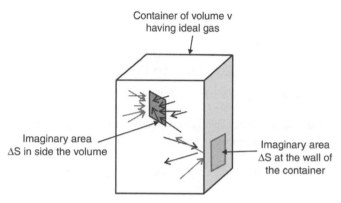

Fig. 1.4   (a) Gas molecules crossing an imaginary area inside the volume

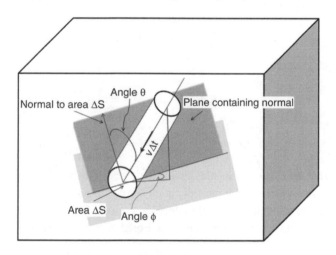

Fig. 1.4   (b) All molecules moving in direction θ, φ with velocity v contained in the cylindrical volume will cross the area ΔS in time Δt

If collisions between molecules contained in the cylindrical volume are neglected, then all the (θ, φ, v) molecules contained within the volume of the cylinder will definitely cross through the area ΔS. It is because all the (θ, φ, v) molecules that are at the other edge of the cylinder, moving with velocity v in direction (θ, φ) will cover a distance $(v\Delta t)$ in time Δt and will cross the area ΔS. The number $N^{(\theta, \varphi, v)}$ of (θ, φ, v) molecules that will pass through the area ΔS is equal to the volume V^{cyl} of the cylinder multiplied by the number density $n^{(\theta, \varphi, v)}$ of (θ, φ, v) molecules. Since $V^{cyl} = (\Delta S \cos\theta).v\Delta t$; therefore,

$$N^{(\theta,\,\varphi,\,v)} = n^{(\theta,\varphi,v)} \cdot \Delta S \cos\theta\, v\, \Delta t \qquad \text{1.3 (b)}$$

Hence,
$$\Phi^{(\theta,\,\varphi,\,v)} = \frac{N^{(\theta,\varphi,v)}}{\Delta S\, \Delta t} = \frac{n^{(\theta,\varphi,v)} \cdot \Delta S \cos\theta\, v\, \Delta t}{\Delta S\, \Delta t}$$

or
$$\Phi^{(\theta,\,\varphi,\,v)} = n^{(\theta,\varphi,v)}\, v \cos\theta \qquad \text{1.3 (c)}$$

or
$$\Phi^{(\theta,\,\varphi,\,v)} = \frac{\Delta n^v}{4\pi}\, v \cos\theta \sin\theta\, \Delta\theta\, \Delta\varphi \qquad \text{1.4}$$

In Eq. 1.4, $\Delta\theta$ and $\Delta\varphi$ are infinitesimally small elements of angle θ and φ and may be replaced by their corresponding differential elements $d\theta$ and $d\varphi$. Further, the angles θ and φ in polar coordinates may vary, respectively, from 0 to $\frac{\pi}{2}$ and 0 to 2π. Integration of Eq. 1.4 over all values of φ will yield the molecular flux of those molecules that have velocities in the range v to $(v + \Delta v)$ coming from direction θ to $(\theta + \Delta\theta)$ with all values of φ (from 0 to 2π).

$$\Phi^{(v,\,\theta)} = \frac{\Delta n^v}{4\pi}\, v \cos\theta \sin\theta\, d\theta \int\limits_0^{2\pi} d\varphi = \frac{\Delta n^v}{2}\, v \cos\theta \sin\theta\, d\theta \qquad \text{1.4 (a)}$$

and the flux Φ^θ of molecules of all velocities coming in direction θ may be obtained by summing Eq. 1.4 (a) over all values of $\Delta n^v v$.

$$\Phi^\theta = \frac{1}{2} \cos\theta \sin\theta\, d\theta \sum \Delta n^v\, v \qquad \text{1.4 (b)}$$

On the other hand

$$\Phi^v = \iint \Phi^{(\theta,\varphi,v)} d\theta\, d\varphi = \frac{\Delta n^v}{4\pi}\, v \int\limits_0^{\pi/2} \cos\theta \sin\theta\, d\theta \int\limits_0^{2\pi} d\varphi$$

or
$$\Phi^v = \frac{\Delta n^v}{4\pi}\, v \left(\frac{1}{2}\right)(2\pi) = \frac{v \Delta n^v}{4} = \frac{1}{4} v\, \Delta n^v \qquad \text{1.5}$$

Equation 1.5 tells that the flux of molecules having their velocities between v and $(v + \Delta v)$, coming from any direction, is one fourth of the multiplication of the velocity v with their number density. If it is required to find the total flux Φ of all molecules, coming from all directions and having all possible velocities, then it may be found by summing Eq.1.5 over all values of velocities. i.e.,

$$\Phi = \frac{1}{4} \sum v\, \Delta n^v \qquad \text{1.6 (a)}$$

There are molecules of velocity say v_1 with number density Δn^{v_1}, v_2 with number density Δn^{v_2}, v_3 with number density Δn^{v_3}, and so on, then expression 1.6 (a) may be expended as;

$$\Phi = \frac{1}{4} \sum [v_1\, \Delta n^{v_1} + v_2\, \Delta n^{v_2} + v_3\, \Delta n^{v_3} + v_4\, \Delta n^{v_4} + \ldots] \qquad \text{1.6 (b)}$$

We multiply both the numerator and the denominator of Eq. 1.6 (b) by $n\left(= \dfrac{N}{V}\right)$ the number density of molecules in the container.

$$\Phi = \frac{n}{4}\Sigma[\frac{v_1\,\Delta n^{v_1} + v_2\,\Delta n^{v_2} + v_3\,\Delta n^{v_3} + v_4\,\Delta n^{v_4} + \cdots}{n}]$$

or $$\Phi = \frac{1}{4}n\bar{v} \qquad\qquad 1.6\ (c)$$

where the average velocity (speed) $\bar{v} = \Sigma[\frac{v_1\,\Delta n^{v_1} + v_2\,\Delta n^{v_2} + v_3\,\Delta n^{v_3} + v_4\,\Delta n^{v_4} + \cdots}{n}]$ 1.6 (d)

or $$\bar{v} = \frac{1}{n}\Sigma\Delta n^{v}\,v \qquad\qquad 1.6\ (e)$$

Equation 1.6 (c) shows that the molecular flux at any place inside the container is proportional to the average or mean velocity, $\bar{v}$.

Substitution of $n\bar{v} = \Sigma\Delta n^{v}\,v$ from Eq. 1.6 (e) into Eq. 1.4 (b), the molecular flux in direction θ may be written as

$$\Phi^{\theta} = \frac{1}{2}\cos\theta\,\sin\theta\,d\theta\,\Sigma\Delta n^{v}\,v = \frac{1}{2}n\bar{v}\,\cos\theta\,\sin\theta\,d\theta \qquad 1.6\ (f)$$

1.1.5   Pressure exerted by an ideal gas

Randomly moving ideal gas molecules when hit the walls of the container get elastically scattered and reflected back. As a result of reflection, the directions of motion of molecules get changed and so also their linear momentum. However, rate of change of momentum is equal to force. Molecules colliding with the container walls exert a force on the wall. The average force exerted per unit area of the walls is equal to the pressure exerted by the gas.

In Fig. 1.5 an area of unit dimension at the wall of the container is shown. The normal to the unit area is also shown in the figure. A group of molecules with velocity v moving in the direction θ with the normal to the unit area hit the unit area and are elastically scattered. The velocity vector of both the incident and the scattered molecules may be resolved in to two mutually perpendicular components as shown in the figure. The tangential component $v\sin\theta$ remains unaltered after scattering as the container walls and hence the surface of the unit area is assumed to be smooth. The normal component $v\cos\theta$ after scattering from the surface becomes $(-v\cos\theta)$. Thus the scattered particles move out making an angle θ with the normal, as shown in the figure. The number of (v, θ) particles hitting the unit area per unit time is equal to the flux $\Phi^{(v,\,\theta)}$. If the mass of each molecule is m, then the change in the linear momentum Δp per unit time of $\Phi^{(v,\,\theta)}$. molecules will be,

$$\Delta p = \{mv\cos\theta - (-mv\cos\theta)\}\Phi^{(v,\theta)} \qquad 1.7\ (a)$$

Hence the force $F^{(v,\,\theta)}$ exerted by $\Phi^{(v,\,\theta)}$ molecules when they are elastically scattered by the unit area on the container wall is given by;

$$F^{(v,\,\theta)} = \{2mv\cos\theta\}\Phi^{(v,\theta)} \qquad 1.7\ (b)$$

But from Eq. 1.4 (a), $\Phi^{(v,\,\theta)} = \dfrac{\Delta n^{v}}{2}v\cos\theta\,\sin\theta\,d\theta$; therefore,

$$F^{(v, \theta)} = \{2mv\cos\theta\}\frac{\Delta n^v}{2} v\cos\theta\sin\theta\, d\theta$$

$$= m\Delta n^v v^2 \cos^2\theta\sin\theta\, d\theta \qquad\qquad 1.7\ (c)$$

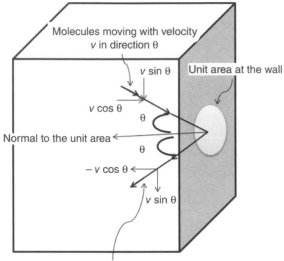

Molecules moving with velocity
v in direction θ

$v\sin\theta$ Unit area at the wall

$v\cos\theta$

θ

Normal to the unit area

θ

$-v\cos\theta$

$v\sin\theta$

(v, θ) molecules after elastic scattering from the unit area at the wall

Fig. 1.5   Gas molecules moving with velocity v in direction θ are elastically scattered
by the unit area at the wall of the container

In order to find the force $F^{(v)}$ exerted on the unit surface at the wall of the container by all molecules with velocity v coming from all directions may be obtained by integrating Eq. 1.7 (c) over θ from 0 to $\frac{\pi}{2}$. Integration becomes easy if the substitution $\cos\theta = t$ is made so that $dt = -\sin\theta\, d\theta$ and the integration in the given limits results in

$$F^{(v)} = \frac{1}{3}mv^2\Delta n^v \qquad\qquad 1.7\ (d)$$

The total force F exerted by all the gas molecules coming from all directions having all possible velocities and hitting the unit area may be obtained by summing Eq. 1.7 (d) over all velocities. You may wonder why we do not integrate the above equation over v as we have done it in case of θ and φ. The reason is that if the expression is integrated over v it will give infinite value as the upper limit for v is ∞. To avoid this we sum it over all values of v.

Force F experienced by the unit area on the container wall by all gas molecules hitting the area

$$= \frac{1}{3}m\Sigma v^2\Delta n^v = \frac{n}{3}m\Sigma \frac{v_1^2\nabla n^{v_1} + v_2^2\nabla n^{v_2} + v_3^2\nabla n^{v_3} + v_4^2\nabla n^{v_4} + ...}{n}$$

Force per unit area is pressure; hence pressure P (= F) may be written as

$$\text{or } P = \frac{1}{3}mn\left(v^2\right)_{average} = \frac{1}{3}mn\overline{v^2} = \frac{1}{3}\rho\ \overline{v^2}, \rho \text{ is the density of the gas} \qquad 1.8$$

where $\overline{v^2} = \left(v^2\right)_{average} = \sum \dfrac{v_1{}^2 \nabla n^{v_1} + v_2{}^2 \nabla n^{v_2} + v_3{}^2 \nabla n^{v_3} + v_4{}^2 \nabla n^{v_4} + \ldots}{n}$ 1.8 (a)

is the mean or average value of the squares of the molecular velocities (or speeds). $(v^2)_{average}$ is also denoted by $\overline{v^2}$. The under-root of $(v^2)_{average}$ is called root-mean square speed and is denoted by v_{rms}, i.e.,

$$v_{rms} = \sqrt{(v^2)_{average}}$$

$$= \sqrt{\sum \dfrac{v_1{}^2 \nabla n^{v_1} + v_2{}^2 \nabla n^{v_2} + v_3{}^2 \nabla n^{v_3} + v_4{}^2 \nabla n^{v_4} + \ldots}{n}} \qquad \text{1.8 (b)}$$

According to Eq. 1.8, the pressure exerted by a gas is proportional the mean square speed $(v^2)_{average}$ or $\overline{v^2}$.

1.1.6   Equation of state of an ideal gas

The mass (M), pressure (P), volume (V) and temperature (T) are the variables required to completely specify a gas. However, all these parameters are not independent. A functional relationship between these parameters like, $f(M, P, V, T) = 0$ is called the equation of state of the substance (gas). In order to reduce the number of parameters, it is sometimes convenient to consider the unit mass of the substance so that the equation of state has only three variables and becomes $f(P, v, T) = 0$, where $v\,(=V/M)$ is the volume per unit mass called the specific volume.

The concept of an ideal gas has originated from some experiments carried out on different gases. Let us define the specific molar (also called molal) volume $\mathbb{V}$ of a gas, to appreciate these experiments. From Avogadro's hypothesis it is known that 1 kilo mole (molecular weight taken in kilogram) of every gas contains Avogadro number A_v of molecules, where A_v has the value 6.03×10^{26}. If a gas has a volume V which contains $\mathbb{N}$ kilo-moles of the gas then the specific molar volume of the gas is

$$\mathbb{V} = \dfrac{V}{\mathbb{N}} \qquad \text{1.9 (a)}$$

Also the number of molecules N in volume V is $N = \mathbb{N} \times A_v = \mathbb{N} \times 6.02 \times 10^{26}$ 1.9 (b)

We now come to the experiments. In these experiments the quantity $\dfrac{P\mathbb{V}}{T}$, where P is the pressure, $\mathbb{V}$ the specific volume and T the absolute temperature, was plotted as a function of the pressure P. The $\dfrac{P\mathbb{V}}{T}$ *vs* P graph for CO_2 is shown in Fig. 1.6. It may be noted in Fig. 1.6, that the data for each temperature T lies on a smooth curve, while for different temperatures there are different curves. One very special feature of the graph is that all curves at different temperatures cut the y-axis at the same point where the value of $\dfrac{P\mathbb{V}}{T}$ is R which is called universal gas constant and has the value R $= 8.3143 \times 10^3$ (J kilomole^{-1} K^{-1}). Further, not only for CO_2 but for all other gases also the curves meet the Y-axis at the same point *R*. This really means that all gases at very low pressure (very dilute gases) are alike and behave in identical manner. This observation that all gases behave in identical manner if the pressure is very low, gave the idea that there might be a hypothetical gas for which $\frac{P\mathbb{V}}{T}$ is constant and is equal to R at all pressures. The $\frac{P\mathbb{V}}{T}$ vs P curve for such a gas will be a straight line, as shown by a dotted horizontal line in Fig. 1.6. The equation of state of such a gas will be a

$$\frac{P\mathbb{V}}{T} = R, \text{ or } P\mathbb{V} = RT \qquad \qquad 1.9 \text{ (c)}$$

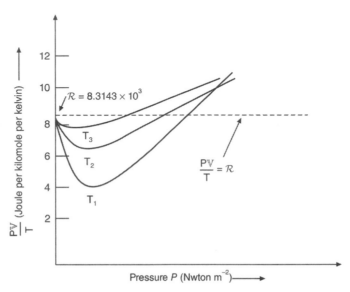

Fig. 1.6 Experimental data on the variation of $\frac{P\mathbb{V}}{T}$ with pressure P for CO_2 at three different temperatures $T_1 > T_2 > T_3$

Substituting the value of specific molar volume $\mathbb{V}$ from Eq. 1.9 (a) in Eq. 1.9 (c) we get the equation of state for an ideal gas as

$$P\frac{V}{\mathbb{N}} = RT \text{ or } PV = \mathbb{N}RT$$

$$PV = \mathbb{N}A_V \frac{1}{A_v} RT = N\,Tk_B \qquad \qquad 1.10$$

In Eq. 1.10, $k_B = \dfrac{R}{A_v} = \dfrac{8.3143 \times 10^3 \left(\text{Joule perkilomol per Kelvin} \right)}{6.03 \times 10^{26} \left(\text{per kilomol} \right)} = 1.38 \times 10^{-23}$ Joule per

Kelvin (JK^{-1}) is a constant, called the Boltzmann constant.

The equation of state of an ideal gas may be written as

$$PV = Nk_B T \qquad \qquad 1.11 \text{ (a)}$$

or $\qquad\qquad\qquad\qquad\qquad PV = \mathbb{N}RT \qquad \qquad 1.11 \text{ (b)}$

where N and $\mathbb{N}$ are, respectively, the number of molecules and the number of kilomole in volume V of the gas. Equations 1.11 (a) and (b) provide the thermodynamic definition of an ideal gas.

We now come back to Eq. 1.8 for the pressure exerted by a gas that has been derived assuming the molecular kinetic theory.

$$P = \frac{1}{3} mn \overline{v^2}$$

In the above equation, $n = \dfrac{N}{V}$, where N is the total number of gas molecules in volume V.

Substituting this value for n, we get;

$$P = \frac{1}{3} m \frac{N}{V} \overline{v^2}$$

or
$$PV = \frac{1}{3} Nm \,\overline{v^2} \qquad\qquad 1.12$$

Equation 1.12 represents the equation of state for an ideal gas derived from the molecular kinetic theory, and may be compared with Eq. 1.11 (a), that gives;

$$Nk_B T = \frac{1}{3} Nm \,\overline{v^2}$$

or
$$T = \frac{1}{3} \frac{m}{k_B} \overline{v^2} \; = \frac{1}{3} \frac{2}{k_B} \left(\frac{1}{2} m \overline{v^2} \right) = \frac{2}{3} \frac{1}{k_B} E_{tran,kin}^{mean} \qquad\qquad 1.13$$

Equation 1.13 gives the kinetic theory interpretation of temperature T. According to which, **the absolute temperature is proportional to the mean square speed** $\overline{v^2}$ **of gas molecules or to the mean kinetic energy of translational motion of molecules** $E_{trans,kin}^{mean}$ $\left(= \frac{1}{2} m \overline{v^2} \right)$. It may, however,

be emphasized that **temperature T, according to the kinetic theory, is a joint property of all the gas molecules**. It is not a property of one single molecule, because the average or mean kinetic energy of translational motion of gas molecules depends on the kinetic energies of all the molecules. Total kinetic energy of all the gas molecules, which is equal to the mean kinetic energy multiplied by the number of molecules of the gas, is called the internal energy of the gas and is denoted by U_{int}, i.e.,

$$U_{int} = N \times E_{tran,kinc}^{mean} = \frac{3}{2} Nk_B T = \frac{3}{2} NRT \qquad\qquad 1.13\,(a)$$

The concept of internal energy of a gas plays important role in thermodynamic description of gases. Equation 1.13 may be written as,

$$\left(\frac{1}{2} m \overline{v^2} \right) = \frac{3}{2} k_B T \qquad\qquad 1.14$$

Since the right hand side of the above equation depends only on temperature, irrespective of the mass of the molecule, the mean kinetic energy of molecules for all different gases at same temperature is same. For example, if there are oxygen, carbon dioxide, nitrogen, etc., each with molecules of different masses, are kept at the same absolute temperature, then the mean kinetic energy of the molecules of the three gases will be same. This means;

$$\frac{1}{2} m_1 \overline{v_1^2} \; = \frac{1}{2} m_2 \overline{v_2^2} = \frac{1}{2} m_3 \overline{v_3^2} = \frac{3}{2} k_B T \qquad\qquad 1.14(a)$$

where m_1, m_2, m_3 and $\overline{v_1^2}$, $\overline{v_2^2}$, $\overline{v_3^2}$ are, respectively, the masses and mean square speeds of the molecules of the three gases. Since molecules of different gases have different masses, it is obvious that the mean square speeds of the molecules of different gases at the same temperature also have different numerical values, such that Eq. 1.14(a) holds good.

Further, for a given gas,
$$\overline{v^2} = \frac{3k_B T}{m}$$
1.14 (b)

Here m is the mass of the gas molecule, k_B Boltzmann constant and T the temperature of the gas in Kelvin.

1.1.7   Degrees of freedom and the law of equipartition of energy

A material body may have two types of energies, namely, the kinetic energy and the potential energy. As is known, the kinetic energy is due to some kind of motion of the body. It may be the translational motion, rotational motion, vibrational motion, or combination of all the three. In general, therefore, the total kinetic energy of a body may be written as the sum of the energies of translational, rotational and vibrational motions. Each component of the total energy depends on a few mutually independent parameters. The number of such independent parameters that are required to completely define the energy of the system is called the degrees of freedom for that component of energy. We may understand the concept of the degrees of freedom by taking an example. Let us first consider a point particle of mass M (only possible in imagination) moving with some speed v in three-dimensional space. The velocity v may be resolved into three components v_x, v_y, and v_z, each of which can have any value independent of each other. Thus for translational motion in 3-D space the degrees of freedom for the kinetic energy are three. Further, the translational kinetic energy is equal to $\frac{1}{2} M v^2 =$

$\frac{1}{2} M \left(v_x^2 + v_y^2 + v_z^2 \right)$; which shows that the kinetic energy depends on the square of the three degrees

of freedom (v_x, v_y, v_z). If the system we are considering is not a point particle but is made up of two particles, like the molecule of a diatomic gas, and is moving in 3-D space, then again the problem may be reduced to the motion of a single particle in center of mass (CM) frame. In general, therefore, it may be said that any translational motion in three dimensional space will have three degrees of freedom and the kinetic energy associated with this translational motion will be a function of the square of the parameters (v_x, v_y, v_z) defining the degrees of freedom.

In the derivation of the expression for the pressure exerted by an ideal gas, only the translational motion of gas molecules has been considered. Therefore, each gas molecule has three degrees of freedom. The average kinetic energy of gas molecules, from Eq. 1.14, is given by;

$$\left(\frac{1}{2} m \overline{v^2} \right) = \frac{3}{2} k_B T$$

Since each gas molecule has three degrees of freedom with equal likelihood (that means that v_x, v_y or v_z have equal preference and none is more preferred than the other), the average kinetic energy associated with each degree of freedom may be obtained by dividing the right hand side of the above expression by three;

Average kinetic energy per degree of freedom $= \frac{1}{2}k_B T$ 1.15

The rule that $\frac{1}{2}k_B T$ of energy is associated with each degree of freedom is called the **law of equipartition of energy**. In the section on quantum thermodynamics it will be shown that the law of equipartition of energy is applicable only to those energies that are quadratic function of the parameters that define the degrees of freedom.

Monatomic molecule

A rigid monatomic molecule has three degrees of freedom for translational motion. It may also develop a rotational motion on collision with other molecules and the wall of the container. If the angular velocity of rotation is ω and the moment of inertia of the molecule about the axis of rotation is $\mathbb{I}$, then the rotational kinetic energy of the molecule E_{kin}^{rot} is given by,

$$E_{kin}^{rot} = \frac{1}{2}\mathbb{I}\omega^2 = \frac{1}{2}\mathbb{I}\left(\omega_x^2 + \omega_y^2 + \omega_z^2\right)$$ 1.16 (a)

Here, ω_x, ω_y, ω_z are, respectively, the x, y and z components of the angular frequency. It may be observed that for a single rigid body the rotational kinetic energy also has three degrees of freedom and that the energy is a quadratic function of the parameters $(\omega_x, \omega_y, \omega_z)$ defining the degrees of freedom. Hence the law of equipartition of energy is also applicable to rotational energy. The rotational energy of a rigid body depends on its moment of inertia about the axis of rotation, which in turn depends on the mass and the distribution of mass around the axis of rotation. In the case of a rigid molecule made up of only one atom (monatomic), the moment of inertia is very small and, therefore, the rotational kinetic energy is negligible as compared to the energy of the translational motion and may be neglected. It may be pointed out that rotational motion of molecules was not considered in the kinetic theory of gases while deriving the expression for the average kinetic energy of the molecule given by Eq. 1.15. The average kinetic energy of gas molecules was derived assuming that molecules have only translational motion and, therefore, only three degrees of freedom. A rigid molecule cannot vibrate and, therefore, does not have any vibrational energy. As such the total effective degrees of freedom of a monatomic rigid molecule f^{mon} is just the translational degrees of freedom f_{tran}

$$f^{mon} = f_{tran} = 3$$ 1.16 (b)

Diatomic molecule

Molecules with more than one atom may have significant magnitudes of rotational and vibrational energies along with the energy of translational motion. For example a diatomic molecule, as shown in Fig. 1.7 (a) may independently rotate along the three mutually perpendicular axes aa', bb' and cc'. However, the moment of inertia of the diatomic molecule which is a dumbbell shaped body, along the axis aa' is much smaller as compared to the its moment of inertia for rotations about the other two axes, bb' and cc'. As such a diatomic molecule can have two degrees of freedom for rotational motion,

or $$f_{rot}^{dia} = 2$$ 1.16 (c)

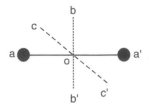

Fig. 1.7 (a) The moment of inertia of the dumbbell shaped diatomic molecule is negligible for its rotation about *aa'* axis as compared to the rotations about *bb'* and *cc'* axes.

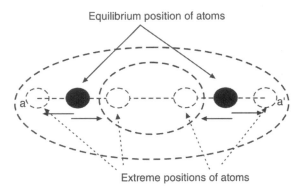

Fig. 1.7 (b) Vibration of atoms in a diatomic molecule

In a diatomic molecule the two atoms of the molecule are held at some equilibrium distance X from each other due to the molecular force between them. If the atoms are displaced from their equilibrium position by an amount ΔX, they undergo vibratory motion around their equilibrium position. Collisions with other gas molecules or with the walls of the container may setup vibratory motion of the molecule. The energy in vibratory motion oscillates between kinetic and potential forms. The energy is totally kinetic when atoms of the molecule pass through their equilibrium positions and is totally potential at the two extremes where the atoms come to a momentary rest before reversing the direction of motion. At some intermediate point between the extreme displacement and the equilibrium position, the energy is partly kinetic and partly potential. However, both the potential and the kinetic energies are proportional to the square of the displacement, $(\Delta X)^2$, which defines the degrees of freedom. Thus, there is one degree of freedom for the kinetic energy and another degree of freedom for the potential energy. As such the total degrees of freedom for vibrational motion f_{vib}^{dia} for a diatomic molecule are 2.

$$f_{vib}^{dia} = 2 \qquad \text{1.16 (d)}$$

The total number of degrees of freedom of a diatomic molecule due to translational, rotational and vibrational motions is given by,

$$f_{total}^{dia} = f_{tra} + f_{rot}^{dia} + f_{vib}^{dia} = 3 + 2 + 2 = 7 \qquad \text{1.16 (e)}$$

Further, the law of equipartition of energy is applicable to all the three components of the total energy.

Molecules of higher atomicity

The degrees of freedom for translational motion of any molecule, irrespective of its atomicity, are always 3. But the number of degrees of freedom for rotational and vibrational modes depends not only on the atomicity but also on the shape of the molecule. For example the degrees of freedom for a linear triatomic molecule are 3 for translation, 2 for rotation and 2 for vibration, a total of 7. But if the molecule is non-linear then its degree of freedoms will be 3 for translation, 3 for rotation and 3 for vibration. We thus see that the exact number of degrees of freedom of a higher atomicity molecule depends on its shape and the number of atoms in the molecule. However, in general, it may be said that the number of degrees of freedom increases with the atomicity of the molecule.

There may be energies such that they are not quadratic functions of the parameter that defines their degrees of freedom. For example, the molecules of a gas in a container also have gravitational potential energy. The height h of the molecule from the ground level determines the gravitational potential energy of the molecule, and, therefore, h is the parameter that defines the degrees of freedom for the gravitational potential energy. But the gravitational potential energy E_{pot}^{grav} is given by

$E_{pot}^{grav} = \mathrm{mgh}$, where m is the mass of the molecule and g the acceleration due to gravity at the place of observation and h the height of the particle from a reference height.

We observe that E_{pot}^{grav} is not a quadratic function of h, the parameter that defines the degrees of freedom and hence the law of equipartition of energy is not applicable to the gravitational potential energy.

In general, therefore, if f denotes the total number of degrees of freedom for each molecule of a gas, then according to the law of equipartition of energy, the total kinetic energy of such a molecule at temperature T will be $\left(\frac{1}{2} f k_B T\right)$. We denote this total kinetic energy of each molecule by $u(T)$. It

may be observed that the total kinetic energy, $u(T)$ of the molecule, which the molecule may have due to the translational, rotational and/or vibrational motions is a function of the absolute temperature T and the degrees of freedom of the molecule. If the total number of gas molecules in volume V is N, then the total energy of these N molecules at temperature T, $U(T)$ will be,

$$U(T) = Nxu\left(T\right) = Nx\frac{1}{2} fk_B T = \frac{1}{2} f \frac{N}{A_v}\left(A_v k_B\right)T = \frac{1}{2} f \mathbb{N} RT \qquad 1.17$$

Here $\mathbb{N}\left(= \dfrac{N}{A_v}\right)$ is the number of kilomoles (in SI system and moles in CGS system) in volume

V and $R(=A_v k_B)$ the universal gas constant. $U(T)$ is the sum of the kinetic energies of all molecules contained in volume V of the gas and is called the internal energy of the gas. As may be seen from Eq. 1.17, the internal energy of a given mass of the gas depends on the degrees of freedom f of the molecules and the absolute temperature T.

1.1.8   Work done by an expanding gas

As shown in Fig. 1.8, an ideal gas is contained in an insulated container fitted with an insulated and leak proof piston which may smoothly slide without any friction. The gas molecules hit all the sides of the container including the piston and exert a pressure (= force per unit area) on them. The walls of the container are rigid and fixed, therefore, do not move. The piston, however, is moveable and moves outward. We consider the system when the piston is moving outward with a velocity v_p, which is much smaller than the mean velocity $\bar{v}$ of the gas molecules in the container. Let V be the volume of the container having N molecules of the gas at temperature T Kelvin. The motion of the piston is so slow that the thermal equilibrium in the container is maintained. Thermal equilibrium means that the temperature of the gas throughout the container remains same.

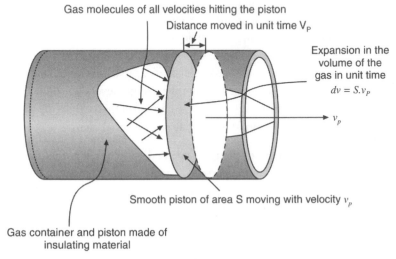

Fig. 1.8   Ideal gas contained in an insulated container fitted with smooth and frictionless insulating piston moving outward with velocity, v_p

We consider a group of molecules moving toward the piston with velocity v in the direction θ with respect to the normal to the piston. The normal component of the velocity of this bunch will be $v \cos \theta$ with respect to the stationary walls of the container. However with respect to the piston the relative velocity of the normal component will be $(v \cos \theta - v_p)$, as both piston and the molecule are moving in the same direction. In elastic scattering of a gas molecule by the moving piston, the normal component of the relative velocity of the molecule with respect to the piston is conserved. Therefore, in elastic scattering with the moving piston the normal component of the velocity will be conserved, only its direction of motion will be reversed. Therefore, the normal component of the velocity of the (v, θ) group with respect to the piston after elastic scattering with it, will be $[-(v \cos \theta - v_p)]$. The negative sign before the curly bracket means that the direction of the normal component of the velocity of the molecule is now opposite to the direction of v_p. The relative velocity of the container walls with respect to the piston is $(-v_p)$. Therefore, the speed of the scattered normal component with respect to the container walls is $\left[-\left\{\left(v \cos \theta - v_p\right) - v_p\right\}\right] = \left[-\left(v \cos \theta - 2v_p\right)\right]$. It may, however be kept in

mind that the tangential component of the (v, θ) bunch $((v \sin \theta)$ does not change on scattering because of the assumption that the surface of the piston is perfectly smooth. The kinetic energy of these molecules will get changed as a result of elastic scattering by the moving piston. The change in the kinetic energy will be due to the change of the magnitude of the normal component of the velocity after scattering.

Loss ΔE_{kin} in the kinetic energy of the (v,θ) molecule on elastic scattering with the moving piston

$$\Delta E_{kin} = \frac{1}{2}m\left(v\cos\theta\right)^2 - \frac{1}{2}m\left(v\cos\theta - 2v_p\right)^2 = \frac{1}{2}\left[4mv_pv\cos\theta + 4v_p^2\right] \approx 2mv_pv\cos\theta \quad 1.18$$

In the final expression v_p^2 term is neglected as v_p is assumed to be much smaller than v.

The loss in kinetic energy of (v,θ) group per unit time per unit area of the piston, ΔE_{kin}^{bunch}

$$\Delta E_{kin}^{bunch} = \Delta\Phi^{(v,\theta)} \times \Delta E_{kin}$$

We now replace the value of $\Delta\Phi^{(v,\theta)}$ from Eq. 1.4 into the above expression to get,

$$\Delta E_{kin}^{bunch} = \frac{\Delta n^v}{2}v\cos\theta\,\sin\theta\,d\theta \times 2mv_pv\cos\theta = mv_pv^{2"}n^v\sin\theta\cos^2\theta\,d\theta \quad 1.18\text{ (a)}$$

The total kinetic energy loss per unit time per unit area of the piston ΔE_{kin}^{total} may be calculated

by integrating Eq. 1.18 (a) over θ from 0 to $\frac{\pi}{2}$ and summing over $v^2\Delta n^v$.

$$\Delta E_{kin}^{total} = \frac{1}{3}mv_p\Sigma v^2\Delta n^v = \frac{1}{3}mv_p\left(n\overline{v^2}\right) = \left(\frac{1}{3}mn\overline{v^2}\right)v_p = Pv_p \quad 1.18\text{ (b)}$$

The loss of kinetic energy by the gas molecules per unit time may be obtained by multiplying Eq. 1.18 (b) by the area S of the piston,

The total kinetic energy lost per unit time ΔE_{kin}^{total} by the molecules of the gas in pushing the piston

$$\Delta E_{kin}^{total} = Pv_pS = PdV \quad 1.18\text{ (c)}$$

where, $dV = (S\,v_p)$ is the change in the volume of the gas per unit time. On the other hand, the loss in kinetic energy per unit time of the gas molecules is also equal to Force (= P S) multiplied with the distance travelled (v_p) in unit time, i.e., the work done per unit time by the force exerted by the molecules on the piston.

In summing up, it may be said that the work done by the gas against the piston per unit time is equal to the decrement in the kinetic energy of the molecules in this time. As a result of the loss of kinetic energy, the average kinetic energy of all molecules of the gas decreases, which results in the lowering of the temperature of the gas.

Free expansion of a gas

Let us once again look at the piston. On one side of the piston (on the right hand side of piston in Fig. 1.8) there is atmospheric pressure P_a and on the other side the gas pressure P, which is obviously greater than P_a. Basically, the gas has expended against the constant atmospheric pressure P_a and the gas has done work in pushing the piston against the atmospheric pressure. The expansion of the gas, if not stopped, will continue till the gas pressure P becomes equal to the atmospheric pressure P_a. Now for a moment let us assume that on the right hand side of the piston there is vacuum and no atmosphere (or atmospheric pressure). In that case there will be no force or pressure that will oppose the expansion of the gas and so no work will be done by the gas molecules in expansion; consequently, no loss of kinetic energy or the temperature of the gas will take place. The expansion of a gas in vacuum is called the *free expansion* of the gas. No loss of kinetic energy of gas molecules takes place in free expansion and if the container is insulated, there is no change in the average kinetic energy of gas molecules. The temperature of the gas remains constant.

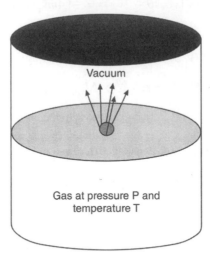

Fig. 1.9   An ideal gas at temperature *T* and pressure *P* is contained in an insulated container and is separated by a diaphragm from the vacuum on the other side. The gas undergoes free expansion if a hole is made in the diaphragm.

In Fig. 1.9, an ideal gas at temperature *T* and pressure *P* is kept in an insulated container and is separated from the vacuum on the other side by a diaphragm. If a hole is made in the diaphragm, the gas will rush out or expand in vacuum. This is an example of free expansion and no loss of kinetic energy of the gas molecules will take place in free expansion. The temperature of the gas in the insulated container will also remain same.

1.1.9   Thermal capacity

When some thermal or heat energy is given to a body either its temperature changes or its temperature remains constant but its state changes. For example, when water is heated initially its temperature increases and once the temperature reaches 100°C, further rise of temperature stops and water undergoes phase change into steam at 100°C.

In the situation when no phase change takes place, the thermal or heat energy supplied to a certain mass of a substance goes in changing its temperature. If ΔQ is the amount of heat energy absorbed by the substance (also termed as the heat flow into the substance) and ΔT the change in its temperature, the quantity $\dfrac{\Delta Q}{\Delta T}$ is called the mean thermal capacity of the substance and is denoted by $\overline{C}$.

$$\overline{C} = \frac{\Delta Q}{\Delta T}$$ 1.19 (a)

The thermal capacity *C* of the substance is defined by,

$$C = \lim_{\Delta T \to 0} \frac{\Delta Q}{\Delta T}$$ 1.19 (b)

The MKS unit of thermal capacity is 1 joule per kelvin (1 J K^{-1}). From the definition of the thermal capacity it is apparent that its numerical value will depend not only on the material but also on the amount of the substance. Thermal capacity of 1 kilogram of the material is called the specific thermal capacity. Similarly, the thermal capacity of 1 kilomole of the substance is called the specific kilo molar or specific kilo molal thermal capacity. The specific thermal capacities, both per kilogram or per kilomole (or per mole) are represented by the small letter c and are characteristics of the substance. The units of the specific capacity and specific molal capacity in MKS system are respectively, J kg^{-1} K^{-1} and J kilomol $^{-1}$ K^{-1}.

Since for a given change in temperature ΔT, the amount of heat flow, ΔQ, may be negative, zero or positive, the thermal capacity for a system may have any value. Further, the amount of heat flow also depends on whether the volume of the substance is kept constant during the heat flow or the external pressure on the specimen is kept constant. The specific thermal capacity (or the specific molal thermal capacity, as the case may be) at constant volume is denoted by c_v and the one at constant pressure by c_p.

What happens to the heat flow?

When heat, a form of energy, flows into a substance it increases the internal energy of the substance. If the substance is a gas, according to the kinetic theory, the heat energy increases the kinetic energy of gas molecules. This results in the increase of their average kinetic energy. Since average (or mean) kinetic energy of gas molecules is a measure of the gas temperature, the temperature of the gas rises. Let us assume that we have 1 kilomole of an ideal gas in a container of volume v at temperature T K and pressure P. The molal thermal capacity (or capacitance) of the gas at constant volume c_v is equal in magnitude to the thermal energy (say H Joule) supplied to this 1 kilomole of the gas to raise its temperature from T to $(T + 1)$ K, keeping the volume of the gas v constant. The total heat energy H that flows into the gas is used only in increasing the velocities of gas molecules so that their average kinetic energy got increased to a value corresponding to the temperature $(T + 1)$ K. In the case of c_p the heat energy (say H') is supplied at constant pressure to raise the temperature of 1 kilomol of the gas by 1 K. It is obvious that if the pressure of the gas is kept constant the volume of the gas will increase. Let dv be the increase in the volume of the gas. The heat energy H' performs two tasks: (i) it increases the mean kinetic energy of gas molecules so that the temperature of the gas increases by 1 K, just like the case of c_v. This will need H Joule of heat energy; (ii) but in addition a part of energy (= $H' - H$) performs mechanical work, dW, in the expansion of the gas volume by the amount dv against the constant pressure. It is, therefore, obvious that c_p is more than c_v by the amount dW, the work done in the expansion (of 1 kilomol) of the gas volume v by dv. For an ideal gas, we know from Eq. 1.10.

$$PV = \mathbb{N}RT,$$

here $\mathbb{N}$ is the number of kilomoles in volume V of the gas.

If, $\mathbb{N} = 1$ (kilomole), and the volume of the gas is v, the above equation of state of the ideal gas becomes;

$$Pv = RT$$

On differentiation the above expression, keeping P constant, one gets;

$$P\,dv = R\,dT$$

If the rise in temperature $dT = 1$ K, we get

$$P\,dv = R$$

But $(P\,dv)$ is the work done (dW) by the gas at constant pressure in the expansion of its volume by dv.

Hence, $c_p = c_v + dW = c_v + R$

or $c_p - c_v = R$ 1.19 (c)

Equation 1.19 (c), called **Mayer's relation**, tells that for an ideal gas the difference between the molal thermal capacities at constant pressure and constant volume is equal to the universal gas constant R.

1.1.10   Molal thermal capacities and their ratios

It is known (Eq. 1.17) that the total kinetic energy $U(T)$ of all the molecules of a gas at temperature T is given by,

$$U(T) = \frac{1}{2} f\, \mathbb{N} R T$$ 1.20 (a)

Here f is the number of degrees of freedom of the gas, $\mathbb{N}$ number of moles of the gas and T the temperature. The total kinetic energy of molecules of 1 kilomol of the gas at temperature T, $U^{mol}(T)$ may be obtained from Eq. 1.20 (a) by putting $\mathbb{N} = 1$,

$$U^{mol}(T) = \frac{1}{2} f\, R T$$ 1.20(b)

Similarly, the total kinetic energy of the molecules of 1 kilomole of the gas at temperature $(T + 1)$, $U^{mol}(T + 1)$, is,

$$U^{mol}(T + 1) = \frac{1}{2} f\, R(T + 1)$$ 1.20 (c)

The difference in energy $[U^{mol}(T + 1) - U^{mol}T]$ is equal to the energy supplied to 1 kilomole of an ideal gas at constant volume to raise its temperature by 1 K, which is equal to c_v, the molal specific thermal capacity at constant volume. Hence,

$$c_v = \frac{1}{2} f\, R \ \text{ and } \ c_p = \frac{1}{2} fR + R = \left(\frac{1}{2} f + 1\right) R = \left(\frac{f+2}{2}\right) R$$ 1.20 (d)

and the ratio, $\dfrac{c_p}{c_v}$, denoted by γ, is given by

$$\gamma = \frac{(f+2)}{f}$$ 1.20 (e)

1.2 Test of the Kinetic Theory

1.2.1 Kinetic theory and the gas laws

Boyle's, Charles's and Gay-Lussac's laws

The equation of state of an ideal gas derived from the kinetic theory and given by Eq. 1.12 is,

$$PV = \frac{1}{3} Nm \overline{v^2}$$

The right hand side of this equation contains the mean square speed $\overline{v^2}$ which is a function of the gas temperature T. N and m are constants for a given gas; hence one may write,

$$PV = KT$$

where K is constant for a gas

Boyle's law ($P \propto \frac{1}{V}$, at constant temperature); Charles's law ($V \propto T$, at constant pressure); and Gay-Lussac's law ($P \propto T$, at constant volume) follow easily from the equation of state of the ideal gas obtained using the kinetic theory.

Dalton's law of partial pressures

Dalton's law of partial pressures says that the pressure P exerted by a mixture of N_1 molecules of gas-1, N_2 molecules of gas-2, N_3 molecules of gas-3, and so on, contained in volume V at temperature T is equal to the sum of the partial pressures $P_1, P_2, P_3, \ldots$ exerted by each individual gas separately held in volume V at temperature T. In order to derive this law from the kinetic theory, we refer to Eq. 1.12;

$$PV = \frac{1}{3} Nm \overline{v^2}$$

which may be written as,

$$P = \frac{2}{3} \frac{1}{V} \left[Nx \left(\frac{1}{2} m \overline{v^2} \right) \right] = \frac{2}{3} \frac{1}{v} \left[Nx \text{ mean kinetic energy of molecules} \right]$$

or
$$P = \frac{2}{3} \frac{1}{V} E^{kin}_{all\ molecules}$$

In above equation E_{kin} is the total kinetic energy of all the N molecules of the gas contained in volume V at temperature T K. When gases are kept separately in volume V at temperature T, then,

$$P_1 = \frac{2}{3} \frac{1}{V} E^{kin}_1 ; \quad P_2 = \frac{2}{3} \frac{1}{V} E^{kin}_2, \quad P_3 = \frac{2}{3} \frac{1}{V} E^{kin}_3$$

Here $P_1, P_2, P_3, \ldots$, called the partial pressures, are, respectively, the pressures exerted by N_1 molecules of gas-1 when it alone is contained in volume V at temperature T, N_2 molecules of gas-2 when it alone is contained in volume V at temperature T and so on. E^{kin}_1, E^{kin}_2, E^{kin}_3, $\ldots$ are, respectively, the total kinetic energies of all molecules of gas-1, gas-2, etc. at temperature T. The sum of all partial pressures may be written as,

$$\Sigma P_i = P_1 + P_2 + P_3 + \dots = \frac{2}{3}\frac{1}{V}\left[E_1^{kin} + E_1^{kin} + E_1^{kin}\dots\right] = \frac{2}{3}\frac{1}{V}E_{all\,the\,molecules}^{kin}$$

It may now be observed that the sum of partial pressures is equal to $\frac{2}{3}\frac{1}{V}E_{all\,the\,molecules}^{kin}$, where

$E_{all\,the\,molecules}^{kin}$ is the sum of the total kinetic energies of N_1 molecules of gas-1 at temperature T, N_2 molecules of gas-2 at temperature T, N_3 molecules of gas-3 at temperature T, ... etc. It is to be kept in mind that the kinetic energies of gas molecules depend only on the temperature and not on the volume in which the molecules are kept. If N_1 molecules of type-1, N_2 molecules of type-2, N_3 molecules of type-3, etc. are all put together in volume V at temperature T, then also the total kinetic energy of each type of molecules will remain the same as it was when they were kept individually in volume V at temperature T. Therefore, $E_{all\,the\,molecules}^{kin}$ is the total kinetic energy of all the molecules $(N_1 + N_2 + N_3, \dots)$ of all the gases in the mixture of gases at temperature T kept in volume V.

The pressure exerted by the mixture of gases

$$P = \frac{2}{3}\frac{1}{V}\left(\text{Total kinetic energy of all the gas molecules}\right)$$

$$= \frac{2}{3}\frac{1}{V}E_{all\,the\,molecules}^{kin}$$

or $$P = \Sigma P_i$$

Thus Dalton's law of partial pressures may be derived from the kinetic theory of gases

Graham's law

Another important gas law is Graham's law of effusion and diffusion. Effusion rate is the rate at which a gas escapes through a pinhole into vacuum, while diffusion rate is the rate at which two gases mix. Rate of effusion (or diffusion) is the amount (either mass or volume) of a given gas that effuses (or diffuses) in unit time. Graham's law says that the rates of effusion and of diffusion are inversely proportional to the square-root of the densities or the square-root of the molar masses of the gases. If D_1 and D_2 represent the densities of two gases and MM_1 and MM_2, respectively, their molar masses, then according to Graham's law,

$$\frac{\text{rate of effusion or diffusion of gas} - 1}{\text{rate of effusion or diffusion of gas} - 2} = \frac{\sqrt{D_2}}{\sqrt{D_1}} = \frac{\sqrt{MM_2}}{\sqrt{MM_1}}$$

Let us talk about the density of gases. According to Avogadro's hypothesis, one kilomole of all gases contains A_v (= 6.03×10^{26}) molecules and occupy same (22.4 kiloliter) volume at standard temperature and pressure (STP). It means that the densities of different gases are proportional to their molar masses, represented by MM. Therefore, the square-root of density is proportional to the square-root of molar masses of the gases. This means that,

$$\frac{\sqrt{D_2}}{\sqrt{D_1}} = \frac{\sqrt{MM_2}}{\sqrt{MM_1}}$$

The effusion or diffusion rates of different gases kept at the same temperature and pressure will be proportional to the root-mean square speed v_{rms} of the gases. Gases kept at same temperature and pressure have same mean kinetic energies, i.e.;

$$\frac{1}{2} m_1 \overline{v_1^2} = \frac{1}{2} m_2 \overline{v_2^2}$$

Where m_1 and m_2 are the molecular masses and $\overline{v_1^2}$ and $\overline{v_2^2}$ the mean square speeds of the gas molecules. It follows from the above equation,

$$\frac{\overline{v_1^2}}{\overline{v_2^2}} = \frac{m_2}{m_1} \text{ and } \sqrt{\frac{\overline{v_1^2}}{\overline{v_2^2}}} = \sqrt{\frac{m_2}{m_1}}$$

or $\dfrac{\text{rate of effusion or diffusion of gas} - 1}{\text{rate of effusion or diffusion of gas} - 2} = \dfrac{(v_{rms})_1}{(v_{rms})_2} = \sqrt{\dfrac{m_1 x A_v}{m_2 x A_v}} = \dfrac{\sqrt{MM_2}}{\sqrt{MM_1}}$

Where we have used the fact that the molecular weight multiplied by Avogadro's number is equal to the molar mass, i.e. $(m \times A_v) = MM$

Therefore, the kinetic theory of gases is quite successful in deriving all gas laws.

1.2.2   The molar specific heat capacities

The ratio of the molal specific heat capacities, γ, for a large number of gases has been determined experimentally. A comparison of the experimental γ values with the values obtained theoretically using Eq. 1.20 (e) may be used to test the validity of the kinetic theory.

According to the kinetic theory, (Eq. 1.20 d), the ratio of the two molal specific thermal capacities for monatomic gases may be given as,

$$\gamma_{theo}^{mon} = \frac{f^{mon} + 2}{f^{mon}}$$

The degrees of freedom f^{mon} for a monatomic gas are 3, and hence

$$\gamma_{theo}^{mon} = \frac{5}{3} = 1.66 .$$

The experimental values of the ratio γ for some monatomic gases are listed in Table 1.1 (a).

Table 1.1   (a) Experimental values of γ for some monatomic gases

Gas	Experimental value of $\gamma = \dfrac{c_p}{c_v}$ for monatomic gases
He	1.66
Ne	1.64
Kr	1.69
Xe	1.67

As may be seen from Table 1.1 (a), there is very good agreement between the experimental and theoretical values of γ for monatomic gases.

According to the kinetic theory, the number of degrees of freedom for a diatomic gas f^{dia}, has a value 7, with 3 degrees of freedom for translational, 2 degrees of freedom each for rotational and vibrational motions. The magnitude of, γ_{theo}^{dia}, the ratio of molal thermal capacities for diatomic gases according to kinetic theory, comes out to be

$$\gamma_{theo}^{dia} = \frac{9}{7} = 1.285$$

But the experimental values of the ratio γ_{exp}^{dia} as listed in Table 1.1 (b) are closer to 1.4. If it is assumed that in case of diatomic gases, two degrees of freedom, either rotational or of vibrational motion (and not both of them) along with the three degrees of freedom of translational motion, i.e., a total of 5 degrees of freedom take part, then one may obtain, $\gamma_{theo}^{dia} = \frac{7}{5} = 1.4$, a value that is very close to the experimental values. It is a fact that vibrational degrees of freedom are excited at higher energies and it is just possible that near room temperatures where most of the measurements are done, these are not excited.

Table 1.1   (b) Experimental values of the ratio γ for diatomic gases

Gas	Experimental value of $\gamma = \dfrac{c_p}{c_v}$ for diatomic gases
H_2	1.40
O_2	1.40
N_2	1.40
CO	1.42
NO	1.43

The ratio γ, according to the kinetic theory, may be written as,

$$\gamma = \frac{f+2}{f} = 1 + \frac{2}{f}$$

It means that, γ, irrespective of the atomicity of the gas, is always greater than 1. Further, the number of degrees of freedom f increases with the atomicity of the molecule and, therefore, the numerical value of γ should decrease with the increase of the molecular atomicity. Both these facts are found to be true. No gas has a value of γ less than one and its value is found to decrease with the atomicity as is clear from Tables 1.1 (a) and 1.1 (b).

1.2.3   Specific molal heat of solids and Dulong and Petit's law

In case of solids only the vibrational motion of the molecules is possible. Each molecule can have 3-degrees of freedom for vibrational motion along the three axes. The energy of vibrational motion in each direction consists of two components, the kinetic energy and potential energy. Thus in each direction there are two degrees of freedom, one for kinetic energy and one for potential energy. Since the vibrational motion of a molecule of a solid material is possible independently in three directions, the total number of degrees of freedom for each molecule of a solid becomes $f^{solid} = 3 \times 2 = 6$. The molal specific thermal capacity at constant volume for a solid, c_v^{solid}, according to Eq. 1.20 (d), is

therefore, $\dfrac{f^{solid}}{2} R = 3\,R$. It may be observed that the kinetic theory predicts the same value of c_v^{solid}

(= 3R) for solids as is given by the empirical law of Dulong and Petit that says, "At temperatures that are not too low, the molal specific heat capacities at constant volume for all pure substances in solid state are very nearly equal to 3R".

1.2.4   Drawbacks of the kinetic theory

Although the kinetic theory of gases has been successful in explaining gas laws and correctly predicting the magnitudes of the molal specific heat capacities and their ratios, not only for gases but also for solids, but it fails to explain the temperature dependence of c_v for gases. As a matter of fact, according to the kinetic theory c_v or c_p for gases should be independent of temperature. Experiments, however, show that except for monatomic gases, c_v for all other gases of higher atomicity generally increase with the increase of the temperature and decreases with the lowering of the temperature. In the case of hydrogen which remains a gas at 20 K, the value of c_v changes from $\dfrac{5}{2} R$ to $\dfrac{3}{2} R$. Similarly, if one tries to deduce the value of the degrees of freedom f for a gas by substituting the experimental value of γ in Eq. 1.20 (e), often one gets a fractional value of f. A fractional value of the degrees of freedom is physically meaningless, as either there should be at least one degree of freedom or none.

Kinetic theory also fails in explaining why Dulong and Petit's law holds in the case of metals. In case of metals there are large number of free electrons and, therefore, a piece of metal may be considered to be a combination of positive ions, capable of vibrating independently in three mutually perpendicular directions and a cloud of free electrons behaving like a monatomic gas. The total number of degrees of freedom for the system is, therefore, 6-for the vibrational motion of the ions plus 3- for the monatomic electron gas, i.e., in all 9. According to the kinetic theory (Eq. 1.20 d) the molar specific thermal capacity at constant volume for the system is, therefore, 9/2 R. However, the experimental value of the specific molal thermal capacity of metals is also of the order of 3R, that is the value for all other non-conducting materials.

These discrepancies, as will be seen later, were finally resolved by the quantum mechanical treatment of the subject.

1.3   Velocity Distribution of Gas Molecules

Space distribution of molecules and the molecular velocity distribution functions

It was Daniel Bernoulli who, in the year 1738, for the first time considered the rapid motion of gas molecules as the cause of the pressure exerted by a gas. Bernoulli applied the laws of Newtonian physics to the collisions between the molecules and between molecules and container walls. He was of the view that the motion of each gas molecule may have to be followed in time for the complete understanding of gas properties. Maxwell, however, realized that the relevant microscopic information does not require the knowledge of the position and velocity of each gas molecule at every instant, but just the two distribution functions, namely, the space distribution function of molecules and the molecular velocity distribution function, are enough to understand the behavior of a gas. The space distribution function gives what percentage of molecules are in a certain part of the container at each instant of time, while the velocity distribution function tells what percentage of molecules have velocities within a certain range, at each instant of time. Maxwell further argued that for a gas in thermal equilibrium the space distribution function of molecules is independent of time. If minor effects due to gravity are neglected, the gas molecules will be uniformly distributed in the container. Therefore, the only unknown quantity is the velocity distribution function. In order to understand velocity distribution function, we first define the velocity space. Velocity space is an imaginary three dimensional space having axes along the three mutually perpendicular directions, v_x, v_y, v_z. Suppose a typical gas molecule has the velocity $\vec{V}$ with components V_x, V_y, and V_z. In the velocity space this molecule may be represented by the point P_1 in Fig. 1.10. Thus at a given instant all the molecules of the gas in the container may be represented by points in the velocity space forming a cloud of points. In principle every point of this cloud will be dancing with time because of the collisions between the molecules and between the molecules and the wall. Thus the shape of the cloud pattern will go on changing with time. Again, because of the very large number of molecules even in relatively very small parts of the space within the container, Maxwell argued that it is reasonable to assume that on an average the cloud pattern remains unaltered with time and the density of molecules both in the ordinary space and in the velocity space remains uniform.

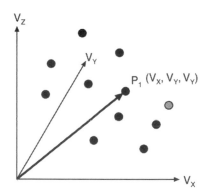

Fig. 1.10   Cloud of points in velocity space representing the velocities of different molecules of the gas

We assume that there are N molecules of the gas contained in volume V at temperature T K. Let N_1 molecules out of the total N have the X-components of their velocities between v_x and $(v_x + dv_x)$. The fraction defined as $f(v_x)dv_x = \dfrac{N_1}{N} \cdot$ '$f(v_x)dv_x$', may also be looked as the probability of finding a molecule with the X-component of its velocity in the range v_x to $(v_x + dv_x)$. Since the total probability of finding a molecule with the X-component of velocity with any value between $-\infty$ to $+\infty$ is 1,

$$\int_{-\infty}^{\infty} f(v_x) dv_x = 1 \qquad\qquad 1.21\ (a)$$

Since there is nothing special with direction X, there will be similar functions $f(v_y)dv_y$ and $f(v_z)$ dv_z for the Y and Z-directions with the corresponding conditions,

$$\int_{-\infty}^{+\infty} f(v_y) dv_y = \int_{-\infty}^{+\infty} f(v_z) dv_z = 1 \qquad\qquad 1.21\ (b)$$

Since v_x, v_y and v_z are all independent, the probability P that a molecule may have velocity between $\left(\vec{v_x} + \vec{v_y} + \vec{v_z}\right)$ and $\left[\left(\vec{v_x} + d\vec{v_x}\right), \left(\vec{v_y} + d\vec{v_y}\right), \left(\vec{v_z} + d\vec{v_z}\right)\right]$ may be obtained by multiplying the individual probabilities,

$$P = f(v_x) f(v_y) f(v_z) dv_x\, dv_y\, dv_z \qquad\qquad 1.21\ (c)$$

The number of molecules dN with velocity v and $(v + dv)$ is, therefore,

$$dN = Nf(v_x) f(v_y) f(v_z) dv_x\, dv_y dv_z \qquad\qquad 1.22$$

The most important concept involved in this derivation is the cleaver argument made by Maxwell. He argued that since each direction is equivalent, the velocity distribution function should be a function of the total velocity v. A new velocity distribution function $G = F\left(\sqrt{v_x^2 + v_y^2 + v_z^2}\right) dv$ which is a function of total velocity $v = \left(\sqrt{v_x^2 + v_y^2 + v_z^2}\right)$, may be defined that gives the probability that a molecule has the velocity between v and $(v + dv)$. It is obvious that P and G represent the same probability but P is a function of three independent factors, each depending on one of the three velocity components while G depends on the total velocity. Hence, equating P and G one gets,

$$F\left(\sqrt{v_x^2 + v_y^2 + v_z^2}\right) dv = f(v_x) f(v_y) f(v_z) dv_x\, dv_y dv_z \qquad\qquad 1.23$$

We take logarithm of the both sides,

$$\ln F(v) = \ln f(v_x) + \ln f(v_y) + \ln f(v_z)$$

On partially differentiating both sides of Eq. 1.23 with respect to v_x, one gets,

$$\frac{F'}{F(v)} \frac{1}{2} \frac{1}{\sqrt{v_x^2 + v_y^2 + v_z^2}} 2v_x = \frac{f'}{f}$$

or
$$\frac{F'}{F(v)v} = \frac{f'}{v_x f} \qquad\qquad 1.24\ (a)$$

In Eq. 1.24 (a) quantities with prime represent the first derivatives.

Similarly, differentiation of Eq. 1.23 partially with respect to v_y and v_z will yield,

$$\frac{F'}{vF} = \frac{f'}{v_x f} = \frac{f'}{v_y f} = \frac{f'}{v_z f} = -2k \qquad\qquad 1.24\ (b)$$

In Eq. 1.24 (b) terms that are functions of a different independent variables, are equal to each other. This is possible only if they are equal to a constant. Let the constant be –2k.

On integration Eq. 1.24 (b) gives,

$$F(v) = Ae^{-kv^2} = A\,e^{-k\left(v_x^2+v_y^2+v_z^2\right)}$$

$$= A^{1/3}e^{-kv_x^2}A^{1/3}e^{-kv_y^2}A^{1/3}e^{-kv_z^2} \qquad \text{1.24 (c)}$$

This gives $f(v_x)dv_x = A^{1/3}e^{-kv_x^2};\ f(v_y)dv_y = A^{1/3}e^{-kv_y^2};\ f(v_z)dv_z = A^{1/3}e^{-kv_z^2}$ 1.24 (d)

Here A is a constant of integration.

Coming back to Eq. 1.22 we may write,

$$dN = Nf(v_x)f(v_y)f(v_z)dv_x\,dv_y dv_z$$

$$= N F(v)dv_x\,dv_y dv_z \qquad \text{1.25 (a)}$$

Here $F(v) = Ae^{-kv^2}$ and the function $F(v)dv_x\,dv_y\,dv_z$ satisfy the normalization condition

$$\iiint F(v)\,dv_x dv_y dv_z = 1 \qquad \text{1.25 (b)}$$

Here $(v_x\,dv_y\,dv_z)$ represents an element of volume (d^3v) in the velocity space which in spherical polar coordinates may be written as,

$$d^3v = v^2dv\sin\theta\,d\theta\,d\phi = v^2dv\,d\Omega$$

Here $d\Omega = \sin\theta\,d\theta\,d\phi$ is the element of solid angle which on integration over all angles gives a factor of 4π. So that $d^3v = 4\pi v^2 dv$. Substitution of this value of the volume element in Eq. 1.25 (a), gives the velocity distribution function,

$$F(v)dv = A\,e^{-kv^2}(4\pi v^2 dv) = 4\pi Av^2e^{-kv^2}\,dv \qquad \text{1.26 (a)}$$

As may be seen from the above equation, the velocity distribution function contains two unknown constants, A and k. We need two equations to determine the values of these constants. The first relation may be obtained using the normalization condition,

$$\int_0^\infty F(v)\,dv = 1 \text{ or } \int_0^\infty 4\pi Av^2e^{-kv^2}\,dv = 1 \text{ or } 4\pi A\int_0^\infty v^2e^{-kv^2}\,dv = 1 \quad \text{1.26 (b)}$$

or
$$4\pi A\cdot\frac{\sqrt{\pi}}{4}k^{-3/2} = 1 \text{ or } A\pi^{3/2}k^{-3/2} = 1 \qquad \text{1.27}$$

Here we have made use of the property of the definite integral $\int_0^\infty x^2e^{-kx^2}\,dx = \frac{\sqrt{\pi}}{4}k^{-3/2}$

For the second relation between A and k, we make use of the definition of the mean square velocity $\overline{v^2}$ that is defined as,

$$\overline{v^2} = \int_0^\infty v^2 F(v)\,dv \qquad \text{1.28}$$

So,
$$\overline{v^2} = 4\pi A\int_0^\infty \left(v^2e^{-kv^2}\right)dv$$

$$= 4\pi A \frac{3\sqrt{\pi}}{8} k^{-5/2} = \frac{3\pi^{3/2}}{2} A k^{-5/2} \qquad \text{1.29 (a)}$$

We made use of the relation $\int\limits_{0}^{\infty} x^4 e^{-kx^2} dx = \frac{3\sqrt{\pi}}{8} k^{-5/2}$

However, from Eq. 1.14 (b) we know that; $\overline{v^2} = \frac{3k_BT}{m}$, where k_B is the Boltzmann's constant, T

the temperature of the gas (in Kelvin) and m the mass of the molecule of the gas. Substituting this value of the mean square velocity in Eq.1.29 one gets

$$\frac{3k_BT}{m} = \frac{3\pi^{3/2}}{2} A k^{-5/2} \quad \text{or} \quad \frac{k_BT}{m} = \frac{\pi^{3/2}}{2} A k^{-5/2} \qquad \text{1.29 (b)}$$

From Eqs.1.29 (b) and 1.27 we get,

$$k = \frac{m}{2Tk_B}; A = \left(\frac{k}{\pi}\right)^{3/2} = \left(\frac{m}{2\pi k_BT}\right)^{3/2} \qquad \text{1.30}$$

The velocity distribution functions become;

$$F(v) = \left(\frac{m}{2\pi k_BT}\right)^{3/2} 4\pi v^2 e^{-\left(\frac{\epsilon}{k_BT}\right)} \quad \text{where } \epsilon = \frac{1}{2} m v^2 \qquad \text{1.31}$$

and

$$f(v_x) = \left(\frac{m}{2\pi k_BT}\right)^{1/2} e^{-\frac{\epsilon_x}{k_BT}} \quad \text{where } \epsilon_x = \frac{1}{2} m v_x^2 \qquad \text{1.32}$$

The number dN of molecules with speeds between v and $v + dv$ may now be given as,

$$dN = NF(v)dv = N\left(\frac{m}{2\pi k_BT}\right)^{3/2} 4\pi v^2 e^{-\left(\frac{\epsilon}{k_BT}\right)} dv$$

$$= N\left(\frac{m}{2\pi k_BT}\right)^{3/2} 4\pi v^2 e^{-\left(\frac{mv^2}{2k_BT}\right)} dv$$

and

$$\frac{dN}{dv} = N\left(\frac{m}{2\pi k_BT}\right)^{3/2} 4\pi v^2 e^{-\left(\frac{mv^2}{2k_BT}\right)} \qquad \text{1.33}$$

$\frac{dN}{dv}$, represented by Eq. 1.33 is called the Maxwell–Boltzmann speed distribution function. The distribution function is zero when v is zero. This means that no (or negligibly small number) molecule is at rest. The plot of the function $\frac{dN}{dv}$ against the speed v is shown in Fig. 1.11. The graph first increases with the increase of v essentially because of the v^2 term, reaches a maximum value and then decreases as the exponential term decreases faster than the v^2 term. The shape of the distribution function also depends on temperature T. The curve becomes broader and the position of the maximum shifts to the higher velocity as the temperature increases. However, the total area under the curve remains constant, always equal to N, the total number of gas molecules.

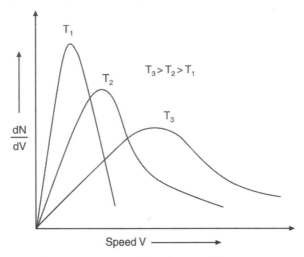

Fig. 1.11   Speed distribution functions at different temperatures

Characteristic speeds of gas molecules

There are four characteristic speeds associated with the Maxwell–Boltzmann speed distribution.
The first is the mean square speed $\overline{v^2}$ that is given by $\overline{v^2} = \dfrac{3k_BT}{m}$. The second is the root-mean

square speed, $v_{rms} = \sqrt{\overline{v^2}} = \sqrt{\dfrac{3k_BT}{m}} = 1.732\sqrt{\dfrac{k_BT}{m}}$, and the third characteristic speed is the mean

speed $\overline{v}$. The mean speed is defined by

$$\overline{v} = \int_0^\infty v\,F(v)\,dv = \int_0^\infty \left(\frac{m}{2\pi k_BT}\right)^{3/2} 4\pi v^3 e^{-\left(\frac{mv^2}{2k_BT}\right)}\,dv$$

$$= 4\pi \left(\frac{m}{2\pi k_BT}\right)^{3/2} \int_0^\infty v^3 e^{-\left(\frac{m}{2k_BT}\right)v^2}$$

or
$$\overline{v} = 4\pi \left(\frac{m}{2\pi k_BT}\right)^{3/2} \frac{1}{2(\frac{m}{2k_BT})^2} = \sqrt{\frac{8k_BT}{\pi m}} = 1.596\sqrt{\frac{k_BT}{m}}$$

We have made use of the fact that the value of the definite integral $\int_0^\infty x^3 e^{-cx^2} = \dfrac{1}{2c^2}$.

The fourth characteristic speed is v_m, the speed of maximum number of molecules or the most
probable speed that corresponds to the maximum of $F(v)$. To obtain the value of v_m we differentiate
$v^2 e^{-\left(\frac{m}{2k_BT}\right)v^2}$ with respect to v^2 and equate it to zero.

$$\frac{d}{d\left(v^2\right)}\left[v^2 e^{-\left(\frac{m}{2k_BT}\right)v^2}\right] = 0$$

or
$$e^{-\left(\frac{m}{2Tk_B}\right)v^2} - \frac{m}{2k_BT}v^2 e^{-\left(\frac{m}{2k_BT}\right)v^2} = 0$$

or
$$v_m = \sqrt{\frac{2k_BT}{m}} = 1.414\sqrt{\frac{k_BT}{m}}$$

Summing up we may write:
$$\overline{v^2} = \frac{3k_BT}{m} \qquad \text{1.34 (a)}$$

$$v_{rms} = \sqrt{\frac{3k_BT}{m}} = 1.732\sqrt{k_B\frac{T}{m}} \qquad \text{1.34 (b)}$$

$$\overline{v} = \sqrt{\frac{8k_BT}{\pi m}} = 1.596\sqrt{\frac{k_BT}{m}} \qquad \text{1.34 (c)}$$

$$v_m = \sqrt{\frac{2k_BT}{m}} = 1.414\sqrt{\frac{k_BT}{m}} \qquad \text{1.34 (d)}$$

Hence, $v_{rms} > \overline{v} > v_m$

The most probable speed v_m, the mean speed $\overline{v}$ and the rms speed v_{rms} of gas molecules are shown in Fig. 1.12.

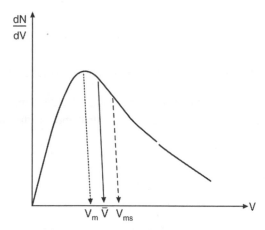

Fig. 1.12 Most probable, the mean and the root-mean-square speeds

1.4 Isothermal and Adiabatic Processes

Isothermal Process: The process in which temperature is kept constant either by adding or taking away heat from a system, is called isothermal process. Slow expansion of a gas contained in a perfectly conducting cylinder fitted with a smooth and perfectly conducting piston is an example of isothermal expansion. Since the cylinder and piston are conducting and the expansion is slow, the temperature of the whole system, including the gas contained in the cylinder, remains constant, which is equal to the temperature of the surroundings.

Adiabatic process: On the other hand if the expansion occurs suddenly so that the system (cylinder + piston + gas contained inside) does not have enough time to come to the temperature of the surroundings, the process is called Adiabatic. In an adiabatic process no heat energy is allowed to enter or leave the system. This may be achieved either by making the process very fast (sudden) or by keeping the system isolated from the surroundings by enclosing it in an insulator shield. Thus in adiabatic processes the total heat contents of the system remain constant though the temperature of the system may change. While in isothermal processes the total heat contents of the system may change, the system may absorb heat from the surroundings or lose heat to the surrounding but the temperature of the system remains fixed.

In isothermal changes for an ideal gas $PV (= RT)$ remains constant as the temperature T does not change. However, in adiabatic processes PV^γ remains constant, where γ is the ratio of the two specific heats of the gas. The graph between P and V for an isothermal change is called an isotherm and that for an adiabatic change an adiabatic. We now calculate the slopes for the two curves. Let us consider the an isotherm at a temperature T_1 so that,

$$PV = RT_1 = \text{constant } K_1,$$

hence $P = \dfrac{K_1}{V}$ and $\dfrac{dP}{dV} = -\dfrac{k_1}{V^2} = -\dfrac{P}{V}$

The slope of the isotherm is therefore, $-\dfrac{P}{V}$. Similarly, for an adiabatic

$$PV^\gamma = \text{constant say, } K_2 \text{ and, therefore, } P = \frac{K_2}{V^\gamma} \text{ and } \frac{dP}{dV} = -\gamma\frac{K_2}{V^{\gamma+1}} = -\gamma\frac{P}{V}$$

Hence, the slope of an adiabatic is $-\gamma\dfrac{P}{V}$. We see that an adiabatic curve falls off much faster than

an isotherm, as is shown in Fig. 1.13. Further, isotherms at different temperatures will be parallel to each other and will not cut each other. However, an isotherm and an adiabatic may cross each other.

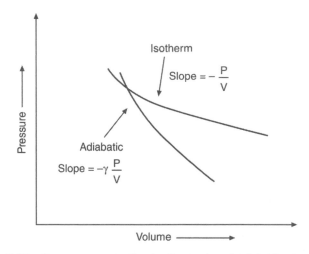

Fig. 1.13   Curves representing isothermal and adiabatic changes

1.4.1   Work done by an ideal gas in isothermal expansion

Suppose an ideal gas at temperature $T\ K$ expands isothermally from an initial volume V_i to a final volume V_f. The work done dW in expansion by amount dV is given by

$$dW = P\ dV \qquad\qquad 1.35$$

But for an ideal gas $\qquad\qquad PV = RT \text{ or } P = RT/V$

Substituting this value of P in Eq. 1.35 one gets

$$dW = RT\frac{dV}{V}$$

Integrating this expression, the total work done is

$$W = RT \int_{V_i}^{V_f} \frac{dv}{V}$$

Since in isothermal changes temperature T remains constant, it may be taken out of the integration as also R, the gas constant.

or $\qquad\qquad W = RT \ln \dfrac{V_f}{V_i} = RT \times 2.303 \log_{10} \dfrac{V_f}{V_i} \qquad\qquad$ 1.36 (a)

Since in isothermal case, $P_i V_i = P_f V_f$ Eq. 1.36 may also be written as

$$W = RT \times 2.303 \log_{10} \frac{P_i}{P_f} \qquad\qquad 1.36\ (b)$$

1.4.2   Work done by an ideal gas in adiabatic expansion

Suppose an ideal gas at pressure P expands by an infinitesimally small volume dV adiabatically. The work done by the gas in this small expansion is again given by Eq. 1.36, i. e., $dW = P\ dV$. As a result of this expansion the pressure P of the gas changes in such a way that PV^γ remains constant, say K. So that,

$$PV^\gamma = K \text{ or } P = \frac{K}{V^\gamma} \qquad\qquad 1.37$$

or $\qquad\qquad dW = P\,dV = K\dfrac{dV}{V^\gamma} \text{ and } W = K\displaystyle\int_{V_i}^{V_f} V^{-\gamma} dV$

Here V_i and V_f are, respectively, the initial and the final volumes of the gas. The above expression on integration gives

$$W = K\left(\frac{1}{1-\gamma}\right)\left[V^{(1-\gamma)}\right]_{V_i}^{V_f} = K\left(\frac{1}{1-\gamma}\right)\left[V_f^{1-\gamma} - V_i^{1-\gamma}\right]$$

Since γ is always greater than 1, we make the first parenthesis positive and absorb the negative sign in the second parenthesis where both terms are also multiplied by K.

$$W = \left(\frac{1}{\gamma - 1}\right)\left[KV_i^{1-\gamma} - KV_f^{1-\gamma}\right]$$

But $K = P_iV_i^\gamma = P_fV_f^\gamma$, when these values of K are substituted in the above equation we get,

$$W = \left(\frac{1}{\gamma - 1}\right)\left[P_iV_i - P_fV_f\right] \qquad\qquad 1.38 \text{ (a)}$$

Also $P_iV_i = RT_i$ and $P_fV_f = RT_f$, here T_i and T_i are, respectively, the initial and final temperatures of the gas. Substituting these values in Eq. 1.38 (a) one gets

$$W = \frac{R}{\gamma - 1}\left[T_i - T_f\right] \qquad\qquad 1.38 \text{ (b)}$$

Expressions 1.38 (a) and (b) give the work done by an ideal gas in adiabatic expansion.

Solved Examples

1. A box of sides L_x, L_y and L_z has N_x gas molecules each of mass M moving in X-direction with speed U_x, N_y molecules moving with speed U_y in Y-direction and N_z molecules moving with speed U_z in Z-direction. (a) Calculate the pressure exerted by the gas molecules on the sides of the box. (b) Under what conditions the pressure on all the sides will be equal. (c) In the case when pressures are same on all sides and $N_x = N_y = N_z$, derive a relation between the root- mean-square speed v_{rms} and U_x.

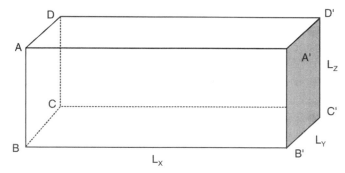

Fig. 1.14   Box containing N_x, N_y and N_z molecules of the gas moving in X, Y and Z directions

Solution:

(a) Let us consider the motion of N_x molecules each of mass M in the direction X. These molecules will collide with either the face ABCD or A'B'C'D' after travelling a distance $2L_x$. Therefore, the number of collisions that N_x molecules will suffer per unit time, $N_{coll} = \frac{U_x}{2L_x}$. Each molecule on one collision with the wall of the container will suffer a change in

momentum $\Delta P = MU_x - \left(-MU_x\right) = 2MU_x$.

The rate of change of momentum (change in momentum in unit time) of a single molecule = Number of collisions per unit time × change in momentum per collision = $N_{coll} \times \Delta P$. Hence,

force exerted on the face ABCD (or A'B'C'D') of the box by a single molecule = rate of change of momentum = $2MU_x \times \dfrac{U_x}{2L_x} = \dfrac{MU_x^2}{L_x}$.

The force F_x exerted on face ABCD by N_x molecules moving in direction $X = N_x \dfrac{MU_x^2}{L_x}$.

Pressure P_x on face ABCD (or A'B'C'D') = $\dfrac{F_x}{\text{Area of face ABCD}\left(= L_y L_z\right)} = \dfrac{N_x MU_x^2}{L_x L_y L_z} =$

$\dfrac{N_x MU_x^2}{V}$.

Here $V (= L_x L_y L_z)$ is the volume of the box.

Similarly, $P_y = \dfrac{N_y MU_y^2}{V}$ and $P_z = \dfrac{N_z MU_z^2}{V}$

(b) For pressure to be same on all faces, i.e. $P_x = P_y = P_z$, the condition becomes,

$$N_x U_x^2 = N_y U_y^2 = N_z U_z^2$$

(c) Further, if pressure on all the faces is same and $N_x = N_y = N_z$, then,

$$U_x^2 = U_y^2 = U_z^2 \text{ and } v_{rms}^2 = U_x^2 + U_y^2 + U_z^2 = 3U_x^2$$

or $v_{rms} = \sqrt{3}\, U_x$

2. An ideal gas is contained in a box of volume 49.86 m³ at a pressure of 100 kPa and temperature of 27°C. Calculate the kilomole, number of gas molecules, the mean kinetic energy of gas molecules and the internal energy of the gas. If the gas is helium, calculate the rms speed of the molecules. Given that the value of the gas constant R is 8.31 J/mol/K.

Solution:
The absolute temperature of the gas in the box $T = 273 + 27 = 300$K, Volume of the gas $V = 49.86$ m³. The pressure of the gas $P = 100 \times 10^3$ Pa. Using the ideal gas law $PV = \mathbb{N} RT$, where $\mathbb{N}$ is the number of moles, one gets

$$100 \times 10^3 \times 49.86 = \mathbb{N} \times 8.31 \times 300 \text{ or } \mathbb{N} = 2000$$

So the number of moles of the gas in the box = 2000 = 2 kilomole

But the number n of molecules = A_v (Avogadro's number) × $\mathbb{N}$ (number of moles)

The number of molecules in 2000 mol = $6.02 \times 10^{23} \times 2000 = 12.04 \times 10^{26}$.

The number of gas molecules in the box = 12.04×10^{26}.

Average kinetic energy of gas molecules = $\frac{3}{2}k_B T$ = 1.5 × 1.38 × 10^{-23} × 300 J = 621.0 × 10^{-23} J

Internal energy of the gas = sum of the kinetic energies of all molecules = Number of gas molecules × average kinetic energy of molecules = 621 × 10^{-23} × 12.04 × 10^{26} J = 74.77 × 10^5 J.

The rms speed $v_{rms} = \sqrt{\dfrac{3k_B T}{m}} = \sqrt{\dfrac{3 \times 1.38 \times 10^{-23} \times 300}{4 \times 1.66 \times 10^{-27}}}$ = 1.36 × 10^3 ms^{-1}

3. If a helium filled balloon takes 10 hours to reduce its size by 5%, how much time will be taken by oxygen filled similar balloon to reduce its size by the same amount? Assume that both helium and oxygen molecules pass through the unseen micro holes in the balloon with equal opportunity.

Solution:

The effusion rates are inversely proportion to the root of the masses of the molecules. Hence

$$\frac{\text{effusion time of oxygen for 5\% reduction}}{\text{effusion time of Helim for 5\% reduction}} = \sqrt{\frac{\text{mass of oxygen molecule}}{\text{mass of helium molecule}}} = \sqrt{\frac{32}{4}} = 2.83$$

The time taken by oxygen balloon to reduce 5% in size = 10 × 2.83 = 28.3 hours.

4. Prove that for an adiabatic process PV^γ = constant, for an ideal gas.

Solution:

When some energy, dQ is given to an isolated system having one mol of a gas, a part of this energy $(C_v\, dT)$ goes in increasing the internal energy of the gas and the part $(P\, dV)$ is used up in doing work in the expansion of the volume. Here C_v is the thermal capacity at constant volume of the gas, dT the rise in the temperature T of the gas and dV the increase in the volume V of the gas at initial pressure P. So, one may write

$$dQ = C_v\, dT + P dV$$

In case of adiabatic process $dQ = 0$;

therefore, from the above expression one gets,

$$dT = -\frac{P\, dV}{C_v} \tag{A}$$

Also for an ideal gas $PV = RT$

Differentiating this we get,

$$P\, dV + V\, dP = R\, dT = (C_p - C_v)\left(-\frac{P\, dV}{C_v}\right);$$

Here the value of $R = C_p - C_v$ and the value of dT from Eq. (A) has been substituted. Collecting the terms of dV, one gets

$$\left[P + P\left(\frac{C_p - C_v}{C_v}\right)\right]dV + V\, dP = 0 \text{ or } P\{1 + (\gamma - 1)\}dV + V\, dP = 0 \text{ or } \gamma\, P dV + V\, dp = 0$$

Dividing the above expression by PV, one may get

$$\frac{dP}{P} + \gamma \frac{dV}{V} = 0 \qquad\qquad (B)$$

Equation (B) on integration gives,

$$\ln P + \gamma \ln V = \text{constant say } K$$

or $PV^\gamma = K'$, here K' is some other constant.

Hence proved.

5. The speed distribution functions for the two cases are defined as: (a) $\Delta N_v = K_1 \Delta v$ and (b) $\Delta N_v = K_2 v \Delta v$ in the limits $V_0 > v > 0$, where ΔN_v is the number of particles with speed between v and $v + \Delta v$. If the total number of particles in each system is N_0, calculate: (i) the values of constants K_1 and K_2 (ii) the mean speed $\bar{v}$ and (iii) the rms speed $v_{rms} = \left(\overline{v^2}\right)^{1/2}$ for the two cases.

Solution:

The velocity distribution functions may be written in their differential forms as:

(a) $dN = K_1\, dv$ and (b) $dN = K_2 v\, dv$ with the condition that the velocity v can have the minimum value 0 and maximum value V_0.

(i) To find the values of the constants K_1 and K_2, we make use of the normalization condition and integrate the distribution functions in the limits v going from zero to V_0 and put them equal to N_0 i.e.,

(a) $K_1 \int_0^{V_0} dv = N_0$ or $K_1 V_0 = N_0$; hence $K_1 = \dfrac{N_0}{V_0}$

(b) $K_2 \int_0^{V_0} v\, dv = N_0$ or $K_2 \dfrac{1}{2} V_0^2 = N_0$; hence $K_2 = \dfrac{2N_0}{V_0^2}$

(ii) The mean value of a parameter x of a function $f(x)$, is given by

$$\bar{x} = \frac{\int_{x_{min}}^{x_{max}} x f(x)\, dx}{\int_{x_{min}}^{x_{max}} f(x)\, dx} , \text{ we use this method to find the mean speed and the rms speed.}$$

(a) $\bar{v} = \dfrac{K_1 \int_0^{V_0} v\, dv}{K_1 \int_0^{V_0} dv} = \dfrac{V_0}{2}$

(b) $\bar{v} = \dfrac{K_2 \int_0^{V_0} v^2\, dv}{K_2 \int_0^{V_0} v\, dv} = \dfrac{2V_0}{3}$

(iii) To determine the rms speed, we first find the mean of v^2 and then take the square root of this mean value.

(a) $\overline{v^2} = \dfrac{K_1 \int_0^{V_0} v^2\, dv}{K_1 \int_0^{V_0} dv} = \dfrac{V_0^2}{3}$, hence $v_{rms} = \dfrac{1}{\sqrt{3}} V_0 = 0.577 V_0$

(b) $\overline{v^2} = \dfrac{K_2 \int_0^{V_0} v^3\, dv}{K_2 \int_0^{V_0} v\, dv} = \dfrac{V_0^2}{2}$, hence $v_{rms} = \dfrac{1}{\sqrt{2}} V_0 = 0.707 V_0$

6. A hollow spherical chamber of radius R is filled with an ideal gas at temperature T. What fraction of the gas molecule will hit the portion of the chamber wall defined by $\theta = 30°$, $d\theta = 0.5°$, $\phi = 45°$, and $d\phi = 0.5°$.

Solution:

The number density $n^{(\theta,\ \phi)}$ of molecules moving in direction θ and $\theta + d\theta$, and ϕ and $\phi + d\phi$ is given by (see Eq. 1.3)

$$n^{(\theta,\ \phi)} = \dfrac{n}{4\pi} \sin\theta\, d\theta\, d\phi$$

Here n is the total number density of molecules (moving in any direction). Further in case where trigonometric function like $\sin\theta$ is used, the angle θ is in degree, while for the case of small angles like $d\theta$ and $d\phi$, the angle is to be taken in radians. It is because trigonometric functions are used to represent the ratio of the sides of triangles while for small angles like $d\theta$, $\sin(d\theta)$ is taken equal to $d\theta$ which is true only when $d\theta$ is in radians. We, therefore, convert $0.5°$ into radians.

$$0.5° = \dfrac{3.14 \times 0.5}{180}, \text{ and } \sin 30 = 0.5$$

The required fraction, f, of the molecules moving in the direction specified by the given angles is,

$$f = \dfrac{n^{(\theta,\phi)}}{n} = \dfrac{1}{4\pi} \sin\theta\, d\theta\, d\phi = \dfrac{1}{4 \times 3.14} \times 0.5 \times \dfrac{3.14 \times 0.5}{180} \times \dfrac{3.14 \times 0.5}{180} = 3.03 \times 10^{-6}$$

Thus an average fraction 3.03×10^{-6} of the total molecules in the chamber will be moving in the specified direction.

7. (a) Using the following data, calculate (*i*) the values of the most probable, the mean and the rms speeds of air molecules; (*ii*) the number density of air molecules in the atmosphere

(b) An evacuated copper container of volume 1 m³ with a hole of area 10^{-14} m² is placed in the atmosphere. Assuming that all the air molecules that hit the hole enters the container, how many air molecules will enter the container in 30 minutes and what pressure they will produce inside the container?

Data: Atmospheric pressure $= 1 \times 10^5$ N/m²; Atmospheric temperature $= 27°C$; Average atomic weight of air molecule $= 29$. Assume that no gas molecule leaves the evacuated container.

Solution:

(a) (*i*) The three characteristic speeds are given by the relations,

$$\overline{v} = 1.596\sqrt{\dfrac{k_B T}{m}}, \quad v_m = 1.414\sqrt{\dfrac{k_B T}{m}} \text{ and } v_{rms} = 1.732\sqrt{\dfrac{k_B T}{m}};$$

We, therefore first calculate the value of $\sqrt{\dfrac{k_B T}{m}} = \sqrt{\dfrac{1.38 \times 10^{-23} \times 300}{29 \times 1.66 \times 10^{-27}}} = 2.93 \times 10^2 \, \text{m/s}.$

Here we have converted the temperature in K; mass of the air molecule (in kg) and the value of Boltzmann constant also in MKS system. Hence,

$$\bar{v} = 467.62 \frac{m}{s}; \, v_m = 414.30 \frac{m}{s}; \, v_{rms} = 507.5 \text{ m/s}$$

(ii) To calculate the number density n of air molecules in the atmosphere we make use of the relation $P = P$ (see Eq.1.8) and substitute the values of

$$P = 1 \text{ atm} = 1.013 \times 10^5 \text{ N/m}^2; \, m = 29 \times 1.66 \times 10^{-27} \text{ kg and } \overline{v^2} = (507.5 \text{ m/s})^2.$$

To get $n = \dfrac{3 \times 1.013 \times 10^5}{29 \times 1.66 \times 10^{-27} \times (507.5)^2} = 2.45 \times 10^{25} \text{ m}^{-3}$

In order to calculate how many molecules enter the container per second, we have to calculate the molecular flux of the atmosphere Φ that gives the number of air molecules per unit area per second in the atmosphere. For that we use Eq. 1.7

$$\Phi = \frac{1}{4} n \bar{v} = \frac{1}{4} \times 2.45 \times 10^{25} \times 467.62 = 286.42 \times 10^{25}.$$

The number of molecules hitting the hole per second $N_{hole} = n \times$ area of the hole $= 286.42 \times 10^{25} \times 10^{-14} = 286.42 \times 10^{11}$ per second.

The number of molecules entered the container in 30 min $= N_{hole} \times 30 \times 60 = 5.16 \times 10^{16}$

The number density of molecules inside the container of volume 1m^3 after 30 minutes is 5.16×10^{19} molecules per unit volume. The temperature of the container is the same as that of the atmosphere i.e., 300 K. The rms speed of the molecules inside the container will be same as that outside in the atmosphere, hence again from Eq. 1.8,

Pressure inside the container $P_{inside} = $ Atmos.pressure $\dfrac{\text{number density in the container}}{\text{number density in atmosphere}}$

$$= \text{Atmos.pressure} \times \frac{5.16 \times 10^{16}}{286.42 \times 10^{25}}$$
$$= 1.013 \times 10^5 \times 1.80 \times 10^{-11} \text{ N/m}^2$$
$$= 1.82 \times 10^{-6} \text{ N/m}^2.$$

The pressure inside the container is 1.82×10^{-6} N/m^2.

8. Estimate (i) the number of molecules in the air in a room of volume 100 m^3 (ii) their energy per kilomole, take room temperature as 300 K (iii) quantity of heat to be added to warm 1 kilomole of air at 1 atm from 0 to 20 °C.

Solution:

(i) Since 1 kilomole of any gas occupies a volume of 22.4×10^3 l and has 6.02×10^{26} molecules, the number of molecules in a room of volume 100 m^3 is

$$= \frac{100 \times 10^3}{22.4 \times 10^3} \times 6.02 \times 10^{26} = 26.88 \times 10^{26} \quad (1 \text{ m}^3 = 10^3 \text{ l})$$

(*ii*) Assuming that air is largely diatomic gas, energy per kilomole

$$= \frac{5}{2} RT = \frac{5}{2} \times 8.31 \times 10^3 \times 300 = 6.23 \times 10^6 \text{ J}$$

(*iii*) $\Delta Q = C_p \Delta T = \frac{7}{2} \times R \times \Delta T = 3.5 \times 8.34 \times 10^3 \times 20 = 58.38 \times 10^4 \text{ J}$

Problems

1. (a) At what temperature the rms-speed of helium gas molecules will be 1360 m/s? (b) Calculate the speed of oxygen molecules at 2400 K.

2. At what temperature (K) the molecules of an ideal gas will have average kinetic energy of 0.025 eV?

3. When a gas expands adiabatically, its Kelvin temperature decreases by a factor of 1.32 and its volume increases by a factor of 2. Calculate the degrees of freedom of the gas.

4. 1×10^{20} molecules of oxygen are filled in a container at temperature $27°$ C. Calculate the number of oxygen molecules with velocities between 199 m/s and 201 m/s.

5. 1.3 kg of oxygen is filled in an insulating box of volume 1 m³. Calculate the temperature of the oxygen if the pressure inside the box is 1×10^5 N/m².

6. 84 g of nitrogen-14 gas is held in an insulating container at $37°$C, what is the total internal energy of the gas?

7. At what Kelvin temperature the rms speed of the gas molecules is half of the value at STP (or NTP)

8. A gas of density 0.3 g/liter is kept at a pressure of 300 mm of mercury. What is the rms speed of the gas molecules?

Short Answer Questions

1. What property of a gas is reflected in its temperature?

2. What is meant by the random distribution of molecular velocities in kinetic theory of gases?

3. Discuss one major success and one major drawback of the kinetic theory.

4. Write an expression for the total internal energy of one kilomol of an ideal gas.

5. Draw a rough sketch for the velocity distribution function of an ideal gas molecule and indicate the locations of the most probable, average and rms velocities.

6. On a P–V graph draw rough sketches for an adiabatic and an isothermal process indicating the value of the slopes for the two curves.

7. Explain why C_p is greater than C_v.

8. What happens to the internal energy of a gas in free expansion? Give reason for your answer.

9. When a gas molecule collides elastically with a slowly moving wall what quantity is conserved? Calculate the energy loss of the molecule per collision in this case.

10. Write expressions for (i) the number density of gas molecules moving in direction defined by $\theta \,\&\, (\theta + \Delta\theta)$ and $\phi \,\&\, (\phi + \Delta\phi)$ and (ii) the total molecular flux.

11. Under what condition the law of equipartition of energy holds? Is the law applicable to the distribution of gravitational potential energy of gas molecules?

12. Draw a rough sketch for an isothermal process on a T–P graph.

13. Explain why the pressure of an ideal gas increases when it is heated at constant volume.

Long Answer Questions

1. List the assumptions of the kinetic theory of gases and discuss their validity.

2. Using kinetic theory drive an expression for the molecular flux of an ideal gas and hence establish a relation between the pressure P exerted by the gas, the density ρ of the gas and, $\overline{v^2}$, the mean square velocity of gas molecules, $P = \dfrac{1}{3}\rho\overline{v^2}$.

3. What is an ideal gas? Show how the equation of state for an ideal gas may be derived from the kinetic theory of gases.

4. What is meant by the degrees of freedom? Obtain the law of equipartition of energy using the kinetic theory approach. Spell out the condition for the application of the law of equipartition.

5. Derive molecular velocity distribution function using kinetic theory approach and find the value of the most probable velocity.

6. What are the two types of molar specific heats and why C_p is greater than C_v? Obtain Mayer's relation between C_p and C_v of an ideal gas. Discuss the variation of the ratio C_p/C_v with the atomicity of the gas molecules.

7. Clearly distinguish between adiabatic and isothermal transformations. Free expansion of a gas is an adiabatic or an isothermal process? Drive expressions for the work done in the two types of transformations.

Multiple Choice Questions

Note: In some cases more than one alternative may be correct. All correct alternatives must be picked for the complete answer in such cases.

1. Air contains molecules of nitrogen and of oxygen. At a given temperature which of the following quantity is same for the two molecules?
   (a) Mass   (b) mean kinetic energy
   (c) molecular flux   (d) rms-speed

2. The ratio of molar specific heats C_p/C_v for an ideal gas
   (a) Increases with the atomicity of the gas   (b) decreases with the atomicity of the gas
   (c) does not depend on the atomicity   (d) is always greater than 1.

3. Work done by a gas in adiabatic expansion from initial conditions P_i, V_i T_i, to the final conditions P_f, V_f, T_f, is

(a) $\dfrac{P_i V_i^{\gamma}}{1 - \gamma}\left(V_f^{1-\gamma} - V_i^{1-\gamma}\right)$

(b) $R V_i \times 2.303 \log_{10} \dfrac{P_i}{P_f}$

(c) $\dfrac{R}{\gamma - 1}\left[T_i - T_f\right]$

(d) $\left(\dfrac{1}{\gamma - 1}\right)\left[P_i V_i - P_f V_f\right]$

4. Which of the terms has/ have a constant value for adiabatic process?

(a) $T^{\gamma-1}V$

(b) $TV^{\gamma-1}$

(c) PV^{γ}

(d) $P^{\gamma}V^{\gamma-1}$

5. The average speed of CH_4 molecules (molar mass 16) at 300 K is 412 m/s; the average speed of helium (He) molecules (molar mass 4) at the 300 K will be

(a) 103 m/s

(b) 206 m/s

(c) 824 m/s

(d) 1648 m/s

6. The kinetic energy (in kilo joule per mole) of 1 mole nitrogen molecules at 300 K is

(a) 11.22

(b) 7.48

(c) 3.74

(d) 1.82

7. The specific heat of a gas in case of an isothermal process is

(a) Zero

(b) positive

(c) negative

(d) infinite

8. The ratio of the rms speeds of the molecules of helium to that of the hydrogen is 5/7. If the temperature of helium sample is 273 K, the temperature of hydrogen sample is

(a) 27 K

(b) 100 K

(c) 150 K

(d) 273 K

9. If C_p and C_v, respectively, denote the molar specific heat at constant pressure and constant temperature and R the universal gas constant, then C_v is given by

(a) $\dfrac{\gamma R}{\gamma - 1}$

(b) $\dfrac{R}{\gamma - 1}$

(c) $\dfrac{\gamma - 1}{R}$

(d) $\sqrt{\dfrac{R}{\gamma - 1}}$

10. An ideal gas has the following parameters: volume V, pressure P, Kelvin temperature T, molecular mass m and density ρ. If k denotes the Boltzmann constant, then ρ is given by;

(a) $\dfrac{Pm}{kT}$

(b) $\dfrac{P}{kTV}$

(c) $\dfrac{P}{kTm}$

(d) $mkTV$

11. The slope of an adiabatic curve is

(a) $-\gamma \dfrac{P}{V}$

(b) $+\gamma \dfrac{P}{V}$

(c) $-\gamma \dfrac{V}{P}$

(d) $-\dfrac{P}{V}$

12. The slope of an isothermal curve is
    (a) $-\gamma \dfrac{P}{V}$
    (b) $+\gamma \dfrac{P}{V}$
    (c) $-\gamma \dfrac{V}{P}$
    (d) $-\dfrac{P}{V}$

13. The volume of the given mass of an ideal gas is isothermally reduced to half of its original value, the ratios of the initial and final values of the number density and the average kinetic energy of gas molecules will, respectively, be,
    (a) 2, 2
    (b) 1, 1
    (c) 2, 1
    (d) 1, 2

14. Tick the correct statements about the kinetic theory
    (a) It could explain the gas laws
    (b) It could explain the Dulong's law of specific heat of solids
    (c) It could explain the specific heat of conductors
    (d) It could not explain the atomicity dependence of γ, the ratio of C_p to C_v

15. On reducing the temperature of a fixed amount of an ideal gas, the value of the most probable velocity and the total area of the velocity distribution curve respectively,
    (a) Increases, decreases
    (b) remains same, decreases
    (c) decreases, increases
    (d) decreases, remains same

Answers to Problems and Multiple Choice Questions

Answer to problems

1. (a) 300 K (b) 1360 m/s
2. 289 K
3. 5
4. 2.3×10^{17}
5. 293 K
6. 1.16×10^4 J
7. 68.25 K
8. 632 m/s

Answers to multiple choice questions

1. (b)	2. (b) and (d)	3. (a), (c) and (d)	4. (b) and (c)
5. (c)	6. (c)	7. (d)	8. (d)
9. (b)	10. (a)	11. (a)	12. (d)
13. (c)	14. (a) and (b)	15. (d)	

Revision

1. Element of surface area on a sphere of radius r in spherical polar coordinates

$$\Delta S = r^2 \sin\theta \, \Delta\theta \, \Delta\varphi$$

2. The molecular flux $\Phi = \dfrac{1}{4} n v^{avg} \left(= \dfrac{1}{4} n \bar{v} \right)$

3. Pressure exerted by a gas: $P = \dfrac{1}{3} mn \left(v^2 \right)_{average} = \dfrac{1}{3} mn \overline{v^2} = \dfrac{1}{3} \rho \overline{v^2}$, where ρ is the density of

   the gas.

4. Equation of state for an ideal gas: $PV = \mathbb{N}\mathcal{R}T$, here $\mathbb{N}$ is the number of kilomole or mole of the gas as the case may be, contained in volume V.

   $PV = \mathbb{N}A_V \dfrac{1}{A_V} \mathcal{R}T = N k_B T$, here N is the total number of gas molecules contained in volume

   V, and $N = \mathbb{N}A_V$ where A_V is Avogadro's number. A_v has a value 6.02×10^{23} per mole and 6.02×10^{26} per kilomole. The Boltzmann constant $k_B = \mathcal{R}/A_V$ has a value of 1.381×10^{-23} Joule/ Kelvin.

5. The Kelvin temperature T of the ideal gas is related to the average kinetic energy of translational motion of gas molecules $E_{tran,kin}^{mean}$

$$T = \dfrac{1}{3}\dfrac{m}{k_B} \overline{v^2} = \dfrac{1}{3}\dfrac{2}{k_B}\left(\dfrac{1}{2} m \overline{v^2}\right) = \dfrac{2}{3}\dfrac{1}{k_B} E_{tran,kin}^{mean}$$

6. The number of mutually independent parameters required to completely specify the energy of a system is called the degrees of freedom of the system.

7. Law of equipartition of energy: The rule that $\dfrac{1}{2} k_B T$ of energy is associated with each degree

   of freedom is called the **law of equipartition of energy**. In the section on thermodynamics it will be shown that the law of equipartition of energy is applicable only to those energies that are quadratic function of the parameters that define the degrees of freedom. The average internal

   energy of each gas molecule at Kelvin temperature T, $u(T) = \dfrac{1}{2} k_B T f$, where f is the degrees of

   freedom of the molecule.

8. The internal energy of a gas at Kelvin temperature T is represented by U(T), then

$$U(T) = Nx\,u\left(T\right) = N \times \dfrac{1}{2} f k_B T = \dfrac{1}{2} f \dfrac{N}{A_v}\left(A_v k_B\right)T = \dfrac{1}{2} f \mathbb{N}\mathcal{R}T$$

9. In elastic scattering of a gas molecule by the moving piston, the normal component of the relative velocity of the molecule with respect to the piston is conserved. Loss ΔE_{kin} in the kinetic energy

   of the (v, θ) molecule on elastic scattering with the piston moving with velocity v_p

$$\Delta E_{kin} = \dfrac{1}{2} m \left(v \cos\theta\right)^2 - \dfrac{1}{2} m \left(v \cos\theta - 2v_p\right)^2 = \dfrac{1}{2}\left[4muv \cos\theta + 4v_p^2\right] \approx 2m v_p\, v \cos\theta$$

   The total loss of kinetic energy ΔE_{kin}^{total} by the gas molecules hitting the piston is given by

$$\Delta E_{kin}^{total} = P v_p S = P dV$$

10. The expansion of a gas in vacuum is called the *free expansion* of the gas. No loss of kinetic energy of gas molecules takes place in free expansion and if the container is insulated there is no change in the average kinetic energy of gas molecules. The temperature of the gas remains constant.

11. When some heat energy ΔQ flows into a body and its temperature (Kelvin) increases by ΔT, then thermal capacity C of the body is defined by,

$$C = \lim_{\Delta T \to 0} \frac{\Delta Q}{\Delta T}$$

12. $c_p - c_v = R$

    The above equation is called **Mayer's relation** and tells that for an ideal gas the difference between the molal thermal capacities at constant pressure and constant volume is equal to the universal gas constant, $\mathcal{R}$.

13. The ratio, $\dfrac{c_p}{c_v}$, which is denoted by γ, is given by,

$$\gamma = \frac{(f+2)}{f}$$

Test of kinetic theory

14. Kinetic theory may reproduce all the gas laws including the Graham's law of partial pressures. The pressure exerted by the mixture of gases $P =$

$$\frac{2}{3}\frac{1}{V}(\text{Total kinetic energy of all the gas molecules}) = \frac{2}{3}\frac{1}{v}E_{all\ the\ molecules}^{kin}$$

    or $P = \Sigma P_i$

15. Graham's law of diffusion or effusion. Effusion rate is the rate at which a gas escapes through a pinhole into vacuum, while diffusion rate is the rate at which two gases mix. Rate of effusion (or diffusion) is the amount (either mass or volume) of a given gas that effuses (or diffuses) in unit time. Graham's law says that the rates of effusion and of diffusion are inversely proportional to the square-root of the densities or the square-root of the molar masses of the gases. If D_1 and D_2 represent the densities of two gases and MM_1 and MM_2, respectively, their molar masses, then according to Graham's law,

$$\frac{\text{rate of effusion or diffusion of gas} - 1}{\text{rate of effusion or diffusion of gas} - 2} = \frac{\sqrt{D_2}}{\sqrt{D_1}} = \frac{\sqrt{MM_2}}{\sqrt{MM_1}}$$

    Also, $$\frac{\text{rate of effusion or diffusion of gas} - 1}{\text{rate of effusion or diffusion of gas} - 2} = \frac{(v_{rms})_1}{(v_{rms})_2} = \sqrt{\frac{m_1 x\, A_v}{m_2 x\, A_v}} = \frac{\sqrt{MM_2}}{\sqrt{MM_1}}$$

16. The ratio of the molar specific heat capacities c_p/c_v which is denoted by γ, according to the kinetic theory, may be written as,

$$\gamma = \frac{f+2}{f} = 1 + \frac{2}{f}$$

It means that, γ, irrespective of the atomicity of the gas, is always greater than 1. Further, the number of degrees of freedom f increases with the atomicity of the molecule and, therefore, the numerical value of γ should decrease with the increase of the molecular atomicity. Both these facts are found to be true. No gas has a value of γ less than one and its value is found to decrease with the atomicity.

17. The kinetic theory may also be applied to the solids, and it predicts the molal specific heat capacity at constant volume for solids to be $3R$. The same value is predicted by the empirical law of Dulong and Petit.

18. Drawbacks of Kinetic theory: (*i*) It fails to explain the variation of molal specific thermal capacity with temperature (*ii*) The molar thermal capacity of conducting materials according to the kinetic theory should be $9/2\,R$. But they are around $3\,\mathcal{R}$, the value characteristic of non-conducting solids.

19. Velocity distribution function

$$dN = NF(v)\,dv = N\left(\frac{m}{2\pi k_B T}\right)^{3/2} 4\pi v^2 e^{-\left(\frac{\epsilon}{k_B T}\right)}dv = N\left(\frac{m}{2\pi k_B T}\right)^{3/2} 4\pi v^2 e^{-\left(\frac{mv^2}{2k_B T}\right)}dv$$

and

$$\frac{dN}{dv} = N\left(\frac{m}{2\pi k_B T}\right)^{3/2} 4\pi v^2 e^{-\left(\frac{mv^2}{2k_B T}\right)}$$

20. Characteristic speeds: *rms* speed v_{rms}, mean or average speed $\bar{v}$ and the most probable speed v_m

$$\overline{v^2} = \frac{3k_B T}{m}$$

$$v_{rms} = \sqrt{\frac{3k_B T}{m}} = 1.732\sqrt{\frac{k_B T}{m}}$$

$$\bar{v} = \sqrt{\frac{8k_B T}{\pi m}} = 1.596\sqrt{\frac{k_B T}{m}}$$

$$v_m = \sqrt{\frac{2k_B T}{m}} = 1.414\sqrt{\frac{k_B T}{m}}$$

Hence, $v_{rms} > \bar{v} > v_m$

21. Isothermal and Adiabatic processes: The slope of an isotherm is $-\dfrac{P}{V}$ and of an adiabatic $-\gamma\dfrac{P}{V}$

22. Work done by an ideal gas in isothermal expansion

$$W = RT \times 2.303 \log_{10}\frac{P_i}{P_f}$$

23. Work done by an ideal gas in adiabatic expansion

$$W = \left(\frac{1}{\gamma - 1}\right)\left[P_i V_i - P_f V_f\right] = \frac{R}{\gamma - 1}\left[T_i - T_f\right]$$

Ideal to a Real Gas, Viscosity, Conductivity and Diffusion

2.0   The Ideal Gas

In the earlier chapter properties of an ideal gas were discussed. It was also mentioned that any gas at very low pressure when its density is very small, may be treated as an ideal gas. It essentially means that the laws of ideal gas may be applied to any gas when much larger volume of space is available to each gas molecule in comparison to its size. It is, however, evident that an ideal gas is only a conceptual gas that does not exist in real terms.

2.1   Difference between an Ideal Gas and The Real Gas

Finite molecular size and attraction between molecules

A real gas differs from an ideal gas in two respects: first, the molecules of a real gas are not point particles but have a finite size. This means that the actual volume available to the gas molecules for their motion is restricted by the amount of the volume occupied by the molecules themselves. Secondly, in the case of a real gas the gas molecules attract each other. The force of molecular attraction, called the Van der Waals force, originates from the net electrostatic force of attraction between the electron cloud of one constituent atom and the nucleus of the other atom of the molecule minus the force of repulsion between the nuclei and the electron clouds of the atoms in the molecule. Since electrostatic forces have infinite range, the net electrostatic force of attraction exceeds beyond the molecular dimensions. The Van der Waals force may also be looked as the net resultant of the forces of attraction between the electron cloud of one molecule and the nuclei of the other molecule and the forces of repulsion between the electron clouds and nuclei of the two molecules as shown in Fig. 2.1. This leaked or residual force of attraction (Van der Waals force) is responsible for the molecular attraction in real gases. Since an attractive force may be derived from a negative potential, the potential between two real gas molecules may be represented as shown in Fig. 2.2.

As is shown in Fig. 2.2 , Van der Waals potential becomes repulsive when the two molecules come closer to a certain distance between them and remains negative over a small range of mutual separation. If the separation between the molecules becomes too large the force of attraction becomes negligible. Repulsive core is essential to avoid the collapse of the molecules into each other. The repulsive core indicates that the molecules behave like hard spheres if separation between them is small. When two hard spheres just touch each other, the potential between them becomes infinite repulsive. As may be seen from Fig. 2.2, the range of Van der Waals molecular force is very small. A molecule is therefore, attracted by only a few molecules around it, not by the molecules much farther away. The molecules of a real gas immediately next to the container walls experience an unbalanced pull back towards the inside when the gas is kept in a container. It is assumed that no force of attraction exists between the molecules of the gas and the molecules of the container walls. This unbalanced backward pull reduces the pressure exerted by the gas on the container walls.

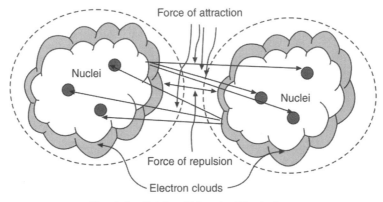

Fig. 2.1   Origin of Van der Waals force

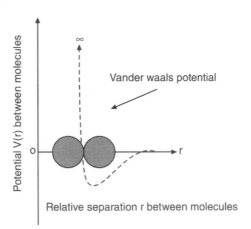

Fig. 2.2   Van der Waals potential

In view of the above-mentioned differences between a real and an ideal gas, no real gas is either expected or obeys the ideal gas equation, $PV = \mathbb{N}RT$, for $\mathbb{N}$ moles of the gas. It is only under the extreme conditions of low pressure and high temperature that the above-mentioned ideal gas equation may be applied to a real gas, and that too, over a limited range of pressure and temperature. An important experimental observation in this regard is that the ideal gas equation may be applied to a real gas with about 1% accuracy to the pressure and temperature range where the specific molar volume v is >5 liter/mole for diatomic gases and >20 liter/mole for other gases.

Several attempts have been made to develop equation of state for real gases, mostly by modifying either semi-empirically or empirically the ideal gas equation. Such attempts resulted in the development of: 1. Cubic equations of state, 2. Virial equations of state and 3. Compressibility factor equation.

Van der Waals equation of state for real gases is a typical example of cubic equations. A cubic equation gives three roots for specific volume for each set of pressure and temperature, at least one of which is real, and can thus explain the co-existence of different phases.

2.2   Modification of Ideal Gas Equation: Van der Waals Equation of State

The equation of state for an ideal gas is written as

$$PV = \mathbb{N}RT \qquad\qquad 2.1$$

Here P is the gas pressure, V the volume, $\mathbb{N}$ the number of moles of the gas contained in volume V, R the gas constant and T the Kelvin temperature of the gas. The gas equation per mole may be obtained by dividing both sides of Eq. 2.1 by the number of moles $\mathbb{N}$ and putting specific molar volume $v = \dfrac{V}{\mathbb{N}}$ to get,

$$Pv = RT \qquad\qquad 2.2$$

2.2.1   Reduction in pressure

In the case of ideal gas, the pressure exerted by the gas molecules was calculated assuming that the molecules are free point particles and have no attraction between them. In a real gas, however, molecules attract each other and, hence, the force with which the molecules hit the walls of the container is less than that in the case of the ideal gas. As a result the pressure exerted by the real gas will be smaller than the pressure exerted by the ideal gas. In the case of the real gas, therefore, one must use $(P + \Delta P)$ instead of P in Eq. 2.2. Here ΔP is the amount of reduction in the pressure due to the molecular attraction. We will now try to estimate the value of ΔP. In fact the reduction term ΔP is made up of two components $(\Delta P)_p$, the reduction due to the number density of molecules and the other $(\Delta P)att$ due to molecular attraction. Therefore, the net reduction factor $\Delta P = (\Delta P)p(\Delta P)att$.

Let the total number of real gas molecules present in volume V be N so that the number density of molecules $n = \dfrac{N}{V}$. It is obvious that $N = \mathbb{N} \times A_v$, where A_v is Avogadro's number.

Figure 2.3 shows a layer of real gas molecules on the extreme right having unit area that is about to hit the shaded wall of the container. When molecules contained in this unit area say, N_1 will get scattered by the container wall, they will exert a force on the wall that will be equal to the pressure. The pressure exerted will be more if the number N_1 is large. But N_1 will be proportional to the number density n of the molecules. Reduction in pressure $(\Delta P)p$ will in turn be proportional to the pressure,

Hence, $\qquad\qquad (\Delta P)p \propto P \propto n.$ $\qquad\qquad 2.3$

Next we consider the effect of the force of attraction between the molecules. As is shown in Fig. 2.3, molecules contained in a layer of unit area just behind and adjacent to the extreme layer will exert force of attraction on the molecules that are going to hit the container wall. Other molecules, because of the short range of Van der Waals force, will not interact with the molecules that are going to hit the wall. Reduction in the pressure due to molecular attraction $(\Delta P)_{att}$ will be proportional to the number of molecules contained in unit area of the adjacent layer; this number on an average will be equal to N_1 and will be proportional to the number density n.

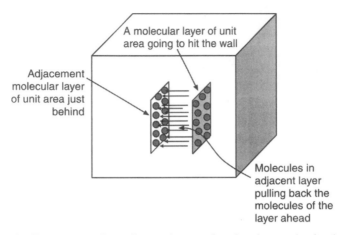

Fig. 2.3 Two successive adjacent layers of molecules each of unit area

Thus, $(\Delta P)_{att.} \propto n$ 2.4

The net reduction $\Delta P = (\Delta P)_p \cdot (\Delta P)_{att} \propto n^2 = \dfrac{kN^2}{V^2} = \dfrac{k.\left((\mathbb{N}A_V)^2\right)}{V^2} = \dfrac{k.A_V^{\,2}}{\left(\dfrac{V}{\mathbb{N}}\right)^2}$

or $\Delta P = \dfrac{a}{v^2}$ 2.5

In Eq. 2.5, $a = k.(A_v)^2$, k is a constant, A_v is Avogadro number and $v = \dfrac{V}{\mathbb{N}}$ is specific molar volume.

Thus the corrected expression for the pressure of the real as is $\left(P + \dfrac{a}{v^2}\right)$

2.2.2 Correction in volume

As discussed above, the pressure term in ideal gas Eq. 2.2 needs to be replaced by $(P + \dfrac{a}{v^2})$ in case

the equation is to be used for a real gas. Next we consider the change in the second term v for the real gas. It is obvious that in the case of a real gas the whole total specific molar volume v will not be available to the gas molecules for their motion. Let us assume that a volume 'b' per mole is not available for the motion of the gas molecules because of the finite size of the gas molecules. The available volume per mole is, therefore, $(v - b)$, and the equation of state, per mole of a real gas becomes,

$$\left(P + \dfrac{a}{v^2}\right)(v - b) = RT$$ 2.6

Equation 2.6 was first derived by Van der Waals and is called Van der Waals equation of state for a real gas.

We now try to estimate the reduction term b in specific molar volume.

Let us concentrate on a gas molecule of radius r, denoted by 'A' in Fig. 2.4 (a). We try to estimate how much volume around this molecule is such that no other molecule can come within that volume. This volume is called the exclusion volume. As is indicated by the dotted big sphere, other molecules, like the molecule denoted by 'B', (Fig. 2.4 a) may come around the molecule A in a spherical volume of radius $3r$. If the gas molecules are assumed to be rigid spheres then in the volume of spherical shell bounded by the radius r and radius slightly less than $3r$, no other molecule can come. This is shown in Fig. 2.4 (b). However, molecules are not really rigid and therefore, they may collapse into each other to some extent, like the rubber balls. The mean value of r and $3r$ is $2r$. Hence it is reasonable to assume that the average radius of the sphere of exclusion is $2r$, as shown in Fig. 2.4(a). Thus, on an average a volume $V_{exc} = \frac{4}{3}\pi(2r)^3$ around each molecule is excluded for the motion of other molecules.

In one mole of the gas there are A_v (Avogadro number) molecules. If we associate an exclusion volume V_{exc} with each of the A_v molecules than we will be over estimating the total exclusion volume, as the exclusion volume is the volume around one molecule in which no other molecule can intrude. Hence it is enough if the exclusion volume is considered only around half of the total molecules present in a given volume of the gas. The excluded volume in one mole of the gas $b = \frac{A_v}{2}V_{exc} = \frac{1}{2}[A_v V_{exc}]$.

Therefore,

$$b = \frac{1}{2}A_v\frac{4}{3}\pi(2r)^3 = 4x\left[A_v\frac{4}{3}\pi(r)^3\right] \qquad 2.7$$

It is interesting to note from Eq. 2.7 that in one mole of the gas, the volume not available for the motion of the molecules (b) is four times the total volume of all the (A_v number) molecules present in one mole.

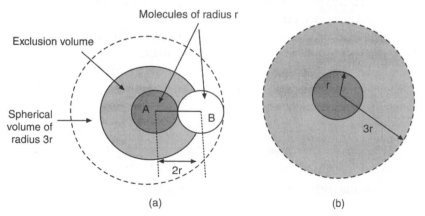

(a)   (b)

Fig. 2.4   (a) The sphere of exclusion has a radius 2r; (b) Concentric spherical shell volume with inner radius r and outer radius just less than 3r is excluded for any other molecule.

2.2.3 Real gas behavior and Van der Waals equation of state

Thomas Andrews studied the behavior of real gases in detail. He measured the change in the volume of a fixed amount of a gas at a fixed temperature with pressure. Curves showing the variation of volume V of a fixed amount of a gas with pressure P at a fixed temperature T are called isotherms. Andrews carried out his classic experiments using the apparatus outlined in Fig. 2.5.

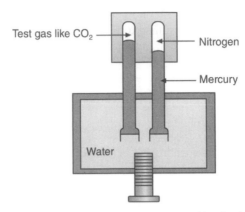

Fig. 2.5 Outline of the apparatus used by Andrews

As shown in Fig. 2.5, the pressure on the test gas as well as on nitrogen may be changed by the screw at the base of the water tank. Nitrogen was taken as the standard gas for the measurement of pressure P. The capillaries used in these experiments were strong enough to sustain pressures up to 10^7 Pa (Pascal). The results of these experiments may be described taking the typical case of carbon dioxide gas.

As may be observed in Fig. 2.6 (a), the family of isotherm curves may be divided into two distinct classes, one below the critical isotherm where unsaturated vapors, liquid plus saturated vapors and pure liquid phases exist in different regions of the isotherms. For carbon dioxide the critical temperature is 30.9 °C. Below critical temperature denoted by T_c, the horizontal part of the isotherm where liquid and saturated vapor phases co-exist, decreases with the increase of the isotherm temperature. Above the critical isotherm only one phase, the gaseous phase, exists. Further, Fig. 2.6 (b) shows that all gases at lower temperature and moderate pressure considerably deviate from ideal gas behavior and it is only at high temperature and low pressure gases behave like an ideal gas. The following important observations may be made from the above figures.

1. In order to liquefy a gas by the application of pressure alone, it must first be cooled below its critical temperature. In other words a gas cannot be converted into liquid no matter how much pressure is applied to the gas if the temperature of the gas is above its critical temperature.

2. At critical point the critical isotherm has a point of inflexion, which means that both $\dfrac{dP}{dv}$ and $\dfrac{d^2P}{dv^2}$ are zero at the critical point.

The critical temperature T_c, the critical pressure P_c, the critical volume V_c, and the boiling point B_p of some substances are given in Table 2.1.

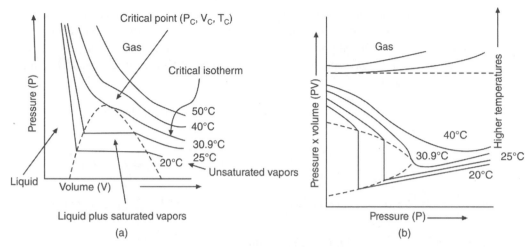

Fig. 2.6 (a) CO_2 isotherms at different temperatures; (b) PV verses P curves for CO_2

Table 2.1 Critical point parameters and boiling point for some substances

Substance	Critical temperature. T_c (K)	Critical pressure P_c (N/m²)	Critical volume V_c (m³/kilomole)	Boiling point B_p (K)
Water	647.4	209.0	0.056	373.4
Carbon dioxide	304.2	73.0	0.094	195.2
Oxygen	154.8	50.2	0.078	90.4
Hydrogen	33.3	12.8	0.065	20.4

Coming back to the Van der Waals equation of state and its application to a real gas, one may re-write it for a mole of a gas as

$$Pv^3 - (RT + Pb)v^2 + av - ab = 0 \qquad 2.8$$

Equation 2.8 is cubic in specific molar volume v and, therefore, for given values of pressure P and temperature T, it may have three roots for v, at least one of which must be real. The theoretical isotherms drawn according to Van der Waals equation of state are shown in Fig. 2.7

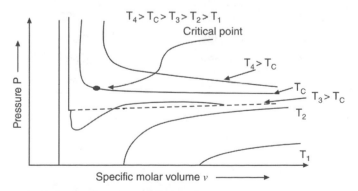

Fig. 2.7 Isotherms as per Van der Waals equation of state

As expected, the theoretical curves of Fig. 2.7 resemble to a large extent the experimental isotherms of Fig. 2.6(a). The requirement that the first and second derivatives of pressure with the specific molar volume v must vanish at the critical point (observation no.2 above) provides a mean of co-relating the critical parameter of a real gas with the constants 'a' and 'b' of Van der Waals equation. Van der Waals equation may be written as;

$$P = \frac{RT}{(v-b)} - \frac{a}{v^2}, \text{ that on differentiation with respect to } v \text{ gives;}$$

$$\frac{dP}{dv} = -\frac{RT}{(v-b)^2} + \frac{2a}{v^3} \quad \text{and} \quad \frac{d^2P}{dv^2} = \frac{2RT}{(v-b)^3} - \frac{6a}{v^4}$$

Putting $\dfrac{dP}{dv} = \dfrac{d^2P}{dv^2} = 0$ at critcal point where $v = v_c$, $P = P_c$, and $T = T_c$ it is easy to get

$$v_c = 3b, \; T_c = \frac{8a}{27bR}, \text{ and } P_c = \frac{a}{27b^2} \qquad\qquad 2.9$$

Thus from the experimentally measured values of the critical volume and critical temperature for different substances it is possible to get the values of constants 'a' and 'b' for these substances. The magnitude of the constant b is related to the molecular radius of the substance through Eq. 2.7. It is possible to determine the size or the molecular radius using the critical point data.

$$b = \frac{v_c}{3} \text{ and from Eq. 2.7 radius of the moleculer} = \left(\frac{v_c}{16\pi A_V}\right)^{1/3} \qquad\qquad 2.10$$

Here v_c and A_v are, respectively, the critical molar volume and Avogadro's number.

The numerical values of the constants 'a' and 'b' for different gases may be determined in two ways: 1. by fitting the experimental isotherms for each gas at different temperatures and thus to obtain the best values for 'a' and 'b'; 2. experimentally determining the values of critical constants V_c, P_c and T_c for the gas and then determining the value of a and b using Eq. 2.9. The values of the constants for the same gas determined from these methods generally differ from each, in some cases substantially. A typical set of values of a and b for some gases are given in Table 2.2.

Table 2.2   A typical set of a and b values for some gases

Substance	a (J m³ / kilomole²)	b (m³ / kilomole)
H_2O	580.0	0.032
CO_2	366.0	0.043
O_2	138.0	0.032
H_2	24.8	0.027

Van der Waals equation is the simplest cubic equation. More involved cubic equations of state have been proposed by many different authors. Some examples are:

Soave–Redlich–Kwong equation: $P = \dfrac{RT}{(v-b)} - \dfrac{a(T)}{v(v+b)}$ \qquad 2.11 (a)

Peng–Robinson equation:
$$P = \frac{RT}{(v-b)} - \frac{a(T)}{v(v+b)+b(v-b)}$$
2.11 (b)

Patel–Teja equation:
$$P = \frac{RT}{(v-b)} - \frac{a(T)}{v(v+b)+c(v-b)}$$
2.11 (c)

Most of these equations are obtained by changing the strengths of the terms corresponding to molecular attraction and the hard core repulsion in Van der Waals equation.

2.2.4 Reduced gas variables and Van der Waals correspondence principle

P, V and T are the three variables of a gas, any two of which may be treated as independent and the value of the third depends on the other two. The ratios of the gas variables to their respective critical values are called the reduced variables and are generally denoted by putting a subscript r. Thus P_r $= \dfrac{P}{P_C}$, $V_r = \dfrac{V}{V_C}\left(\text{or } v_r = \dfrac{v}{v_C} \text{ as the case may be}\right)$, and $T_r = \dfrac{T}{T_C}$. It may be realized that relative variables are only numbers or fractions that tells how many times the actual values of the gas parameters are bigger or smaller than the corresponding critical value. Van der Waals observed that the reduced gas parameters for different gases follow the same cubic gas equation even at moderate pressures. *This similarity in the behavior of all gases when expressed in terms of their reduced variables is called Van der Waals principle of corresponding states.* Since different cubic gas equations represent deviation of the real gas from the ideal gas, the Van der Waals principle of corresponding states essentially means that the reduced variables give a better measure of the deviation of a gas from the ideal gas, in comparison to the absolute variables. In other words, the deviation of the gas from ideality is determined essentially by the reduced parameters and not by the absolute parameters. It may, therefore, be said that the real nature of the gas is contained in the reduced parameters P_r, V_r, and T_r and not in P, V, and T.

2.3 Virial Equation of State

The virial equation of state is an extension of the real gas equation by including terms of higher order either in $\dfrac{1}{v}$ or $\dfrac{1}{P}$. The two virial equations of state for one mole of the gas may be written as:

$$Pv = RT\left[1 + \frac{B(T)}{v} + \frac{C(T)}{v^2} + \frac{D(T)}{v^3}\cdots\right]$$
2.12 (a)

or
$$Pv = RT\left[1 + \frac{B'(T)}{P} + \frac{C'(T)}{P^2} + \frac{D'(T)}{P^3}\cdots\right]$$
2.12 (b)

Here B(T), C(T), D(T), …, etc., and B'(T), C'(T), D'(T) …, etc., are called virial coefficients and are temperature(T) dependent constants for each gas. It may be noted that B(T) and B'(T); C(T) and C'(T), …, etc., are different constants for a given gas but related to each other.

The word virial is derived from Latin and means 'force'. The real gas differs from an ideal gas because of the attractive intra molecular force and the repulsive force between the molecules when they are very near to each other. The molecular forces are, therefore, interplay of repulsive and attractive forces, which perhaps is signified by the term virial. The virial equations are popular because of the constants involved are readily obtained by perturbative treatment as from statistical mechanics. Further, the values of the virial coefficients can also be obtained from the linear fitting of experimental data. It may, however be added that the virial equations cannot describe the situation when different phases of the substance coexist.

2.3.1 Boyle's temperature

Boyle's temperature is the temperature at which a real gas behaves like an ideal gas. It happens when the second term in Eq. 2.12 (a) becomes zero and all the other succeeding terms become negligible. At Boyle's temperature, the attractive forces between the molecules and the repulsive forces between the molecular cores are exactly balanced. Formally, the Boyle temperature $T_{Boy} = \dfrac{a}{Rb}$.

2.4 Compressibility Factor

The ratio of the molar volume of a real gas to that of an ideal gas at same temperature and pressure is defined as the compressibility factor of the gas, and is denoted by Z.

$$Z = \frac{v}{v_{ideal}} = \frac{Pv}{RT} \text{ (since } v_{ideal} = \frac{RT}{P} \text{)} \qquad 2.13$$

The pressure dependence of Z (at fixed temperature) for different gases is shown in Fig. 2.8 (a), while the pressure dependence of Z at different temperatures for nitrogen gas is shown in Fig. 2.8 (b).

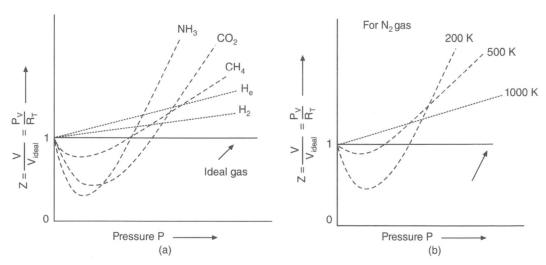

Fig. 2.8   (a) Compressibility factor Z (at fixed temperature) as a function of pressure for different gases; (b) Compressibility factor Z for nitrogen gas as a function of pressure at different temperatures

Van der Waals, however, showed that different curves of Fig. 2.8 (a) may collapse in a single curve if reduced parameters are used in place of their absolute values. This shows that Z is function of (V_r, P_r, and T_r) that has the same forms for all gases. The universal or generalized Z-values applicable to all gases, for different values of P_r and T_r are shown in Fig. 2.9.

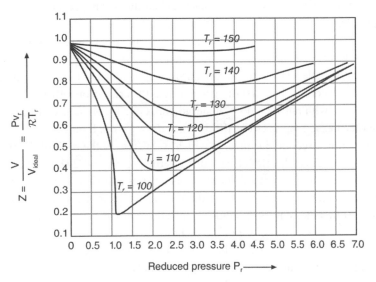

Fig. 2.9 Compressibility factor *Z* as a function of reduced pressure *Pr*, and reduced temperature *Tr*

Generalized Z curves again show the validity of principle of correspondence and the importance of reduced parameters over their absolute values. As may be observed from Fig.2.9, the generalized Z initially decreases and then increases with the reduced pressure P_r for all values of reduced temperature T_r. This typical behavior of generalized Z may be understood in terms of the interplay between the attractive and repulsive forces between the gas molecules. At moderately high pressure the molar volume of the ideal gas $v_{ideal}\left(=\dfrac{RT_r}{P}\right)$ decreases while the collision rate between gas

molecules increases. Increase in the collision rate brings into picture the repulsive forces between molecules more often and to have a noticeable effect. Decrease of v_{ideal} and noticeable effect of repulsive forces increases Z. On the other hand at low pressures gas molecules have more freedom to move and have lesser rate of collisions. Further the molar volume of the ideal gas is also not too small. Predominance of attractive forces and relatively larger value of v_{ideal} results in a smaller value of $Z(= \dfrac{v}{v_i})$.

The compressibility of a gas at its Boyle's temperature is 1, as at this temperature $v = v_{ideal}$.

2.5   Collisions Between Real Gas Molecules

It is a common experience that when a bottle of some perfume is opened, the smell of the perfume takes some time to spread all around. This delay is due to the collisions of the perfume molecules with the molecules of the air and with its own molecules. Each collision deviates the molecule from its original direction of motion as a result the molecule traverses a zigzag path reducing the displacement of the molecule from its initial position and considerably reducing the effective velocity of the molecule in a particular direction. A molecule suffers several collisions and the average distance travelled by a molecule between two successive collisions is called the mean free path and is denoted by l or λ. Another equivalent way of defining the mean free path is to consider a group of molecules and calculate the mean distance after which the molecules of the group make their first collision. In the following we will develop an expression for the mean free path.

We assume that the molecules of the gas are hard spheres, like a billiard ball, and do not flex like tennis balls, when they collide. The first thing to define is the criteria of collision. Suppose molecules of two different gases with radii r_1 and r_2 undergo collision. A collision will occur only when the centers of the molecules are separated by the distance $d = (r_1 + r_2)$. It may thus be seen that collision is defined by the centre to centre distance, d, which is the sum of the radii of the two molecules. If the collision is between the molecules of the same gas then $r_1 = r_2 = d/2$ (Fig. 2.10 a). The important point to note is that the centre to centre distance d that ensures collision does not depend individually on the radii of the two molecules but only on their sum. For example, if one of the molecule has a radius $(d-a)$ and the other molecule is of radius 'a', then the centre to center distance when they collide is again d (Fig. 2.10 b). Therefore, if one of the colliding molecules is treated as a point particle and the other molecule of radius d, then also the collision will take place if centre to centre distance is d as shown in Fig. 2.10 (c). The condition for collision between two molecules does not change if their radii are changed in such a way that the sum of their radii remains same. Just from the point of convenience we assume that one of the colliding molecules is of radius d and the other is a point particle. Again for simplicity, we call the molecule of radius d as the incident molecule and all other point like molecules as target molecules. Now all the gas molecules are in motion generally with different speeds as given by Maxwell speed distribution. Since it is not possible to track all molecules moving with different speeds, one assign an average speed $\bar{v}$ $\left(= \sqrt{\dfrac{8 k_B T}{\pi m}}\right)$ to the gas molecules.

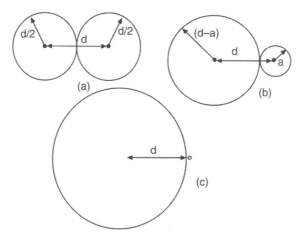

Fig. 2.10   The criteria for collision does not change if the centre to centre distance between the molecules is same

One may also assign an average relative velocity $\overline{v_r}$ to the molecule of radius d with respect to all other molecules assumed to be point particles. When a relative velocity is assigned to the molecule of radius d, it means that all other molecules that are assumed to be point like, are stationary. An expression for $\overline{v_r}$ will be derived later. The incident molecule of radius d projects an area πd^2 and any other molecule hitting this area will undergo a collision with the incident molecule. The quantity πd^2 is called the microscopic collision cross section and is denoted by σ. As shown in Fig. 2.11, the incident molecule will cover in unit time a volume $\mathbb{V} = \pi d^2 \overline{v_r}$ of the gas and all other gas molecules contained in this volume, denoted by points, will collide with the incident molecule of radius d. If n (=N/V) is the number density of the gas molecules than the number of gas molecules in volume $\mathbb{V}$ is $= \pi d^2 \overline{v_r} n$

. Therefore, the number of collisions per unit time, also called the frequency f_{col} of collisions, will be $f_{col} = \pi d^2 \overline{v_r} n$. Thus in unit time the distance travelled by the incident molecule is $\overline{v_r}$ and the number of collisions that have taken place in unit time and in travelling a distance $\overline{v_r}$ is $\pi d^2 \overline{v_r} n$. Therefore,

$$\text{Average distance between two successive collisions} = \lambda = \frac{\text{total distance travelled}}{\text{number of collisions}}$$

or
$$\lambda = \frac{\overline{v_r}}{\pi d^2 \overline{v_r} n} = \frac{1}{\pi d^2 n} = \frac{1}{\sigma n} \qquad 2.14$$

The quantity $(\sigma . n)$ is called the macroscopic cross section and has the dimensions of inverse length, while σ the microscopic cross section has the dimensions of area or length square.

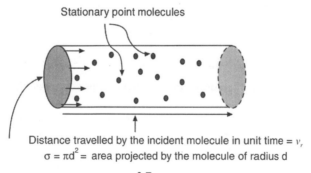

Stationary point molecules

Distance travelled by the incident molecule in unit time = v_r
$\sigma = \pi d^2$ = area projected by the molecule of radius d

Fig. 2.11   Cylinderical volume $\mathbb{V} = pd^2 . \overline{v_r}$ swept by the incident molecule in unit time

The reciprocal of collision frequency is the average collision time, i.e., the average time between two successive collisions

$$t_{col} = \frac{1}{f_{col}} = \frac{1}{\pi d^2 \overline{v_r} n} \qquad 2.15$$

2.5.1   Refinements of mean free path

The mean free path may be corrected for the actual distribution of molecular velocities. Clausius assumed that all gas molecules travel with the same velocity and he showed that the mean free path may be given as,

$$\lambda_{Cla} = \frac{3}{4}\frac{1}{n\sigma} = 0.75\,\lambda$$

On the other hand if one takes the molecular velocity distribution as predicted by Maxwell–Boltzmann, then it may be shown that

$$\lambda_{Maxwell} = 0.707\,\lambda$$

As may be observed, the correction terms in both the above cases are small and for all practical applications $\lambda = \frac{1}{n\sigma}$ is a reasonable value for the mean free path.

2.5.2 Average relative velocity (speed) of molecule $\bar{v}_r$

It is reasonable to assume that all gas molecules are moving in different directions with average velocity (or speed) $\bar{v}$. The velocities of two such molecules with their directions of motion making an angle θ are represented in Fig. 2.12 by vectors **AB** and **AC**. The relative velocity vector **CA,** represented by v_r is also shown in the figure. It is easy to see that,

$$v_r^2 = \bar{v}^2 + \bar{v}^2 + 2\bar{v}\cos\theta = 2\bar{v}^2 + 2\bar{v}\cos\theta$$

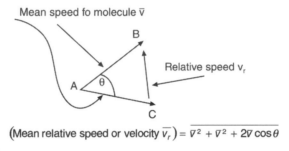

(Mean relative speed or velocity $\bar{v}_r$) $= \overline{\bar{v}^2 + \bar{v}^2 + 2\bar{v}\cos\theta}$

Fig. 2.12   Vector diagram of relative speed or velocity

To take the average value of the above expression one has to take the average of the $\cos\theta$ term that gives zero over the complete cycle of θ. Second term of the above expression, therefore, vanishes. Hence,

$$\bar{v}_r = \sqrt{2}\,\bar{v} \qquad\qquad 2.16$$

By substituting the above value of the average relative velocity, the collision frequency and collision time becomes,

$$f_{col} = \sqrt{2}\pi d^2\,\bar{v}\,n = \sqrt{2}\,\sigma\,n\bar{v}$$

and

$$t_{col} = \frac{1}{\sqrt{2}\pi d^2\,\bar{v}\,n} = \frac{1}{\sqrt{2}\,\sigma\,n\bar{v}} = \frac{\lambda}{\sqrt{2}\,\bar{v}} \qquad\qquad 2.17$$

It may be noted that the mean free path λ is not affected by the value of the mean relative velocity.

2.6 The Survival Equation

In the previous section while deriving the expression for the mean free path it was assumed that the molecule with effective radius d is moving with the mean relative velocity with respect to the other stationary point like molecules. However, the process of collision does not alter if it is assumed that the molecule of assumed radius d is stationary and the assumed point like molecules are moving with respect to it with relative average velocity $\overline{v_r}$ $(=\sqrt{2}\,\overline{v}\,)$.

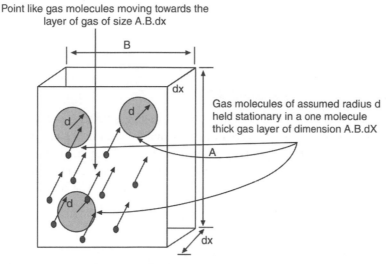

Fig. 2.13 Assumed point like molecule moving towards a layer of gas of dimension A.B.dX

Figure 2.13 shows a thin rectangular slice of the gas of height A, length B and thickness dX. The thickness dX is assumed to be of the order of the thickness of a single molecule. This assumption is made so that gas molecules (of radius d) do not overlap and shadow one another. Let us further assume that N number of point like molecules are moving towards the gas slice with mean relative velocity $\overline{v_r}$. Some, say ΔN out of the total N, incident point like molecules hit the area blocked by the molecules of assumed radius d. These ΔN incident molecules will be thrown out of the incident beam of N point like molecules because of the collisions they will suffer on hitting the area blocked by the target molecules. The remaining incident molecules will pass through the slice of the gas through the un-blocked area. The ratio of the molecules that have scattered out of the incident beam to the total incident molecules, $\dfrac{\Delta N}{N}$, may be given by the ratio of the area blocked by molecules (of radius d) in the gas slice to the total area of the gas slice. Each molecule of radius d projects an area $\sigma = \pi d^2$ and the total area blocked by these molecules = Number of molecules in the slice of the gas $\times$ σ. If n is the number density of the gas molecules, then the number of molecules in the slice of the gas is equal to the $n \times$ the volume of the gas slice $= n.ABdX$. Hence the total blocked area $= \sigma.n.ABdx$. Therefore,

$$\frac{\Delta N}{N} = \frac{\text{total blocked area}}{\text{total area of the gas slice}}$$

$$= \frac{\text{area projected by one molecule} \times \text{number density} \times \text{volume of the slice}}{A.B}$$

$$= \frac{\sigma.n.ABdx}{AB} = \sigma.n \, dX \qquad\qquad 2.18$$

Now ΔN that gives the number of incident molecules that are scattered away by collision with target molecules, may also be looked as the reduction in the number of incident molecules as they pass through the slice of the gas. Since ΔN represents reduction in number it is assigned a negative sign. Further, if N is sufficiently large, $-\Delta N$ may be replaced by $-dN$. With these modifications Eq. 2.18 becomes,

$$-dN = N \sigma n \, dx \qquad\qquad 2.19$$

Integration of Eq. 2.19 gives,

$$N = N_0 e^{-\sigma n x} \qquad\qquad 2.20$$

Equation 2.20 tells that the number of molecules N that pass through a slice of gas of thickness x without undergoing any collision decreases exponentially with the thickness x or the distance travelled by the molecules. Here N_0 is the number of gas molecules at the front of the gas slice where $x = 0$. Equation 2.20 is called the **survival equation.** It is easy to observe that the exponential decay rate is faster if the macroscopic collision cross-section (σn) is large. The survival equation may also be expressed in terms of the mean free path $\lambda \left(= \dfrac{1}{\sigma n}\right)$ as,

$$N = N_0 e^{-\frac{x}{\lambda}} \qquad\qquad 2.21$$

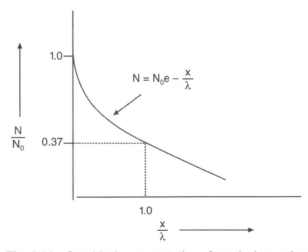

Fig. 2.14   Graphical representation of survival equation

Figure 2.14 gives the graphical representation of the survival equation. The X- and the Y- axes in the graph are plotted in dimensionless parameters $\dfrac{x}{\lambda}$ and $\dfrac{N}{N_0}$ respectively. Mean free path λ, as its name suggests, is the mean value of the distances between two successive collisions of the molecules. It is obvious that in some collisions the distances between two collisions are larger than λ and in some other collisions the distance is smaller than λ. Figure 2.14 tells that in 37% of collisions the distance between successive collisions is smaller than λ and in the remaining 63% it is larger than λ. Collision parameters of some substances are given in Table 2.3. The molecular separation, the average distance between the two molecules of a substance, depends on temperature and has different value for different substances. However, at room temperature the order of molecular separation is $\approx$ 30×10^{-10} m. It may thus be seen that the mean free path is much larger than the molecular separation. Typical magnitudes of collision parameters are shown in Fig. 2.15.

Table 2.3   Collision parameters of some substances at 20°C and 1 atm pressure

Substance	Collision diameter $(10^{-10}$ m$)$	Mean free path λ $(10^{-10}$ m$)$	Collision frequency $(10^9$ s$^{-1})$
H_2	2.73	1240	14.3
He	2.18	1910	6.6
N_2	3.74	656	7.2
O_2	3.57	716	6.2
Ar	3.62	700	5.7
CO_2	4.56	441	8.6

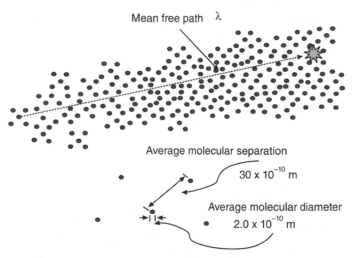

Fig. 2.15   Relative magnitudes of collision parameters

Since mean free path is defined as the mean distance between two successive collisions, if a molecule in the gas is picked up randomly and it is asked how much average distance earlier it would have suffered a collision in past, the answer is one mean free path earlier. Similarly, if it is asked as to how much average distance from now it will suffer a collision in future, the answer is again after an average distance of one mean free path. This essentially means that the future distribution of free paths and the past distributions of free paths with respect to a given instant are identical.

2.7 Average Normal Distance (or Height) $\bar{y}$ above or below an Arbitrary Plane at which a Molecule made its Last Collision before Crossing the Plane

Figure 2.16 shows a container having some gas. The molecules of the gas are moving randomly, as shown by arrows on the molecules. Some of these molecules will cross an imaginary arbitrary plane in the gas from above to downward, while almost an equal number of molecules, on an average, will move through the plane from down to the upward. We desire to calculate the average distance normal to the plane, say $\bar{y}$, above or below the plane where the molecules crossing the plane in either direction suffered their last collision. This distance is important because after the last collision the momentum and energy, etc., of the molecule does not change because of the collisions. In order to calculate $\bar{y}$ we consider the case of an ideal gas and concentrate on the flux Φ^θ of molecules approaching the imaginary plane from direction θ (see Fig. 2.16). Some important characteristics of the molecules contained in this flux are: 1. All the molecules contained in this flux are moving making an angle θ with the direction normal to the imaginary plane. 2. Molecules contained in this flux must be colliding with each other as a result of which some molecules may be scattered out from this bunch. However, on an average, almost an equal number of molecules with parameters identical to those molecules that have been scattered out are replaced in the bunch from other scattering events elsewhere in the gas. The number of molecules and the structure of the Φ^θ bunch do not change. 3. From the definition of the mean free path, it follows that the molecules of the Φ^θ bunch may have suffered their last collision at the mean slant distance λ from the imaginary plane in direction θ. The component of the mean free path in the direction of normal is the distance $y = \lambda \cos\theta$. We want to find out the average or mean value of $y (= \lambda \cos\theta)$ for all molecules that are approaching the plane from all possible values of θ. Therefore, to obtain $\bar{y}$ we need to carry out the following mathematical steps of calculating the mean value of the quantity

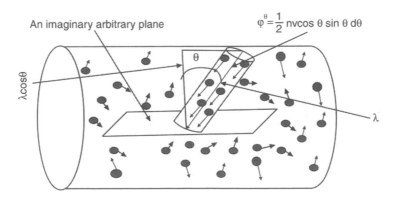

Fig. 2.16   Molecular flux in direction θ approaching the imaginary plane

$$\overline{y} = \frac{\int_0^{\frac{\pi}{2}} \lambda \cos\theta \; \Phi^\theta d\theta}{\int_0^{\frac{\pi}{2}} \Phi^\theta d\theta} = \frac{\int_0^{\frac{\pi}{2}} \lambda \cos\theta \; \Phi^\theta d\theta}{\text{Total molecular flux}} = \frac{\frac{1}{2} \lambda n\overline{v} \int_0^{\frac{\pi}{2}} \sin\theta \; \cos^2\theta d\theta}{\frac{1}{4} n\overline{v}} = \frac{2}{3}\lambda \qquad 2.22$$

In deriving the above equation the values of total flux $= \frac{1}{4} n\overline{v}$ and the molecular flux in direction

θ, $\Phi^\theta = \frac{1}{2} n\overline{v} \; \sin\theta \cos\theta \; d\theta$, as derived in chapter 1 have been used. Equation 2.22 tells that all

molecules approaching the imaginary plane, from above or below, suffer their last collision at a mean

distance $\frac{2}{3}\lambda$ from the imaginary plane. The properties like momentum, kinetic energy and the

direction of motion of the molecules change only on collision, and the molecules suffer their last

collision at a mean distance $\frac{2}{3}\lambda$ from the plane, Therefore, the molecules reaching the imaginary

plane will have the values of energy, momentum and direction of motion same as they had at a

distance $\frac{2}{3}\lambda$ from it. This fact will be used in driving expressions for transport properties.

2.8 Transport Properties of Gases

Thermal conductivity, viscosity and diffusion are called the transport properties of fluids. As will be shown in this section, these properties arise from the transport of, respectively, the kinetic energy, momentum and mass through the layers of the gas when respectively, temperature, velocity or density gradient exists in the gas (fluid) layers. Though in this section we will discuss these properties for an ideal gas, however, the results may well be extended to the case of real gases and other fluids.

2.8.1 The thermal conductivity

It is known that heat or thermal energy across a piece of matter may be transported from one place to another by conduction, convection and radiation. If the substance is a gas or liquid, the heat transfer by convection plays an important role. However, presently we are interested in the property of the thermal conductivity of gases, and, therefore, we do not consider the convection and radiation modes of thermal energy transfer.

Figure 2.17 shows a gas contained in a vessel. Suppose two metallic plates are placed in the gas such that the upper plate bbbb is at a higher temperature than the lower plate cccc. There will be a flow of thermal energy current H from top towards the bottom, where H is the thermal energy per unit area per second (units: 1 Jm^{-2}s^{-1}) passing downwards (in −y direction). A temperature gradient $\frac{dT}{dy}$, increasing upwards (in + y direction), will also be established in the gas between the two plates.

The thermal energy current *H* will be proportional to the temperature gradient and, therefore,

$$H = \kappa \left(-\frac{dT}{dy} \right) \qquad 2.23$$

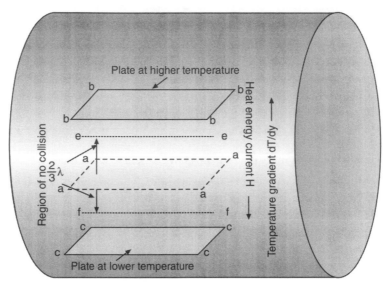

Fig. 2.17   Flow of heat energy in a gas

The negative sign in Eq. 2.23 is due to opposite directions of H and dT/dy. The constant of proportionality κ, called the coefficient of thermal conductivity or simply conductivity, has units of $(= 1\dfrac{Jm^{-2}s^{-1}}{K/m} = 1Jm^{-1}K^{-1}s^{-1})$ 1 Joule per meter per Kelvin per second in MKS system. It is now desired to develop an expression for κ in terms of the properties of the gas.

We consider an imaginary surface aaaa between the two plates in the gas shown by dotted lines in Fig. 2.17. Let the temperature of the gas at this surface be T_0 K. As already shown earlier, there will be a region around this surface of height $\dfrac{2}{3}\lambda$ (where λ is the mean free path of the gas molecules) in which molecules will have no collisions. The temperature of the gas at a distance $\dfrac{2}{3}\lambda$ from the imaginary surface aaaa upwards, i.e., at surface ee will be $T_e = (T_0 + \dfrac{2}{3}\lambda\dfrac{dT}{dy})$ and downwards at surface ff it will be $T_f = (T_0 - \dfrac{2}{3}\lambda\dfrac{dT}{dy})$ .Further, the temperature of molecules coming from above to the surface aaaa will remain T_e and that coming from down will be T_f. No change in the temperature of molecules will take place in collision free region between planes ee and ff.

According to the kinetic theory, the kinetic energy of molecules is determined by the temperature. The kinetic energy E_{kin} of one mole of an ideal gas is equal to $c_v T$, where c_v is the molar heat capacity at constant volume and T the temperature of the gas in kelvin. If c_v is divided by the number of molecules in one mole (i.e., Avogadro's number A_v), one may get a quantity c_v^m $(=\dfrac{c_v}{A_v})$ which may be called the mean thermal capacity at constant volume of each molecule and, therefore, the average kinetic energy of each molecule, E_{kin}^m at temperature T may be written as,

$$E_{kin}^m = c_v^m T \qquad\qquad 2.24$$

It is now clear that every molecule crossing the surface aaaa coming from above will transport a

kinetic energy $\left(E_{kin}^m\right)_{down} = c_v^m\, T_e = c_v^m\left(T_0 + \dfrac{2}{3}\lambda\dfrac{dT}{dy}\right)$ downwards, while each molecule crossing the

surface coming from below will transport a kinetic energy $\left(E_{kin}^m\right)_{up} = c_v^m T_f = c_v^m\left(T_0 - \dfrac{2}{3}\lambda\dfrac{dT}{dy}\right)$

upwards. The number of molecules that cross unit area of surface aaaa per unit time from above and

from below is equal to the molecular flux $\Phi = \dfrac{1}{4}n\bar{v}$. The net kinetic energy flow per unit area per

second of the surface aaaa is equal to the thermal current H, Hence,

$$H = \Phi\left[\left(E_{kin}^m\right)_{up} - \left(E_{kin}^m\right)_{down}\right] = -\frac{1}{4}n\bar{v}\,c_v^m\left[\frac{4}{3}\lambda\frac{dT}{dy}\right] = -\frac{1}{3}c_v^m n\bar{v}\lambda\frac{dT}{dy} \qquad 2.25$$

Comparing Eqs. 2.23 and 2.25, one gets,

$$\kappa = \frac{1}{3}c_v^m n\bar{v}\lambda = \frac{1}{3}c_v^m\frac{\bar{v}}{\sigma} \qquad 2.26$$

In Eq. 2.26 κ is the coefficient of thermal conductivity of the gas, c_v^m the average thermal capacity

at constant volume of each gas molecule, n the number density of the gas, λ the mean free path, $\bar{v}$

the mean velocity of the gas molecules and $\sigma\left(= \frac{1}{n\lambda}\right)$ the microscopic collision cross section.

The values of the coefficient of thermal conductivity calculated using Eq. 2.26 agree reasonably
well with the experimental values for those pressures where the mean free path is not comparable
or larger than the dimensions of the gas container. At very low pressures where the mean free path
becomes comparable to the dimensions of the gas container, the concept of mean free path loses its
meaning and the theory based on collisions between gas molecules does not hold.

Equation 2.25 may also be written as,

$$H. = -\frac{1}{3}n\bar{v}\lambda\frac{d\left(c_v^m T\right)}{dy} \qquad 2.26\ (a)$$

That gives $\kappa = \dfrac{1}{3}n\bar{v}\lambda$ provided the gradient of the molecular kinetic energy $\left(c_v^m T\right)$ is taken into

consideration. Equation 2.26 (a) further tells that the gas possesses the property of thermal conductivity
because of the transport of molecular kinetic energy across gas layers.

2.8.2   The viscosity

Viscosity is a property shown by fluids when they undergo shearing flow in which each adjacent
layers of the fluid move parallel to each other but with different velocities. The adjacent layers of the
fluid apply a sort of a frictional force on each other opposing their motion.

Figure 2.18 shows a container having a gas at some temperature T K. The molecules of the gas
have random motion and velocities. As shown in the figure, the gas is made to flow from left to right
(in X-direction) by maintaining a small pressure difference $(P_1 - P_2)$ between the two ends of the
container. We assume that the pressure difference does not disturb the equilibrium of the system.

The velocities of the gas molecules in the X-direction are small and are super imposed on random velocities of relatively much larger magnitudes. We consider a fixed and stationary plate at the bottom of the container and another plate at some vertical height from the bottom plate. Parallel layers of the gas are moving in the x-direction with velocities in X-direction increasing with the height of the layer from the bottom. This is indicated by the increasing lengths of arrows in the figure and velocity gradient $\frac{dv}{dy}$ normal to the direction of flow. Some gas molecules of a given layer get scattered to the adjacent layers above and below it. Similarly, molecules from the adjacent layers, both above and below a given layer get scattered to the middle layer. Since molecules in adjacent layers have velocities of different magnitudes in X-direction, the scattered molecules bring in unequal amounts of linear momentums. For example, a molecule which has been scattered from an upper layer will carry a larger linear momentum in X-direction and will try to accelerate the lower layer in X-direction. While a molecule scattered from a lower layer to an upper layer will carry a smaller linear momentum in X- direction than all other molecules in the layer and will try to reduce the linear momentum of the upper layer in X-direction. This exchange of molecules in adjacent layers develops a dragging effect. The slower lower layer try to reduce the relative velocity of the middle layer and the faster upper layer tries to accelerate and drag the lower layer.

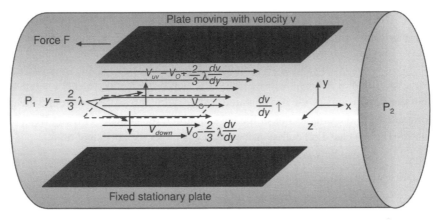

Fig. 2.18   Successive layers of the gas moving parallel to each other with increasing velocity

It appears surprising, at the first instance, that how layers of a gas, the molecules of which undergo elastic collisions, may develop a friction like force opposing the motion of adjacent layers. However, as explained above, the apparent friction force between the adjacent gas layers is the outcome of the transport of unequal linear momentum across layers of the gas.

In Fig.2.18, if left free, the upper plate will move in the +ve X-direction with the flow of the gas. In order to keep the upper plate stationary a force, F, will have to be applied along the –ve X-direction. The force F is

$$F \propto \text{Area of the plate } A.\frac{dv}{dy},$$

Here, $\frac{dv}{dy}$ is the velocity gradient in a direction normal to the direction of flow.

or
$$\frac{F}{A} = \eta \frac{dv}{dy} \qquad 2.27$$

Here η is the constant of proportionality and is called the coefficient of viscosity or simply the viscosity of the gas. The MKS unit for the coefficient of viscosity is 1 N m^{-2}s. The corresponding CGS unit is 1 dyne cm^2 s called 1 Poise = 10 N m^{-2}s.

Again we consider an imaginary surface aaaa, shown by dotted lines in the centre of Fig. 2.18. Let the X-velocity of the gas layer at surface aaaa be v_0. Gas molecules suffer their last collision at a distance $\frac{2}{3}\lambda$ above and below the imaginary surface and there is a collision free region up to a distance of $\frac{2}{3}\lambda$ all around the imaginary surface aaaa. The molecules at the upper edge of the collision free region have velocity in X-direction $v_{up} = v_0 + \frac{2}{3}\lambda\frac{dv}{dy}$ while those at the lower edge $v_{down} = v_0 - \frac{2}{3}\lambda\frac{dv}{dy}$. All molecules approaching the imaginary surface from above have the same velocity v_{up} and all molecules coming to the surface aaaa from below the velocity $v_{up} = v_{down}$. The number of molecules crossing a unit area of surface aaaa per second from above and from below is equal to the flux Φ of the gas molecules. If m is the mass of one gas molecule, then the momentum transported in the forward direction per unit area per second of surface aaaa, by molecules from above $\bar{M} \downarrow$ is given by

$$\bar{M} \downarrow = \Phi m v_{up} = \frac{1}{4}n\bar{v}m\left(v_0 + \frac{2}{3}\lambda\frac{dv}{dy}\right) \qquad 2.28$$

Similarly, the linear momentum transported to the unit area of surface aaaa per second by molecules coming from below $\bar{M} \uparrow$ is given by,

$$\bar{M} \uparrow = \Phi m v_{down} = \frac{1}{4}n\bar{v}m\left(v_0 - \frac{2}{3}\lambda\frac{dv}{dy}\right) \qquad 2.29$$

The net momentum transported per unit area per second $\bar{M} = (\bar{M} \downarrow - \bar{M} \uparrow)$,

$$\bar{M} = \frac{1}{3}n\bar{v}m\lambda\frac{dv}{dy} \qquad 2.30$$

But change in momentum per second is force and hence, $\bar{M} = \dfrac{Force}{Area} = \dfrac{F}{A}$

Therefore,
$$\frac{F}{A} = \frac{1}{3}n\bar{v}m\lambda\frac{dv}{dy} \qquad 2.31$$

Comparing Eqs. 2.27 and 2.31, one gets,

$$\eta = \frac{1}{3}n\bar{v}m\lambda = \frac{1}{3}\frac{m\bar{v}}{\sigma} \qquad 2.32$$

In Eq. 2.32, η is the viscosity, n the number density (N/V), $\bar{v}$ mean molecular velocity, m the mass of the molecule, λ the mean free path and $\sigma\left(= \frac{1}{n\lambda}\right)$ the microscopic collision cross section for gas molecules.

Equation 2.31 may also be written in a slightly modified form as below,

$$\frac{F}{A} = \frac{1}{3} n \bar{v} \lambda \frac{d(mv)}{dy}$$
2.32 (a)

That gives,
$$\eta = \frac{1}{3} n \bar{v} \lambda$$
2.32 (b)

Provided the gradient of momentum (mv) instead of velocity v, is taken into consideration. Further, Eq. 2.32 (a) tells that a gas possesses viscosity because of the transport of linear momentum across layers of the gas.

Equation 2.32 indicates that the viscosity does not depend on the gas pressure or the density of the gas. However, viscosity does depend on the temperature of the gas through $\bar{v}$ as given below.

Since, $\bar{v} = \left(\frac{8k_B T}{\pi m}\right)^{1/2}$, therefore, $\eta = \frac{1}{3}\left(\frac{8k_B}{\pi}\right)^{1/2} \frac{(mT)^{1/2}}{\sigma}$
2.33

Equation 2.33 indicates that in case the above derivation is correct then the viscosity of a given gas should be proportional to $\sqrt{T}$ and for different gases at the same temperature it should be proportional to $\frac{\sqrt{m}}{\sigma}$. The experimentally measured values of viscosities for gases at different temperatures have shown that both of these predictions are true with in experimental uncertainties. The experimental values of the viscosity are often used to determine the microscopic collision cross section $\sigma (= \pi d^2)$ and hence the collision diameter d of the gas molecules.

One can determine the ratio of the thermal conductivity κ to the viscosity η for a given gas using the Eqs. 2.26 and 2.32 as,

$$\frac{\kappa}{\eta} = \frac{\dfrac{1}{3}\dfrac{c_v^m \bar{v}}{\sigma}}{\dfrac{1}{3}\dfrac{m\bar{v}}{\sigma}} = \frac{c_v^m}{m} = \frac{c_v}{m.A_v} = \frac{c_v}{M}$$
2.34

In Eq. 2.34 c_v is the molar thermal capacity at constant volume, M the molecular weight, m the mass of one molecule of the gas, and c_v^m mean thermal capacity at constant volume of each gas molecule. It also follows from Eq. 2.34 that the ratio $\frac{\kappa M}{\eta c_v} = 1$ for all gases, however, experimental values of the ratio is found to vary from 1.5 to 2.5 for different gases showing that the order of magnitude of the ratio is correctly predicted by the theory. The discrepancy in the absolute magnitude may be assigned to the fact that the gas molecules are not hard spheres but are more like tennis balls which flexes on collision.

2.8.3 The coefficient of diffusion

It is a common experience that a puff of smoke thrown out by a smoker slowly dissolves out in the air. Though no large scale motion of particles is observed but slowly the molecules of smoke diffuse into the air and the air molecules into the puff of smoke. This slow motion of molecules from the region of higher concentration to the region of lower concentration is called diffusion. Diffusion is a very general process observed in gases, liquids and even in solids. The basic reason of diffusion is the random thermal motion of molecules; however, the rate of diffusion depends on several factors including the masses of the molecules, the temperature of the system, and the concentration gradient of the molecules. If there is a system where only one type of molecules are present but the concentration of molecules is different in different parts of the system, then molecules from the region of higher concentration diffuse to the region of lower concentration and the process is called self-diffusion. On the other hand if there is a system which contains two different species of molecules concentrated into different parts of the system, then also diffusion takes place and molecules of one kind diffuse into the molecules of the other kind. In principle it is easy to detect experimentally the process of diffusion of different types of molecules but theoretical treatment of the process is much involved. The diffusion of identical molecules from the region of higher concentration to the region of lower concentration is experimentally difficult to detect, because it is not possible to distinguish between a molecule that has reached a particular point by diffusion and the one that was already present at that point. However, the theoretical treatment of self-diffusion is much simpler. A compromize may be made by studying the diffusion of not exactly identical but very nearly similar molecules. For example if there are two isotopes of an element, one stable and the other radioactive (which can be tagged through its radioactive properties) and the two isotopes are concentrated in different parts of a system, then it will be possible to follow experimentally the diffusion of the tagged molecules.

Let us consider that a gas consisting of tagged and the normal molecules is kept in a container such that initially the tagged molecules are in the upper part and the normal ones in the lower. We further assume that the temperature in the container is same throughout. As a result of diffusion tagged molecules from above will move downwards and after some time a gradient $\dfrac{dn^{tag}}{dy}$ in the concentration of tagged molecules in y-direction will be established. Here n^{tag} is the number density of the tagged molecules. The flux $F\downarrow$ of the tagged molecules, i.e., the number of tagged molecules per unit area per second, diffusing downwards is given by,

$$F\downarrow = -D\frac{dn^{tag}}{dy} \qquad\qquad 2.35$$

Here D is called the coefficient of diffusion. The MKS unit for D is $1\ m^2s^{-1}$.

We consider the situation after some time when sufficient number of tagged molecules have mixed with the normal ones and imagine a surface aaaa in the middle of the container. Tagged molecules from both the above and also from the lower sides will be crossing this imaginary surface at every instant. However the number of tagged molecules crossing the surface from above will be larger than the number of molecules crossing from below. It is because of their larger concentration upwards and the finite concentration gradient. The average number of tagged molecules crossing from above will become equal to those crossing from below when the concentration gradient vanishes, i.e., when the molecules are uniformly distributed all over and diffusion has stopped. We are, however, considering the situation when diffusion is still taking place. Let n_0^{tag} be the number density of tagged molecules at the surface aaaa. Molecules coming from above would have suffered their last collision at a distance $\frac{2}{3}\lambda$ above the surface where their number density n_{up}^{tag} will be

$$n_{up}^{tag} = n_0^{tag} + \frac{2}{3}\lambda\frac{dn^{tag}}{dy} \qquad\qquad 2.36$$

Since these molecules do not suffer any further collisions till they reach the surface aaaa, their number density does not change when they cross the surface.

Similarly, the number density of molecules coming from below n_{down}^{tag} will be

$$n_{down}^{tag} = n_0^{tag} - \frac{2}{3} \lambda \frac{dn^{tag}}{dy}$$

2.37

and the number density will not change till they cross the surface.

The flux of tagged molecules crossing surface aaaa down wards $F^S \downarrow$ is given by

$$F^S \downarrow = \frac{1}{4} \bar{v} \, n_{up}^{tag}$$

and the upwards flux of molecules crossing from below is given by

$$F^S \uparrow = \frac{1}{4} \bar{v} \, n_{down}^{tag}$$

The net down ward flux $F^{net} \downarrow$ is given by

$$F^{net} \downarrow = \frac{1}{4} \bar{v} \left(n_{down}^{tag} - n_{up}^{tag} \right) = -\frac{1}{3} \bar{v} \, \lambda \frac{dn^{tag}}{dy}$$

2.38

A comparison of Eq.2.39 with Eq.2.35 gives,

$$D = \frac{1}{3} \bar{v} \lambda = \frac{\bar{v}}{3n\sigma}$$

2.39

Eq. 2.38 may also be written as,

$$F^{net} \downarrow = -\left(\frac{1}{3} \bar{v} n \lambda \right) \frac{d\left(n^{tag} / n \right)}{dy}$$

2.40

and now the expression for the coefficient of self-diffusion D becomes

$$D = \left(\frac{1}{3} \bar{v} n \lambda \right)$$

2.41

Provided the concentration gradient is taken in terms of the relative number density $\left(n^{tag} / n \right)$ of tagged molecules. Further, Eq. 2.40 signifies that the property of diffusion results from the difference in the relative concentration of molecular number density in different parts of the system.

2.9   Application of Kinetic Theory to Free Electrons in Metals: Success and Failure

Only three years after the discovery of electron Paul K. L. Drude, a German physicist, applied the concepts of kinetic theory of gases to the free electrons of metals. Although electrons are charged and free electron density in metals is larger than the number density of gas molecules at room temperature, still Drude applied the kinetic theory that was essentially developed for neutral dilute gases. The basic assumptions of Drude's classical theory may be summarized as;

1. Free electrons in metal collide with the lattice of positive ions and electron–electron collisions are neglected. Between collisions electrons move in a straight line when no external field is present. The electromagnetic interactions between electron–electron and electron–positive ion are neglected.

2. The mean free time between collisions is τ and the probability of collision per unit time is $\frac{1}{\tau}$; the probability of having a collision in infinitesimal time dt is $\frac{dt}{\tau}$. The mean free time τ is independent of the position of the free electron as well as its velocity.

3. After each collision the free electron forgets about its previous history, i.e., about its velocity, direction of motion, etc., and emerges in a random direction with speed characteristic of the temperature of the region where collision has occurred, hotter the region higher the speed. In other words, free electrons achieve thermal equilibrium by collision with positive ion lattice.

With these assumptions Drude's theory successfully explained Ohm's law and electrical conductivity of metals. However, the theory failed to explain why free electrons do not contribute to the constant volume heat capacity of metals.

2.9.1   Lattice–electron collisions

Drude model assumes that free electrons in a metal collide only with the positive ion lattice and ignores electron–electron collisions. Further, it assumes that after every collision electron acquires the speed characteristic of the temperature at the location of the collision and moves with that speed in any random direction. Lattice electron collisions are due to this random motion. Let us consider a piece of metal of length L and area of cross section S in thermal equilibrium at temperature T. If v_e denotes the random speed of the electron just after a collision with the lattice, then applying the principle of equipartition,

$$\frac{1}{2}m_e v_e^2 = \frac{3}{2}k_B T \text{ or } v_e = \sqrt{\frac{3k_B T}{m_e}} \qquad 2.42$$

Here m_e is the mass of an electron. At room temperature (T $\approx$ 300 K) v_e is of the order of 10^5 m/s.

If the mean time between two successive collisions is denoted by τ, often called the relaxation time, and the mean distance travelled by the free electron between collisions (the mean free path) is denoted by λ_e, then,

$$\tau = \frac{\lambda_e}{v_e} \qquad 2.43$$

An important point to note is that although electron–lattice collisions occur because of speed v_e but the average value of velocity $\overline{v_e}$ is zero on account of its being random in direction, i.e.,

$$<v_e> = \overline{v_e} = 0 \qquad 2.44$$

2.9.2   D.C. electrical conductivity of metals from Drude's theory

The direct current density j in a metallic conductor is defined as the electric current per unit time that passes through a unit area held normal to the direction of the flow of current. If $-e$ denotes the charge, v_e the velocity and n_e the number density (number per unit volume) of free electrons in a metal, then the current density due to the motion of free electrons may be given by,

$$j = -e n_e \overline{v_e} \qquad\qquad 2.45$$

When there is no externally applied electrical field across the metal, after every collision electrons emerge with random direction of motion and with speeds corresponding to the temperature of the metal at the location of collision. Because of the random directions of motion of free electrons the average value of electron velocity $\overline{v_e} = 0$. Hence the current density j is zero.

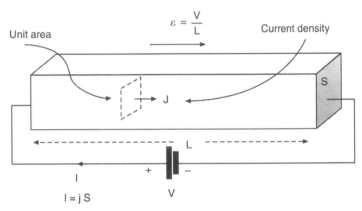

Fig. 2.19 Flow of current through a piece of conductor when an electric field is applied across it

Suppose an external electric field ε is produced across the metallic conductor by connecting a source of potential difference V across its two ends as shown in Fig. 2.19. Each free electron will experience a force $F = -e\varepsilon$ in a direction opposite to the applied electric field and a corresponding acceleration $a_e = \dfrac{-e\varepsilon}{m_e}$. If τ denotes the average time between two collisions then a free electron will gain a velocity $v = \dfrac{-e\varepsilon\tau}{m_e}$ due to the applied electric field during its motion between two successive collisions. Please note that just at the end of a collision the velocity due to the electric field was zero and just before the next collision it has become v. The mean velocity of the free electron due to the applied electric field between two successive collisions is called the drift velocity and is denoted by v_d. It is given as $v_d = \left(\dfrac{0 + v_m}{2}\right) = \dfrac{-e\varepsilon\tau}{2m_e}$. The drift velocity is not random and is, therefore, superimposed over the random velocity, v_e, of electrons. Thus, under the influence of an external electric field the average value of the current density is given by

$$j = -e\, n_e\, \overline{v_e} + \left(-e\, n_e\, v_d\right) = 0 + \left(-e\, n_e\, \dfrac{-e\varepsilon\tau}{2m_e}\right)$$

or

$$j = \dfrac{n_e e^2 \tau}{2m_e}\varepsilon \qquad\qquad 2.46$$

But,
$$j = \sigma\varepsilon = \frac{1}{\rho}\varepsilon \qquad\qquad 2.47$$

where, σ and ρ are, respectively, the conductivity and resistivity of the metal. Hence,

$$\rho = \frac{2m_e}{n_e e^2 \tau} \qquad\qquad 2.48$$

All other quantities except τ are known in Eq. 2.48, switching the values of known quantities τ is estimated to be of the order of 1×10^{-14}s. Taking $\underline{v}_e \approx 10^5$ m/s and using Eq. 2.43, one may estimate λ_e to be of the order of 1×10^{-9} m, which is the order of intermolecular distance. As such Drude's theory apparently gives the correct order of magnitude for the mean free path of free electrons in metals.

Multiplying the two sides of Eq. 2.47 by the area of cross section S of the conductor one gets,

$$Sj = \frac{n_e e^2 \tau S}{2m_e}\varepsilon = \frac{n_e e^2 \lambda_e S}{2m_e v_e}\frac{V}{L} = \frac{S}{\rho L}V$$

Putting $Sj = I$, one gets

$$V = RI, \text{ where } R = \frac{\rho L}{S} \text{ is the resistance of the specimen.}$$

Thus assuming that the free electrons in a metal behave like the molecules of an ideal gas, Drude successfully derived Ohm's law and showed that the resistance of the metallic conductor originates from electron–lattice collisions.

However, Drude's assumption that free electrons in metals behave like molecules of an ideal gas completely failed in explaining the experimental fact that the constant volume molar heat capacity of metals at room temperature is of the order of 3R, the value predicted by Dulong-Petite for non-metallic solids. The value 3R for molar specific heat capacity for non-metallic solids comes from the three dimensional (vibratory and or rotatory) motion of the molecules of the solid which are fixed at their locations. In the case of metallic solids the positive ions may, therefore, contribute 3R to the molar specific heat. The free electrons, if they are like the molecules of an ideal gas must also contribute to the molar heat capacity in addition to 3R. The molar heat capacity of metals should, therefore, be much larger than 3R. This failure of Drude's classical theory has been successfully explained by the quantum thermodynamics, as will be shown later.

Solved Examples

1. Hydrogen gas is contained in a cubical box of side 0.1m at 1.0 atm pressure and 300 K temperature. Given that the mass of hydrogen molecule is 3.34×10^{-27} kg and the critical molar volume V_c, $= 0.065$ m^3/kilomole, calculate, (i) rms, average and average relative speeds of molecules (ii) number density (iii) radius of the molecule (iv) microscopic and macroscopic collision cross sections (v) collision frequency and mean free path and (vi) Average number of collisions suffered by each container wall per second.

Solution:

(i) $V_{rms} = \left(\dfrac{3k_B T}{m}\right)^{1/2} = \left(\dfrac{3 \times 1.38 \times 10^{-23} \times 300}{3.34 \times 10^{-27}}\right)^{1/2} = 19.3 \times 10^2$ m/s)

$speed\, \overline{v} = \left(2.55 \dfrac{k_B T}{m}\right)^{1/2} = 17.7 \times 10^2$ m/s ;

Average relative speed $\overline{v}_r = \sqrt{2}\,\overline{v} = 25.0$ m/s

(ii) $P = \dfrac{1}{3}nm\overline{v^2}$, hence $n = \dfrac{3P}{m\overline{v^2}} = \dfrac{3 \times 1.01 \times 10^5}{3.34 \times 10^{-27} xv_{rms}^2} = 2.4 \times 10^{25}$ m^{-3}

(iii) Radius of the molecule $r = \left(\dfrac{v_c}{16\pi A_V}\right)^{1/3} = \left(\dfrac{65.0 \times 10^{-3}}{16 \times 3.14 \times 6.02 \times 10^{26}}\right)^{1/3} = 1.30 \times 10^{-10}$ m

(iv) Microscopic cross section $\sigma = \pi(2r)^2 = 3.14 \times (2 \times 1.30 \times 10^{-10})^2 = 21.2 \times 10^{-20}$ m^2 per molecule

Macroscopic cross section $= n\sigma = 2.4 \times 10^{25}$ m$^{-3} \times 21.2 \times 10^{-20}$ m$^2 = 50.9 \times 10^5$ m^{-1}

(v) Mean free path $\lambda = \dfrac{1}{n\sigma} = \dfrac{1}{50.9 \times 10^5} = 1.97 \times 10^{-7}$ m;

Collision frequency $f_{col} = \sqrt{2}\,\overline{v}\,n\sigma = \sqrt{2} \times 17.7 \times 10^2 \times 50.9 \times 10^5\, s = 1.3 \times 10^{10}$ collisions per sec

Mean collision time $\tau = \dfrac{1}{f_{col}} = 7.7 \times 10^{-11}$ s

(vi) Average number of collisions suffered by one wall of the container: A cube has six identical faces and the gas pressure at each face is same, hence it is reasonable to assume that on average n/6 molecules per unit volume moves with average speed $\overline{v}$ in the direction of each face of the cube. If A is the area of each face then $\left(\dfrac{n}{6}\overline{v}\,A\right) molecules$ will hit the face per second. The number of hits per second on each face $= \left(\dfrac{n}{6}\overline{v}\,A\right) = 42 \times 10^{25}$

2. An insulated container is divided into two parts by a diaphragm that has a hole of dimensions D. The two parts are filled with the same gas but are kept at temperatures T_1 and T_2 by connecting them to the two heat sources. If λ_1 and λ_2 are the mean free paths on the two sides, discuss the conditions for achieving steady state when (i) $D \gg \lambda_1$ and $D \gg \lambda_2$ (ii) $D \ll \lambda_1$ and $D \ll \lambda_2$

Solution:

Since heat is continuously supplied to the two parts of the container the temperatures of the two parts will always remain different, T_1 and T_2 and there will never be a thermal equilibrium. However, a steady state will eventually reach when the net flow of molecules through the hole will be zero, i.e., as many molecules per unit time pass from one side to the other exactly same number will cross from the other side to the first.

(*i*) If the size of the hole i.e., D is larger than both the free paths λ_1 and λ_2 (condition *i*) there will be macroscopic flow of gas molecules and steady state will reach when the collision rate of molecules on the two sides becomes equal. The collision rate or frequency of collision is given by $d^2 \bar{v_r} n$, where d = diameter of the molecule, $\bar{v_r} = \sqrt{2}\,\bar{v}$ the mean relative velocity of molecules, $\bar{v}$ the average velocity of the molecule and n the number density. Since same gas is kept in two parts d is same but n and $\bar{v}$ will have different values for the two parts. So for steady state

$$\pi d \sqrt{2}\,\bar{v_1}\,n_1 = \pi d \sqrt{2}\,\bar{v_2}\,n_2$$

or

$$\frac{n_1}{n_2} = \frac{\bar{v_2}}{\bar{v_1}} = \frac{\sqrt{\dfrac{8k_B T_2}{\pi m}}}{\sqrt{\dfrac{8k_B T_1}{\pi m}}} = \sqrt{\frac{T_2}{T_1}}$$

Therefore, in this case steady state will reach when the ratio of the number of molecules on the two sides will be equal to the square root of the inverse ratio of their temperatures.

(*ii*) In the second case when the dimension of the hole is much less than the mean free paths, molecules will not be able to pass through the hole because of their mutual collisions. However, they will be pushed through the hole because of the pressure difference on the two sides. The steady state will reach when pressures on the two sides of the hole becomes equal. But pressure $P = \dfrac{1}{3} n m v_{rms}^2$ and for steady state in the second case,

$$n_1 \left(v_{rms}\right)_1^2 = n_2 \left(v_{rms}\right)_2^2 \quad \text{or} \quad \frac{n_1}{n_2} = \frac{\left(v_{rms}\right)_2^2}{\left(v_{rms}\right)_1^2} = \frac{T_2}{T_1}$$

It may be observed that in the first case it is the square root of the ratio of two temperatures while in the second case it is simply the ratio of two temperatures that decides the ratio of the number of molecules on the two sides of the hole.

3. In example-1 above, if it is assumed that hydrogen gas in the container is in thermal equilibrium and that there is no macroscopic flow of the gas, molecules of the gas move from one location to another because of their mutual collisions that are characterized by the mean free path $\lambda (= 1.97 \times 10^{-7}$ m) and the mean velocity $\bar{v} (= 17.7 \times 10^2$ m/s). It is now required to calculate the average time that a particular molecule will take to move to a distance say, L = 5 m.

Solution:

Let us assume that a given molecule moves a distance L in X number of collisions. Since collision is a random process,

$$L^2 = X \lambda^2 \tag{2.3.1}$$

The required time $\qquad\qquad T = X\,\tau_{col} = X\dfrac{\lambda}{\sqrt{2}\,\bar{v}}$ $\qquad\qquad$ (2.3.2)

Substituting the value of X from Eq. (2.3.1) into Eq. (2.3.2)

$$T = X\,\tau_{col}$$

$$= X\dfrac{\lambda}{\sqrt{2}\,\bar{v}}$$

$$= \dfrac{L^2}{\sqrt{2}\,\bar{v}\,\lambda}$$

$$= \dfrac{25}{\sqrt{2}\times 17.7\times 10^2 \times 1.97\times 10^{-7}} = 5.0\times 10^5 \text{ s}$$

4. (a) Under what approximation the ratio of pressure to the coefficient of viscosity for a gas is equal to molecular collision frequency? (b) For a particular gas the molecular collision frequency at atmospheric pressure and 300 K temperature is 5×10^9 per second. What is the value of the coefficient of viscosity for the gas under the above approximations? (c) Using the value of coefficient of viscosity obtained above obtain he diameter of the molecule if the mass of the molecule is 4×10^{-27} kg.

Solution:

(a) $P = \dfrac{1}{3} n\,m\,v_{rms}^2$ and $\eta = \dfrac{1}{3} n m \bar{v}\lambda$

Hence, $\dfrac{P}{\eta} = \dfrac{v_{rms}^2}{\bar{v}\lambda}$, If one makes the approximation that the *rms* velocity is equal to the mean velocity which is also equal to the relative molecular velocity v_r, i.e., $v_{rms} = \bar{v} = v_r$,

then $\dfrac{P}{\eta} = \dfrac{\bar{v}}{\lambda} = \dfrac{v_r}{\lambda}$ = molecular collision frequency

(b) Using the above approximation $\eta = \dfrac{P}{f_{col}} = \dfrac{1.013\times 10^5}{5\times 10^9} = 2.06\times 10^{-5}\ Nm^{-2}s$

(c) Also, $\qquad\qquad \eta = \dfrac{1}{3} mn\bar{v}\lambda = \dfrac{1}{3}\bar{v}\dfrac{1}{\sigma}$ or $\sigma = \dfrac{1}{3}\bar{v}\dfrac{m}{\eta}$

but $\qquad\qquad \sigma = \pi d^2$

or $\qquad\qquad d^2 = \dfrac{\bar{v}}{3\pi\eta} = \dfrac{\left(m\dfrac{8k_BT}{\pi}\right)^{1/2}}{3\times\pi\times 2.06\times 10^{-5}} = 1.7\times 10^{-20}\ m^2$

and diameter of the molecule $d = 1.36 \times 10^{-10}$ m.

5. A container is divided into two equal parts 'a' and 'b' each of volume V by a partition. Initially part 'b' is empty and part 'a' is filled with a gas that has n_0 molecules per unit volume. A small hole of area s (s>> mean free path) is then made in the partition so that macroscopic flow of gas into part b takes place. The hole is small enough so that temperature of parts 'a' and 'b' remains same and thermal equilibrium is maintained at all times. Derive a relation for the number of gas molecules in part 'a' at some time t after the opening of the hole in terms of the mean velocity $\bar{v}$ of molecules.

Solution:

Since temperature of the two parts is same the mean velocity $\bar{v}$ will have same value for part-a

and part-b and will remain constant throughout as $\bar{v} = \sqrt{\dfrac{8k_BT}{\pi m}}$. Initially the total number of

molecules in part-a is ($V n_0$) and at some later time 't' let there be Vn_t molecules in part-a and $(V(n_0 - n_t))$ molecules in part-b. The number density of molecules in part-b at instant 't' is $(n_0$

$- n_t)$. At instant 't' a flux $\left(\dfrac{sn_t\bar{v}}{4}\right)$ is moving through the hole of area s towards part-b and

similarly a flux $\left[\dfrac{s(n_0 - n_t)\bar{v}}{4}\right]$ is moving from part-b through the hole towards part-a. The

net change in the flux of molecules in part-a at instant 't' may be given as,

$$\frac{d}{dt}\left[Vn_t\right] = -\left(\frac{sn_{t\bar{v}}}{4}\right) + \left[\frac{s(n_0 - n_t)\bar{v}}{4}\right] \tag{S-2.5.1}$$

or
$$\frac{dn_t}{dt} = -\frac{sn_t\bar{v}}{2V} + \frac{sn_0\bar{v}}{4V}$$

or
$$\frac{dn_t}{dt} + \frac{s\bar{v}}{2V}n_t - \frac{sn_0\bar{v}}{4V} = 0 \tag{S-2.5.2}$$

Equation (S-2.5.2) is a first order differential equation which has the following solution for boundary condition that at $t = 0$, $n_t = n_0$,

$$n_t = \frac{n_0}{2}\left[1 + e^{-\left(\frac{s\bar{v}t}{2V}\right)}\right] \tag{S-2.5.3}$$

Equation (S-2.5.3) gives the required relation. If both sides of this equation are multiplied by $\frac{1}{3}mv_{rms}^2$, one gets:

$$P_t = \frac{P_0}{2}\left[1 + e^{-\left(\frac{s\bar{v}t}{2V}\right)}\right]$$

The above relation tells how the pressure in part-a changes with time.

6. Gases are often used as insulation material. Discuss the dependence of the insulating properties of a gas on its molecular weight.

Solution:

The insulation properties of any material depend on the coefficient of thermal conductivity κ $= \frac{1}{3}c_v^m \frac{\bar{v}}{\sigma}$, here c_v^m is the constant volume heat capacity of each molecule, $\bar{v}$ the mean velocity of the molecule and A is the microscopic cross section $= \pi d^2 = \pi(2r)^2$ where r is the radius of the molecule. Now, $\bar{v} \propto \sqrt{\frac{T}{M}}$ where M is the Molecular weight. Also, assuming that the molecule is a sphere of radius r and density ρ, $M = \frac{4}{3}\rho\pi r^3$ and $r \propto M^{1/3}$. Hence $\kappa \propto \sqrt{\frac{T}{M}}.M^{-\left(\frac{2}{3}\right)}$ or

$\kappa \propto T^{1/2}\,M^{-7/6}$ Thus molecular weight is the most important parameter of the molecule that decides its coefficient of conductivity. Further, the coefficient of thermal conductivity depends on $T^{1/2}$ i.e., on the square root of the temperature.

7. A group of nitrogen molecules at temperature 300 K starts simultaneously and travel in a particular direction in the volume of nitrogen gas, all with the same velocity equal to the average speed $\bar{v}$. Assuming that the mean free path of molecules is 5×10^{-2} m, calculate the time till which 60% of nitrogen molecules in the group suffers no scattering.

Solution:

The weight m of each nitrogen molecule $= \dfrac{28}{6.02 \times 10^{26}}$ kg ,

The average speed $\bar{v} = \sqrt{\dfrac{2.55 \times k_B \times T}{m}} = \sqrt{\dfrac{2.55 \times 1.38 \times 10^{-23} \times 300 \times 6.02 \times 10^{26}}{28}}$

$$= 4.76 \times 10^2 \text{ m/s}$$

Let us assume that after travelling a distance X the number of molecule that does not suffer scattering is 60%. Hence, from survival equation,

$$0.6 = e^{-\left(\frac{X}{\lambda}\right)} = e^{-\left(\frac{X}{5\times10^{-2}}\right)}$$

But from the table of exponential functions, $e^{-0.51} = 0.6$,

Therefore, $\left(\dfrac{X}{5\times10^{-2}}\right) = 0.51$

or $X = 0.51 \times 5 \times 10^{-2} = 2.55 \times 10^{-2}$ m

Time 't' taken in travelling the distance 2.55×10^{-2} m with average speed $\bar{v} = 4.76 \times 10^2$ m/s

$$t = \dfrac{2.55 \times 10^{-2}\text{ m}}{4.76 \times 10^2\text{ m/s}} = 5.35 \times 10^{-5}\text{ s}$$

8. (a) What percent of molecules, on average, will travel a distance of λ, 2λ, and 3λ before suffering their first collision, λ being the mean free path of the molecules? (b) How many molecules out of 100000 will travel a distance more than 2λ and less than 3λ?

Solution:

(a) According to survival equation $\dfrac{N}{N_0} = e^{-\frac{x}{\lambda}}$, where N is the number of molecules that have

travelled a distance X without suffering any collision out of the initial N_0. The percent of molecules that have not suffered any collision till $X = \lambda$, $X = 2\lambda$ and $X = 3\lambda$ are respectively,

$e^{-\frac{\lambda}{\lambda}} \times 100 = e^{-1} \times 100 = 0.367 \times 100 = 36.7\%$; $e^{-2} \times 100 = 13.5\%$; $e^{-3} \times 100 = 4.97\%$.

(b) Number of molecules out of 1000 that will travel a distance more than $2\lambda = 1 \times 10^5 \times e^{-2}$

$$= 13500$$

Also the number of molecules that will travel a distance more than $3\lambda = 1 \times 10^5 \times e^{-3}$

$$= 4970$$

Hence the number of molecules that suffered scattering between 2λ and 3λ distance

$$= 13500 - 4970 = 8530$$

9. A thin walled bottle of volume V contains N_0 ideal gas molecules at time $t = 0$. The bottle is kept in vacuum and its temperature is maintained to a fixed value T. A small hole of area S is made in the wall of the bottle. Assuming that no molecule from vacuum will ever enter the bottle, obtain an expression for the number of molecules in the bottle at some subsequent instant t in terms of the average speed $\bar{v}$, area S of the hole and volume V of the bottle.

Solution:

The average flux of molecules in the bottle at any point is $\dfrac{n\bar{v}}{4} = \dfrac{\dfrac{N}{V}\bar{v}}{4} = \dfrac{N\bar{v}}{4V}$

Let N_t denote the number of molecules in the bottle at instant 't'.

Number of molecules leaving at instant t through the hole of area S per second $= -\dfrac{dN_t}{dt} = s\dfrac{N_t\bar{v}}{4V}$

or $\dfrac{dN_t}{N_t} = -\dfrac{S\bar{v}}{4V}dt$ or $\ln N_t = -\dfrac{S\bar{v}}{4V}t + \text{Constant}$

Putting the boundary condition $N_{t=0} = N_0$, one gets the desired result.

$$N_t = N_0 e^{-\left(\frac{S\bar{v}}{4V}\right)t}$$

10. Derive an expression for the intensity of the molecular beam leaking into vacuum from a gas container through a small hole of area S in terms of the gas pressure P, temperature T, area S and the mass m of the molecule.

Solution:

Molecular beam intensity I may be given as,

$$I = \text{number of molecules coming out per unit time} = \frac{S n \bar{v}}{4} \tag{2.10.1}$$

$$\text{Pressure } P = \frac{1}{3} nmv_{rms}^2 = \frac{1}{3} n \left(2 x \frac{1}{2} mv_{rms}^2 \right) = \frac{2}{3} n \left(\frac{3}{2} k_B T \right) = n k_B T$$

or

$$n = \frac{P}{k_B T} \, ;$$

Also

$$\bar{v} = \sqrt{\frac{8 \pi k_B T}{m}}$$

Substituting the above values of $\bar{v}$ and n in Eq. (2.10.1), one gets

$$I = \frac{1}{4} S x \frac{P}{k_B T} x \sqrt{\frac{8 \pi k_B T}{m}} = \frac{SP}{\sqrt{mT}} \sqrt{\frac{\pi k_B}{2}} \tag{2.10.2}$$

Hence, $I \propto S; I \propto P; I \propto \dfrac{1}{\sqrt{T}}; I \propto \dfrac{1}{\sqrt{m}}$

Equation (2.10.2) gives the desired relation.

Problems

1. At 1.0 atm. pressure and 300 K the mean free path for a diatomic gas is 3.0×10^{-8} m and the mean velocity of gas molecules is 450 m/s. calculate the value of the coefficient of viscosity and the coefficient of thermal conductivity of the gas.

2. A molecular beam is produced by leaking the gas through a small hole of area S into the vacuum from a gas container at constant temperature T and pressure P. What will happen to the intensity of the molecular beam when the temperature, the pressure and the size of the hole are all increased by a factor of four?

3. The critical temperature and critical pressure for a diatomic gas are, respectively, 304.1 K and 73 atm. Calculate the radius of the molecule of the gas.

4. One kilomole of CO_2 gas is held in a container of volume 10 m³ at 519 K temperature. Assuming that the radius of CO_2 molecule is 1.6×10^{-10} m determine (i) the average speed (ii) mean free path (iii) collision frequency for molecules.

5. Two gases A and B with molecular masses 3×10^{-26} kg and 5×10^{-26} kg and diameters respectively, 1.9×10^{-10} m and 2.3×10^{-10} m are mixed together and are held in equilibrium at 300 K temperature and 10^{-4} times the atmospheric pressure. Obtain the (a) number density for the molecules, (b) the exclusion radii for the collision of A-A molecules, B-B molecules and A-B molecules, (c) the mean velocities for A molecules and B molecules

6. A $\frac{1}{8}$ litre bottle contains Nitrogen gas at 10^{-5} atmospher pressure and 300 K temperature. The gas is allowed to leak through a hole of 1 micron radius into vacuum. Calculate (i) the initial number density of molecules in the bottle (ii) the average speed of gas molecules (iii) the time it will take the gas molecules to reduce to one fifth of their initial number.

7. A mixture of Oxygen and Hydrogen is kept in a bottle of 1/8 liter volume at 300 K and 10^{-5} atmosphere pressure. Calculate the time the two gases will take in reducing to 1/10 of their original amounts when they are allowed to leak through a hole of radius 1.0 micron into vacuum.

8. A beam containing 10^6 molecules of gas A of radius 2.5×10^{-10} m is projected on gas B, molecular radius 1.8×10^{-10} m, and number density 1×10^{23} m^{-3}. (i) How many molecules of A will survive after travelling a distance 2.58×10^{-5} m in gas B? (ii) what distance the beam will travel in gas B before half of its molecules are scattered?

9. The atomic weight of copper is 64 and its density is 8.9×10^3 kg m^{-3}. Calculate the density of free electrons at room temperature in a piece of copper of mass 3 kg.

10. Silver has a density of 10.5 cm^{-3}, atomic weight 108; and resistivity $1.59 \times 10^{-8}\,\Omega - m$. Calculate (i) the number density of free electrons (ii) random velocity of electron at 300 K (iii) the relaxation time (iv) mean free path (v) mean drift velocity for potential gradient of 0.5 volt per meter.

Short Answer Questions

1. Differentiate between the random and the drift velocities of free electrons in metal and briefly outline their roles.
2. Derive an expression for the mean drift velocity of free electrons in metals
3. Write a short note on Van der Waals force.
4. Obtain correction term for the reduction in pressure due to molecular attraction in case of real gases.
5. What is meant by the "sphere of exclusion" in molecular collisions?
6. Show that the volume not available for the motion of molecules is four times the total volume of molecules in one mole of a real gas.
7. Discuss Van der Waals' correspondence principle.
8. Write survival equation (without derivation) and discuss its importance.
9. Define mean free path and collision frequency for a gas and write expressions for them.
10. Define coefficients of (i) thermal conductivity (ii) viscosity and (iii) diffusion for gases.
11. What is compressibility of a real gas and what is its physical significance?

Long Answer Questions

1. Derive Van der Waals equation of state for a real gas and show that critical volume $v_c = 3b$, where symbols have their usual meaning.

2. Discuss the collision of gas molecules and define collision diameter, microscopic and macroscopic cross sections, collision frequency, mean free path and hence derive the survival equation for gas molecules.

3. Show that the mean height from an arbitrary plane in the volume of a gas at which the gas molecules suffered their last collision before reaching the plane is given by $\frac{2}{3}\lambda$, where λ is the mean free path.

4. Show that a gas possesses the property of thermal conductivity because of the transport of molecular kinetic energy across gas layers.

5. Derive an expression for the coefficient of viscosity of a gas and hence show that according to the kinetic theory the ratio $\dfrac{\kappa\, M}{\eta\, C_V} = 1$ for all gases. Symbols have their usual meanings.

6. Discuss the process of diffusion in gases and obtain expression for the coefficient of self-diffusion. Further, show that the relative concentration gradient is responsible for diffusion.

7. What assumptions were made by Drude while applying the concepts of kinetic theory to the free electrons in a metal? Show how Drude's classical picture was successful in driving Ohm's law but failed in explaining the constant volume heat capacity of metals.

Multiple Choice Questions

Note: Some of the following problems may have more than one correct alternative. All correct alternatives must be marked for complete answer and full marks

1. If P_i & P_r are, respectively, the pressures exerted by the same amounts of an ideal gas and an real gas then,
   (a) $P_i > P_r$ (b) $P_i = P_r$
   (c) $P_i < P_r$ (d) no definite relation between P_i & P_r

2. N molecules each of diameter 'd' are held in a volume V. The actual volume available to the molecules for their motion is,
   (a) $\left(V - \dfrac{16}{3}\pi d^3\right)$ (b) $\left(V - \dfrac{8}{3}\pi d^3\right)$

   (c) $\left(V - \dfrac{4}{3}\pi d^3\right)$ (d) $\left(V - \dfrac{2}{3}\pi d^3\right)$

3. If b denotes the volume not available for the motion of molecules per mole, A_v Avogadro's number and P_c, v_c, & T_c the critical constants of a gas then the radius r of the molecule is,
   (a) $\left(\dfrac{3b}{16\pi A_v}\right)^{1/3}$ (b) $\left(\dfrac{3RT_c}{128\pi P_c A_v}\right)^{1/3}$

   (c) $\left(\dfrac{8b}{16\pi A_v}\right)^{1/3}$ (d) $\left(\dfrac{16\pi A_v}{3b}\right)^{1/3}$

4. A beam containing 1000 molecules was projected on a gaseous target and it was observed that 2570 molecules were scattered out of the beam after travelling a distance of 10×10^{-8} m in the gas. The mean free path of incident molecules in the gaseous medium is,
   (a) 5×10^{-8} m   (b) 15×10^{-8} m
   (c) 20×10^{-8} m   (d) 25×10^{-8} m

5. If λ, d, and s, respectively, denote the mean free path, the molecular diameter and the mean molecular separation, then;
   (a) $s \ll \lambda \gg d$ (b) $s \approx \lambda \gg d$
   (c) $s > \lambda < d$ (d) $s > \lambda \gg d$

6. A potential difference of V volt is applied across a metallic rod of length L and area of cross section S. If e is the charge, m_e the mass and λ the mean free path of free electrons then their mean drift velocity is,
   (a) $\dfrac{eV\tau}{2m_e SL}$ (b) $\dfrac{eV\tau}{2m_e L}$

   (c) $\dfrac{eV\tau}{2m_e SL}$ (d) $\dfrac{eVS\tau}{2m_e L}$

7. A container is divided into two equal halves A and B by a thin partition. Part A is filled with an ideal gas at pressure P and temperature T while the other part B is evacuated. A very small hole of dimension smaller than the mean free path of the gas molecules is then made in the partition and the two parts of the container are kept at the same temperature T. The system will attain a steady state when,
   (a) frequency of collision of molecules is same in A and B
   (b) number density of molecules is same in A and B
   (c) gas pressure is same in A and B
   (d) since temperature is kept same, A and B will always be in steady state.

8. A molecular beam is allowed to leak from a gas container at temperature T and pressure P through a small hole of size S into vacuum. The beam intensity I will be proportional to,
   (a) S   (b) $T^{1/2}$
   (c) $T^{-1/2}$ (d) P

9. An ideal gas of molecular mass m is held in a bottle of volume V at pressure P and temperature T. The number density of gas molecules is proportional to,
   (a) V^{-1} (b) m^{-1}
   (c) T^{-1} (d) P

Answers to Numerical and Multiple Choice Questions

Answer to problems

1. 5.62×10^{-6} Nsm^{-2}; 3.73×10^{-3} $Jm^{-1}s^{-1}K^{-1}$
2. The intensity will increase by a factor of eight.
3. 1.6×10^{-10} m

4. (*i*) 500 m/s (*ii*) 5.2×10^{-8} m (*iii*) 13.7×10^9 per second
5. (a) 2.48×10^{21} m^{-2} (b) 1.9×10^{-10} m; 2.3×10^{-10} m; 2.1×10^{-10} m; (c) 593 m s^{-1}; 459 ms^{-1}
6. (*i*) 2.4×10^{20} m^{-3} (*ii*) 476.2 m s^{-1} (*iii*) 5.38×10^5 s
7. Oxygen 8.2×10^5 s and Hydrogen 2.0×10^5 s
8. (a) 2.2×10^5 (b) 1.2×10^{-5} m
9. 8.34×10^{28} m^{-3}
10. (*i*) 5.8×10^{28} m^{-3} (*ii*) 7.70×10^{-14} s (*iii*) 1.17×10^5 ms^{-1} (*iv*) 9.0×10^{-9} m (*v*) 3.95×10^{-4} ms^{-1}

Answers to multiple choice questions

1. (c)
2. (d)
3. (a), (b)
4. (c)
5. (a)
6. (c)
7. (c)
9. (a), (c), (d)
10. (c), (d)

Revision

1. The molecules of a real gas attract each other and also have finite size. The pressure exerted by a real gas is, therefore, to be corrected for the molecular attraction and the volume for the reduction in it due to the finite size of the molecules as a result of which whole of the volume is not available to the molecules for their motion. Van der Waals estimated these correction terms and modified the ideal gas equation of state for application to real gases as;

$$\left(P + \frac{a}{v^2}\right)(v - b) = RT$$

2. The volume around a molecule in which no other molecule can enter or penetrate is called the volume of exclusion. The average radius of sphere of exclusion is 2r. The excluded volume in one mole of the gas is denoted by '*b*' and is given by $b = \frac{16}{3} A_v \pi r^3$, which is four times the total volume of all molecules present in one mole.

3. Pressure vs Volume diagram for all gases show a general behavior, below critical point (temperature T_c, volume v_c and pressure P_c) gases may stay both in liquid and vapor phases but at critical point and above it gases have only vapor phase. Since *PV* curve has a point of inflexion at the critical point both $\frac{dp}{dv}$ and $\frac{d^2p}{dv^2} = 0$ that gives $v_c = 3b$, $T_c = \frac{8a}{27bR}$, and $P_c = \frac{a}{27b^2}$

4. Using the above relations one may determine the molecular radius from the critical volume as,

$$r = \left(\frac{v_c}{16\pi A_V}\right)^{1/3}$$

5. The ratio of the gas variables P, *v*, *T* to their respective critical values are called reduced variables and are denoted by $P_r = \frac{P}{P_c}$, $v_r = \frac{v}{v_c}$, $T_r = \frac{T}{T_c}$. Van der Waals observed that the reduced gas parameters for different gases follow the same gas equation. This similarity in the behavior of

all gases when expressed in terms of their reduced variables is called Van der Waals principle of corresponding states. This leads to the conclusion that the deviation of a gas from ideality is determined essentially by the reduced parameters and not by absolute parameters.

6. Viral equation of state is an extension of the real gas equation by including terms of higher order either in $\frac{1}{v}$ or $\frac{1}{P}$, such as

$$Pv = RT\left[1 + \frac{B(T)}{v} + \frac{C(T)}{v^2} + \frac{D(T)}{v^3}.....\right] \text{ or } Pv = RT\left[1 + \frac{B'(T)}{P} + \frac{C'(T)}{P^2} + \frac{D'(T)}{P^3}...\right]$$

The coefficients B(T), C(T) and B'(T), C'(T), etc., are different from each other and are temperature dependent. Further the successive terms of viral equations decrease sharply. The temperature at which the second term of viral equation B(T) or B'(T) is zero is called Boyle's temperature, because at this temperature gas behaves like an ideal gas. Boyle's temperature is given by $T_b = \frac{a}{Rb}$.

7. The compressibility of a gas is denoted by Z and is equal to the ratio of the actual gas volume to the volume of an ideal gas under same conditions. Further, Z = 1 at Boyle's temperature. It has been shown by Van der Waals that the compressibility Z behaves in identical way for all gases if it is represented in terms of the reduced parameters, i.e., $Z = \frac{v}{v_{ideal}} = \frac{Pv_r}{RT_r}$.

8. The condition for the collision of two rigid spheres of radii r_1 & r_2 is that the distance of separation between their centers $d = r_1 + r_2$. In case of a collision between two identical molecules $d = 2r$. The quantity πd^2 that is equal to the collision cross section per molecule is called the microscopic cross section and is denoted by σ. The quantity $n\sigma$ that gives the total collision cross section for all molecules in one mole of the gas is called the macroscopic cross section. The average distance between two successive collisions of gas molecules is called mean free path, denoted by λ is given by $= \frac{1}{n\sigma}$. The number of collisions per second, called the collision frequency f_{col} is given by $f_{col} = \pi n d^2 \overline{v_r}$. The reciprocal of the collision frequency is called the average time between two collisions $t_{col} = \frac{1}{\pi n d^2 v_r}$. The relative velocity v_r is the relative velocity of a molecule assuming that all other molecules are at rest. It is related to the average velocity by the relation $v_r = \sqrt{2}\,\overline{v}$.

9. The number of molecules N out of the initial number N_0 survived after travelling a distance x through a gas of number density n is given by the following equation called the survival equation, $N = N_0 e^{-\sigma nx} = N_0 e^{-x/\lambda}$

10. The average normal distance above or below any plane in a gas at which the a molecule suffers its last collision before crossing the plane, denoted by $\overline{y} = \frac{2}{3}\lambda$

11. The heat energy current H through a gas is proportional to the temperature gradient $\frac{dT}{dy}$ and is in a direction opposite to the temperature gradient, i.e., $H = \kappa\left(-\frac{dT}{dy}\right)$, the constant of

proportionality κ is called the coefficient of thermal conductivity or simply thermal conductivity. The MKS unit for κ is $1 Jm^{-1} K^{-1} s^{-1}$. It can be shown that

$$\kappa = \frac{1}{3} c_v^m n \bar{v} \lambda = \frac{1}{3} c_v^m \frac{\bar{v}}{\sigma}$$

Here c_v^m the average thermal capacity at constant volume of each gas molecule, n the number density of the gas, λ the mean free path, $\bar{v}$ the mean velocity of the gas molecules and $\sigma \left(= \frac{1}{n\lambda} \right)$ the microscopic collision cross section. Further, H may also be given as,

$$H. = -\frac{1}{3} n\bar{v}\lambda \frac{d\left(c_v^m T \right)}{dy}$$

where $\dfrac{d\left(c_v^m T \right)}{dy}$ gives the rate of change of kinetic energy per molecule. This tells that the gas possesses the property of thermal conductivity because of the transport of molecular kinetic energy across gas layers.

12. Viscosity is a property shown by fluids when they undergo shearing flow in which each adjacent layer of the fluid move parallel to each other but with different velocities. The adjacent layers of the fluid apply a sort of a frictional force on each other opposing their motion. The force F that must be applied to keep a plate stationary is given by

$$\frac{F}{A} = \eta \frac{dv}{dy}$$

where A is the area of the plate and $\dfrac{dv}{dy}$ is the velocity gradient. The constant of proportionality η is called the coefficient of viscosity or simply viscosity of the gas. Its MKS unit is $1 \text{ N m}^{-2}\text{s}$. The corresponding CGS unit is 1 dyne $cm^2 s$, called 1 Poise $=10 \text{ N m}^{-2}$ s. Further,

$$\eta = \frac{1}{3} n\bar{v}m\lambda = \frac{1}{3}\frac{m\bar{v}}{\sigma}$$

Also,

$$\eta = \frac{1}{3}\left(\frac{8k_B}{\pi} \right)^{1/2} \frac{(mT)^{1/2}}{\sigma}$$

and

$$\frac{\kappa}{\eta} = \frac{\frac{1}{3}\frac{c_v^m \bar{v}}{\sigma}}{\frac{1}{3}\frac{m\bar{v}}{\sigma}} = \frac{c_v^m}{m} = \frac{c_v}{m.A_v} = \frac{c_v}{M}$$

A gas possesses viscosity because of the transport of linear momentum across layers of the gas. The ratio $\dfrac{\kappa M}{\eta c_v} = 1$ for all gases, however, experimental values of the ratio is found to vary from 1.5 to 2.5 for different gases showing that the order of magnitude of the ratio is correctly predicted by the theory. The discrepancy in the absolute magnitude may be assigned to the fact that the gas molecules are not hard spheres but are more like tennis balls which flexes on collision.

13. The slow motion of molecules from the region of higher concentration to the region of lower concentration is called diffusion. The flux F $\downarrow$ of the tagged molecules, i.e., the number of tagged molecules per unit area per second, diffusing downwards is given by,

$$F \downarrow = -D \frac{dn^{tag}}{dy}$$

Here D is called the coefficient of diffusion. The MKS unit for D is $1 \ m^2 s^{-1}$. Further,

$$D = \frac{1}{3}\bar{v}\lambda = \frac{\bar{v}}{3n\sigma}$$

The property of diffusion results from the difference in the relative concentration of molecular number density in different parts of the system.

14. Drude, a German physicist, applied the concepts of kinetic theory of gases to the free electrons of metals. Although electrons are charged and free electron density in metals is larger than the number density of gas molecules at room temperature, still Drude applied the kinetic theory that was essentially developed for neutral dilute gases. The basic assumptions of Drude's classical theory may be summarized as;

15. Free electrons in metal collide with the lattice of positive ions and electron-electron collisions are neglected. Between collisions electrons move in a straight line when no external field is present. The electromagnetic interactions between electron-electron and electron-positive ion are neglected.

16. The mean free time between collisions is τ and the probability of collision per unit time is $\frac{1}{\tau}$; the probability of having a collision in infinitesimal time dt is $\frac{dt}{\tau}$. The mean free time τ is independent of the position of the free electron as well as its velocity.

17. After each collision the free electron forgets about its previous history, i.e., about its velocity, direction of motion etc., and emerges in a random direction with speed characteristic of the temperature of the region where collision has occurred, hotter the region higher the speed. In other words free electrons achieve thermal equilibrium by collision with positive ion lattice.

Drude's theory successfully explained Ohm's law and electrical conductivity of metals. However, the theory failed to explain why free electrons do not contribute to the constant volume heat capacity of metals. Drude derived the following expression for the current density J in a metallic conductor, $j = \frac{n_e e^2 \tau}{2m_e}\varepsilon$, and for the resistivity $\rho = \frac{2m_e}{n_e e^2 \tau}$.

Assuming that free electrons in a metal behave like the molecules of an ideal gas Drude successfully derived Ohm's law and showed that the resistance of the metallic conductor originates from electron – lattice collisions.

However, Drude's assumption that free electrons in metals behave like molecules of an ideal gas completely failed in explaining the experimental fact that the constant volume molar heat capacity of metals at room temperature is of the order of 3R, the value predicted by Dulong-

Petite for non-metallic solids. The value 3R for molar specific heat capacity for non-metallic solids comes from the three dimensional (vibratory and or rotatory) motion of the molecules of the solid which are fixed at their locations. In the case of metallic solids the positive ions may, therefore, contribute 3R to the molar specific heat. The free electrons, if they are like the molecules of an ideal gas must also contribute to the molar heat capacity in addition to 3R. The Molar heat capacity of metals should, therefore, be much larger than 3R. This failure of Drude's classical theory has been successfully explained by the quantum thermodynamics.

Thermodynamics: Definitions and the Zeroth Law

3.0   Introduction

Thermodynamics is the science that has developed from the observations of the behavior of macroscopic systems. Keen observation and analysis of macroscopic systems led to some very general laws that form the basis of thermodynamics. It is worth mentioning that in most cases studies were made for some entirely different purpose but the analysis of the data from different experiments indicated some order and similarities that led to general laws of thermodynamics. For example, experiments carried out by Benjamin Thompson, (also called Count Rumford) on production of heat while boring holes in gunmetal for making cannons, for the first time clearly showed that heat is a form of energy that may be produced from work. Later, James P. Joule, British scientist, again through experiments established a relation between the mechanical work and the heat energy. The experiments of both Rumford and Joule contributed to the formulation of the first law of thermodynamics. In a similar way, French Engineer Nicolas Leonard Sadi Carnot, working for the improvement of the efficiency of steam engines, developed the well know Carnot cycle which laid the basis for the second law of thermodynamics. It may also be mentioned that the importance and physical significance of some of the common observations were realized much later, an example is the zeroth law of thermodynamics which was formulated much after the first three laws. The zeroth law, that defines the state of thermal equilibrium, was considered to be more fundamental than the already formulated three laws and was, therefore, assigned number zero in the hierarchy of thermodynamic laws.

Like all other branches of science, thermodynamics also has its own terminology. Some of the terms frequently used in thermodynamics are defined here.

3.1   System, Boundary and Surroundings

Since it is not possible to observe whole of the universe at a time it is generally a part of the universe that is studied. The portion or part of the universe (which is under observation) enclosed by a *boundary* is called the *system*. The boundary of the system divides the universe into two parts, the system and the *surroundings* or the environment. In other words, system and surroundings make the universe.

The system may be anything: a solid, liquid, gas, or plasma, or a mixture of all these. It might be a distribution of charges or magnetic poles or radiations, i.e., photons in vacuum, etc. A system is called *closed* when the total mass contained in the system does not change. It means that in a closed system neither additional mass may enter through the boundary nor any mass from the system can leave through the boundary. However, energy may leave or enter a closed system. A system is said to be an *isolated system* if neither mass nor energy may leave or enter the system through its boundary. When a system may exchange matter (mass) and energy with its surroundings, the system is called an *open system*. The fact is that most of the systems in nature are open systems. However, for the easy of analysis, one may approximate a system as closed or isolated under suitable boundary conditions.

The boundary separating the system from the environment may be a *real* boundary, like the walls of a container that holds a liquid or a gas as the system. Boundary may be *imaginary*; for example, if attention is paid to a certain portion of a fluid flowing through a tube and that portion is considered as the system, then the progress of the system, bounded by imaginary boundary may be followed in imagination through the motion of the fluid. A boundary, through which energy may pass from system to the surroundings or vice-versa, is called a *diathermic* boundary. Boundaries made out of conductors are diathermic. On the other hand, if energy cannot pass through the boundary it is called *adiabatic* or *insulating* boundary. It is, however, a fact that no adiabatic or insulating boundary is perfectly insulating. A system enclosed in a diathermic boundary may exchange energy with its surroundings but both diathermic and adiabatic boundaries do not allow matter to pass through them.

3.2   State Parameters or Properties that define a System

A set of values of experimentally measurable quantities defines the state of a system. The physically measurable quantities are called the *state parameters* or *state properties*. The main advantage of the state parameters is that their knowledge makes it possible to duplicate an exactly identical system anywhere.

The state parameters or properties may be divided into two types: the one that are proportional to the mass of the system, are called *extensive parameters* and the other ones that do not depend on the mass of the system are called *intensive parameters*. Total energy E and total volume V of a system are examples of extensive properties of a system, while pressure, temperature, and density are examples of intensive parameters. The extensive properties are *additive* and are also called the *capacity factors*. The *specific value* of an extensive property is obtained by dividing the total value by the mass of the system. Specific values are generally represented by lower case letters, while the values of total extensive properties by upper case or capital letters. For example, the extensive parameter volume of a system is denoted by V and the corresponding specific volume $\frac{V}{m} = v$ by the small letter v. It may be noted that the density of the system, denoted by $= \frac{m}{V}$, is reciprocal of the specific volume v.

Further, specific quantities are independent of mass and, therefore, become intensive property of the system. Intensive properties are *not additive*; for example, if one liter of milk of density ρ_1 is added to five liter of some other fluid of density ρ_2, the total volume (extensive property) will be $1 + 5 = 6$ liter (additive) but the density (an intensive property) of the mixture $\rho \neq \rho_1 + \rho_2$ (non-additive). Intensive properties are also called *intensity factors*

It follows from Avogadro's hypothesis that one kilomole (molecular weight in kilogram) of any substance contains Avogadro number $A_v (= 6.02 \times 10^{26})$ of molecules. In case the mass of the substance is taken equal to molecular weight in gram (in CGS system) the value of Avogadro number A_v is 6.02 $\times 10^{23}$. The point is that the mass of the substance may also be specified in terms of the number of kilomole (or mole) of the substance. For example 1 kilomole of O_2 is equal to 32 kg of it and that 1 kilomole of CO_2 corresponds to 44 kg of the gas. Specific molar (or specific kilomolar) volume is also denoted by lowercase v, as the specific volume per gram or per kilogram. However, it is either explicitly mentioned or may be understood from the context whether the symbol v in a given expression refers to the volume per unit mass or to volume per kilomole (or mole). The advantage of specifying mass in units of kilomole/mole lies in the fact that the number of molecules in the specified mass becomes equal to the Avogadro number for all substances. As is obvious, 1 kilomole of different substances will have different masses, but the number of molecules will be same for all substances.

The main advantage of using specific parameters, instead of their absolute magnitudes, in thermodynamic equations is that the equations become independent of the mass of the system.

The parameters or quantities like pressure, volume, mass, total energy, and temperature are associated with each system and are called system parameters or properties. However, there are some other state parameters also about which we shall talk later. In thermodynamics we also come across quantities that are not the properties of the system. An example is the energy lost or gained (received) by a system (enclosed in a diathermal boundary) from its surroundings. Energy flow between two systems is not the property of any of the two systems, though it depends on the properties of both of them. Energy flow and work done on a system or by the system are not system properties. These are operations or processes.

A system may have many state properties. The *minimum number* of state properties (or parameters) required to *completely specify* a system is called *state variables* or *state functions*. The number of state variables depends on the system. Let us assume that a particular state of a given system is completely defined by three state variables; for example, a mass m of CO_2 gas contained in a perfectly insulating boundary of volume V, at temperature T and pressure P may be completely specified by the numerical values P_1, T_1, and $v_1 \left(= \dfrac{V}{m} \right)$ of the pressure, temperature, and specific volume. We call this state of the gas by A. Another set of state variables P_2, T_2, and v_2 will specify a different state, say B, of the same system. If we now consider a three dimensional space with three axes representing, respectively, the state variables pressure P, temperature T and specific volume v, then state A of the system may be represented by point A having the coordinates $A(P_1, T_1, v_1)$ and the state B by point $B(P_2, T_2, v_2)$ on the graph as shown in Fig. 3.1. In general, therefore, if a system has N-number of state functions, then each state of the system may be represented by a point in an N-dimensional space made up of N-state variables.

The state variables or state functions are properties of a thermodynamic state of a system and do not depend on the way the state is achieved. For example, if by some process the system changes from state A to state B, then the state variables $(P_2, T_2,$ and $v_2)$ of state B do not depend on the process by which the system is brought from A to B, they remain the properties of the final state B of the system. Starting from the initial state A the final state B may be reached through several paths, for example paths 1, 2, and 3 in Fig. 3.1 all leads to the same final state B. As we will see later, either work is done on the system or by the system in taking a system from an initial state to a final state. The different paths 1, 2, 3 etc., show different *processes* or *paths* by which the system in initial state A may be brought to the final state B and, in general, the amount of work done is different for each path.

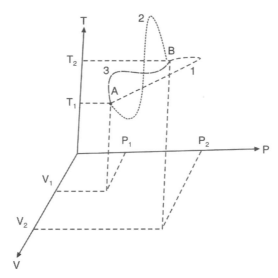

Fig. 3.1   State of a system may be represented by a point in a N-dimensional space, where N is the number of state variables.

The process or path by which the system from initial state A reaches the final state B may be considered to be made up of a large number of successive small changes in pressure dP, in temperature dT, and in specific volume dv as the system is taken along the specific path. Using the method of differential calculus it may be shown that,

$$\int_{P_1}^{P_2} dP = (P_2 - P_1) = \text{difference in the final and initial value of the pressure}$$

$$\int_{T_1}^{T_2} dT = (T_2 - T_1) = \text{difference in the final and initial value of the temperature}$$

$$\int_{V_1}^{V_2} dV = (V_2 - V_1) = \text{difference in the final and initial value of the volume}$$

In above equations the integration is done over the path, which may be path 1, or 2 or 3 or any other. The important point to note is that the result of integration is independent of path and depends only on the initial and final values of P, T, and v. The variables, integrals of which do not depend on the path of integration, are called *exact differentials*. *State functions are exact differentials.* Let us assume that a variable Z (X, Y), is a function of other variables, say X and Y. If Z is an exact differential, then it can be shown that;

$$\frac{\partial^2 Z}{\partial X \partial Y} = \frac{\partial^2 Z}{\partial Y \partial X} \qquad\qquad 3.1$$

Equation 3.1 shows that the second partial derivative of Z, (i.e., $\frac{\partial^2 Z}{\partial X \partial Y}$) does not depend on the order of differentiation. Thus, another property of *exact differentials* and of *state functions* is that their *second order derivatives do not depend on the order of differentiation.* In our example, if pressure P is treated to be a function of temperature T and specific volume v, then:

$$\frac{\partial^2 P}{\partial v \partial T} = \frac{\partial^2 P}{\partial T \partial v}; \ \frac{\partial^2 T}{\partial P \partial v} = \frac{\partial^2 T}{\partial v \partial P} \ \text{and} \ \frac{\partial^2 v}{\partial P \partial T} = \frac{\partial^2 v}{\partial T \partial P} \qquad 3.2$$

3.3 Some State Functions

Most frequently used and directly measureable state parameters are: pressure, volume, temperature, and energy.

3.3.1 Pressure

We have seen that a gas exerts pressure on the walls of the container, which according to the kinetic theory is due to the collision of gas molecules with the container walls. Pressure is force per unit area, and in MKS system its unit is Newton per square meter, written as $N \ m^{-2}$ in short. Newton is MKS unit of force. Pressure may be expressed in several other units like;

$$1 \ \text{Pascal (1Pa)} = 1N \ m^{-2} = 1kg \ m^{-1} \ s^{-2}$$

$$1 \ \text{mm of Hg} = 1 \ \text{torr} = 133.22 \ Pa$$

$$1 \ \text{bar} = 1 \times 10^5 \ Pa$$

$$1 \ \text{atmosphere (1 atm)} = 760 \ \text{torr} = 1.01325 \times 10^5 \ Pa = 1.01325 \times 10^5 \ Nm^{-2}.$$

3.3.2 Volume

Volume is a measure of the space occupied by the system. The volume of gaseous system depends on the temperature and the pressure to a large extent. The volume of liquid systems depends to a lesser extent on the pressure and temperature, while the volume of solids depends to the least extent on temperature and temperature. In MKS system the volume is measured in cubic meter, written in short as m^3. Liter and cubic centimeter (cc) are other commonly used units of volume,

$$1 \ \text{Liter} = 1 \times 10^{-3} \ m^3 \ \text{and} \ 1 \ cc = 1 \times 10^{-6} \ m^3.$$

3.3.3 Temperature

Temperature is an abstract concept that may be related to the feeling of relative coldness or hotness of bodies. According to the kinetic theory, temperature of a system is a measure of the average kinetic energy of the molecules of the system. However, in thermodynamics, the microscopic structure of the system is not taken into consideration and, therefore, the definition of temperature based on molecules, the microscopic constituent of the matter, may not be accepted. Further, temperature plays a dominant role in thermodynamics as it is intimately related to the flow of energy from one system to the other. We shall, therefore, study temperature in more details.

3.3.3.1 *Thermal equilibrium*

Thermodynamic concept of temperature is based on a very general observation that is exhibited by all macroscopic systems. It is a common observation that systems left to themselves, show that their parameters change with time, initially at a faster rate then slowly as time increases and finally after a considerable period of time the system reaches a state when there is no further change in system parameters. One simple example is a cup of hot tea. If left as such, the temperature of the tea falls, initially rapidly, then slowly and after a considerable period of time when the temperature of tea becomes same as the temperature of the surroundings, there is no further change in the temperature of the tea. If observed carefully, the mass and the volume of the tea in cup will also change because of the evaporation but will finally become constant. The state where no further changes in state variables take place is defined as the *state of thermal equilibrium*. Temperature, in thermodynamics, characterizes the state of thermal equilibrium. In thermal equilibrium it is not only assumed but may be experimentally verified that the temperature of all parts of the system is same. Further, if two systems in contact with each other (through a diathermic boundary) are in state of thermal equilibrium, i.e., no parameter of either of the system is changing with time then the temperatures of the two systems will be same. Another equivalent statement may be: if the temperatures of the two systems are same, they will be in thermal equilibrium if put in contact. Thermal equilibrium between two systems in contact means that no thermal or radiation energy is exchanged between the two systems.

Yet another fact about thermal equilibrium, that may be experimentally verified, is that if a system A is in thermal equilibrium separately with two systems B and C, then systems B and C will also be in thermal equilibrium. Experimental verification of the above statement may be obtained by putting A in contact with B and observing that no change in parameters of both A and B takes place. Next put A and C in contact and again no change in system parameters of both A and C will be observed. Finally, when B and C will be put in contact then also no change in the parameters of B and C will be observed. In order to verify that temperatures of A, B and C are equal, one may employ a device that is sensitive to temperature. For example, a capillary filled with some liquid will show the same height of liquid column when put successively in systems A, B and C.

3.3.3.2 *Zeroth law of thermodynamics*

The above mentioned behavior of systems in thermal equilibrium was known for long but perhaps Ralph H. Fowler for the first time in 1935 named it the Zeroth law of thermodynamics. The zeroth law of thermodynamics states: *If a system A is separately in thermal equilibrium with two other systems B and C, then systems B and C will also be in thermal equilibrium and the temperatures of systems A, B and C will be the same.*

On the surface it appears that thermal equilibrium between B and C as enumerated in zeroth law of thermodynamics, is the obvious outcome of the thermal equilibrium between A and B and between A and C. However, it is not true. The equivalence of the temperature of B and C is true only in the case of thermal equilibrium. To make this point clear, we consider a primary cell shown in Fig. 3.2.

A primary cell consists of an electrolyte in which two rods of dissimilar metals are immersed. The metallic rods work as electrodes. When the electrodes are not connected externally, the two electrodes are separately in electric equilibrium with the electrolyte. It is because of this electric equilibrium that no electrical changes, like the flow of current or generation of free ions etc., are taking place in the electrolyte. If a law like the zeroth law of thermodynamics also exists for electric equilibrium, then the two electrodes that are separately in electrical equilibrium with the electrolyte should also be in electric equilibrium with each other. But they are not. As soon as the electrodes are connected externally, a current flows through the circuit, indicating that the electrical potentials of the two electrodes are not equal and, therefore, the electrodes are not in electrical equilibrium with each other. This observation illustrates that the zeroth law is specific for thermal equilibrium. It may also be verified that before externally connecting the electrodes when there was thermal equilibrium between the two electrodes and the electrolyte, the temperatures of all the three were also same, as required from the zeroth law of thermodynamics.

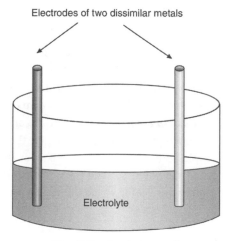

Electrodes of two dissimilar metals

Electrolyte

Fig. 3.2   A primary cell

In thermodynamics, the use of the zeroth law is implicit whenever a measurement of temperature is done. It is because of this reason that this law of thermal equilibrium has been assigned zeroth number, ahead of the first law.

As already mentioned, every system if left to itself, will eventually reach a state of thermal equilibrium. The rate of attaining thermal equilibrium depends on the type of the boundary enclosing the system. If the boundary is perfectly diathermal, the system will attain thermal equilibrium with its surroundings faster. On the other hand if the boundary is perfectly adiabatic, the system will never attain thermal equilibrium with the surroundings. It may, however, be emphasized that no boundary is either perfectly adiabatic or diathermal. The best practical diathermic boundary may be a very thin sheet of silver or copper while the best practical adiabatic boundary may be a thick layer of cotton–wool. It is worth mentioning that a system may attain two different types of thermal equilibrium: (i) thermal equilibrium within the system, when all parts of the system have same temperature, and (ii) thermal equilibrium with the surroundings. A system enclosed in a perfectly adiabatic boundary will of course attain thermal equilibrium within the system but will never be able to attain thermal equilibrium with surroundings.

3.3.3.3   *Measurement of temperature*

Any property of a substance that depends on the temperature may be used to measure temperature. This property is called thermometric property. Expansion of liquids with temperature, dependence of thermo-emf on temperature, change of resistivity of metals with temperature, change of the volume of the gas at constant pressure with temperature, change in pressure of the gas at constant volume with temperature, etc., have all been used as thermometric property in making thermometers. Two important steps are involved in making a thermometer. First is to choose the thermometric property and second to calibrate the thermometer using some temperature as standard temperature. In a thermometer change in the thermometric property is then correlated with the temperature. Suppose by η_1 and η_2 we denote, respectively, the change in the thermometric property corresponding to the two values of the temperatures ϑ_1 and ϑ_2, then

$$\frac{\eta_1}{\eta_2} = \frac{\vartheta_1}{\vartheta_2} \qquad\qquad 3.3$$

A thermometer can be calibrated by keeping it in contact with some natural system that remains at a fixed temperature. Calibrating systems often used for the purpose are: (*i*) a mixture of ice and water at 1 atm pressure, called the *ice point of water (IPW)*; (*ii*) a mixture of water and water vapors at 1 atm pressure, called the *boiling point of water (BPW)*; (*iii*) a mixture of water, water vapors and ice in thermal equilibrium, called the *triple point of water (tri)*. In principle the boiling point or the freezing point of any material at a fixed pressure may be used for the calibration as well. However, by international agreement the triple point of water is now taken as the calibrating standard. If the triple point temperature is assigned the value ϑ_{tri} and the corresponding change in thermometric property is η_{tri} then any other temperature ϑ for which the change in the thermometric property is η may be given as,

$$\vartheta = \vartheta_{tri} \frac{n}{\eta_{tri}} \qquad\qquad 3.4$$

When attempts were made to measure a given temperature ϑ, say the boiling point of Nitrogen or of water at 1 atm pressure, by different thermometers, like the platinum resistance thermometer and thermometer based on the thermo-emf, using Eq. 3.4, it was found that the two thermometers gave different numerical values for ϑ. It was not unexpected also, because the ratio $\dfrac{n}{\eta_{tri}}$ has different values for different thermometric properties. Further, for the same thermometer the ratio $\dfrac{n}{\eta_{tri}}$ varies non-linearly with the temperature. This is shown in Table 3.1 where the magnitudes of the thermometric properties, i.e., thermo-emf of the copper–constantan thermocouple (denoted by η_1) and resistance of the platinum resistance thermometer (denoted by η_2) and their ratios with the value at triple point at different fixed temperatures; the triple point of water, the boiling point of nitrogen, the boiling point of water and the melting point of tin (Sn) are listed in the table.

Table 3.1   Magnitudes of thermometric properties thermo-emf (η_1) for copper–constantan thermocouple and resistance (η_2) for platinum resistance thermometer and their ratios with the value at triple point, at some fixed temperatures

Fixed temperature	Thermo-emf η_1 mV	Ratio η/η_{tri}	Resistance η_2 Ω	Ratio η_1/η_{tri}
Triple point of water	6.26	1.0	9.83	1.0
Boiling point of N_2	0.73	0.12	1.96	0.2
Boiling point of H_2O	10.05	1.51	13.65	1.39
Melting point of Sn	17.50	2.79	18.56	1.89

It may be observed from Table 3.1 that the ratios $\dfrac{\eta_1}{\eta_{tri}}$ (for copper–constantan thermocouple) and

$\dfrac{\eta_2}{\eta_{tri}}$ (for platinum resistance thermometer) have different magnitudes at the same fixed temperature.

We thus see that such thermometers will give different empirical values for the same temperature in spite of the fact that the calibrating temperature, the triple point of water, is same.

3.3.3.4 *Constant volume gas thermometer*

Constant volume gas thermometer is better than all other thermometers, because the ratio $\dfrac{\eta}{\eta_{tri}}$ for

this type of thermometer does not vary with temperature. However, it is bulky and its operation is a bit complicated. As such this thermometer serves as a laboratory standard and other small and handy thermometers are calibrated using the constant volume gas thermometer. A sketch of a constant volume gas thermometer is shown in Fig. 3.3.

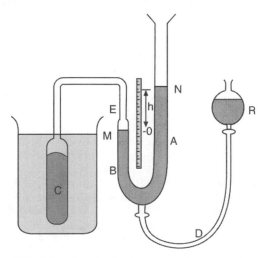

Fig. 3.3   Constant volume gas thermometer

The important components of the constant volume gas thermometer are: the bulb C contains a fixed mass of some gas like hydrogen, nitrogen, etc., at a pre-determined pressure. The bulb is connected through a tube E to mercury monometer EBA. The mercury level in limb B of the monometer U-tube may be adjusted to the desired height by adjusting the height of the mercury reservoir R. Reservoir R is connected to the U-shaped monometer through a flexible rubber tube D. A scale marked in millimeters is provided between the two limbs of the monometer U- tube. A mark M is engraved on the glass tube EB. The zero of the scale is kept aligned with the mark M. Every time before taking any reading the height of the mercury reservoir is so adjusted that the mercury level in limb EB just touches the mark M. This ensures that the volume of the gas contained in the bulb is always same. The body (in the present case the liquid in the beaker), the temperature of which is required to be measured, is kept in contact with the thermometer bulb for sufficient time so that they come in thermal equilibrium and the gas inside the bulb C attains the temperature of the body. The gas in bulb C will either expand or contract on attaining the temperature of the body. As a result, on expansion of the gas the mercury level in arm B will be pushed blow the mark M while on contraction of the gas, it will move above the mark M. In either of the two cases, the mercury level is brought back to mark M by changing the height of the reservoir R. The pressure P on the gas in bulb C, when its volume is kept constant by bringing the mercury level to mark M, is equal to the pressure P_1 on the open end N of the monometer + the pressure ΔP due to the height h of the mercury column. But P_1 is the atmospheric pressure. Hence, P = atmospheric pressure + pressure of height h of the mercury. Thus for each setting of the gas in bulb C to the constant volume (mark M), the pressure on the gas may be determined from the knowledge of the atmospheric pressure and the measured height h of the mercury column. In case of the constant volume gas thermometer any temperature ϑ for which the value of the gas pressure at constant volume is P, and the gas pressure at triple point temperature ϑ_{tri} is P_{tri}, is given by;

$$\vartheta = \vartheta_{tri} \left(\frac{P}{P_{tri}} \right)_{const.Vol.}$$

3.5

The subscript const.vol in Eq. 3.5 refers to the fact that pressures are taken at constant volume of the gas. In principle any gas at any pressure may be filled in bulb C of the gas thermometer. As an example, the data of pressure P and the ratio $\left(\frac{P}{P_{tri}} \right)_{const.Vol.}$ at three fixed temperatures, the boiling point of N_2, the boiling point of H_2O, and the melting point of tin for a hydrogen filled thermometer at two different pressures of hydrogen at triple point of water; P_{tri} = 1.0 atm and P_{tri} = 6.8 atm are recorded in Table 3.2.

Table 3.2   Pressure P and the ratio $\left(\frac{P}{P_{tri}} \right)_{const.Vol.}$ at three different fixed temperatures

for hydrogen thermometer at P_{tri} = 1.0 atm and P_{tri} = 6.8 atm

Fixed temperature	Pressure P (atm)	Ratio P/P_{tri}	Pressure P (atm)	Ratio P/P_{tri}
Triple point of water	P_{tri} =1.0	1.00	P_{tri} =6.8	1.00
Boiling point of N_2	0.29	0.29	1.82	0.27
Boiling point of H_2O	1.37	1.37	9.30	1.37
Melting point of tin	1.85	1.85	12.70	1.87

A comparison of data in columns 3 and 5 clearly show that the ratio $\left(P/P_{tri} \right)_{const\ Vol}$ for the same fixed temperature to a large extent does not depend on the pressure of the gas. This is a definite improvement over the conventional thermometers which use the variations, in the length of some fluid, or thermo-emf, or the resistance etc., as thermometric property.

Next let us investigate the effect of the change of the gas on the absolute values of pressure P (data in column-2 of Table 3.2) and the ratio $\left(P/P_{tri} \right)_{const.vol.}$ when some other gas, like nitrogen, air, oxygen etc., is used in the constant volume gas thermometer. It has been observed that the absolute values and hence the ratios change with the gas, but for a given gas the ratios vary linearly with the value of P_{tri}. The general trends showing the variation of the ratio $\left(\dfrac{P}{P_{tri}}\right)_{const.vol.}$, as a function of the triple point pressure P_{tri}, for different gases are shown in Fig. 3.4(a).

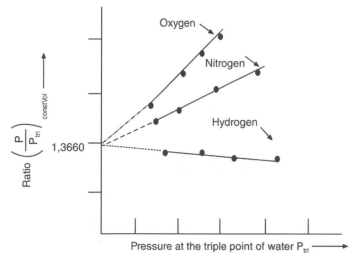

Fig. 3.4 (a) Variation of the ratio $(P/P_{tri})_{const.Vol}$ with the pressure P_{tri} at the triple point of water

In Fig. 3.4 (a), it is important to note that experimental data could be taken only up to a certain low value of P_{tri}, shown in the graph by solid lines with data points. However, when straight line graphs for different gases were extrapolated (extended) backward, shown by dotted lines, they all met the y-axis at the same point where the ratio had the value 1.3660. This extrapolated value of the ratio is denoted by $\lim\limits_{P_{tri}\to 0}\left(\dfrac{P}{P_{tri}}\right)_{const.Vol.}$ and it means that the ratio is obtained by extrapolation in the limit when the pressure P_{tri} approaches zero. Graph in Fig. 3.4 (a) shows that at very low pressure all gases behave in the same way. In order to make the measured value of the temperature by gas thermometer independent of the gas, Eq. 3.5 is modified to the limiting value of the ratio, as given below;

$$\vartheta = \vartheta_{tri}\lim_{P_{tri}\to 0}\left(\frac{P}{P_{tri}}\right)_{const.Vol.} \qquad 3.6$$

With the view to see how the nature of graphs in Fig. 3.4(a) changes with the change of the fixed point temperature, the same three gases were filled in the same thermometer but not at the triple point temperature, instead at the temperature of the ice point of water (IPW). The variation of the ratio (P/P_{IPW}) with the Pressure at the ice point of water P_{IPW}, is shown in Fig. 3.4(b). It may be observed from Fig. 3.4(b), that the basic nature of the graph does not change, only the magnitude of the ratio $\lim\limits_{P_{IPW}\to 0}\left(\dfrac{P}{P_{IPW}}\right)_{constVol}$ has changed to 1.3661. Thus changing the fixed temperature at which gas is filled in the thermometer, changes the magnitude of the extrapolated ratio.

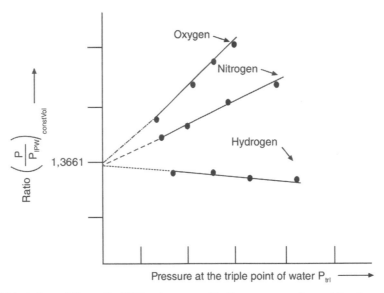

Fig. 3.4 (b) Variation of the ratio $(P/P_{IPW})_{const.Vol}$ with the pressure P_{IPW} at the ice point of water

Suppose the temperatures of the ice point of water and the boiling point of water are, respectively, denoted by ϑ_i and ϑ_b, then using Eq. 3.6 one gets,

$$\frac{\vartheta_b}{\vartheta_i} = \frac{\lim\limits_{P_{tri}\to 0}\left(\frac{P_b}{P_{tri}}\right)_{cons.Vol}}{\lim\limits_{P_{tri}\to 0}\left(\frac{P_i}{P_{tri}}\right)_{cons.Vol}} \quad \text{or} \quad \frac{\vartheta_b}{\vartheta_i} = \lim\limits_{P_{tri}\to 0}\left[\frac{P_b}{P_i}\right]_{constVol} \qquad 3.7$$

or $$\frac{\vartheta_b - \vartheta_i}{\vartheta_i} = \lim\limits_{P_{tri}\to 0}\left[\frac{P_b - P_i}{P_i}\right]_{constVol} \quad \text{or} \quad \vartheta_i = \frac{\vartheta_b - \vartheta_i}{\lim\limits_{P_{tri}\to 0}\left[\frac{P_b - P_i}{P_i}\right]_{constVol}}$$

or $$\vartheta_i = \frac{(\vartheta_b - \vartheta_i)}{\lim\limits_{P_i\to 0}\left(\frac{P_b}{P_i}\right)_{constVol} - 1} \qquad 3.8$$

If the temperature difference between the boiling point and the ice point of water $(\vartheta_b - \vartheta_i)$ is divided into 100 equal parts, and the value of the ratio $\lim\limits_{P_i\to 0}\left(\frac{P_b}{P_i}\right)_{constVol}$ is put as 1.3661 then Eq. 3.8 becomes,

The temperature of the ice point of water $\vartheta_i = \dfrac{100}{1.3661 - 1} = \dfrac{100}{0.3661} = 273.15$ \qquad 3.9

It may be observed that using the experimental value of the ratio $\lim\limits_{P_i\to 0}\left(\frac{P_b}{P_i}\right)_{constVol}$ and assuming that the temperature difference between the boiling point and ice point of water is divided into 100 equal parts, called degree (to match the Celsius scale of temperature) the constant volume gas thermometer gives the numerical value of the ice point as 237.15 degree. The triple point temperature is experimentally found to be 0.01 degree above the ice point temperature, Hence the best value of the triple point temperature ϑ_{tri} is 273.16 degree. Substituting the value of the triple point temperature in Eq. 3.5, any empirical temperature ϑ measured by the gas thermometer may be written as,

$$\vartheta(\text{gas thermometer}) = 273.16 \times \left[\lim_{P_{tri} \to 0} \left(\frac{P}{P_{tri}} \right)_{constVol} \right] K \qquad 3.10$$

Constant volume gas thermometer is the basic instrument for measuring temperature in thermodynamics and temperatures measured by it are generally denoted by T. The unit of temperature is Kelvin, written as K (not degree K). As such for thermodynamic applications Eq. 3.10 may be rewritten as,

$$T = 273.16 \times \left[\lim_{P_{tri} \to 0} \left(\frac{P}{P_{tri}} \right)_{constVol} \right] K \qquad 3.11$$

3.3.3.5 *Other temperature scales*

In thermodynamics temperature is represented in Kelvin. However, some other important scales for temperature measurement are:

(*i*) Celsius (to honor Swedish scientist Anders Celsius), the old name of this scale is Centigrade. The temperature is written in units of °C, meaning degree Celsius. In this scale, the ice point temperature is taken 0.0 °C, the boiling point of water as 100 °C and the temperature difference from ice point to boiling point is divided in 100 equal parts, each called a degree. 1°C = 1 K. The Celsius temperature *t* is related to the thermodynamic temperature *T* by the relation,

$$t \; (in \; degree \; Celsius) = T(in \; Kelvin \; or \; K) - 273.15K \qquad 3.12$$

(*ii*) The Rankine scale (W.J.M. Rankine, Scottish Engineer). In this scale the temperature difference between the boiling point and the ice point is divided into 180 equal parts, each part called Rankine and denoted by 1 R. In Fahrenheit scale (G.D. Fahrenheit, German Scientist) also the temperature difference between the boiling point and the ice point is divided into 180 parts and, therefore, 1 degree Fahrenheit, written as 1 °F is equal to 1 R. 1 Rankine is defined in terms of the thermodynamic scale (Kelvin) by the relation,

$$1R = \frac{5}{9} K \qquad 3.13$$

Equation 3.13 may be used to convert temperatures from Kelvin to Rankines and vice versa.

(*iii*) The following equation gives the relation between the temperature in degree Fahrenheit (°F) and temperature in Kelvin.

$$t \; °F \; (\text{temperature in Fahrenheit}) = \frac{9}{5} \times T \left(\text{temperature in Kelvin} \right) - 459.67 \qquad 3.14$$

Some standard fixed temperatures in different units are given in Table 3.3.

Table 3.3   Some standard temperatures in different units

Temperature	Kelvin (K)	Celsius (°C)	Rankine (R)	Fahrenheit (°F)
Triple point of H_2	13.81	286.96	24.86	−434.81
Ice point of water	273.15	0.00	491.67	32.00
Triple point of water	273.16	0.01	491.69	32.02
Boiling point of water	373.15	100.00	671.67	212.00
Freezing point of silver	1235.08	961.93	2223.14	1763.47
Freezing point of gold	1337.58	1064.43	2407.64	1947.97

3.3.4   Energy

Energy is also an important state function of a system. It is defined as the capacity to do work. A system, in general, may possess two types of energies: (i) *Internal* or *inherent energy* (ii) *External* or *bulk energy*.

3.3.4.1   *Internal energy*

It is the energy possessed by the microscopic structure of the system. Since thermodynamics does not go in the details of the microscopic structure, the nature of the internal energy is not specified in thermodynamics. The only information relevant to thermodynamics is that the internal energy of a system depends on the temperature T of the system. Generally, it is denoted by U(T). However, kinetic theory which assumes that matter is made up of molecules, tells that molecules of a system at temperature T possess kinetic energies and the average kinetic energy of molecules is a measure of temperature T. With this input from the kinetic theory, it may be said that the internal energy of a system U(T) is distributed amongst the molecules of the system as their kinetic energies. Raising the temperature of the system increases the internal energy which means that the velocities of random motions of molecules increase resulting in an increase of their average kinetic energy. The specific internal energy is denotes by u(T) and may be energy per unit mass or per mole of the system.

3.3.4.2   *External or bulk energy*

When the system as a whole is given a motion or is put in some external field, it may acquire bulk kinetic or potential energy. For example, if one considers a cricket ball to be a system, the system is given both the kinetic and the potential energies when the bowler bowls or when the ball is hit by the batsman.

A system may be given energy or energy from a system may be withdrawn by carrying out some operation or process on the system. The first law of thermodynamics deals with such situations. The total energy E_{total} of a system may be written as,

$$E_{total} = U(T) + E_{external}\left(= E_{Bulk\ kinetic} + E_{Bulk\ Potential}\right) + E_{any\ other\ form\ of\ energy} \qquad 3.15$$

Energy is measured in terms of the work and, therefore, the units of energy and work are same. The MKS unit of energy is joule (J). Energy may also be expressed in erg, electron volt (eV), kilo watt hour (kWH), British thermal unit (Btu) and calorie (cal). The conversion factors are given below.

$$1J = I \ N \ m; \ 1 \ eV = 1.602 \times 10^{-19} \ J; \ 1 \ erg = 1 \times 10^{-7} \ J; \ 1 \ cal = 4.1868 \ J; \ 1 \ Btu = 1.0550 \times 10^3 \ J.$$

3.4   Equilibrium

Equilibrium occupies a central place in thermodynamics. A system is said to be in equilibrium if no change in its state functions occur with time. In principle, therefore, it is required to keep the system in observation for infinite time to ensure that no change in system parameters has taken place. Observation of a system for infinite time is possible only in imagination and in practice if system parameter do not change in reasonable time, it is assumed that the system has attained equilibrium. Equilibriums may be of two kinds: (i) *natural equilibrium* (ii) *forced equilibrium* or *steady state*.

3.4.1   Natural equilibrium or equilibrium

Any system left to itself for a sufficiently long time attains equilibrium in a natural way. Several examples of natural equilibrium may be given, like hot tea in a cup left for few hours attains the temperature of the surroundings, reaches the state of equilibrium and remains in it for indefinite time without any further efforts. This type of equilibrium, which does not require any effort or energy to maintain equilibrium once it has been attained, is called natural or simply equilibrium. Energy may be lost or gained during the process of attaining equilibrium when system parameters change, but no loss or gain of energy by the system occurs after the equilibrium has established and system parameters have stabilized.

3.4.2   Forced equilibrium or steady state

The other type of equilibrium, called forced equilibrium or steady state is that in which continuous supply of energy is required to maintain the state of equilibrium. A simple example is an electric oven. Suppose the temperature of the oven is set at 300 °C, and the oven is switched on. After some time the temperature of oven will reach 300 °C and will remain stabilized at this value so long the oven is not switched off. The state of equilibrium attained by the oven at 300°C temperature is forced on the oven by the electrical energy supplied to it. Further, the state of equilibrium stays only so long as the energy supply is maintained. The oven is said to be in steady state when the temperature has reached 300°C and the electric supply is on. Similarly, the temperature of human body stays almost constant at 37°F, showing that human body is in a state of equilibrium with its surroundings. However, this forced equilibrium or steady state of human body is maintained by the energy human body gets from food.

Cosmic radiations hitting the atmosphere of earth from all directions produce the radioactive isotope ^{14}C. The ^{14}C isotope produced by cosmic rays decays as it is radioactive. For more than the last 3000 years the concentration of ^{14}C isotope in atmosphere is almost constant. This steady state of ^{14}C concentration is attained by constant bombardment of atmosphere by the cosmic rays. The decay and the production rates of ^{14}C isotope are exactly equal and, therefore, the concentration of ^{14}C in atmosphere is constant. This is also an example of steady state.

Plants and other living organism continuously absorb ^{14}C isotope from the atmosphere, either by photosynthesis or by inhaling. As a result the ratio of the amounts of stable isotope ^{12}C and the radioactive isotope ^{14}C in the living organism is constant. This is again an example of forced equilibrium or steady state. The stable carbon isotope ^{12}C and the radioactive carbon isotope ^{14}C are in a steady state in the body of a living organism. The number of ^{14}C atoms in a living organism that decays in a given time because of their radioactivity, exactly same number in the give time is replenished by inhaling/or by photosynthesis. And, therefore, in a living organism the equilibrium in the relative concentrations of stable and radioactive isotopes of carbon is maintained. However, after the death of the orgasm, absorption of ^{14}C from the atmosphere stops and as a consequence, the relative concentration of ^{14}C in a dead body decreases with time. This is the basis of carbon dating technique used to determine the age of biological samples. It may again be observed that the steady state in the relative concentrations of the stable and the radioactive carbon isotopes in living organism is maintained by the energy that living organism get from their food.

Summing up it may be said that the maintenance of steady state requires a continuous source of energy, while the maintenance of equilibrium does not need any source. If a system in steady state is left to itself after removing the source of energy, after some time eventually the system will again attain equilibrium. But this time the system will attain natural equilibrium. Steady state basically represents a group the natural equilibrium is a subset of this group.

Apart from thermal equilibrium, which establishes the equality of temperature, there may be other types of equilibriums also. Examples are mechanical equilibrium and chemical equilibrium. Thermal equilibrium is characterized by the equality of temperature, mechanical equilibrium by the minimum of the potential energy and the chemical equilibrium by the law of mass action. *A system is said to be in thermodynamic equilibrium when it is simultaneously in thermal, mechanical and chemical equilibrium.* As will be seen later some thermodynamical functions play the same role in determining the stability of systems as is played by potential energy in determining the stability of mechanical systems.

3.5   Processes

In section 3.2 it was shown that a system in equilibrium may be represented by a point in the N-dimensional space made up of N-system variables. In order to change the state of a system, say from state A to state B, (see Fig. 3.2) some operations have to be performed. *The operations that change the state of the system are called processes and are represented by lines/ curves on the N-dimensional space made of state variables.* These curves/lines join the initial and the final states of the system. In general, a final state of the system may be reached through many different processes, each represented by a different path, (as shown by 1, 2, 3, etc., in Fig. 3.2), in the space made up of state functions. The processes may be of several different kinds:

(a) **Isothermal process**: A processes in which the temperature of the system remains constant. In order to keep the temperature of the system constant, generally, energy is either absorbed by the system from the surroundings or is lost to the surrounding. It is obvious that systems enclosed by diathermic boundary may undergo Isothermal change.

(b) **Isobaric or isopiestic process:** A processes in which the pressure of the system remains constant is called an isobaric process. Expansion of a system against the constant atmospheric pressure is an example of an isobaric process. Generally, energy of the system changes during an isobaric process.

(c) **Isochoric or isovolumic process:** A processes in which the volume of the system remains constant is called an isochoric process. No external work is done on the system or by the system during an isochoric change.

(d) **Adiabatic process:** A. process in which a system enclosed in an adiabatic boundary undergoes some change is called an adiabatic process. No energy leaves or enters the system in an adiabatic process. It will be shown later that another important system parameter called 'entropy' of the system also remains constant during a reversible adiabatic process.

(e) **Quasi-static process:** In the processes considered above some property of the system remains unaltered. However, the speed or the rate at which a process occurred is also important. Change in the state of a system may take place in two different ways:

   (*i*) very slowly, in infinitesimal small steps, such that after each step the system attains equilibrium. Such a process is called a quasi-static process.

   Figure 3.5 shows a system enclosed in a diathermic boundary in equilibrium at temperature T_0. It is required to change the temperature of the system to T_1 ($T_1 > T_0$). One way to do it is to surround the system by another system (surrounding) which is at an infinitely small temperature ΔT above T_0, i.e., at temperature $(T_0 + \Delta T)$. In a short time the system will come in equilibrium with the surroundings and will acquire the temperature $(T_0 + \Delta T)$. Next surround the system by another system at temperature $(T_0 + 2\Delta T)$. The system will then establish equilibrium at temperature $(T_0 + 2\Delta T)$. Next, the system is enclosed by an environment at temperature $(T_0 + 3\Delta T)$. In this way the temperature of the system may be changed slowly in small steps such that the system attains equilibrium at each step. It means that at each step the temperature in every part of the system becomes same as that of surrounding. This slow process in which at each step system attains equilibrium is called the quasi-static process. *It may, therefore, be said that a quasi-static process is made of a succession of equilibrium states.* This is shown in Fig. 3.5 as the system follows the path from left to right, down and from right to left in several small steps. Ultimately the temperature of the last surrounding will be T_1 and the system will reach the desire temperature once it is in thermal equilibrium with the surrounding.

   (*ii*) The other possible and faster way of changing the temperature is to enclose the system in a surrounding at temperature T_1, in a single step, as shown by a big downward arrow at the extreme left of Fig. 3.5. In this type of change the system will not pass through succession of equilibrium states. Initially the part of the system near the boundary will attain the temperature of the surrounding, and then slowly the temperature of the inner part of the system will rise. It will be only at the end of the process when thermal equilibrium will be established. No equilibrium states will occur during the process when temperature of system is changing.

   In the example above quasi-static and non-quasi-static processes for the change of temperature of a system were considered. The same is true for any other state variable like the pressure or the volume etc.

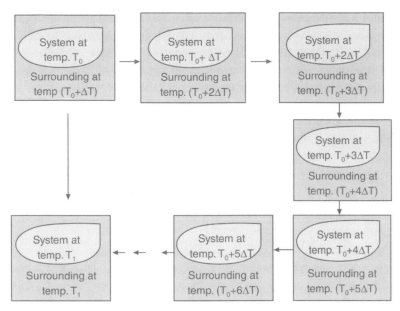

Fig. 3.5   Quasistatic change of temperature of a system from T_0 to T_1

(*f*)   **Reversible and irreversible processes:** A process refers to some change in the value of some state variable of the system. One may assign a direction to the process, say, positive when the magnitude of the system variable increases and negative when it decreases. *The processes in which an infinitesimally small change in the system variable may reverse the direction of the process are called reversible processes.* On the other hand, if *an infinitesimal change in the system variable does not change the direction of the process,* the processes is called irreversible.

Figure 3.6 shows both the irreversible and the reversible processes. In Fig. 3.6 (a) a system in equilibrium at temperature T_0 is shown enclosed by surroundings at temperature T_1, where $T_1 \gg T_0$. The heat energy is, therefore, flowing from the surroundings at higher temperature to the system at lower temperature, indicated by arrows. If now the temperature of the system is increased by an infinitesimal amount ΔT, even then the temperature of the surrounding remains higher than the temperature of the system and the direction of flow of heat energy remains the same, from surroundings to the system, as shown in Fig. 3.6 (b). The process, shown in Fig.3.6 (a) , is, thus, irreversible.

In the lower half of Fig. 3.5 (C), a system in equilibrium at temperature T_0 is shown surrounded by environment at temperature $(T_0 + \frac{\Delta T}{2})$.

Since the temperature of the surrounding is higher than the system, heat energy is going from the surrounding to the system. If now the temperature of the system is increased to $(T_0+\Delta T)$, where ΔT is infinitesimally small, the direction of heat flow will be reversed, the heat energy will now flow from the system to the surrounding, as shown in Fig. 3.6(C). The process represented by Fig. 3.6 (C) is a reversible process. It is simple to follow that *all reversible processes are quasi-static,* because in a reversible process difference in the magnitude of the state variable between the system and the surroundings is very small and, therefore, equilibrium between the system and the surrounding is established at each step. It is, however interesting to note that *all quasi-static processes are not reversible*. Let us consider a system at a considerably higher temperature than the surroundings and bound by a boundary that is not perfectly adiabatic. It means that a small amount of heat energy is slowly lost by the system to the surrounding through the non-perfectly adiabatic boundary. The process of heat loss by the system is quasi-static but not reversible. It is quasi-static because the boundary is partially adiabatic and allows only small amount of heat energy to be lost slowly in each step and during this period thermal equilibrium get established in the system. The process is not reversible because an infinitely small increase in the temperature of the system will not reverse the direction of the heat flow. This is an example of a quasi-static but non-reversible process. A hot cup of tea left in a closed thermos flask is a practical example of quasi-static but non-reversible heat loss process. In order to make a process reversible it is required that there is very small difference between the system variables and the variables of the surrounding.

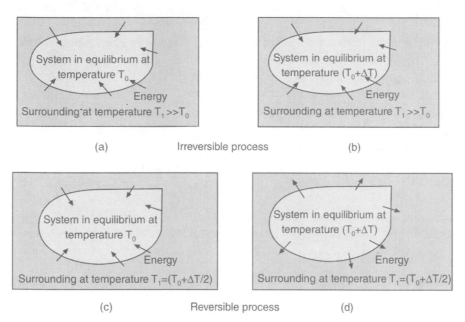

Fig. 3.6   Reversible and irreversible processes

3.6   The Work

Work results from the application of force on a body or a system. It is known from mechanics that force may be either conservative or non-conservative. Work done by a conservative force brings about a change in the potential energy of the body while that of non-conservative force changes the kinetic energy. When a body is simultaneously acted upon by both types of forces, the work–energy principle of mechanics tells the total work done is equal to the sum of the changes in the potential and kinetic energies of the body. The principle of mechanics, stated above, implies that change of either potential or kinetic energy or both must occur when some work is done on or by a system. However, there are situations where the bulk kinetic and potential energies of a system do not change when work is done on or by the system. For example, work is done by a cell in sending current through a resistance but there is no change in the kinetic and potential energies of either the cell or the resistance. Similarly, charging and discharging of a capacitor and magnetization and demagnetization of bodies involve work but no change in the kinetic or potential energies. Many such examples may be cited. Answer to the question as to what happens to the work performed in such cases is provided by thermodynamics.

Work is a form of energy which is different from the internal energy U(T) or the bulk kinetic and potential energies of a system. While internal, bulk kinetic and potential energies are properties of the system (are state variable), the *work done by a system or on a system is not a system property.* Work done on a system or by a system does depend on the initial and the final states of the system but it also depends on the path or the process through which the final state of the system is reached starting from the initial state. It may be realized that work would have been a state variable (or system parameter) had it not dependent on the path. It will be shown later that a special type of work, i.e., adiabatic work between two equilibrium states of a system (of same bulk kinetic and potential energies) is independent of path, and therefore, a system property.

Before we proceed further, it is necessary to define when work is done by a system and when it is done on the system. In mechanics, the work dW (scalar quantity) done by a force F (vector) in displacing the body by a distance dx (vector) is given by the dot (or scalar) product,

$$dW = \vec{F}.\vec{dx} = F\,dx\,\cos\theta \qquad\qquad 3.16$$

In Eq. 3.16, θ is the angle between the direction of force F and the displacement vector dx. θ is zero when dx is in the direction of F, and, therefore, according to Eq. 3.16, dW is positive when the force F and the displacement are in the same direction. However, in thermodynamics, generally, Eq. 3.16 is modified as;

$$-dW = \vec{F}.\vec{dx} = F\,dx\,\cos\theta \qquad\qquad 3.17$$

This means that the work done is positive when the direction of the force is opposite to the direction of displacement. This modification in the sign of the work done has been done to make the work done by the system positive and the work done on the system negative. However, in thermochemistry work done on the system is taken as positive and the work done by system as negative. It may be realized that the sign convention does not change the absolute values.

3.6.1   Work done in changing the volume of a system

We consider a system, the initial and the final states of which are shown in Fig. 3.7(a), respectively, by the continuous and dotted boundary lines. The system is enclosed by the surrounding which exerts a pressure P_{ext} on the system. The system in its initial state exerts a pressure P^i_{int} on the surrounding as shown in the figure. Let us assume that P_{ext} is constant and larger in magnitude than P^i_{int}.

Since the pressure exerted by the surroundings is larger than the initial internal pressure of the system, the volume of the system decreases. For how long does this decrease in the volume of the system will continue? As the system contracts, the pressure inside the system increases and the decrease in the volume will stop when the internal system pressure P^f_{int} becomes equal to the pressure P_{ext}. Two points need to be kept in mind, one, during the process of contraction of the system the internal pressure of the system keeps changing and second, in the final state the external and internal pressures ($P_{ext} = P^f_{int}$) are equal.

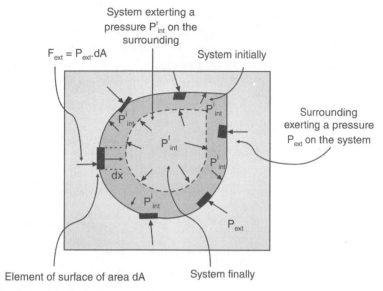

Fig. 3.7 (a) Work done on the system by the pressure of the surrounding

Surroundings perform work on the system in decreasing the volume of the system. Force F exerted by the surrounding per unit area of the boundary of the system is equal to the pressure P_{ext}. If an area dA of the boundary is considered than the force F on this surface area is;

$$\vec{F} = \overrightarrow{P_{ext}} \, .dA \qquad\qquad 3.18$$

The force F has displaced the surface area dA by the amount $\overrightarrow{dx}$ in the direction of F. Here we see that the direction of displacement and the force is same. According to the definition given by Eq. 3.17, the work done by the surroundings in displacing the surface area dA by the distance dx, is negative;

$$-dW = P_{ext}.dA.dx = P_{ext}dV \qquad\qquad 3.19$$

In Eq. 3.19 dV (= $dA.dx$) is the element of the volume. Now the total work W done by the surroundings on the system may be obtained by integrating Eq. 3.19 in the limits V_1 to V_2, where V_1 and V_2 are, respectively, the initial and the final volumes of the system.

$$-\int dW = \int_{V_1}^{V_2} P_{ext} \, dv \text{, and since } P_{ext} \text{ is constant}$$

$$
\begin{aligned}
- W \text{ (work performed by the surrondings on the system)} &= P_{ext} \, (V_2 - V_1) \\
&= -[P_{ext} \, (V_1 - V_2)] \qquad 3.20
\end{aligned}
$$

In the case when we try to fill air in a bicycle tube, the piston of the cycle pump may be considered as the surroundings and the air in the pump below the piston as the system. The piston (surrounding) works on the air (system) in compressing it. A part of this work is converted into heat that makes the pump hot. It is to be noted that processes in which heat is generated corresponds to the case of negate work. In many chemical reactions heat is produced and such reactions are called exoreic or exothermic reactions. For such reactions work is done by the surroundings on the system and according to thermodynamics they are processes of negative work.

Expansion of a system against constant pressure P_{ext} exerted by the surroundings is shown in Fig. 3.7(b). The initial pressure of the system P^i_{int} is more than P_{ext} and the system expands till the final system pressure P^f_{int} becomes equal to P_{ext}. Force $\bar{F}$ acting on a surface area, ds, of the system boundary and the displacement vector $\overrightarrow{dx}$ are shown in the Fig. 3.7(b); they are in opposite directions. Hence, the work done is positive. Therefore,

$$W \text{ (work done by the system on the surrounding)} = P_{ext}(V_2 - V_1) \qquad 3.21$$

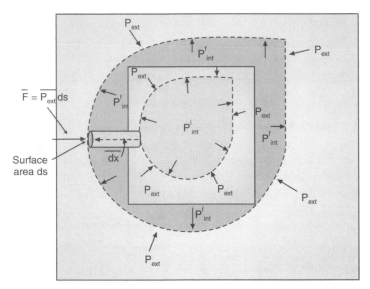

Fig. 3.7   (b) Work done by a system in expansion

It is obvious that in the case when work is done by the system, the energy of the system decreases and the system, if surrounded by diathermal boundary, absorbs energy from the surroundings. Those chemical reactions in which heat is absorbed, correspond to the class where system performs work.

The processes represented in Figs. 3.7(a) and (b) are irreversible, because the system undergoes contraction or expansion rapidly before the establishment of equilibrium. However, if the process is very slow, i.e., at every step the external pressure P_{ext} also varies such that it is always equal to P_{int}, then the process of expansion or contraction will be reversible. For a reversible process of expansion or contraction the magnitude $|P_{ext}|$ will always be equal to $|P_{int}|$ and one may write,

$$|P_{ext}| = |P_{int}| = P$$

and the element of work may be written as,

$$d'W = P\,dV \qquad 3.22$$

It must be emphasized that Eq. 3.22, where P is the pressure of the system, holds only for reversible processes. Further, the dash sign on $d'w$ is used to indicate that $d'w$ is not a state variable (and hence not a complete derivative) and it depends on path.

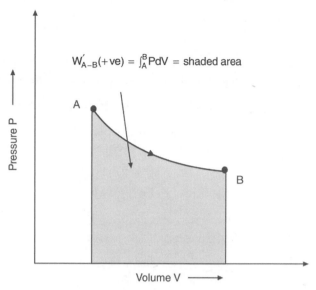

Fig. 3.8   (a) Work done by the system in going from A to B

Figure 3.8 (a) shows the P–V diagram of a system. The system is carried from the initial state represented by point A to a final state represented by point B. In this expansion of the system volume work is done by the system. The work done is positive and is given by,

$$W'_{A \to B}(+ve) = \int_A^B P\,dV = \text{shaded area} \text{ in Fig. 3.8 (a)} \qquad 3.23$$

The system now goes back to the initial state represented by point A. The return path is shown in Fig. 3.8 (b).

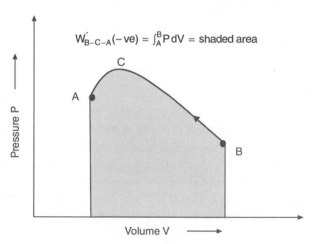

Fig. 3.8   (b) Work done by the surroundings in taking system from B to C to A

In going from point B to A via C, work is performed by the surroundings on the system and is, therefore, negative. The amount of the work done is given by,

$$W'_{B \to C \to A}(-ve) = \int_B^A P \, dV = \text{shaded area in Fig. 3.8(b)} \quad 3.24$$

The net work done in one complete closed path motion from A to B and back to A via C is negative, i.e., the work is done by the surroundings and the amount of net work, that is negative is,

$$W'_{A \to B \to C \to A}(-ve) = \oint P \, dV = \text{shaded area in Fig. 3.8(c)} \qquad 3.25$$

In case the system moves from $A \to C \to B \to A$ the amount of work done will be same as given by Eq. 3.25, but then it will be positive, as the net work will be performed by the system.

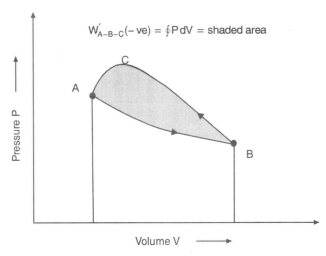

Fig. 3.8   (c) Net work done by the surrounding in taking system from A to B to C and back to A

In case of isobaric processes work done is given by: $W_{isobaric} = P \int_{initial\ value\ of\ V}^{final\ value\ of\ volume} dV$ and in case

of isochoric change of state work done is zero.

3.6.2 Configurational work

In the last section we have seen that the work done in reversible change of the volume of a system is given by $d'W = PdV$, where P is the intensive property and V is the extensive property of the system. In a similar way it can be shown that the work done in charging or discharging of reversible cell is given by $|d'W| = \varepsilon dQ$, where ε (intensive property) is the emf of the cell and dQ (extensive property) the amount of charge deposited in the cell or lost by the cell in the process of charging/ discharging. When the area of a film of some liquid (soap solution) having surface tension σ is increased or decreased by an amount dA, the work done $|d'W|$ is given by $|d'W| = 2\sigma \, dA$. It may be noted that σ is the intensive property of the film and area A an extensive. What we observe here is that in all these examples the work may be written as the product of an intensive property and the differential of an extensive property of the system. Since extensive properties like volume, mass etc., define the configuration of the system, this type of work is called configurational work. If a system has $X_1, X_2,$

X_3... etc., intensive properties and Y_1, Y_2, Y_3, ... etc., the corresponding extensive properties, then the total configurational work may be written as,

$$d'W = X_1dY_1 + X_2dY_2 + X_3dY_3 + ... \qquad 3.26$$

Terms on the right hand side in Eq. 3.26 may be positive or negative depending whether work is performed by the system or on the system.

Configurational work means the work done in changing the configuration of the system. It may be positive, when work is done by the system; and negative if work is done on the system. In some special case it may be zero as well. For example, in the case of free expansion of a gas configurational work is zero.

3.6.2.1 *Free expansion of a gas*

Suppose there is a gas in a container which is separated from the vacuum by a thin diaphragm as shown in Fig. 3.9. Now suppose a hole is made in the diaphragm. The gas will rush out into the vacuum and will fill whole of the container volume. The configurational work $d'W = PdV$, but the pressure P in vacuum is zero and hence, $d'W$ is zero. Therefore in free expansion of a gas, i.e., expansion against zero pressure, the volume of the gas has increased (configuration has changed) without doing any work.

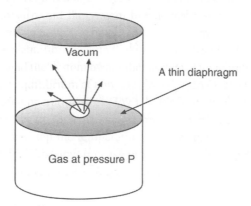

Fig. 3.9 Free expansion of a gas

Work done in a reversible isobaric process $W_{P\ const} = P\left(V_f - V_i\right)$

Work done in a reversible isothermal process $W_{T\ const} = \mathbb{N}RT \times 2.303 \log_{10} \dfrac{V_f}{V_i}$

$$= \mathbb{N}RT \times 2.303 \log_{10} \dfrac{P_i}{P_f}$$

Work done in a reversible adiabatic process $W_{adiabatic} = \left(\dfrac{1}{\gamma - 1}\right)\left(P_i v_i - P_f v_f\right)$

$$= \left(\dfrac{R}{\gamma - 1}\right)\left(T_i - T_f\right)$$

It may be shown that the reversible work $d'W$ (volt coulomb = joule) done in changing the intensity $\in$ (volt per meter) of the electric field in a dielectric slab of dipole moment P (coulomb meter) is given by, $d'W = -\in dP$.

Similarly, the reversible work $d'W$ done in magnetizing a paramagnetic material of volume V (m^3) is given by $d'W = -\mu_0 V H d\mathcal{M}$. Here $\mu_0 = 4\pi \times 10^{-7}$ NA^{-2}, H (A m^{-1}) is the magnetic intensity and $\mathcal{M}$ (A m^{-1}) is the magnetization.

3.6.3 Dissipative work

In a reversible process, the direction of the process reverses if an infinitesimally small change is made in the system variables. An example is the charging and discharging of a reversible cell. Batteries made of series combinations of reversible cells are used in motor cars, scooters and inverters etc. When the emf of such a battery goes down, the battery may be recharged by passing a current through the battery in the reverse direction. The charging and discharging actions of a reversible cell are shown in Fig. 3.10. A reversible cell of emf E is shown at the bottom of the figure. The positive end of the reversible cell is connected through a resistance R to the movable contact b of a rheostat. The end a of the rheostat is connected to the positive end of another cell of emf E_1 which is larger than E. The end c of the rheostat and the negative ends of both cells are connected together. The arrangement of rheostat shown in the figure is called the potential divider arrangement because any potential difference from E_1 to 0 may be obtained between point c and b of the rheostat. Since $E_1 > E$, for some setting of the moveable point the potential of point b will be just equal to the emf E of the reversible cell. At this setting no current will pass through the reversible cell. If point b is moved toward point a by an infinitesimally small amount, a charging current will pass through the resistance R and a reverse chemical reaction in the cell will charge the cell. On the other hand when moveable point b is moved by an infinitesimally small amount towards point c, a current in opposite direction will flow through the resistance R, discharging the reversible cell. Thus it may be observed that the process of charging and discharging of the reversible cell may be reversible.

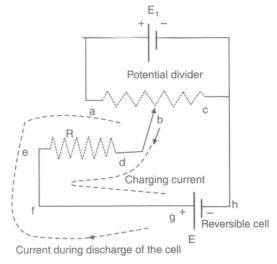

Fig. 3.10   Charging and discharging of a reversible cell

Let us now concentrate on resistance R. A current I passing through a resistance does some work as a result of which heat is generated and the resistance becomes hot. The amount of heat H generated per unit time (rate of doing work) is given by $H = R\,I^2$. Since H depends on the square of the current, it does not depend on the direction of the current. Therefore, energy in the form of heat will be lost by the resistance both when the charging current or the discharge current passes through the resistance. The current performs irreversible work on the resistance and because of the resistance R in the circuit, the charging and discharging processes become irreversible. The resistance R may be the internal resistances of the cells or may be an external resistance. Reversible charging and discharging of the cell is possible only when there are no resistances, external or internal, in the circuit.

Work, like the work done by the current in passing through the resistance, the direction of which cannot be reversed by changing the direction of the process, is called *dissipative work*. A stirrer fan submerged in a liquid will always generate heat no matter in which directions the wings or paddles of the stirrer are made to rotate. The work done by the stirrer is also dissipative. Any work done against the force of friction is yet another example of dissipative work. Electrical system having resistance and mechanical systems with friction are always non-reversible.

3.7   Equations of State and P–v–T Surfaces

A relationship of the type $f\,(\textit{system variables}) = 0$, where f is a function of system variables is called the equation of state of the system. In case of gases intrinsic system variables are pressure P, temperature T and specific volume v. The specific volume may either be volume per unit mass or per mole or per kilomole of the gas. In chapters 1 and 2, properties of an ideal gas were discussed in detail and it was shown that the equation of state for an ideal gas is given by

$$PV = \mathbb{N}RT \text{ or } Pv = RT \qquad\qquad 3.27$$

where, V is the volume of $\mathbb{N}$ moles of the perfect gas at pressure P and temperature T kelvin, v the specific molar volume, $R = 8.3143 \times 10^3$ J kilomole^{-1} K^{-1}, the gas constant, The equation of state of a real gas obtained by Van der Waals is,

$$\left(P + \frac{a}{v^2} \right)(v - b) = RT \qquad\qquad 3.28$$

The constants, a and b, in Eq. 3.28, have different values for different gases.

In principle, equation of state can be written for any system in equilibrium. Important steps in writing down the equation of state for any system are: (i) to identify the intensive and extensive properties of the system (ii) to identify the properties that depend on temperature, and (iii) to write an equation describing the equilibrium of the system. Since temperature plays a key role in thermodynamics, temperature dependence of all system properties must be explicitly included in the equilibrium equation. As an example let us consider the state of equilibrium of a metal wire hanging vertically from a support in the ceiling at the other end of which a body of mass of M kg is attached. You might have used this type of arrangement, shown in Fig. 3.11, in the experiment on the determination of Young's modulus in your laboratory. Suppose before hanging the wire its length was measured as $L_{T_i}^{in}$, when the temperature was T_i. The downward stretching force

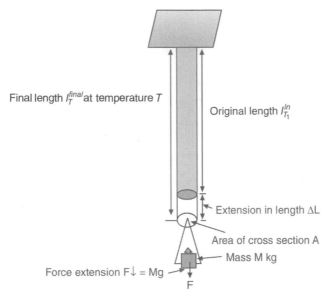

Final length l_T^{final} at temperature T

Original length $l_{T_1}^{in}$

Extension in length ΔL

Area of cross section A

Mass M kg

Force extension $F\downarrow = Mg$

F

Fig. 3.11 Extension in the length of a wire by stretching force

$F = Mg$ will try to increase the length of the wire. However, if the stretching force is within Hook's limits, the extension in the length of the wire will bring into play a restoring force F_{rest} that will be proportional to the increase in the length of the wire ΔL and will act opposite to F. In the equilibrium, the final (total) length L_T^{final} of the wire at temperature T will be given by:

$$L_T^{final} = L_{T_1}^{in} + \Delta L \qquad 3.29$$

Here ΔL is the change in the length of the wire. ΔL will have two components; one due to the change in the temperature T_i when the length of the wire was initially measured, and the other one temperature T when the final length measurement is done. The change in the length of the wire due to the temperature difference is given by $L_{T_1}^{in} x \alpha (T - T_i)$, where α is the coefficient of thermal expansion for the material of wire. It may be noted that this term may be positive when $T > T_i$, and negative if $T < T_i$. The second component will be the extension of the wire due to the elasticity of the wire. If y denotes the young's modulus of elasticity for the material of wire, then the extension in length due to elasticity is given by $\dfrac{L_{T_1}^{in} Mg}{yA}$, where A is the area of cross section of the wire.

Substituting these values for the two types of extensions in Eq. 2.29 we get,

$$L_T^{final} = L_{T_1}^{in} + L_{T_1}^{in} x \alpha (T - T_i) + \frac{L_{T_1}^{in} Mg}{yA} \qquad 3.30$$

Eq. 3.30 is the equation of state for this problem and the intensive and extensive properties are the force $F = Mg$ and the length L, respectively.

It can be shown that the equation of state for a film of surface tension $\sigma(T)$ at temperature T K may be written as $\sigma(T) = \sigma(T_0)\left(\dfrac{T_c - T}{T_c - T_0}\right)$, where $\sigma(T_0)$ is the value of the surface tension at reference temperature T_0 and Tc is the temperature of the fluid at the critical point.

A paramagnetic material put in a magnetizing field of intensity H at temperature T K may be considered to be a thermodynamic system and the equation of state for the same may be written as,

$$M = Cc \frac{H}{T}$$

Here, M is the magnetic moment of the paramagnetic substance and Cc a constant, called the Curie constant. H in above equation of state is the intensive variable while M is extensive.

In a similar way, a dielectric slab placed in an external electric field of intensity E at temperature T (K) develops the dipole moment P that depends both on T and E. The system consisting of the electric field and the slab may be treated as a thermodynamic system, the equation of state for which is,

$$P = \left(A + \frac{B}{T} \right) E$$

Here, A and B are constants. It may be noted that P is an extensive variable and E intensive.

These examples are enough to convince that in principle the rules of thermodynamics may be applied to any macroscopic system.

3.7.1 *P–v–T* surfaces

Three important system variables, particularly in the case of gases, are the pressure P, specific volume v and the kelvin temperature T. These variables are related to each other through the equation of state. Any two of these may be varied independently but the value of the third variable gets fixed, by the equation of state, for the given values of the other two. A three dimensional space may be generated by plotting the values of state variables P, v, and T on three mutually perpendicular axes. In case the system is an ideal gas, all possible values of P, v, and T, allowed by the equation of state, fall on a two dimensional surface in the three dimensional space of system variables. This two dimensional surface, on which each point represents an equilibrium state of the ideal gas, is called the P–v–T surface for an ideal gas.

The shaded surface in Fig. 3.12 is the P–v–T surface for an ideal gas. Any point on this surface corresponds to an equilibrium state of the ideal gas. A line on this surface represents a succession of quasi-static states and hence a process. Isothermal processes are represented by hyperbolas, (Pv = constant) as shown in the figure. Isochors and isobars are straight lines ($P \propto T$) and ($v \propto T$).

P–v–T surface corresponding to Van der Waals equation of state is shown in Fig. 3.13. Projections of some van der Waals isotherms on P–v plane are shown in Fig. 3.14. It may be observed that for isotherms below the critical temperature there are three possible values of volume (points a, b and c in Fig. 3.14) for fixed values of pressure and temperature. Van der Waals equation has only one real root for volume at critical isotherm and isotherms at higher temperatures. Below critical isotherm, the liquid and the vapor or gaseous phases co-exist. However the relative composition of the vapor and liquid phases changes with the temperature. If M_V and M_L, respectively, represent the masses of vapor and liquid phases in equilibrium at temperature T, then mass fraction of the vapor phase X_V and the liquid phase X_L may be given by,

$$X_V = \frac{M_V}{M_V + M_L} \text{ and } X_L = \left(1 - X_V\right) = \frac{M_L}{M_V + M_L} \qquad 3.31$$

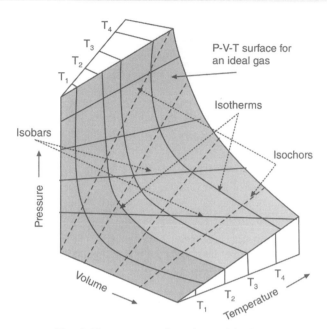

Fig. 3.12   *P–v–T* surface for an Ideal gas

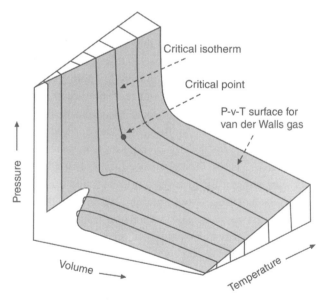

Fig. 3.13   *P–v–T* surface for Van der Waals gas

In engineering terminology X_v is called the 'Quality' of the mixture of the vapor and liquid phases. Other properties of the mixture at a given temperature, may also be expressed in terms of X_v, for example, the composite specific volume of the two phases v may be written as the sum of the specific volumes of the individual phases as,

$$v = X_v v_V + \left(1 - X_V\right)v_L \qquad\qquad 3.32$$

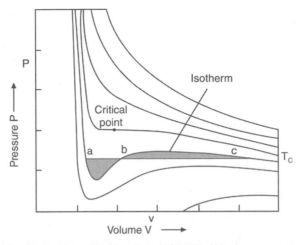

Fig. 3.14 Projection of Isotherms of Van der Waals gas on $P-v$ plane

A $P-v-T$ diagram may be drawn for every substance. Substances that may have all the three phases, vapor, liquid and solid have quite complicated $P-v-T$ surfaces. Some important properties of phase transitions may be studied using the $P-v-T$ surfaces.

3.8   Motion on $P-v-T$ Surface and Differential Calculus

An equilibrium state of a system is given by a point on the $P-v-T$ surface and any quasi-static reversible process by a curve on this surface. When a system moves from one point to another on the $P-v-T$ surface, changes in the values of P, V and T take place. Using the methods of differential calculus it is possible to write these changes in system parameters in terms of the partial derivatives with respect to a single variable keeping the other variable constant. For example, if the volume V is considered to be a function of P and T, i.e., $V = V(P, T)$, then the change in the volume dV of a system when the system moves from one equilibrium state to another may be given as;

$$dV = (dV)_P + (dV)_T = \left(\frac{\partial V}{\partial T}\right)_P dT + \left(\frac{\partial V}{\partial P}\right)_T dP \qquad 3.33$$

Here $(dV)_P$ and $(dV)_T$ are, respectively, the changes in volume at constant pressure and at constant temperature. The former may be written as the partial derivative of V with respect to T at constant pressure $\left(\frac{dV}{\partial T}\right)_P$ multiplied by dT, the change in the value of temperature. Similarly, the term $(dV)_T$ may be written as rate of change of V with pressure at constant temperature $\left(\frac{\partial V}{\partial P}\right)_T$ times the change in pressure dP.

In a similar way, change in specific volume v may be given as,

$$dv = (dv)_P + (dv)_T = \left(\frac{\partial v}{\partial T}\right)_P dT + \left(\frac{\partial v}{\partial P}\right)_T dP \qquad 3.34$$

where, $\left(\dfrac{\partial v}{\partial T}\right)_P$ and $\left(\dfrac{\partial v}{dP}\right)_T$ are, respectively, the rate of change of specific volume with temperature at constant pressure and with pressure at constant temperature. The dT and dP are the total changes, respectively, in temperature and pressure.

3.8.1 The expansivity (β)

The coefficient of volume expansion or expansivity β of a material is defined as the ratio of the rate of change of volume with temperature at constant pressure to the original volume.

$$\beta \equiv \frac{\left(\dfrac{\partial V}{\partial T}\right)_P}{V} = \frac{\left(\dfrac{\partial v}{\partial T}\right)_P}{v} \qquad 3.35$$

In Eq. 3.35, V is the total volume of the system and v the specific volume. Since the rate of change of volume with temperature at constant pressure $\left(\dfrac{\partial v}{\partial T}\right)_P$, depends both on the temperature T and the value of the pressure P, the expansivity of a material depends both on the temperature and the pressure. The unit of expansivity is K^{-1}. In general the expansivity of metals at 1 atm pressure tends to zero at absolute zero and increases slowly with temperature. However, at a fixed temperature, the expansivity of metals decreases with pressure. The expansivity of water at constant pressure of one atmosphere has a negative value (volume decreases) at temperatures below 4 °C, becomes zero at 4 °C and then increases rapidly with temperature.

$$\text{The expansivity of an ideal gas } \beta_{ideal} = \frac{\left(\dfrac{dv}{\partial T}\right)_P}{v} = \frac{1}{T}K^{-1} \qquad 3.36$$

3.8.2 Isothermal compressibility κ

The isothermal compressibility of a material is denoted by κ and is defined by the following equation

$$\kappa \equiv -\frac{1}{V}\left(\frac{\partial V}{\partial P}\right)_T = -\frac{1}{v}\left(\frac{\partial v}{\partial P}\right)_T \qquad 3.37$$

In other words the isothermal compressibility of a material is the ratio of the rate of change of volume with pressure at constant temperature to the original volume. Since volume always decreases with the increase of pressure, $\left(\dfrac{\partial V}{\partial P}\right)_T$ is always negative. A negative sign in the definition of the compressibility has been included to make it a positive quantity. The unit of the compressibility is reciprocal of pressure, that is, $1\ m^2\ N^{-1}$. Like expansivity, the compressibility also depends both on the temperature and the pressure. The compressibility of an ideal gas may be obtained using the equation of state $PV = \mathbb{N}RT$.

$$\kappa_{ideal} = \frac{-\left(\dfrac{\partial v}{\partial P}\right)_T}{v} = \frac{1}{P}\ m^2\,N^{-1} \qquad 3.38$$

3.8.3 Obtaining equation of state from expansivity and compressibility

It follows from Eqs. 3.36 and 3.38 that for an ideal gas

$$\left(\frac{dv}{\partial T}\right)_P = \frac{v}{T} \text{ and } \left(\frac{\partial v}{\partial P}\right)_T = -\frac{v}{P} \qquad 3.39$$

Substituting these values in Eq. 3.34 one gets,

$$dv = \frac{v}{T}\, dT - \frac{v}{P}\, dP$$

or

$$\frac{dv}{v} + \frac{dP}{P} = \frac{dT}{T} \qquad 3.40$$

Equation 3.40, on integration, gives

$$\ln v + \ln P = \ln T \quad \text{or} \quad Pv = \text{constant} . T \qquad 3.41$$

Equation 3.41 is the equation of state for an ideal gas. This shows that the equation of state for a system may be obtained by substituting the values of the expansivity and isothermal compressibility of the system in Eq. 3.33 or 3.34 and integrating it.

Solved Examples

1. Calculate the work that must be done at STP to make room for the products of the octane combustion shown by the equation:

$$2C_8H_{18} + 25O_2 \rightarrow 16CO_2 + 18H_2O$$

Solution:

Increase in the volume of the gas

$$\Delta V = (16 + 18 - 25) = 9 \text{ mole} = 9 \times 22.4 \text{ L} = 9 \times 22.4 \times 10^{-3} \text{ m}^3$$

Pressure P is constant at 1 atm = 1.01325×10^5 Nm^{-2}

Work done in expansion of the volume at constant pressure = $P \times \Delta V$ = $1.01325 \times 9 \times 22.4 \times 10^2$ J

Ans: Work done = 20.43 kilojoules

2. Calculate the minimum work done by 1 kilomole of an ideal gas in isothermal expansion from 22.8 m^3 to 68.4 m^3 at temperature 300 K.

Solution:

The minimum amount of work is done when it is done through a reversible process. The reversible work $d'W$ done by a gas in changing the volume by dV at pressure P is given by $d'W = P\, dV$. In this problem the work is done at constant temperature of 300 K. Since the temperature has remained constant, the pressure of the gas must have varied during the expansion. Since the gas is ideal, $PV = \mathbb{N}RT$, where $\mathbb{N}$ is the number of kilomole and R is the gas constant, given by $R = 8.314 \times 10^3$ J kilomole^{-1}K^{-1}. The value of $\mathbb{N}$ is given as = 1, Hence,

$$P = \frac{RT}{V},$$

Substituting this value of P, we get $d'W = RT \dfrac{dV}{V}$

and $\quad W' = RT \int\limits_{initial\ volume}^{final\ volume} \dfrac{dV}{V} = 8.314 \times 10^3 \times 300 \times \int\limits_{22.8}^{68.4} \dfrac{dV}{V} = 24.942 \times 10^5 \ln \dfrac{68.4}{22.8}$

$\qquad = 24.942 \times 10^5 \times 2.303 \log 3 = 24.942 \times 10^5 \times 2.303 \times 0.477$

Ans: 27.34×10^5 joule kilomole^{-1}

3. Assuming nitrogen to be an ideal gas, calculate the work done in compressing 7 kg of nitrogen from 1.0 atm to 10.0 atm pressure at constant temperature of 600 K.

Solution:

In the present case the work is done on the system and hence it is negative and is given by

$$-W = \int\limits_{V_1}^{V_2} PdV . \tag{A}$$

Here $P_1 = 1.0$ atm; $P_2 = 10$ atm; Number of kilomole $\mathbb{N} = \dfrac{7}{28} = 0.25$; $T = 600$ K; $R = 8.314 \times 10^3$ J kilomole^{-1} K^{-1}

It is given that nitrogen behaves as an ideal gas,

Therefore, $\qquad\qquad PV = \mathbb{N}RT \quad$ or $\quad P = (\mathbb{N}\,RT)/V$

Putting this value in Eq. A, we get

$$-W = \int\limits_{V_1}^{V_2} \dfrac{\mathbb{N}RT}{V}dV = \mathbb{N}RT \int\limits_{V_1}^{V_2} \dfrac{dV}{V} = \mathbb{N}RT \ln \dfrac{V_2}{V_1} \tag{B}$$

In Eq. (B), V_2 and V_1 are the final and initial volumes of the gas, respectively, at pressure $P_2 = 10.0$ atm and $P_1 = 1.0$ atm V_1 and V_2 are not provided in the question. However, for an ideal gas at constant temperature $P_1V_1 = P_2V_2$, and hence, $\dfrac{V_2}{V_1} = \dfrac{P_1}{P_2}$, putting this value in Eq.(B), we gct,

$$-W = \mathbb{N}RT \ln \dfrac{V_2}{V_1} = \mathbb{N}RT \ln \dfrac{P_1}{P_2}\ \mathbb{N}RT \times 2.303 \log_{10} \dfrac{P_1}{P_2},$$

$$= 0.25 \times 8.314 \times 10^3 \times 600 \times 2.303 \times \log \dfrac{1}{10}$$

$$= -28.72 \times 10^5 \text{ joule}$$

Ans: The work done is $W = 28.72 \times 10^5$ joule

4. 0.5 kilomole of a diatomic ideal gas expands adiabatically from 1 liter at 10 atm pressure to 5 liter at 1.0 atm pressure. Who has performed work in this expansion and what is the magnitude of the work? Also calculate the change in the temperature of the gas.

Solution:

The work done by an ideal gas in adiabatic expansion from initial volume V_i, temperature T_i and pressure Pi to final values V_f, T_f and P_f is given by (see chapter 1 section 1.4.1):

$$W = \left(\frac{1}{\gamma - 1}\right)\left(P_i v_i - P_f v_f\right) = \left(\frac{R}{\gamma - 1}\right)\left(T_i - T_f\right) \tag{A}$$

It is given that: $v_f = \dfrac{5 \times 10^{-3}}{0.5} = 1 \times 10^{-2}\ m^3$, $v_i = \dfrac{1 \times 10^{-3}}{0.5} = 2 \times 10^{-3}$, $P_f = 1.0\ \text{atm} = 1.013$

$\times 10^5\ Nm^{-2}$, $P_i = 10\ \text{atm} = 10 \times 1.013 \times 10^5\ Nm^{-2}$; $\gamma = 1.4$(diatomic ideal gas)

Substituting these values in Eq.(A), we get,

$$W = \frac{1}{1.4 - 1}\left(10 \times 1.013 \times 10^5 \times 2 \times 10^{-3} - 1.013 \times 10^5 \times 1 \times 10^{-2}\right)$$

$$= 2.53 \times 10^3\ \text{joule}$$

Also from Eq. (A), change in temperature $\Delta T = \left(T_i - T_f\right) = \dfrac{\left(P_i v_i - P_f v_f\right)}{R} = \dfrac{1.013 \times 10^3}{8.314 \times 10^3} =$

0.122 K

Ans: Since W is positive, work is done by the gas. The amount of work is 53×10^3 joule; Change in temperature is 0.122 K

5. An ideal gas and a block of some metal have same volumes of 1 m³ at 300 K and 1 atm pressure. Pressure on both was increased reversibly and isothermally to 10 atm The isothermal compressibility of the metal is 0.8 atm⁻¹, (a) Draw the P–V diagram for the compression of the gas and the block and (b) calculate the change in the volume of the gas and the block. (c) the number of kilomoles of the gas (d) the work done in compressing the gas and (e) the work done in compressing the block.

Solution:

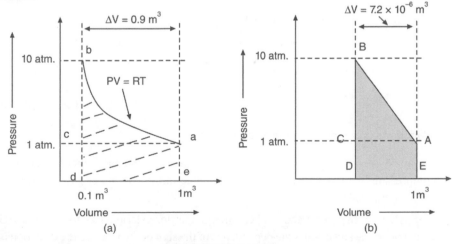

Fig. S-3.5(a) and (b)

(a) The P–V diagrams for the compression of the gas and the block are shown in the Figs. S-3.5 (a) and (b), respectively. The initial positions of the gas and the block is shown by points a and A, respectively, and the final positions by points b and B. The process of compression is Isothermal and reversible therefore, the process goes through the path of minimum work. In case of the gas the gas follows the ideal gas equation $Pv = RT$. In case of the solid block the path of minimum work will be the straight line AB joining the initial and final points.

(b) The change in the volume of the gas: For an ideal gas and isothermal change $P_1V_1 = P_2V_2$ and $V_2 = (P_1/P_2).V_1$. In the case of the gas, $V_2 = (1/10) \times 1 \ \text{m}^3 = 0.1 \ \text{m}^3$.

Therefore, the change in volume of the gas = $(1.0 - 0.1) \ \text{m}^3 = 0.9 \ \text{m}^3$.

For the block it is given that the isothermal compressibility $\kappa = \dfrac{\left(\dfrac{\partial V}{\partial P}\right)_T}{V} = 0.8 \times 10^{-6}$, hence,

$$\left(\frac{\partial V}{\partial P}\right) = 0.8 \times 10^{-6} \times V = 0.8 \times 10^{-6} \times 1 \ \text{m}^3,$$

Therefore, $\partial V = 0.8 \times 10^{-6} \times 1 \times \partial P = 0.8 \times 10^{-6} \times (P_2 - P_1) = 0.8 \times 10^{-6} \times 9 = 7.2 \times 10^{-6} \ \text{m}^3$

Hence change in the volume of the block $\Delta V_{block} = 7.2 \times 10^{-6} \ \text{m}^3$

(c) Let the number of kilomole of the gas be n, then for the ideal gas $PV=nRT$

It is given that: $P = 1.0134 \times 10^5 \ \text{Nm}^{-2}$, $V = 1 \ \text{m}^3$, $T = 300 \ \text{K}$, also $R = 8.314 \times 10^3 \ \text{J kilomole}^{-1}$ K^{-1}. Substituting these values we get, $n = \dfrac{PV}{RT} = \dfrac{1.0134 \times 10^5 \times 1}{8.314 \times 10^3 \times 300} = 0.04$

(d) Let W_1 be the work done in compressing the gas then

$$W_1 = nRT \int PdV = nRT \, 2.303 \log \frac{10 \times 10^5 \times 1.0134}{1 \times 10^5 \times 1.0135} = 0.04 \times 8.314 \times 10^3 \times 300 \times 2.303 \times 1$$

$$= 2.29 \times 10^5 \ \text{joule}$$

(e) Work done in compressing the block W_{block} = area ABCDE in Fig. S-3.5 b

= area of the triangle ABC + area of the rectangle $ACDE$

or $\quad W_{block} = \dfrac{1}{2}\left[\Delta V_2 \times \Delta P\right] + \left[\Delta V_2 \times \Delta P\right]$

$$= \frac{1}{2}\left[7.2 \times 10^{-6} \times 9 \times 1.0134 \times 10^5\right] + \left[7.2 \times 10^{-6} \times 1 \times 1.0134 \times 10^5\right]$$

Work done in compressing the block = 4.01 joule

It may be observed that the work done in compressing the gas is much larger than that in compressing the block. It is because the change in the volume of gas is much larger than that of the block.

6. A system starts from point A and is taken through the closed loop ABCDA five times. Calculate the total work done and discuss the dependence of the sign of the work on the direction of tracing the closed loop.

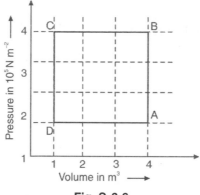

Fig. S-3.6

Solution:

The total work done in taking the system 5-times around the loop

$$= 5 \times \text{Area of the loop}$$
$$= 5 \times (3 \text{ m}^3 \times 3 \times 10^5 \text{ N m}^{-2})$$
$$= 45 \times 10^5 \text{ Nm}$$
$$= 45 \times 10^5 \text{ joule}$$

Total work done is 45×10^5 joules.

Suppose we start from point A and move in clockwise direction, i.e., from A to D than the system is compressed as its volume decreases. Hence work will be done by the surroundings and will be negative. Next in going from D to C no work is done as there is no change in the volume. In going from C to B work is done by the system in expansion. The work done in expansion by the system from C to D is (area of the rectangle *CB41C*) larger than the work done by the surroundings in compression from A to D (area of the rectangle *AD14A*). Hence in clockwise motion the work done will be positive and in counter clock wise direction negative.

7. Figure S-3.7 shows a magnetic material of volume 0.5 m³ starts from point A and is taken around the loop once. Calculate the work done and also when the work will be positive?

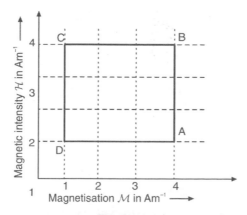

Fig. S-3.7

Solution:

The work done in magnetization of a magnetic substance of volume V is given by:

$$d'W = -\mu_0 \mathcal{H} V \, d\mathcal{M},$$

where, $\mu_0 = 4\pi \times 10^{-7}$ NA^{-2}, $\mathcal{H}$ is the magnetic intensity, V the volume of the speciman and, $\mathcal{M}$ the magnetisation

The magnitude of the work done W in going through the closed path $ABCDA = \mu_0 \mathcal{H} V \, d\mathcal{M}$,

$$W = 4\pi \times 10^{-7} \, \text{NA}^{-2} \left(3 \, \text{Am}^{-1}\right)\left(0.5 \, \text{m}^3\right)\left(3 \, \text{Am}^{-1}\right) = 56.52 \times 10^{-7} \, \text{Nm} = 56.52 \times 10^{-7} \, \text{joule}$$

Work is done by the surroundings in magnetization of the specimen and the work in positive when it is done by a magnetized specimen in demagnetization. In Fig. s-3.7 no work is done along the lines AB and CD. Work is done by the specimen in going from B to C (positive) and by the surroundings (-ve) from D to A. The total work will be positive if the loop is traced in counter clock wise direction.

8. Drive an expression for the work done in changing the area of a soap film of surface tension σ.

   Solution:

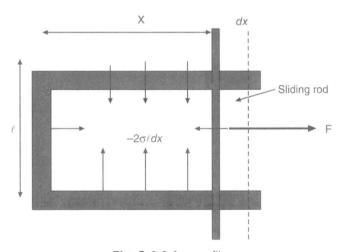

Fig. S-3.8 A soap film

Figure S-3.7 shows a soap film on a U shaped frame with a sliding rod on the right. A force F is applied to extend the rod by a distance dx towards the right. In equilibrium the force F is balanced by the inwards force of surface tension. The surface tension force per unit length is σ. And there are two surfaces, one the front and the other at the back of the film.

Therefore, $F = -2\sigma\ell$ and the work $d'W$ done in moving the rod by a distance dx is,

$$d''W = F \, dx = -2\sigma\ell \, dx = -2\sigma \, dA \tag{A}$$

Here dA is the change in the area of the film.

Expression (A) gives the desired element of work.

8. Show that the work done in isothermal expansion of 1 kilomole of Van der Waals gas from initial volume v_1 to a final volume v_2 is given by, $W = RT \ln \dfrac{v_2 - b}{v_1 - b} + a\left(\dfrac{1}{v_2} - \dfrac{1}{v_1} \right)$.

Solution:

The pressure P in case of one kilomole of the van der Waals gas is given as

$$P = \frac{RT}{v - b} - \frac{a}{v^2}.$$

The work done $d'w$ in isothermal expansion by an amount dv is,

$$d'w = Pdv = \frac{RTdv}{v - b} - \frac{a\,dv}{v^2}$$

The total work in isothermal change in volume from v_1 to v_2 is obtained by integrating the above equation:

$$W = \int_{v_1}^{v_2} d'W = RT \int_{v_1}^{v_2} \frac{dv}{v - b} - a \int_{v_1}^{v_2} \frac{dv}{v^2} = RT \ln\left[v - b\right]_{v_1}^{v_2} - a\left[-\frac{1}{v}\right]_{v_1}^{v_2}$$

or $$W = RT \ln\left[\frac{v_2 - b}{v_1 - b}\right] + a\left[\frac{1}{v_2} - \frac{1}{v_1}\right]$$

9. A thermometer has been made using a liquid, the density of which changes with temperature and is used as the thermometric property. The graph of density verses temperature is shown in Fig. S-3.9. The thermometer gives the same reading d (indicated by the dotted line in the figure) of density when dipped in liquid A and in liquid B separately. Are the two liquids necessarily in thermal equilibrium? If not than what will happen to the level of the liquid in the thermometer if it is dipped in a mixture of the two liquids?

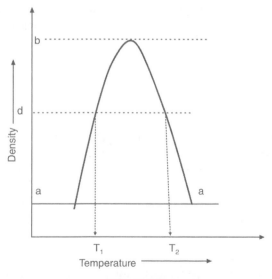

Fig. S-3.9

Solution:

Same density d of the liquid may occur for two different temperatures T_1 and T_2 as shown in the figure. Hence the same reading of the density does not necessarily means that the temperatures of the two liquids A and B are same. It is only at the highest density (corresponding to point b) where there is a unique value of the density and hence of the temperature. For all other values of the densities there will always be two temperatures for the same density.

When the two liquids are mixed, the temperature of the mixture will increase and the density of the liquid will also increase. Since the volume of the liquid in the thermometer is fixed, an increase in the density will mean a decrease in the level of the liquid.

10. A steady current of 0.1 A is passed through a pure resistance of 100 ohm for an hour. Calculate the heat generated. Who has performed the work that generated heat energy? Discuss what will happen if the direction of the current is reversed and the current is passed again for one hour.

Solution:

The source of emf (battery) performs work in passing current through a resistance. The work done in time dt is $d'W = R\,I^2 dt$, and, therefore, the work done in an hour is, $W = RI^2 t = 100 \times (0.1)^2 \times (60 \times 60) = 3600$ joule, So 3.6×10^3 J of energy in the form of heat will be produced. This is irreversible or dissipative work and since the work depends on the square of the current, heat will be produced even when the direction of the current is reversed. In the next one hour also same amount of heat energy (3.6×10^3 J) will be produced.

11. What is meant by the triple point of a substance? Discuss the significance of the triple point of water. Briefly discuss the construction of a triple point cell of water.

Solution:

The pressure and temperature at which the solid, the liquid and the vapor phases of a substance co-exist in thermodynamic equilibrium is called the triple point of the liquid. All thermometers require some reference or standard temperature for calibration of the thermometer scale. Freezing point and the boiling point of water were earlier used as the standard temperatures. However, the values of these standards were difficult to reproduce as they are very sensitive to the conditions of the surrounding environment. Triple point of water, where the ice, liquid water and water vapors co-exist in thermodynamic equilibrium is temperature that does not depend on the conditions of the surroundings and can be easily reproduced in different laboratories using a simple cell- called the triple point cell of water.

Figure S-3.11(b) shows the location of the triple point of water on the P–V graph. A simple triple point cell of water is shown in Fig. S-3.11(a). As may be seen from the figure, it consists of a u-tube which is filled with 99.9999% pure water. The U-tube is then sealed at one end and a vacuum pump is attached to the other end through a side tube. High vacuum of the order of 10^{-4} mm of Hg is produced in the U-tube and the side tube is sealed off. The sealed U-tube containing very pure water is then put in a refrigerator till a layer of ice is formed around the inside edges of the U-tube. The triple point cell is now ready for calibration of other thermometer. Any other thermometer that is to be calibrated is put in the well between the two limbs of the U-tube and is kept in contact with the U-tube for some time. A thin layer of ice melts around the inside walls of the U-tube. At this instant, the liquid water, the solid water (ice) and the water vapors are in thermal equilibrium and the temperature of the triple point cell and of the thermometer in contact with it is at the triple point.

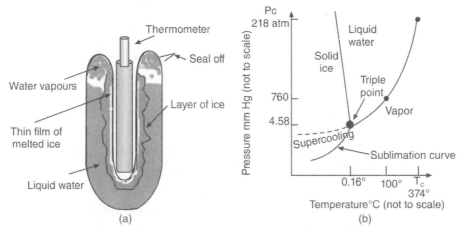

Fig. S-3.11   (a) Triple point cell of water (b) Triple point of water

12. At some temperature and pressure below its critical point, 20 kg of a substance exists in two phases with the mass fraction of vapor phase of 0.75. The specific volumes of saturated vapor and liquid are, respectively, 5 m³/kg and 0.0 5 m³/kg. Determine the mass of the substance in vapor and liquid phases and the total volume of the substance.

Solution:

Mass of liquid phase = $(1 - X_V) \times$ Total mass = $(1 - 0.75) \times 20$ kg = 5 kg

Mass of liquid phase = $X_v \times$ Total mass = 0.75×20 = 15 kg.

Total volume = $M [(1 - X_V)v_L + X_V v_v]$ = 20[0.25 × 0.0 5 + 0.75 × 5]

= 75.25 m³

Ans: The mass of the liquid and vapor phases are, respectively, 5 kg and 15 kg. Total volume = 75.25 m³

13. The temperature of 2 kilomole of a gas increased by 6.0 K when 350 kilojoules of heat energy was supplied to the gas at constant pressure. Calculate the molar heat capacity at constant pressure and also at constant volume.

Solution:

From the definition of the heat capacity at constant pressure it follows that

Heat supplied as constant pressure = Number of kilomole × Cp × rise in temperature (K);

Substituting the values given we get, $350 \times 10^3 = 2 \times Cp \times 6.0$

or $$Cp = \frac{350 \times 10^3}{2 \times 6.0} \text{ J} = 29.17 \times 10^3 \text{ J kilomole}^{-1} \text{ K}^{-1}$$

But $Cv = Cp - R$ = (29.17 × 10³ – 8.314 × 10³) = 20.85 × 10³ J kilomole⁻¹ K⁻¹

Ans: The molar thermal capacities at constant pressure and constant volume are respectively, 29.17×10^3 J kilomole^{-1} K^{-1} and 20.85×10^3 J kilomole^{-1} K^{-1}.

14. The length of the mercury thread in a mercury in glass thermometer is 10 cm when it is in thermal equilibrium with a water triple point cell. What are the values of temperatures in kelvin when the length of mercury thread is 6 cm and 11 cm. If the length of the mercury thread can be read with a precision of 0.01 cm, will it be possible using this thermometer to differentiate between the triple point of water and the melting point of ice? What is the minimum temperature difference that could be read with this thermometer?

Solution:

10 cm length of mercury thread = 273.16 K

Hence 1.0 cm length of mercury thread = 273.16/10 = 27.316 K

Therefore, 6 cm of Hg length = 6 × 27.316 K = 163.90 K

and 11 cm of Hg length = 11 × 27.316 = 300.48 K

The difference between the triple point of water and the melting point of ice is=0.01K. If the mercury thermometer is used than the difference of 0.01 K will be equal to the length of Hg = $\dfrac{0.01}{27.316}$ = 0.000367 cm . However, it is given that the precision of reading the length of mercury thread is only 0.01. So this thermometer will not be able to distinguish between the temperatures of the triple point and the melting point of ice. The minimum difference of two temperatures that could be read with this thermometer is = 0.01 × 27.316 =0.27 K

Problems

1. The readings of a thermometer denoted by t are related to the readings of a constant volume gas thermometer denoted by T_g, through the relation,

$$t = A T_g^3 + B, \text{ here A and B are constants.}$$

Calculate the values of A and B if t =0 for the freezing point of water and 100 for its boiling point.

2. Two thermometer scales A and B defines the triple point of water as 200 A and the other as 350A. What will be the relation between a given temperature measured by A and B?

3. A gas is contained in a vertically held cylinder with a piston at the top on which a heavy weight is placed. 400 J of energy is supplied to the gas and then the system is covered by an adiabatic boundary. If the total mass of the piston and the block is 50 kg and no change in the internal energy of the gas has taken place, calculate the distance by which the piston got displaced, assume that there is no friction between the piston and the cylinder and the value of acceleration due to gravity g at the place is 10 ms^{-2}. Also indicate what type of process has taken place.

4. 2 kg of oxygen is contained in a piston cylinder arrangement at 300 K and 500 kPa pressure. In the first stage the gas was heated so that its volume becomes twice the initial value. In the second stage further heat energy was supplied to the gas keeping the volume constant so that

the final pressure becomes twice the initial pressure. Assuming oxygen to be a diatomic ideal gas calculate: (a) Initial volume of the gas (b) temperature of the gas at the end of the first and the second stages (c) Heat supplied to the gas in first and the second steps (d) Work done by the gas in step one and two.

5. An ideal gas at pressure of 5.4×10^4 Nm^{-2} expands from initial volume V to the final volume of 954 liters and performs 44.36×10^3 J of work. What was the initial volume of the gas?

6. 20 kilomole of an ideal gas at temperature of 107^0C is compressed from 300 liter to 80 liter. Calculate the work done by the gas.

7. 1 kilomole of an ideal gas is contained in a cylinder-piston system at 27^0C and 1.0 atm pressure. The gas was slowly compressed keeping the temperature constant so that the volume of the gas is reduced by 20%. The frictional force between the cylinder walls and the piston is of 10^3 N and the area of cross section of the piston is 0.5 m^2. Calculate (a) the initial and final volumes of the gas (b) the total work done (c) the dissipative work (d) configurational work.

8. 0.2 kilomole of oxygen held in a volume of 1.0 liter was compressed adiabatically and slowly to one fifth of its volume. Assuming oxygen to be a diatomic ideal gas, calculate the work done in the compression.

Short Answer Questions

1. Give one example each of a (i) closed system (ii) open system (iii) adiabatic boundary (iv) diathermic boundary (v) intensive property and (vi) extensive property.

2. Define: (a) a system (b) a process (c) state variables (d) specific molar volume

3. According to gas laws PV = constant. However, when air is pushed into a cycle tube both the pressure and the volume of the gas inside the tube increase simultaneously. How can this be explained.

4. Define isothermal compressibility and expansivity, give their order of magnitudes and units.

5. Show that Van der Waals constants a and b are related to the critical gas constants through the relations: $P_c = \dfrac{a}{27b^2}, v_c = 3b, T_c = \dfrac{8a}{27Rb}$

6. Draw rough sketches for the P–v–T surfaces for an ideal and a real gas.

7. What is the physical significance of P–v–T surface? Discuss.

8. What is a quasi-static process? Give an example of a quasi-static but irreversible process.

9. Clearly distinguish between dissipative and configurational works. Which of the two cannot be reversible? Give one example each of the configurational and dissipative works.

10. Without derivation write expressions for the work done in isobaric, isothermal and adiabatic processes.

11. Discuss the sign of the work when heat is evolved in a reaction.

12. What is meant by the term 'quality' in the texts on engineering thermodynamics?

13. What is the triple point of water and why is it used as a temperature standard?. Briefly discuss the triple point cell for water.

Long Answer Questions

1. State and explain the Zeroth law of thermodynamics. Why this law is so fundamental in thermodynamics? What is meant by thermal and thermodynamic equilibriums? If a system A is in some sort of an equilibrium separately with two other systems B and C, then is it necessary that B and C are also in equilibrium? Explain your answer with suitable example.

2. Distinguish between a reversible and an irreversible process and show that the work done in a reversible process is PdV, where symbols have their usual meanings. Further, show that the work done is a function of the path. Discuss the sign convention adopted for the work in thermodynamics.

3. Drive an expression for the work done in reversible adiabatic and isothermal processes.

4. Briefly discuss the draw backs of conventional thermometers. With the help of a neat diagram explain the working of a constant volume gas thermometer and derive the expression for the empirical temperature determined using this thermometer.

5. What is meant by the equation of state and how one gets $Pv = RT$ as the equation of state for an ideal gas? Do only gases have equations of state? Explain your answer giving one example.

Multiple Choice Questions

Note: In some of the following questions more than one alternative may be correct. Tick all correct alternatives in such cases for complete answer and full marks

1. The expansivity of pure water in unit of K^{-1} at 274.15 K is of the order of
   (a) 50×10^6
   (b) -50×10^6
   (c) 50×10^{-6}
   (d) -50×10^{-6}

2. The empirical temperature T_g measured by a constant volume gas thermometer is given by
   (a) $273.16 \text{ K} \times \lim\limits_{P_{triple}\to 0} \left(\dfrac{P}{P_{triple}}\right)_{const.volume}$
   (b) $273.16 \text{ K} \times \lim\limits_{P_{critical}\to 0} \left(\dfrac{P}{P_{triple}}\right)_{const.volume}$
   (c) $\lim\limits_{P_{triple}\to 0} \left(\dfrac{P}{P_{triple}}\right)_{const.volume}$
   (d) $\lim\limits_{P_{triple}\to 0} \left(\dfrac{P}{P_{triple}}\right)_{const.volume}$

   where symbols have their usual meanings.

3. A constant volume gas thermometer, a Celsius thermometer and a Rankine thermometer all kept in a constant temperature bath show readings 195, X and Y respectively. The approximate values of X and Y are, respectively,
   (a) 78, 351
   (b) 351, 78
   (c) –78, 351
   (d) –351, 78

4. In an isothermal process:
   (a) $\Delta T = 0$
   (b) $PV = \text{constant}$
   (c) $\Delta U = 0$,
   (d) $\Delta W = 0$

5. The volume, pressure, temperature and mass of molecule of an ideal gas are respectively, 2V, 2P , T and m. The density of the gas is: (Here $\Re$ is Boltzmann constant)

(a) $\dfrac{2P}{\Re Tm}$

(b) $m\Re T$

(c) $\dfrac{P}{\Re VT}$

(d) $\dfrac{2Pm}{\Re T}$

6. The volume of an ideal gas changes from Vi to Vf. The work done by the gas is a maximum when the change is,

(a) Isothermal

(b) adiabatic

(c) isobaric

(d) isochoric

7. One end of a metallic rod is kept in a furnace so that the temperature of the rod attains steady state. Which of the following statement (s) about the temperature is (are) true;

(a) it is same throughout the rod

(b) It is constant at each point of the rod

(c) It is non-uniform along the rod

(d) It at a point on the rod is inversely proportional to the distance of the point from the furnace.

8. 2 kg of oxygen is taken along the path ADCBA. Treating oxygen as an ideal gas, work done by the gas is:

(a) $-PV$

(b) $+PV$

(c) $+\pi V^2$

(d) $-\pi V^2$

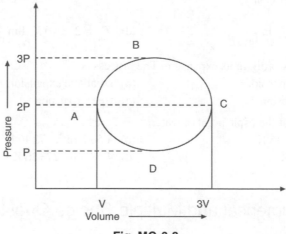

Fig. MC-3.8

9. A system consists of two phases in equilibrium. The specific heat capacity at constant pressure P and the coefficient of thermal expansion of the system are respectively,

(a) 0, 0

(b) 0, ∞

(c) ∞, 0

(d) ∞, ∞

10. A system is taken from initial state a to the final state c via three paths adc work done W_1, ac work done W_2 and abc work done is W_3. Which of the following statement (s) about the work done is (are) correct.

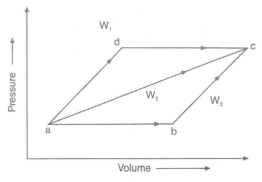

Fig. MC-3.10

(a) $W_1 > W_2 > W_3$ (b) $W_1 > W_2 < W_3$

(c) $W_1 < W_2 < W_3$ (d) $W_1 = W_3 < W_2$

11. 1 torr is a unit of X and is equal to Y, where X and Y are respectively,

(a) volume, 1.333 m^2 (b) density, 13.33 kg m^{-3}

(c) pressure, 133.3 N m^{-2} (d) temperature, 273 K.

12. Empirical temperature T_g obtained by a constant volume gas thermometer at a pressure P and at triple point pressure P_{tri} is given by;

(a) $T_g = 273 \lim\limits_{P_{tri}\to 0}\left(\dfrac{P}{P_{tri}}\right)_V$ (b) $= 273.16\,K \lim\limits_{P_{tri}\to 0}\left(\dfrac{P}{P_{tri}}\right)_V$

(c) $T_g = 273.16\,K \lim\limits_{P_{tri}\to 0}\left(\dfrac{P_{tri}}{P}\right)_V$ (d) $T_g = 273.16\,K \lim\limits_{P_{tri}\to 0}\left(\dfrac{P}{P_{tri}}\right)_P$

13. A gas performs no work in its expansion. The process is:

(a) Isothermal expansion   (b) adiabatic expansion

(c) isobaric expansion   (d) free expansion

14. The coordinates of the triple point of water on P-T surface are,

(a) 4.58 Torr, 273.16 K   (b) 4.58 Nm^{-2}, 273.16 K

(c) 4.58 Pa, 273 K   (d) 4.58 atm, 273.16 K

Answers to Numerical and Multiple Choice Questions

Answer to problems

1. $A = 3.17 \times 10^{-6}$; $B = -65.50$
2. $TA = 0.571\ TB$
3. 0.8 m, isothermal
4. (a) 0.312 m^3, (b) 600 K; 1200 K (c) 7.79×10^5 J; 5.45×10^5 J (d) 1.56×10^5 J, 0
5. 132.5 L
6. -83.53×10^6 J
7. (a) 24.61 m^3; 19.69 m^3 (b) -5.56×10^5 J (c) 9.8×10^3 J (d) -5.46×10^5 J
8. 11.46×10^2 J

Answers to multiple choice questions

1. (d)	2. (a)	3. (c)	4. (a), (b), (c)
5. (d)	6. (c)	7. (b), (c), (d)	8. (d)
9. (d)	10. (a)	11. (c)	12. (b)
13. (d)	14. (a)		

Revision

1. The portion or part of the universe (which is under observation) enclosed by a *boundary* is called the *system*. The system may be anything, a solid, liquid, gas or plasma or a mixture of all these. It might be a distribution of charges or magnetic poles or radiations, i.e., photons in vacuum etc. A system is called *closed* when the total mass contained in the system does not change. A system is said to be an *isolated system* if neither mass nor energy may leave or enter the system through its boundary. When a system may exchange matter (mass) and energy with its surroundings, the system is called an *open* system.

2. The boundary of the system divides the universe into two parts, the system and the *surroundings* or the environment. In other words, system and surroundings make the universe. Boundary may be real or imaginary; *diathermic* that allows energy to pass but not the mass., *adiabatic* that does not allow energy and mass to pass through.

3. A set of values of experimentally measurable quantities defines the state of a system. The physically measurable quantities are called the *state parameters* or *state properties*. The state parameters or properties may be divided into two types: the one that are proportional to the mass of the system, are called *extensive parameters* and the other ones that do not depend on the mass of the system are called *intensive parameters*. The extensive properties are *additive* and are also called the *capacity factors*. Intensive properties are *not additive* and are called *intensity factors*. The *minimum number* of state properties (or parameters) required to *completely specify* a system is called *state variables* or *state functions*. The state variables or state functions are properties of a thermodynamic state of a system and do not depend on the way the state is achieved. State functions are exact differentials. Pressure, volume, temperature and energy are some of the important state functions.

4. The state where no further changes in state variables take place is defined as the *state of thermal equilibrium*.

5. The zeroth law of thermodynamics states: If a system A is separately in thermal equilibrium with two other systems B and C, then systems B and C will also be in thermal equilibrium and the temperatures of systems A, B and C will be the same

6. Conventional thermometers, like the mercury in glass, platinum resistance thermometer, thermometers based on thermo-emf, suffer from the fact that the ratio of the thermometric property at a given temperature to its value at the calibration temperature has different values for different thermometers. As a result different thermometers give different readings for the same temperature.

7. Constant volume gas thermometer is better than all other thermometers, because the ratio $\dfrac{\eta}{\eta_{tri}}$

    for this does not vary with temperature. Any empirical temperature ϑ measured by the gas thermometer may be written as,

$$\vartheta(\text{gas thermometer}) = 273.16 \times \left[\lim_{P_{tri} \to 0} \left(\frac{P}{P_{tri}} \right)_{constVol} \right] K$$

t (in degree Celsius) = T(in kelvin or K) – 273.15 K

Rankine is defined in terms of the thermodynamic scale (kelvin) by the relation,

8. $1R = \dfrac{5}{9}K$

9. $t\,^{0}F$ (temperature in Fahrenheit) = $\dfrac{9}{5} \times T$ (temperature in kelvin) – 459.67

10. *Energy* is also an important state function of a system. It is defined as the capacity of doing work. A system, in general, may possess two types of energies: (i) *Internal* or *inherent energy* (ii) *External* or *bulk energy*

11. *Internal energy:* It is the energy possessed by the microscopic structure of the system. From the kinetic theory it is known that molecules of a substance are always in random motion and that the average kinetic energy of all molecules is proportional to the temperature T (in K). Therefore, the internal energy denoted by U(T) is a function of temperature. Specific internal energy u(T) = U(T)/m where m is the mass of the system. Specific internal energy may also be specified as per kilomole (or per mole).

$$E_{total} = U(T) + E_{external} \left(= E_{Bulk\ kinetic} + E_{Bulk\ Potential} \right) + E_{any\ other\ form\ of\ energy}$$

12. An equilibrium that does not require any effort or energy to maintain itself is called natural or simply equilibrium.

13. The other type of equilibrium, called forced equilibrium or steady state is that in which continuous supply of energy is required to maintain the state of equilibrium.

14. A system is said to be in thermodynamic equilibrium when it is simultaneously in thermal, mechanical and chemical equilibrium.

15. The operations that change the state of the system are called processes and are represented by lines/ curves on the N-dimensional space made of state variables.

16. *Isothermal process*: A processes in which the temperature of the system remains constant. *Isobaric* or *isopiestic process*: A processes in which the pressure of the system remains constant is called an isobaric process. Isochoric or isovolumic process: A processes in which the volume of the system remains constant. *Adiabatic process*: A. process in which a system enclosed in an adiabatic boundary undergoes some change is called an adiabatic process. No energy leaves or enters the system in an adiabatic process. Quasi-static process: a quasi-static process is made of a succession of equilibrium states. *Reversible process*: The processes in which an infinitesimally small change in the system variable may reverse the direction of the process are called reversible processes. *Irreversible process*: an infinitesimal change in the system variable does not change the direction of the process.

17. Work done by a system or on a system is not a system property. Work done on a system or by a system does depend on the initial and the final states of the system but it also depends on the path or the process through which the final state of the system is reached starting from the initial state.

18. In thermodynamics the work done by the system is taken as positive and the work done on the system as negative. (However, in thermochemistry work done on the system is taken as positive and work done by the system as negative)

19. Element of work $d'W$ done in a reversible change of volume is given by

$$d'W = P\,dV$$

Work done in a reversible isobaric process $W_{P\,const} = P\left(V_f - V_i\right)$

Work done in a reversible isothermal process $W_{T\,const} = \mathbb{N}RT \times 2.303\log_{10}\dfrac{V_f}{V_i}$

$$= \mathbb{N}RT \times 2.303\log_{10}\dfrac{P_i}{P_f}$$

Work done in a reversible adiabatic process $W_{adiabatic} = \left(\dfrac{1}{\gamma - 1}\right)\left(P_i v_i - P_f v_f\right)$

$$= \left(\dfrac{R}{\gamma - 1}\right)\left(T_i - T_f\right)$$

If a system has $X_1, X_2, X_3\ldots$ etc., intensive properties and $Y_1, Y_2, Y_3, \ldots\ldots$ etc., the corresponding extensive properties, then the total configurational work may be written as,

$$d'W = X_1 dY_1 + X_2 dY_2 + X_3 dY_3 + \ldots$$

20. No configurational work is done in the free expansion of a gas.

21. *Dissipative work*: Works, like the work done by the current in passing through the resistance, the direction of which cannot be reversed by changing the direction of the process are called dissipative work.

Every thermodynamic system has an equation of state. The equation of state of an ideal gas is $Pv = \mathcal{R}T$, for van der Waals gas $\left(P + \dfrac{a}{v^2}\right)(v - b) = RT$; for a wire stretched by a force:

$$L_T^{final} = L_{T_1}^{in} + L_{T_1}^{in}\alpha\left(T - T_i\right) + \dfrac{L_{T_1}^{in}\,Mg}{\mathcal{Y}A}\,, \text{ for a film of surface tension } \sigma(T) \text{ at temperature } T\text{ K}$$

the equation of state may be written as $\sigma(T) = \sigma(T_0)\left(\dfrac{T_c - T}{T_c - T_0}\right)$, A paramagnetic material put

in a magnetizing field of intensity H at temperature T K may be considered to be a thermodynamic system and the equation of state for the same may be written as,

$$M = Cc\,\dfrac{H}{T}$$

A dielectric slab placed in an external electric field of intensity E at temperature T (K) develops the dipole moment P which depends both on T and E. The system consisting of the electric field and the slab may be treated as a thermodynamic system, the equation of state for which is,

$$P = \left(A + \dfrac{B}{T}\right)E$$

22. In case of an ideal gas, all possible values of P, v, and T, allowed by the equation of state, fall on a two dimensional surface in the three dimensional space of system variables. This two dimensional surface, on which each point represents an equilibrium state of the ideal gas, is called the P- v - T surface for an ideal gas. Similarly, There is a two dimensional curved surface for Van der Waals gas. Isotherms are hyperbolic curves on the $P–v–T$ surface, while isobars and isochors are straight lines.

23. At temperatures below the critical point, substances may co-exist in vapor and liquid phases in equilibrium. The mass fraction of the vapor phase $X_V = \dfrac{M_V}{M_V + M_L}$ is called the 'quality' of the substance in engineering texts on thermodynamics.

24. The values of state parameters change in moving from one equilibrium state to another on the $P–v–T$ surface. These changes in state parameters may be written using the tools of differential calculus. For example, a change dV in the volume may be written as,

$$dV = \left(dV\right)_P + \left(dV\right)_T = \left(\frac{dV}{\partial T}\right)_P dT + \left(\frac{\partial V}{dP}\right)_T dP$$

25. The coefficient of volume expansion or expansivity β of a material is defined as the ratio of the rate of change of volume with temperature at constant pressure to the original volume.

$$\beta \equiv \frac{\left(\dfrac{dV}{\partial T}\right)_P}{V} = \frac{\left(\dfrac{dv}{\partial T}\right)_P}{v}$$

26. *Isothermal compressibility κ:* The isothermal compressibility of a material is denoted by κ and is defined by the following equation,

$$\kappa \equiv -\frac{1}{V}\left(\frac{\partial V}{\partial P}\right)_T = -\frac{1}{v}\left(\frac{\partial v}{\partial P}\right)_T$$

27. It is possible to obtain the equation of state of a system if the expansivity and the isothermal compressibility of the substance are known.

First Law of Thermodynamics and some of its Applications

4.0 Introduction

Generally, the first law of thermodynamics is referred to as another manifestation of the well-known law of conservation of energy. However, in thermodynamics it has a different significance as the law evolved from the observation that *adiabatic work* done in taking a system from an initial state to a final state of same bulk (potential and kinetic) energies is independent of the path and depends only on the properties of the initial and the final states. This sounds strange because in general work is path dependent and furthermore any property that depends only on the initial and final states (and not on the path) must represent some state function. We shall see how this leads to the thermodynamic concept of internal energy of a system.

4.1 Adiabatic Work between Two States of same Bulk Energies

A thermodynamic system may possess bulk potential and kinetic energies. For example, consider a fixed mass of a gas contained in a cylinder piston arrangement and carried in an airplane from one place to another. Taking the gas as our system, it has bulk potential energy due to the height of the airplane above the surface of earth and bulk kinetic energy due to its speed. If during the flight some operations are done on the gas, say it is compressed, the final and the initial states of the system are different but they have the same bulk energies. In the following we consider situations of this type.

It is known that a system may be taken from a given initial equilibrium state (that may have some fixed values of bulk kinetic and potential energies) to a final equilibrium state (having same bulk energies) in many different ways. On the P–v–T surface each equilibrium state is represented by a point and each process by a curve. In general the work done in reaching the final state starting from the same initial state depends on the path and has different value for each path. However, if one selects only those paths or combinations of paths in which either no work is done and/ or those in which work is done under adiabatic conditions, it has been observed that the work done in reaching the final state through all such paths and combinations of paths, is same, independent of the path and depends only on the initial and final states of the system.

Adiabatic work means that the work is done when the system is contained in an adiabatic boundary across which no energy and mass can transfer. For example, a gas contained in a piston cylinder arrangement which is covered from all sides by a perfectly insulated boundary, performs adiabatic work if it expands and adiabatic work is done on the gas if the piston compresses the gas.

In Fig. 4.1 the initial and final equilibrium states of a system are shown by points A and B. We consider the path ACB. It is made of two parts: AC that shows the free expansion of the system. It is a path in which no work is done. Next part CB shows the adiabatic expansion of the system in which work (W_1) equivalent to the area Y, marked with horizontal cross lines, is performed by the system. Thus the total work done in path ACB is equal to or the area Y. Next we consider the path ADB. The part AD shows the adiabatic expansion till the point D such that free expansion from D to B brings the system to the final state B. Again, the total work done (W_2) in the path ADB is equal to the area X marked by horizontal cross lines in the figure. The third path we consider is ADEB. The system is allowed to expand adiabatically from A to E till its volume becomes equal to volume of the system in the final state B. In this adiabatic expansion work done by the system (W_3) is equal to the shaded area ADEifA. Since the work is done by the system, its energy, temperature and pressure must have gone down below their values at the final state B. The system at E is brought to the final state B by pumping energy through dissipative work $-W_{dis}$ done on the system by some external agency. The net adiabatic work done in path ADEB is ($W_3 - W_{dis}$). It may be seen that the system initially at state A is taken to the final state B by three different adiabatic paths, ACB, ADC and AEB. The adiabatic work done in the three paths are respectively, W_1, W_2, and ($W_3 - W_{dis}$). It has been observed that the adiabatic works done in the three different paths W_1, W_2, and ($W_3 - W_{dis}$) are all equal, i.e.,

$$W_1 = W_2 = (W_3 - W_{dis})$$

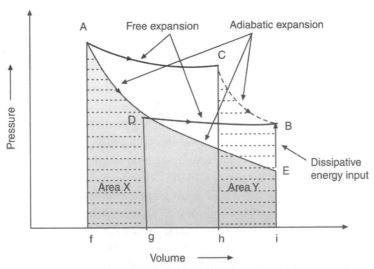

Fig. 4.1   A system in initial state A may reach a final state B via many different paths

4.2   First Law of Thermodynamics and The Internal Energy

The first law of thermodynamics reflects the special property of the adiabatic work between two equilibrium states of same bulk potential and kinetic energies of a system. It says, 'Adiabatic work between two equilibrium states of same bulk potential and kinetic energies is same for all allowed paths'.

It is obvious that the amount of adiabatic work $dW_{adi}^{A \to B}$ will depend on the properties of the initial and the final states A and B, but is independent of the path through which the final state is reached. It means that $dW_{adi}^{A \to B}$ represents and is equal to the change in some state property when system moves from state A to B. This may be compared with the change $\Delta V = (V_B - V_A)$ in the value of the state function volume of the system when the system moves from A to B. ΔV depends on the volume V_A and V_B but not on the path through which the system has reached from state A to state B. Therefore, $dW_{adi}^{A \to B}$ is a perfect differential like dV or dP, etc., and is equal to the change in some state function.

Let us assign a state function U to each equilibrium state of the system that has values U_A and U_B when the equilibrium states of the system are represented by points A and B respectively. We call this state function the internal energy of the system. Internal energy U, like other system variables has different values at different points on the P–v–T surface. Thermodynamics does not define the absolute value of the internal energy U but the change in the internal energy $dU^{A \to B}$ when system moves from state A to final state B is defined by the following relation,

$$dU^{A \to B} \equiv -dW_{adi}^{A \to B} \qquad 4.1$$

or
$$\int_A^B dU^{A \to B} = -\int_A^B dW_{adi}^{A \to B}$$

or
$$U_B - U_A = -\left(W_{adi}^{A \to B}\right)$$

or
$$U_A - U_B = W_{adi}^{A \to B} \qquad 4.2$$

The negative sign in the defining Eq. 4.1 is included to ensure that when adiabatic work is done by the system, i.e., $dW_{adi}^{A \to B}$ is positive, $dU^{A \to B}$ is negative which means that the internal energy decreases. This further means that the adiabatic work is done by a system at the cost of its internal energy. In case the adiabatic work is done on the system, the internal energy of the system will increase. It is obvious from Eq. 4.2 that the units of the internal energy and that of the adiabatic work is the same (1Joule in MKS system).

Figure 4.2 shows two equilibrium states A and B of the system that have same bulk potential and kinetic energies. Solid lines marked a, b and c in Fig. 4.2 represent three different adiabatic paths that lead from initial state A to the final state B. The amount of work done in these paths are denoted, respectively, by W_a, W_b, and W_c. Dotted lines in the figure show three non-adiabatic paths D, E and F with their associated works W_D, W_E and W_F. This is also indicated in the figure that,

$$W_{adi} = W_a = W_b = W_c \equiv (U_A - U_B)$$

and
$$W_D \neq W_E \neq W_F \qquad 4.3$$

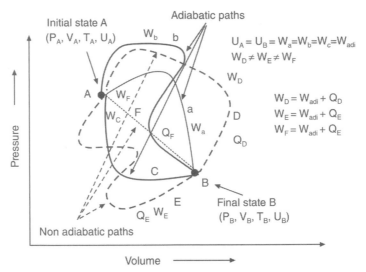

Fig. 4.2   Solid and dotted lines respectively show the adiabatic and non-adiabatic paths connecting the initial and final states. Work done in all adiabatic paths are equal and are equal to the change in the internal energy of the system.

4.3   Non-adiabatic Work and Heat Flow

Non-adiabatic work is done by the system or on the system when the system is in contact with other systems or surroundings with a diathermic boundary through which energy may transfer. As the result of the non-adiabatic work, the temperature of the system either increases or decreases with respect to the surroundings. Therefore, in non-adiabatic process energy in the form of heat flows from the system to the surroundings or from surroundings to the system. The amount of heat flow Q is different both in magnitude and sign for different non-adiabatic paths. If heat flows for the three different non-adiabatic paths in Fig. 4.2 are denoted, respectively, by Q_D, Q_E and Q_F then they are related to the adiabatic work W_{adi} by the following relations,

$$W_D = W_{adi} + Q_D; \; W_E = W_{adi} + Q_E; \; W_F = W_{adi} + Q_F \qquad 4.4$$

In general, therefore, it may be said that the work done in going from an initial state to a final state of equal bulk energies, by a non-adiabatic process $W_{non-adi}^{A \to B}$ is equal to the sum of the adiabatic work $W_{adi}^{A \to B}$ between the same initial and final states and the heat flow $Q_{non-adi}^{A \to B}$ in the non-adiabatic process, i.e.,

$$W_{non-adi}^{A \to B} = W_{adi}^{A \to B} + Q_{non-adi}^{A \to B} \qquad 4.5$$

or

$$Q_{non-adi}^{A \to B} = W_{non-adi}^{A \to B} - W_{adi}^{A \to B} \qquad 4.6$$

Equation 4.6 defines the heat flow and also tells that the unit of heat flow is same as that of work, i.e., 1 joule. But $W_{adi}^{A \to B}$ is equal to $(U_a - U_b)$, the change in the internal energy, therefore;

$$
\underset{\substack{\uparrow \\ \text{Increase in the} \\ \text{internal energy} \\ \text{of the system}}}{(U_B - U_A)} \; = \; \underset{\substack{\uparrow \\ \text{Heat flow} \\ \text{into the} \\ \text{system}}}{Q^{A \to B}_{non-adi}} - \underset{\substack{\uparrow \\ \text{Work done} \\ \text{by the} \\ \text{system}}}{W^{A \to B}_{non-adi}} \qquad 4.7
$$

In Eq. 4.7, $(U_B - U_A)$ is the increase in the internal energy of the system when system goes from state A to state B of same bulk energies, and it is equal to the heat flow $Q^{A \to B}_{non-adi}$ into the system minus the non-adiabatic work $W^{A \to B}_{non-adi}$ done by the system. The differential form of Eq. 4.7 is,

$$dU = dQ - d'W \qquad 4.8$$

In Eq. 4.8 $d'W$ denotes the non-adiabatic work that is path dependent. Further, if $dQ = 0$, $dU = -dW_{adi}$ where $-dW_{adi}$ is the adiabatic work. This means that any work in which heat flow is zero is adiabatic.

Often in literature on thermodynamics, Eq. 4.8 is referred as the 'analytic form of the first law'. However, the fact is that it is a relation that follows from the first law but is not the first law. Relation 4.8 may be called the defining equation for the heat flow. As has already been stated, the first law says that the adiabatic work between two equilibrium states of a system of same bulk potential and kinetic energies is same for all adiabatic paths.

Before proceeding further let us make the following observations about Eq. 4.8.

1. It is applicable to both the reversible and the irreversible processes
2. Out of the three quantities appearing in the equation, dU is path independent but d'W is path dependent. It is possible only if dQ is also path dependent. As a matter of fact it is already indicated in Fig. 4.2 that different amounts of heat flows, Q_D, Q_E, Q_F take place in different non adiabatic paths.
3. In Eq. 4.8, $d'W$ represents a non-adiabatic work, which in general, may be the combination of configurational work XdY and dissipative work W_{dis}.

$$dU = dQ - (d'W_{con} + d'W_{dis}) \qquad 4.9$$

If $d'W$ involves dissipative work, the process cannot be reversible. Therefore, for the case of a reversible non-adiabatic work, $d'W$ must contain only configurational work and no dissipative work. Hence for reversible processes Eq. 4.8 may be modified as,

$$dU = dQ - d'W_{con} = dQ - XdY \quad \text{for reversible processes} \quad 4.10$$

Here, X and Y are, respectively, the intrinsic and the corresponding extrinsic property of the system. For the more specific case, Eq. 4.10 for reversible processes may be written as,

$$dU = dQ - PdV \text{ for reversible processes} \qquad 4.11$$

4. In Eq. 4.9, dU is the change in the internal energy of the system. Any change in the internal energy of the system may be brought about individually or jointly by quantities appearing on the right hand side of Eq. 4.9. For example, if both $d'W_{con}$ and $d'W_{dis}$ are zero, i.e., no configurational and no dissipative work is done then the internal energy of the system may be changed by the flow of heat dQ. This happens when the temperature of a fixed amount of a gas is increased by heating and keeping its pressure and volume constant. Similarly, when a fixed amount of a

gas contained in a frictionless piston cylinder arrangement covered by an adiabatic boundary is slowly compressed, only configurational work is done and the internal energy of the gas changes. The internal energy of a fixed mass of a gas contained in an insulated container fitted with a stirrer may also be changed by doing dissipative work through the stirrer, keeping the pressure and volume of the gas constant. It means that the system senses only the change in its internal energy and does not distinguish whether the change is brought by heat flow, or by configurational work or by dissipative work or jointly by all of them. Eq. 4.9, based on the first law of thermodynamics, explicitly shows that *heat is just like any other form of work, like configurational work or dissipative work so far as the change in the internal energy of a system is concerned.*

Earlier heat was supposed to be an invisible fluid that passes from a hot body to a cold body when they are put in contact. Count Rumford (an American Physicist, named Benjamin Thompson) pointed out that the temperature of metal chips produced while boring gun metal for making barrels of guns, increases due to the work done in boring and in this way perhaps for the first time indicated that there is a direct correlation between mechanical work done and amount of heat produced.

For many years heat was measured in calorie, 1 calorie being the heat required to raise the temperature of 1 gram of pure water from 15.1 to 16.1°C. Earliest accurate measurement of how much mechanical work is equal to how much quantity of heat, called the mechanical equivalent of heat was done by Joule. From his experiment in which rise in the temperature of a known amount of water was recorded when known amount of dissipative work was done, Joule obtained that 1 calorie = 4.19 joules. The presently accepted value of the mechanical equivalent of heat is, 4.1850 J/cal. The name calorie is used for two units of energy. The small calorie, symbol 'cal' is the energy needed to raise the temperature of 1 g of water from 15.1 to 16.1 degree Celsius. The other unit called the large calories, or kilogram calorie, dietary calorie or simply Calorie with capital C or food calories (symbol: Cal) is the energy needed to raise the temperature of 1 kg of water from 15.1 to 16.1 degree Celsius. Other units for heat energy are British Thermal Unit (BTU); 1 BTU = 1055.06 J = 0.252 kilocalorie. In engineering applications a unit called International steam table calorie, in short written as IT calorie, is also used.

$$1 \text{ IT (International steam table) calorie} = \frac{1}{860} \text{ watt} - \text{hour} = \frac{3600}{860} \text{ joules}$$

and   1 IT calorie = 4.1860 joules.

4.4   Phase Transition and Heat of Transformation: the Enthalpy

Matter, in general, is found in three different phases: solid, liquid, and vapor. Under suitable conditions of pressure and temperature, matter may undergo phase transition, for example water at 373 K and 1 atm pressure undergoes phase transformation from liquid phase to vapor phase. Generally, energy in the form of heat is either absorbed or released in phase transitions. Phase transitions usually occur at constant temperature and pressure. Andrew's (a doctor of medicine, professor of chemistry and physicist from Belfast, Ireland) experiments on the phase transition in CO_2 are of special interest. A part of CO_2 isotherms on P–V plane obtained by Andrews is shown in Fig. 4.3.

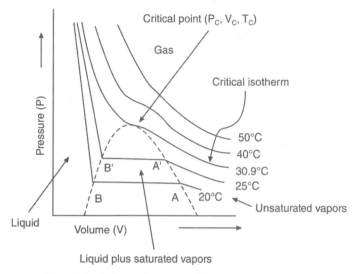

Fig. 4.3 CO_2 isotherms at different temperatures

As may be seen in Fig. 4.3 for all isotherms below the critical isotherm there is a region shown by straight horizontal lines AB, A'B' etc. where phase transitions occur. Let us consider the isotherm at temperature 20°C. On increasing the pressure volume of the gas decreases and at point A in the figure a part of the CO_2 gas becomes liquid. On further increasing the pressure, there is no appreciable change in the pressure but the volume decreases and more CO_2 changes from gaseous to liquid phase. Both the liquid and the gas phase co-exist in the region AB. In the interval A to B an increase in pressure results in the decrease of the volume and in conversion of more and more gas into liquid phase. At point B all the gas is now in liquid phase and since liquid has very small compressibility, further increase in pressure does not decrease volume very much. Behavior shown by CO_2 is typical for gases.

Three important conclusions about the process of phase transformation may be drawn from these curves: (a) *phase transition is always accompanied with change in volume at constant pressure* (b) *since there is always a change in the volume during phase transformation, therefore*, configurational work is always done, either by the system or on the system, during phase transition *and* (c) *since temperature remains constant and work is done during phase transition*, therefore, heat flow to the system or from the system takes place during the process of phase transformation.

The configurational work during phase transition W_{con} is given by,

$$W_{con} = P(V_f - V_i), \text{ where } V_f \text{ and } V_i \text{ are the final and initial volumes} \qquad 4.12$$

If U_i and U_f are the initial and final internal energies of the system, then change in internal energy during phase transformation $\Delta U = U_i - U_f$. On substituting these values in Eq. 4.11, one gets,

$$\left(U_f - U_i\right) = \left(\Delta Q\right)_P - P\left(V_f - V_i\right) \text{ or } \left(\Delta Q\right)_P = \left(U_f + PV_f\right) - \left(U_i + PV_i\right) \qquad 4.13$$

In Eq. 4.13, $(\Delta Q)_P$ is the amount of heat flow at constant pressure from or into the system during the process of phase transformation. It is obvious that $(\Delta Q)_P$ will depend on the mass or moles contained in the system. The specific value of $(\Delta Q)_P$, i.e., heat flow per unit mass or per mole is called the specific or molal heat of transformation (also called latent heat) and is denoted by the small letter ℓ. Further the solid, liquid, and gaseous phases are denoted, respectively, by the subscripts 1, 2 and 3. As an example, ℓ_{12} means specific heat of transformation from the solid to the liquid phase, process called fusion and ℓ_{23} refers to the specific heat of transformation from liquid phase to vapor phase, process called vaporization, In a likewise manner ℓ_{13} is the specific heat of transformation from solid to vapor phase, the process called sublimation. The MKS unit for specific heat of phase transition is J kg^{-1} or J kilomole^{-1}. It may also be given in units of kcal/kg, or Btu/lb. The inter conversion of units may be done using the following conversion factors,

$$1 \times 10^3 \text{ J kg}^{-1} = 0.43 \text{ } Btu \, lb^{-1} = 0.24 \text{ kcal kg}^{-1} \qquad 4.14$$

Dividing both sides of Eq. 4.13 by mass of the system or the number of moles contained in the system, one gets;

$$\ell = \left(u_f + Pv_f \right) - \left(u_i + Pv_i \right) \qquad 4.15$$

In Eq. 4.15, ℓ is the specific heat of transformation, u_1, u_2, the initial and final specific internal energies and v_1, v_2 the initial and final specific volumes of the system, respectively.

The term $(u + Pv)$ or $(U + PV)$ often appears in thermodynamics and is called the *specific enthalpy* or *enthalpy* of the system. Specific enthalpy is denoted by h and enthalpy by the capital letter H. Equation 4.15 may be written as,

$$\ell = h_f - h_i \qquad 4.16$$

It follows from Eq. 4.16, that *the specific heat of phase transformation is equal to the change in the specific enthalpy of the system*. The statement may be generalized to '*heat flow in any isobaric reversible process is equal to the change in enthalpy*'.

Specific heat for fusion and vaporization along fusion and boiling temperatures for some substances are listed in Table 4.1.

Table 4.1   Specific heats of fusion and vaporization and the temperatures of fusion and boiling point of some substances

Substance	Spec. heat of fusion (kJ/kg)	Temp. fusion (K)	Spec. heat of vaporization (kJ/kg)	Temp. of boiling (K)
Water	334	273	2258	373
Ethanol	109	159	838	195
Chloroform	74	209	254	335
Mercury	11	234	294	630
Hydrogen	60	14	449	20
Oxygen	14	54	213	90
Nitrogen	25	63	199	77

4.5   Heat Flow at Constant Pressure and at Constant Volume

We come back to Eq. 4.11, according to which for reversible processes

$$dU = dQ - PdV$$

or
$$dQ = dU + PdV \qquad\qquad 4.17$$

Eq. 4.17 gives a relation between the heat flow dQ, change in internal energy dU and configurational work PdV for a reversible process. Now there could be two possibilities: (a) the work is done at constant pressure P. Heat flow $(\Delta Q)_P$ at constant pressure is given by Eq. 4.13 as,

$$(\Delta Q)_P = \left(U_f - U_i\right) + PdV \qquad\qquad 4.18$$

$$= \left(U_f + PV_f\right) - \left(U_i + PV_i\right)$$

This may also be written as,

$$(\Delta Q)_P = H_f - H_i = \Delta H \qquad\qquad 4.19$$

Therefore, heat flow at constant pressure is equal to the change in the enthalpy of the system.

Next we consider the case (b) heat flow in a reversible process at constant volume, i.e., dV =0, In this case no configurational work is done and,

$$(\Delta Q)_V = (U_f - U_i) = \text{change in the internal energy} \qquad\qquad 4.20$$

It follows from Eqs. 4.18 and 4.20 that,

$$(\Delta Q)_P = \left(\Delta Q\right)_V + PdV = \left(\Delta Q\right)_V + \Delta\left(PV\right) \qquad\qquad 4.21$$

In Eq. 4.21 $\Delta(PV)$ is the change in the value of PV at constant pressure which is equal to PdV. In case the system is an ideal gas and the processes are reversible, one has,

$PV = \mathbb{N}RT$ and $\Delta\left(PV\right) = \Delta\left(\mathbb{N}RT\right)$, substituting this value of $\Delta(PV)$ in Eq. 4.21, one gets:

$$(\Delta Q)_P = \left(\Delta Q\right)_V + \Delta\left(PV\right) = \left(\Delta Q\right)_V + \Delta\left(\mathbb{N}\,RT\right) \qquad\qquad 4.21A$$

In Eq. 4.21A $\mathbb{N}$ is the number of moles or kilomoles of the ideal gas in the system. Equation 4.21 gives a relation between the heat flow in reversible processes in case of an ideal gas at constant pressure and at constant volume and is important from the view point of thermochemistry.

4.6   Heat Capacities at Constant Pressure C_P and at Constant Volume C_V

It follows from the definition of C_P and C_V that,

$$C_P = \frac{(\Delta Q)_P}{dT} = \left(\frac{\partial H}{\partial T}\right)_P \qquad\qquad 4.22$$

and
$$C_V = \frac{(\Delta Q)_V}{dT} = \left(\frac{\partial U}{\partial T}\right)_V \qquad 4.23$$

Hence,
$$C_P - C_V = \left(\frac{\partial H}{\partial T}\right)_P - \left(\frac{\partial U}{\partial T}\right)_V \qquad 4.24$$

But, $H = U + PV$, hence, $\left(\frac{\partial H}{\partial T}\right)_P = \left(\frac{\partial U}{\partial T}\right)_P + P\left(\frac{\partial V}{\partial T}\right)_P \qquad 4.25$

Therefore,
$$C_P - C_V = \left(\frac{\partial U}{\partial T}\right)_P + P\left(\frac{\partial V}{\partial T}\right)_P - \left(\frac{\partial U}{\partial T}\right)_V \qquad 4.26$$

However,
$$U = U(T, V) \text{ and so } dU = \left(\frac{\partial U}{\partial T}\right)_V dT + \left(\frac{\partial U}{\partial V}\right)_T dV \qquad 4.27$$

and
$$V = V(P, T) \text{ and, therefore, } dV = \left(\frac{\partial V}{\partial P}\right)_T dP + \left(\frac{\partial V}{\partial T}\right)_P dT \qquad 4.28$$

Putting the value of dV from Eq. 4.28 in Eq. 4.27, one gets:

$$dU = \left(\frac{\partial U}{\partial T}\right)_V dT + \left(\frac{\partial U}{\partial V}\right)_T \left[\left(\frac{\partial V}{\partial P}\right)_T dP + \left(\frac{\partial V}{\partial T}\right)_P dT\right]$$

or
$$dU = \left[\left(\frac{\partial U}{\partial T}\right)_V + \left(\frac{\partial U}{\partial V}\right)_T \left(\frac{\partial V}{\partial T}\right)_P\right]dT + \left[\left(\frac{\partial U}{\partial V}\right)_T \left(\frac{\partial V}{\partial P}\right)_T\right]dP \qquad 4.29$$

The internal energy U may also be written as a function of P and T, i.e.,

$$U = U(T, P) \text{ and therefore, } dU = \left(\frac{\partial U}{\partial T}\right)_P dT + \left(\frac{\partial U}{\partial P}\right)_T dP \quad 4.30$$

Comparing the coefficients of dT and dP in Eq. 4.29 and Eq. 4.30, one gets:

$$\left(\frac{\partial U}{\partial T}\right)_P = \left[\left(\frac{\partial U}{\partial T}\right)_V + \left(\frac{\partial U}{\partial V}\right)_T \left(\frac{\partial V}{\partial T}\right)_P\right]$$

and
$$\left(\frac{\partial U}{\partial P}\right)_T = \left[\left(\frac{\partial U}{\partial V}\right)_T \left(\frac{\partial V}{\partial P}\right)_T\right] \qquad 4.31$$

Substituting the value of $\left(\frac{\partial U}{\partial T}\right)_P$ from Eq. 4.31 in Eq. 4.26, one gets:

$$C_P - C_V = \left(\frac{\partial U}{\partial T}\right)_P + P\left(\frac{\partial V}{\partial T}\right)_P - \left(\frac{\partial U}{\partial T}\right)_V$$

$$= \left[\left(\frac{\partial U}{\partial T}\right)_V + \left(\frac{\partial U}{\partial V}\right)_T \left(\frac{\partial V}{\partial T}\right)_P\right] + P\left(\frac{\partial V}{\partial T}\right)_P - \left(\frac{\partial U}{\partial T}\right)_V$$

.or
$$C_P - C_V = \left[\left(\frac{\partial U}{\partial V}\right)_T + P\right]\left(\frac{\partial V}{\partial T}\right)_P \qquad 4.32$$

The dimensions of the term $\left(\frac{\partial U}{\partial V}\right)_T$ are $\left[\frac{MLT^2 x L}{L^3}\right] = \left[\frac{MLT^2}{L^2}\right] = \left[\frac{\text{Force}}{\text{area}}\right] = [\text{Pressure}]$

Since, the dimensions of $\left(\dfrac{\partial U}{\partial V}\right)_T$ or $\left(\dfrac{\partial u}{\partial v}\right)_T$ are that of pressure the term $\left(\dfrac{\partial u}{\partial v}\right)_T$ is called the *internal*

pressure. In thermodynamics this is just a term having dimensions of pressure but according to the kinetic theory the internal pressure originates from the molecular attraction. In case of ideal gas, molecules of which to not exert any force of attraction or repulsion on each other $\left(\dfrac{\partial u}{\partial v}\right)_T$ must be zero.

Therefore, from Eq. 4.32, for an ideal gas

$$C_P - C_V = P\left(\frac{\partial V}{\partial T}\right)_P \qquad\qquad 4.33$$

But for an ideal gas $PV = \mathbb{N}RT$, and $\left(\dfrac{\partial V}{\partial T}\right)_P = \dfrac{\mathbb{N}R}{P}$

Hence, $C_P - C_V = \mathbb{N}R$ or $c_p - c_v = R$ $\qquad\qquad 4.34$

Here $\mathbb{N}$ is the number of moles, R, the gas constant, c_p, c_v, respectively, the molal specific heat capacities at constant pressure and constant volume.

4.7 Systems with Three State Variables P, *v* and T

We now consider only those systems which have only three state functions, that may be pressure P, specific volume v and temperature T. Internal energy U (or specific internal energy u) of the system and the enthalpy H (or specific enthalpy h) are also state variables that are functions of P, v and T. Out of the three state variables any two can be taken as independent variables and the third one becomes a dependent variable, as it is related to the other two by the equation of state. In this section we will derive some important differential equations taking three different combinations (T, v), (T, P) and (P, v) as pairs of independent variables.

Before proceeding further, following important properties of partial derivatives may be mentioned. If a function $f(x, y, z)$ depends on variables x, y and z and $f(x, y, z) = 0$, then the following two identities hold good.

$$\left(\frac{\partial x}{\partial y}\right)_z = \frac{1}{\left(\dfrac{\partial y}{\partial x}\right)_z} \qquad\qquad 4.34A$$

and $\qquad \left(\dfrac{\partial x}{\partial y}\right)_z \left(\dfrac{\partial y}{\partial z}\right)_x \left(\dfrac{\partial z}{\partial x}\right)_y = -1 \qquad\qquad 4.34B$

4.7.1 Temperature T and specific volume *v* as independent variables

We start with Eq. 4.11 $dU = dQ - PdV$ which is true for reversible process and rewrite it in terms of the specific quantities as,

$$dq = du + Pdv \qquad\qquad 4.35$$

Treating u to be a function of (T, v), one may write

$$du = \left(\frac{\partial u}{\partial T}\right)_v dT + \left(\frac{\partial u}{\partial v}\right)_T dv \qquad 4.36$$

Substituting the value of du from Eq. 4.36 in Eq. 4.35 and rearranging terms one gets,

$$dq = \left(\frac{\partial u}{\partial T}\right)_v dT + \left(P + \left(\frac{\partial u}{\partial v}\right)_T\right)dv \qquad 4.37$$

For an isochoric reversible processes $dv = 0$ and second term in Eq. 4.37 vanishes. Therefore,

$$(dq)_v = \left(\frac{\partial u}{\partial T}\right)_v dT, \text{ but } \underline{(dq)}_v = c_v dT, \text{ so } \left(\frac{\partial u}{\partial T}\right)_v dT = c_v dT \qquad 4.38$$

Substituting the value of $\left(\frac{\partial u}{\partial T}\right)_v dT$ in Eq. 4.37, one gets

$$dq = c_v dT + \left(P + \left(\frac{\partial u}{\partial v}\right)_T\right)dv \qquad 4.39$$

For an isothermal reversible process dT is zero and Eq. 4.39 becomes,

$$(dq)_T = \left[P + \left(\frac{\partial u}{\partial v}\right)_T\right]dv = Pdv + \left(\frac{\partial u}{\partial v}\right)_T dv \qquad 4.40$$

Eq. 4.40 tells that the heat flow into the system at constant temperature is partly used in doing work (Pdv) and partly in raising the internal energy of the system $\left(\frac{\partial u}{\partial v}\right)_T dv$. One may ask why specific heat capacity at constant temperature c_T is not defined while similar other quantities c_v and c_P are defined. The reason is that if it is defined by the equation $c_T dT = (dq)_T$, then if there is no rise of temperature, i.e., $dT = 0$, even then unlimited amount of heat flow may happen that may go in doing work and, therefore, c_T will have an infinite value. Since c_T may have infinite value it is not defined.

For an adiabatic reversible process $dq = 0$, Eq. 4.39 becomes,

$$(c_v dT)_{adi} = -\left[\left\{P + \left(\frac{\partial u}{\partial v}\right)_T\right\}dv\right]_{adi} \qquad 4.41$$

4.7.2   Pressure P and specific volume v as independent variables

Taking the specific internal energy u to be a function of P and v, i.e., $u = u(P, v)$ we have

$$du = \left(\frac{\partial u}{\partial P}\right)_v dP + \left(\frac{\partial u}{\partial v}\right)_P dv \qquad 4.42$$

Similarly, taking T to be a function of P and v, i.e., $T = T(P,v)$, one gets

$$dT = \left(\frac{\partial T}{\partial P}\right)_v dP + \left(\frac{\partial T}{\partial v}\right)_P dv \qquad 4.43$$

But from Eq. 4.36 we have, $du = \left(\frac{\partial u}{\partial T}\right)_v dT + \left(\frac{\partial u}{\partial v}\right)_T dv$

Substituting the value of dT from Eq. 4.43 in above equation, one gets

$$du = \left(\frac{\partial u}{\partial T}\right)_v \left[\left(\frac{\partial T}{\partial P}\right)_v dP + \left(\frac{\partial T}{\partial v}\right)_P dv\right] + \left(\frac{\partial u}{\partial v}\right)_T dv$$

or

$$du = \left(\frac{\partial u}{\partial T}\right)_v \left(\frac{\partial T}{\partial P}\right)_v dP + \left[\left(\frac{\partial u}{\partial T}\right)_v \left(\frac{\partial T}{\partial v}\right)_P + \left(\frac{\partial u}{\partial v}\right)_T\right] dv \qquad 4.44$$

Comparing Eq. 4.42 and Eq. 4.44, one gets,

$$\left(\frac{\partial u}{\partial P}\right)_v = \left(\frac{\partial u}{\partial T}\right)_v \left(\frac{\partial T}{\partial P}\right)_v \qquad 4.45$$

and

$$\left(\frac{\partial u}{\partial v}\right)_P = \left[\left(\frac{\partial u}{\partial T}\right)_v \left(\frac{\partial T}{\partial v}\right)_P + \left(\frac{\partial u}{\partial v}\right)_T\right] \qquad 4.46$$

From Eq. 4.38, $\left(\dfrac{\partial u}{\partial T}\right)_v = c_V$. Substituting this value in Eq. 4.45, one gets

$$\left(\frac{\partial u}{\partial P}\right)_v = c_v \left(\frac{\partial T}{\partial P}\right)_v \quad \text{and} \quad c_v = \left(\frac{\partial u}{\partial T}\right)_v \qquad 4.47$$

4.7.3   Temperature T and pressure P as independent variables

Taking specific enthalpy as a function of T and P, i.e., $h = h(T, P)$ one gets,

$$dh = \left(\frac{\partial h}{\partial T}\right)_P dT + \left(\frac{\partial h}{\partial P}\right)_T dP \qquad 4.48$$

Also

$$h = u + pv; \quad \text{and} \quad dh = du + pdv + vdP \qquad 4.49$$

But from the relation $dq = du + pdv$, Eq. 4.49 becomes,

$$dh = dq + vdp \quad \text{or} \quad dq = dh - vdP \qquad 4.50$$

Substituting the value of dh in Eq. 4.50 from Eq. 4.48, one gets

$$dq = \left(\frac{\partial h}{\partial T}\right)_P dT + \left[\left(\frac{\partial h}{\partial P}\right)_T - v\right] dP \qquad 4.51$$

In an isobaric reversible process $dP = 0$ and so,

$$(dq)_P = \left(\frac{\partial h}{\partial T}\right)_P dT, \text{ but } (dq)_P = c_P dT$$

Hence, $c_P = \left(\dfrac{\partial h}{\partial T}\right)_P$ \qquad 4.52

Also, expressing $h = h(P, v)$, one gets

$$dh = \left(\frac{\partial h}{\partial P}\right)_v dP + \left(\frac{\partial h}{\partial v}\right)_P dv \qquad 4.53$$

We use the relation given by Eq. 4.50, $dq = dh - vdP$ and substitute the value of dh from Eq. 4.53 to get,

$$dq = \left(\frac{\partial h}{\partial v}\right)_P dv + \left[\left(\frac{\partial h}{\partial P}\right)_v - v\right]dP \qquad 4.54$$

For reversible isobaric process $dP = 0$, and therefore,

$$(dq)_P = \left(\frac{\partial h}{\partial v}\right)_P (dv)_P \qquad 4.55$$

But $(dq)_P = c_P(dT)_P$. Substituting this value in Eq. 4.55, one gets

$$c_P(dT)_P = \left(\frac{\partial h}{\partial v}\right)_P (dv)_P \quad \text{or} \quad c_P\left(\frac{\partial T}{\partial v}\right)_P = \left(\frac{\partial h}{\partial v}\right)_P \qquad 4.56$$

Again from Eq. 4.54, it follows that for reversible isochoric process dv =0 and hence,

$$(dq)_v = \left[\left(\frac{\partial h}{\partial P}\right)_v - v\right]_v (dP)_v \quad \text{but} \quad (dq)_v = c_v(dT)_v; \text{hence}$$

$$\left[\left(\frac{\partial h}{\partial P}\right)_v - v\right]_v = c_v\left(\frac{dT}{dP}\right)_v = c_v\left(\frac{\partial T}{\partial P}\right)_v \qquad 4.57$$

Substituting the value of $\left[\left(\frac{\partial h}{\partial P}\right)_v - v\right]$ from Eq. 4.57 and the value of $\left(\frac{\partial h}{\partial v}\right)_P$ from Eq. 4.56 back in Eq. 4.54, one gets,

$$dq = c_P\left(\frac{\partial T}{\partial v}\right)_P dv + c_v\left(\frac{\partial T}{\partial P}\right)_v dP \qquad 4.58$$

In case of the reversible adiabatic process $(dq)_T = 0$, hence from Eq. 4.58 one gets,

$$c_P\left(\frac{\partial T}{\partial v}\right)_P (dv)_{adi} = -c_v\left(\frac{\partial T}{\partial P}\right)_v (dP)_{adia}$$

or
$$c_P\left(\frac{\partial T}{\partial v}\right)_P = -c_v\left(\frac{\partial T}{\partial P}\right)_v \left(\frac{\partial P}{\partial v}\right)_{adi} \qquad 4.59$$

But from identity 4.34B, $\left(\frac{\partial T}{\partial v}\right)_P = -\left(\frac{\partial P}{\partial v}\right)_T \left(\frac{\partial T}{\partial P}\right)_v$. Substituting this value of $\left(\frac{\partial T}{\partial v}\right)_P$ in Eq. 4.59, one gets:

$$c_P\left(\frac{\partial P}{\partial v}\right)_T = c_v\left(\frac{\partial P}{\partial v}\right)_{adi} \qquad 4.60$$

4.7.4 (h, T, and P) as state variables

As has been pointed out earlier, we are considering systems that may be completely defined by only three state functions or variables. So far we have considered P, v and T to be the three state variables and taking a pair of them as independent variables important thermodynamic relations in the form of partial derivatives of state variables have been obtained. However, it is possible to choose any other three system parameters to be the state functions, for example the combination of (h, T, and P) or (u, T, and v) may as well be chosen as state functions. Let us assume that the three variables (h, T, and P) defines the system and that they are related through the equation of state $f(h, T, P) = 0$. Using identities 4.34A and 4.34B, one may get,

$$\left(\frac{\partial h}{\partial T}\right)_P \left(\frac{\partial T}{\partial P}\right)_h \left(\frac{\partial P}{\partial h}\right)_T = -1$$

But from Eq. 4.52, $\left(\dfrac{\partial h}{\partial T}\right)_P = c_P$, and, therefore, the above equation gives,

$$\left(\frac{\partial h}{\partial P}\right)_T = -c_P \left(\frac{\partial T}{\partial P}\right)_h \qquad 4.61$$

4.7.5 (u, T and v) as state variables

Also, if one considers (u, T and v) as state variables that are related by the equation of state $f(u, T$ and $v) = 0$, then from identity 4.34B it follows that,

$$\left(\frac{\partial u}{\partial T}\right)_v \left(\frac{\partial T}{\partial v}\right)_u \left(\frac{\partial v}{\partial u}\right)_T = -1$$

Now using identity 4.34A, the above equation may be written as,

$$\left(\frac{\partial u}{\partial v}\right)_T = -\left(\frac{\partial u}{\partial T}\right)_v \left(\frac{\partial T}{\partial v}\right)_u$$

But from Eq. 4.47 $c_v = \left(\dfrac{\partial u}{\partial T}\right)_v$, and therefore,

$$\left(\frac{\partial u}{\partial v}\right)_T = -c_v \left(\frac{\partial T}{\partial v}\right)_u \qquad 4.62$$

4.8 Gay-Lussac–Joule Experiment

Purpose of the experiment: In thermodynamics change in the internal energy of a system is defined by the first law but apparently there is no direct method to measure the internal energy, U, or the specific internal energy, u, of a system. Further, it is also not known how the internal energy depends on the volume or the pressure of the system. The related property is the internal pressure $\left(\dfrac{\partial u}{\partial v}\right)_T$, which according to the kinetic theory arises from the inter-molecular attraction and must be zero for ideal gases and should have a non-zero value for real gases. However, from Eq. 4.62, $\left(\dfrac{\partial u}{\partial v}\right)_T = -c_v \left(\dfrac{\partial T}{\partial v}\right)_u$, the internal pressure may be determined from the rate of change of temperature with volume at constant specific internal energy u, $\left(\dfrac{\partial T}{\partial v}\right)_u$. Now $du = dq - Pdv$, and, therefore, du will be zero, or u will be constant when both dq and the configurational work, Pdv, are zero but there is a change in the volume of the system. It may be pointed out that configurational work is zero in free expansion of a gas. Similarly, from Eq. 4.61, $\left(\dfrac{\partial h}{\partial P}\right)_T = -c_P \left(\dfrac{\partial T}{\partial P}\right)_h$, the rate of change of h(or u) with pressure may be determined from the rate of change of temperature with pressure of a system of constant enthalpy.

For the first time Gay-Lussac (Joseph Louis Gay-Lussac, French chemist and physicist*)* and later Joule carried out similar experiments to investigate the volume and pressure dependence of the internal energy. The experimental layout used by Gay-Lussac is shown in Fig. 4.4. As may be seen, two vessels one with air at higher pressure and the other one evacuated to a very low pressure are connected by a stopcock and the system is held in a water bath. A thermometer is kept immersed in the water to record any change in the water temperature. The air at high pressure undergoes free expansion on opening the stopcock, and no configurational work is done by the expanding gas. It was expected that in case the internal energy depends on the volume or the pressure of the gas, it will change in the free expansion and as such heat flow from the system to the surrounding water or vice-versa will take place. However, in the experiment conducted both by Gay-Lussac and later by Joule no change in water temperature was recorded. It was, therefore, concluded that internal energy does not depend on pressure or volume of the system. However, later it was realized that the experiment was not sensitive enough to record small changes in the temperature of water. It was because the thermal capacity of water was very large and for a given small heat flow $dQ = C_{water}\, dT$, which means that if C_{water} is large, dT will be very small and the thermometer may not be able to record it. In this experiment it was erroneously concluded that the internal pressure is zero for air also, which is a mixture of non-ideal gases. While in fact it should be zero for only ideal gas. Since the experiments by Gay-Lussac and by Joule were not conclusive, it may be that an additional assumption is made for ideal gas that $\left(\dfrac{\partial u}{\partial v}\right)_T = 0$.

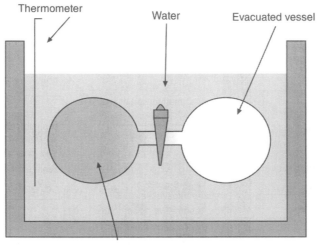

Fig. 4.4 Experimental setup of Gay-Lussac

It follows from Eq. 4.62 that in case $\left(\dfrac{\partial u}{\partial v}\right)_T = -c_v \left(\dfrac{\partial T}{\partial v}\right)_u = 0$ for an ideal gas, then $\left(\dfrac{\partial T}{\partial v}\right)_u = 0$

The quantity $\left(\dfrac{\partial T}{\partial v}\right)_u$ is called the *Joule coefficient* and is represented by η,

i.e., $$\eta \equiv \left(\dfrac{\partial T}{\partial v}\right)_u \tag{4.63}$$

η is zero for an ideal gas but is non-zero for real gases.

4.9   Internal Energy of an Ideal Gas

Since $\left(\dfrac{\partial u}{\partial v}\right)_T = 0$, for an ideal gas, the specific internal energy of an ideal gas is a function only of

temperature T, i.e., $u = u(T)$ for an ideal gas. But we know that

$\left(\dfrac{\partial u}{\partial T}\right) = c_v$ and since u is a function only of temperature T, the partial derivatives may be replaced

by the full derivatives, so

$$\frac{du}{dT} = c_v \quad \text{or} \quad u = c_v T + K_{int} \text{ (constant)} \tag{4.64}$$

where K_{int} is the constant of integration and the value of it is not known. In order to eliminate K_{int}, one defines the difference in specific internal energies when a system with specific internal energy u_0 at temperature T_0 moves to another equilibrium state of specific internal energy u at temperature T.

$$(u - u_0) = c_v(T - T_0) \quad \text{and} \quad u = u_0 + c_v(T - T_0) \tag{4.65}$$

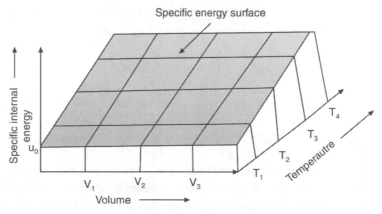

Fig. 4.5   Specific internal energy surface for an ideal gas as a function of volume and temperature

Like the P–v–T surface for equilibrium processes, a specific internal energy surface for an ideal gas as a function of volume and temperature may also be drawn as shown in Fig. 4.5.

Equation 4.65 is a relation between the specific internal energy of an ideal gas with temperature. In the case of an ideal gas internal energy does not depend on the pressure or the volume. Equation 4.65 is called the energy equation of the ideal gas. However, for a substance in general with three system parameters, the internal energy U (or u) may be written a function of any pair of system variables, that is $u = f(P, v)$ or $u = f(P, T)$ or $u = f(v, T)$. The function f that gives the relation between the internal energy of the substance with the system parameters is called the energy equation of the substance. Equation of state $f(P, v, T) = 0$ and the energy equation $U = f(P, v)$ provide complete information about the system.

4.10   Joule–Thomson or Porous Plug Experiment

In Gay-Lussac–Joule experiment change in temperature of the water surrounding the system during the free expansion of the gas could not be recorded due to the large thermal capacity of the surrounding water. Joule and Thomson (William Thomson, who later became Lord Kelvin, British mathematician and physicist) devised another experiment in which a gas was made to pass through a porous plug from higher pressure to a lower pressure under adiabatic conditions. This experiment is often referred to as the porous plug experiment. A porous plug as the name suggests, is like an assembly of very fine capillary tubes held parallel to each other, the size of each capillary being of the order of the mean free path of the gas molecules so that only one molecule at a time may pass through each capillary. In practice a lump of cotton wool with fine porous serves the purpose. A schematic diagram of the porous plug experiment is shown in Fig. 4.6. The porous plug, in the figure divides the insulated adiabatic boundary into two parts A and B, each fitted with a frictionless piston and a thermometer. In part A there is a gas which is pushed into the porous plug by a piston with pressure P_1. Let the initial volume of the gas on side A be V_1 and there be no gas on side B, which means that the piston on side B is initially in contact with the porous plug. Let the pressure exerted by the piston on side B be P_2, and $P_1 > P_2$. After some time finally all the gas from side A will be pushed on to side B. The volume of the gas on side B may be, say, V_2, and since $P_1 > P_2$, $V_2 > V_1$. It means that the gas while passing through the porous plug has undergone adiabatic expansion. Further, since only one gas molecule at a time can pass through the capillaries in the porous plug, the gas molecules are wire-drawn, which means that gas molecules are pulled apart from each other. In this process work is done against the force of attraction between gas molecules.

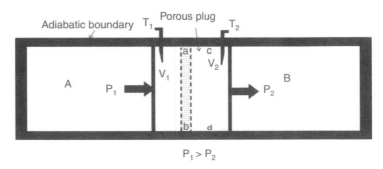

$P_1 > P_2$

Fig. 4.6   Schematic diagram of porous plug experiment

The work W_a is performed by the surrounding piston on the gas in part A and is given by,

$$W_a = -P_1(dV) = -P_1 V_1 \qquad 4.66$$

In part B, gas pushes the piston at constant pressure P_2 and performs the work W_b given by,

$$W_b = P_2(dV) = P_2 V_2 \qquad 4.67$$

Net work done $dW = (P_2 V_2 - P_1 V_1)$ \qquad 4.68

The system is enclosed in an adiabatic boundary hence there is no heat flow in the system from the surroundings or into the surroundings, i.e., $dQ = 0$. The work done dW must be at the coast of the internal energy of the gas, i.e.,

$$dU = dW \text{ or } U_1 - U_2 = P_2 V_2 - P_1 V_1 \text{ or } U_1 + P_1 V_1 = U_2 + P_2 V_2$$

or $$H_1 = H_2 \qquad 4.69$$

It may thus be seen that the process that the gas has undergone in the porous plug experiment is isenthalpic, i.e., the enthalpy (or the specific enthalpy h) of the system remains constant.

The results of porous plug experiment may be summarized as follows.

1. For all real gases it was found that the T_2 is different from T_1, i.e., the temperature of the gas changes when it passes from high pressure side to the low pressure side through the porous plug. At room temperature, for most of the gases, except H_2 and He, there was a fall of temperature. However, the rise or fall of the temperature depends on the values of the temperature T_1, and the pressures P_1 and P_2.

2. For a given gas and for fixed values of P_1 and T_1 on changing the pressure P_2 to different values $P_2^a, P_2^b, P_2^c, \ldots$ etc, respectively, different values of temperature $T_2^a, T_2^b, T_2^c, \ldots$ etc., were recorded. When the points (P_2^a, T_2^a), (P_2^b, T_2^b), (P_2^c, T_2^c) etc., were plotted on a P–T graph they fell on a smooth curve that has a maximum, indicated by M in Fig. 4.7

3. In the next step the pressure and temperature on side A were changed to different sets of values (P_1', T_1'); (P_1'', T_1''); (P_1''', T_1'''); etc., and for each set step-2 was repeated to get the corresponding sets of the values of pressure and temperature $(P_2^{a'}, T_2^{a'})$, $(P_2^{b'}, T_2^{b'})$, $(P_2^{c'}, T_2^{c'})$; $(P_2^{a''}, T_2^{a''})$, $(P_2^{b''}, T_2^{b''})$, $(P_2^{c''}, T_2^{c''})$; $(P_2^{a'''}, T_2^{a'''})$, $(P_2^{b'''}, T_2^{b'''})$, $(P_2^{c'''}, T_2^{c'''})$ etc. Family of curves obtained from plotting the above points on a P–T graph are shown in Fig. 4.7.

4. The family of curves, one for each setting of (P_1, T_1), represents curves of constant enthalpy. Enthalpy remains constant on each curve but has different values for different curves. Each of these curves shows a point of maximum value denoted by M on the curve. In Fig. 4.7 points of maximum value for different curves are joined by a dotted curve called the inversion curve. On the left hand side of the dotted curve the slope of each isenthalpic curve is positive, while on the right hand side it is negative. Thus the slope of the isenthalpic curve changes sign at the point of maximum value M and, therefore, these points are called the points of inversion and the curve joining the points of inversion (dotted curve) the inversion curve. The physical significance of the inversion curve lies in the fact that cooling will be produced when system

crosses the inversion curve from right to left along any isenthalpic curve. For example, in porous plug experiment, the gas initially at pressure and temperature (P_1'', T_1'') goes from point A to point B, or A′ to B or A to B′ etc., (in Fig. 4.7) then cooling will be produced, but if it goes from B to A or A to A′ or A′ to A, or B to B′ etc., heating will be produced. This is important from the point of view of liquefying gases.

5. All isenthalpic curves were found to become flat and horizontal at low pressure and high temperatures.

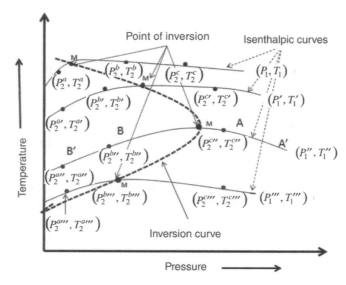

Fig. 4.7   Graphs showing temperature inversion and constant enthalpy curves in porous plug experiment

4.10.1 Joule–Thomson or Joule–Kelvin coefficient

The slope of the isenthalpic curve $\left(\dfrac{\partial T}{\partial P}\right)_H$ is defined as Joule–Thomson (or Joule–Kelvin) coefficient and is represented by the Greek letter μ (mu). Therefore,

$$\mu \equiv \left(\frac{\partial T}{\partial P}\right)_H \qquad\qquad 4.70$$

Since isenthalpic curves tend to become horizontal at low pressure and high temperature, conditions under which all real gases tend to behave as ideal gas, it may be assumed that the Joule–Thomson coefficient for ideal gas, μ_{ide} is zero,

$$\mu_{ide} = \left(\frac{\partial T}{\partial P}\right)_H^{ide} = 0 \qquad\qquad 4.71$$

and from Eq. 4.61 $\left[\left(\dfrac{\partial h}{\partial P}\right)_T = -c_P \left(\dfrac{\partial T}{\partial P}\right)_h\right],$

For an ideal gas, $\left(\dfrac{\partial h}{\partial P}\right)_T = 0$ 4.72

Now treating h to be a function of P and T, one may write,

$dh = \left(\dfrac{\partial h}{\partial P}\right)_T + \left(\dfrac{\partial h}{\partial T}\right)_P$ and from Eq. 4.72 $\left(\dfrac{\partial h}{\partial P}\right)_T = 0$. It means that in case of ideal gas h is a

function of T only and $\left(\dfrac{\partial h}{\partial T}\right)_P = \dfrac{dh}{dT}$. Also, from Eq. 4.52 $\left[c_P = \left(\dfrac{\partial h}{\partial T}\right)_P\right]$, hence,

$$dh = c_p dT \text{ and } \int_{h_0}^{h} dh = c_P \int_{T_0}^{T} dT \text{ or } (h - h_0) = c_P(T - T_0)$$

or $$h = h_0 + c_P(T - T_0)$$ 4.73

Like specific internal energy u the specific enthalpy h for ideal gas is a function only of the temperature and in the rectangular coordinate system of (h, v, T) one can draw a two dimensional (h, T) surface just like the (u, T) surface (Fig. 4.5).

4.11 Reversible Adiabatic Process for an Ideal Gas

Isenthalpic processes referred in porous plug experiment were adiabatic but not reversible. It is because the system has not passed through a succession of quasi-static states.

For reversible adiabatic processes it follows from Eq. 4.60 that

$$c_P\left(\dfrac{\partial P}{\partial v}\right)_T = c_v\left(\dfrac{\partial P}{\partial v}\right)_{adi}$$

or $$\left(\dfrac{\partial P}{\partial v}\right)_{adi} = \dfrac{c_P}{c_v}\left(\dfrac{\partial P}{\partial v}\right)_T = \gamma\left(\dfrac{\partial P}{\partial v}\right)_T$$ 4.74

where $\dfrac{c_p}{c_v} = \gamma$. Also for an ideal gas $P = \dfrac{RT}{v}$ and $\left(\dfrac{\partial P}{\partial v}\right)_T = -\dfrac{RT}{v^2} = -\dfrac{P}{v}$ 4.75

Substituting the value of $\left(\dfrac{\partial P}{\partial v}\right)_T$ from Eq. 4.75 into Eq. 4.74 one gets,

$$\left(\dfrac{\partial P}{\partial v}\right)_{adi} = -\gamma\dfrac{P}{v} \text{ or } \dfrac{dP}{P} = -\gamma\dfrac{dv}{v}$$ 4.76

Integrating Eq. 4.76 one gets,

$$\ln P + \gamma \ln v = \ln K \text{ or } Pv^\gamma = K \text{ (constant of integration)}$$ 4.77

Therefore, for an ideal gas $Pv = RT$ and for reversible adiabatic processes $Pv^\gamma = K$; therefore, for reversible adiabatic processes

$$Tv^{\gamma-1} = \dfrac{K}{R} = K' \text{ (another constant)}$$ 4.78

Similarly it can be shown that for reversible adiabatic processes on an ideal gas

$$TP^{-(\gamma-1)} = \text{constant}$$ 4.79

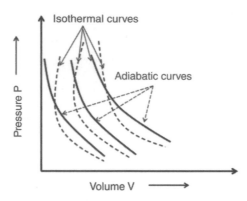

Fig. 4.8 Projections of reversible isothermal and adiabatic processes on P-V plane

Typical curves for reversible isothermal and adiabatic processes in case of an ideal gas on P–V surface are shown in Fig. 4.8. The dotted curves represent a family of isotherms while the family of adiabatic curves is represented by solid curves. It is obvious that for a given system two isotherms and two adiabatic curves cannot cross each other, but adiabatic curves and isotherms for the same system may cut each other as shown in the figure.

The specific work (work per mole) in a reversible adiabatic expansion of an ideal gas from specific volume v_1 to v_2 may be given by,

$$w_{adi} = \int\limits_{v_1}^{v_2} Pdv = \int\limits_{v_1}^{v_2} \frac{K}{v^\gamma} dv = \frac{1}{(\gamma - 1)}(P_1 v_1 - P_2 v_2) \qquad 4.80$$

Since in adiabatic process no flow of heat takes place, the work in expansion is done by the system at the coast of the internal energy and hence,

$$w_{adi} = u_1 - u_2 \qquad 4.81$$

If the system is an ideal gas, $u = c_v T + K'$ (constant) and therefore,

$$w_{adi} = c_v (T_1 - T_2) \qquad 4.82$$

4.12 Carnot Cycle

Carnot (Nicolas Leonard Sadi Carnot, French military engineer and physicist) may be called the founder of thermodynamic reasoning. In an effort to find the parameters on which the efficiency of an engine depends, Carnot looked into the physics of the processes involved in the running of an engine without really bothering about their mechanical aspects. Thus, Carnot concentrated on the underlying principles of physics and developed the science of thermodynamic reasoning. He showed that the working of an engine may be described by four step closed cycle of operations called the Carnot's cycle.

A Carnot's cycle works on a system or a substance, called the working substance, that may be anything, including solids, liquids, gases, mixture of these, magnetic materials, dielectric material, etc. It may be proved that the basic quantities involved in the cycle do not really depend on the state of the working substance that may even under go phase change during the operation of the cycle. The cycle consists of four steps or processes, two isothermal and two adiabatic. These processes operate on the working system in such a way that the system reverts back to the initial state after the four operations. Since the operation of Carnot's cycle is independent of the properties of the working substance, we take ideal gas as the working substance.

The initial state of the working substance (ideal gas) is shown by point 'a' (P_a, v_a, T_2) in the P–v diagram 4.9 (a). The four steps of the Carnot cycle are as follows:

Step-1: The ideal gas is allowed to expand reversibly and isothermally. This may be achieved by keeping the gas in thermal contact with a heat reservoir or a constant temperature bath at temperature T_2. During this expansion the working substance, which in this case is ideal gas, performs work which we denote by w_a. If the gas was not in contact with a constant temperature heat reservoir, the temperature of the gas would have decreased below T_2 in expansion. However, because of the thermal contact with the heat reservoir at temperature T_2, some heat flows from the reservoir into the system and the temperature of the working substance remains constant at T_2. Let us denote the amount of heat flow into the system as Q_2. At the end of step-1, the system is denoted by point 'b' on the P–v graph of Fig. 4.9 (a).

In step-1, the volume of the working substance increases from v_a to v_b; pressure changes from P_a to P_b, but the temperature remains constant at T_2. It means that $T_a = T_b = T_2$. The system performs work w_a and heat Q_2 flows into the system.

Step-2: In step-2, the working substance undergoes adiabatic expansion till it reaches the point 'c' (P_c, v_c, T_1) on the P–v graph of Fig. 4.9 (a). To achieve adiabatic expansion, the thermal contact between the heat reservoir and the system must be replaced by an adiabatic boundary so that no heat flow from the gas or to the gas from the surroundings may take place. Work will be performed by the ideal gas in its expansion and its temperature will fall to the value T_1 which is less than T_2. The work done by the system in step-2 may be denoted by w_b. The final temperature at the end of step-2 is denoted by $T_1 = T_c$.

In step-2 the gas undergoes adiabatic expansion, performs a work w_b on the surroundings, its specific volume increases from v_b to v_c, pressure changes from P_b to P_c and the temperature decreases to T_1 which is less than T_2. No flow of heat takes place in step-2.

Step-3: The system is again put in thermal contact with a constant temperature bath at temperature $T_1 = T_c$. Now the system consisting of ideal gas is reversibly and isothermally compressed till it reaches the state shown by point 'd' in Fig. 4.9 (a). The point 'd' is chosen in such a way that adiabatic compression from there, in step-4, brings the system back to the initial state denoted by 'a'. During isothermal compression work w_c is done on the system by the surroundings and, therefore, the temperature of the gas will rise. However, the gas is in thermal contact with a constant temperature bath at temperature T_1 and so the temperature of the system remains constant at T_1 and some heat flows from the gas to the heat reservoir. By definition the temperature of a constant temperature bath does not change when some heat flows in or out of the bath. The heat flow out of the system in step-3 is denoted by Q_1.

In step-3, the system is isothermally and reversibly compressed, work w_c is done by the surroundings on the system, system temperature remains fixed at T_1 and amount of heat Q_1 flows out of the system.

Step-4: In this step the thermal contact between the constant temperature bath and the gas is removed and the system is covered by an insulating adiabatic boundary. The gas is then adiabatically and reversibly compressed till it reaches back to the initial state denoted by 'a'. During this step the temperature of the gas rises from T_1 back to T_2, volume and pressures change, respectively, from v_d to v_a and P_d to P_a and a work w_d is performed on the system by the surroundings.

In step-4, the system is reversibly and adiabatically compressed till it attains its initial state. Work w_d is performed on the system by the surroundings. No heat flows from or to the system in this step.

Figure 4.9 (a) shows the path of the operations performed on the working substance in (P, v) plane. However, the path for the same Carnot cycle may also be shown in (T, v) plane as is done in Fig. 4.9 (b). As a matter of fact any thermodynamic operation may be represented in a two dimensional plane made up of any two state functions.

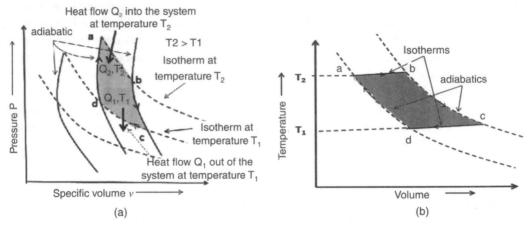

Fig. 4.9   (a) Carnot cycle operations are shown by the shaded area; (b) Carnot cycle in *T–V* plane

The flow chart shown in Fig. 4.10 summarizes the operation of Carnot cycle.

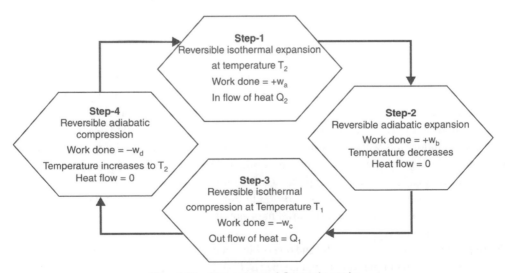

Fig. 4.10   Flow chart of Carnot's cycle

4.12.1 Analysis of Carnot cycle

In step-1, system performed the work w_a and absorbed heat Q_2 from the surrounding heat reservoir. In case the system was isolated from the heat reservoir, the temperature of the system would have fallen to a value below T_2 and the work done by the system would have been at the coast of the decrease in the internal energy. However, the system was thermally connected with the heat reservoir that provided the heat flow Q_2 and maintained the temperature at T_2. Since for ideal gas, internal energy depends only on temperature and the temperature is kept constant by the heat reservoir. It is, therefore, clear that the heat flow Q_2 is equal in magnitude to the work w_a done by the system in isothermal expansion.

$$|w_a| = |Q_2| \tag{4.83}$$

Similarly, it may be argued that,

$$|w_c| = |Q_1| \tag{4.84}$$

Net work done by the system in one cycle of operation $w = (w_a + w_b) - (w_c + w_d)$ 4.85

Heat absorbed by the system $= Q_2$ 4.86

Work done by the system in step-1, $w_a = Q_2 = \int_{v_a}^{v_b} P dv = RT_2 \ln \dfrac{v_b}{v_a}$ 4.87

Further, the work done on the system in step-3, $w_c = Q_1 = \int_{v_c}^{v_d} P dv = RT_1 \ln \dfrac{v_c}{v_d}$ 4.88

Points 'b' and 'c' lies on the same adiabatic curve, therefore,

$$T_2 v_b^{\gamma-1} = T_1 v_c^{\gamma-1} \tag{4.89}$$

Similarly, points 'd' and 'a' lie on the same adiabatic, and so,

$$T_1 v_d^{\gamma-1} = T_2 v_a^{\gamma-1} \tag{4.90}$$

It follows from Eq. 4.89 and Eq. 4.90,

$$\frac{v_b}{v_a} = \frac{v_c}{v_d} \tag{4.91}$$

and from Eq. 4.87 and Eq. 4.88, it follows that,

$$\frac{Q_2}{Q_1} = \frac{T_2}{T_1} \tag{4.92}$$

The special feature of Carnot cycle is that heat flow into the system takes place at the higher temperature T_2 and heat flow out of the system at the lower temperature T_1. Further, in case of the ideal gas as the working substance the ratio Q_2 to Q_1 is same as the ratio of T_2 to T_1.

Work done in the adiabatic steps of the Carnot cycle may also be calculated as follows

Work done by the system in step-2, $w_b = \int_{v_b}^{v_c} P dv = \int_{v_b}^{v_c} \dfrac{dv}{v^\gamma} = \dfrac{R}{(\gamma - 1)}(T_2 - T_1)$ 4.93

Work done on the system in step-4, $w_d = \int\limits_{v_d}^{v_a} Pdv = -\dfrac{R}{(\gamma - 1)}(T_2 - T_1)$ 4.94

It may be observed that work done by the system in adiabatic step-2, w_b is exactly equal in magnitude but opposite in sign to the work w_d performed on the system in step-4. Therefore, the net work done by the system in one cycle

$$w = w_a - w_c = Q_2 - Q_1$$ 4.95

A Carnot cycle operates between a high temperature heat source (heat reservoir at temperature T_2) and a lower temperature heat sink (constant temperature bath at temperature T_1). A quantity Q_2 of heat is taken from the source, a part of this is converted in to mechanical work w and the rest $Q_1 = (Q_2 - w)$ is rejected at the sink.

Two most important properties of Carnot cycle are:

1. The working of Carnot cycle is totally independent of the properties of the working substance
2. The ratio of the heat in-flow Q_2 from the heat source at higher temperature T_2 to heat out-flow Q_1 at the heat sink of temperature T_1 depends only on the ratio of the temperatures T_2 and T_1, $\left(Q_2\Big/Q_1 = T_2\Big/T_1 \right)$ and is independent of the nature and properties of the working substance.

4.12.2 Heat engine

Heat engine is a machine that absorbs heat and performs some mechanical work. Most of the heat engines employ cyclic operations based on Carnot cycle. In a Carnot cycle the working substance returns back to the initial state after each cycle. The thermal efficiency η of a heat engine is defined as the ratio of the mechanical work done to the heat absorbed per cycle. In case of engines based on Carnot cycle, the thermal efficiency may be given as,

$$\eta = \frac{\text{mechanical work performed}}{\text{heat absorbed}}$$

$$= \frac{w}{Q_2} = \frac{Q_2 - Q_1}{Q_2} = 1 - \frac{Q_1}{Q_2} = 1 - \frac{T_1}{T_2} = \frac{T_2 - T_1}{T_2}$$ 4.96

Equation 4.96 tells that thermal efficiency is always less than one, unless Q_1 is zero, i.e., all the heat absorbed is converted into work. It will be shown later that this is against the second law of thermodynamics.

4.12.3 Refrigerator

A refrigerator removes heat from a system at lower temperature in each cycle of its operation. This is done at the coast of mechanical work performed by some external agency. Heat removed from the system at lower temperature plus the heat equivalent to the mechanical works done by the external agency, is ejected at the higher temperature source. In thermodynamic sense a refrigerator also works on Carnot cycle operated in reverse order. It may be noted that operation in reverse order is possible as each stage of Carnot cycle is reversible.

In step-1 of the refrigerator Carnot cycle, the working substance at higher temperature T_2 expands adiabatically and performs work, w_d; in step-2 it expands isothermally, absorbs Q_1 amount of heat at the lower temperature T_1. Work w_c is performed by the system in step-2. The system is made to perform adiabatic compression in step-3 by performing work w_b on the system. Finally, in the last step-4, the system undergoes isothermal compression and ultimately attains the initial state. Work w_a is performed on the system by the surroundings/ external agency and heat $Q = Q_1 + w$, where $w = [(w_a + w_b) - (w_c + w_d)]$ flows out of the system at the higher temperature T_2.

The performance of a refrigerator is judged in terms of its coefficient of performance (COP) denoted by c and defined as,

$$c \equiv \frac{\text{Amount of heat removed from the lower temperature}}{\text{Work performed by the external agency or surroundings}} = \frac{Q_1}{w} \qquad 4.97$$

A good refrigerator is one that removes large amount of heat from the lower temperature, cools fast and to a lower temperature and only a small amount of work is required to be done by the external agency. In fridges an electrical motor generally performs the work required for heat transfer from lower to higher temperature. The temperature of the environment at the back of an air conditioner or a fridge is generally higher than other areas. It is because of the heat taken from the freezer of the refrigerator/air conditioner is rejected at the back.

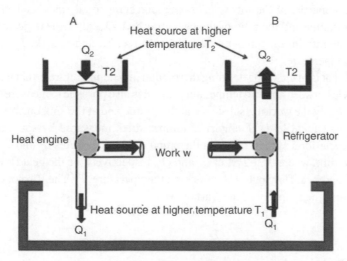

Fig. 4.11   Schematic diagram of heat engine and refrigerator

A schematic diagram of a heat engine and a refrigerator is shown in Fig. 4.11. A heat engine, (Fig. 4.11 A), absorbs a quantity of heat Q_2 from the heat source at higher temperature T_2. This is shown by a pipe of large diameter in the figure. A part of this energy is converted into mechanical work, w, and the remaining part, $Q_1 = (Q_2 - w)$, is rejected in the heat sink at lower temperature T_1. The diameters of the pipes showing the heat absorbed, Q_2, the work, w, and the heat, Q_1, rejected at the sink are proportional to their magnitudes. A more efficient heat engine should convert a larger part of the absorbed heat into work and only a small fraction of the absorbed energy should be rejected at sink. Therefore, for an efficient engine the bottom pipe in the figure representing Q_1 should be as thin as possible. A circle in the center of the figure symbolizes that the system is based on Carnot cycle.

Part B of Fig. 4.11 shows a refrigerator. In this case mechanical work, w, is performed by some external agency (like an electric motor) which makes it possible to draw heat Q_1 from the sink at lower temperature. The heat equal to the sum of w and Q_1 is then rejected at the heat source at higher temperature T_2. For an efficient refrigerator the pipe representing the work should be of small diameter, while the pipe representing the heat Q_1 withdrawn from the sink should have the large diameter. The diameter of the pipe representing the heat rejected at the heat source at higher temperature T_2 will be equal to the sum of the diameters of the other two pipes,

4.13 Thermodynamic Temperature

In chapter-3 we studied various temperature scales, some based on the thermometric properties of substances and the one measured by the constant volume gas thermometer. Temperature measured by a gas thermometer is to a large extent independent of the properties of the gas used, provided the pressure is low. At low pressure and high temperature all gases tend to behave like ideal gas. We shall now discuss another temperature scale, called the thermodynamic temperature. Temperature on this scale is absolutely independent of the properties of the substance used for the measurement and, therefore, it is also called the absolute or natural temperature. The temperature of this natural scale is based on the properties of the Carnot cycle.

Two important properties of Carnot cycle, its operation being totally independent of the properties of the working substance and the ratio of the heat absorbed, Q_2, to the heat ejected at the sink, Q_1, depends only on the ratio of the temperatures of the source, T_2, and of sink, T_1, are used to define thermodynamic temperature.

An assumption that is made while defining thermodynamic temperature is that it is always possible to run a Carnot cycle between any two temperatures and it is also possible to measure the heat absorbed and the heat rejected by the working substance at the source and at the sink during the Carnot cycle.

Let us denote by capital letter T empirical temperatures measured by some thermometer and draw the Carnot cycle for a system between the empirical temperatures T_a and T_d, respectively, of the source and the sink, where $T_a > T_d$. Let Q_a and Q_d, respectively, be the heat flow into the system at temperature T_a and heat flow out of the system at temperature T_d. The Carnot cycle is shown in Fig. 4.12. Then, from the property of the Carnot cycle,

$$\frac{Q_d}{Q_a} = \frac{T_d}{T_a}$$

It means that $\frac{Q_d}{Q_a}$ is some function only of T_d and T_a, i.e.,

$$\frac{Q_d}{Q_a} = F(T_d, T_a), \text{ where } F \text{ is some function.} \qquad 4.98$$

Next we consider the Carnot cycle run between temperatures T_a and an intermediate temperature T_f. This Carnot cycle that will go along the path abefa, heat inflow into the system during isothermal expansion (ab in Fig. 4.12) will be Q_a, and the heat outflow during the isothermal compression (denoted in Fig. 4.12 by ef) will be Q_f. Again, from the property of the Carnot cycle,

$$\frac{Q_f}{Q_a} = \frac{T_f}{T_a} = F(T_f, T_a) \qquad 4.99$$

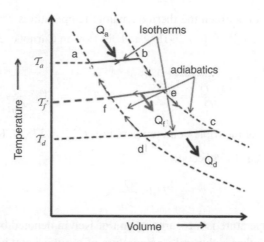

Fig. 4.12 Carnot cycle in *T–V* plane

Similarly, for the Carnot cycle ecdf

$$\frac{Q_d}{Q_f} = \frac{T_d}{T_f} = F(T_d, T_f)$$ 4.100

Though we are not interested in knowing the exact value of the function F but we are interested in knowing the temperature dependence of the function. From Eqs. 4.98, 4.99 and 4.100, we may write;

$$\frac{Q_f}{Q_a}\frac{Q_d}{Q_f} = F(T_f, T_a)F(T_d, T_f) = \frac{Q_d}{Q_a} = F(T_d, T_a)$$

or $$\qquad F(T_f, T_a)F(T_d, T_f) = F(T_d, T_a)$$ 4.101

Equation 4.101 tells that T_f dependence of function F vanishes when $F(T_f, T_a)$ is multiplied by $F(T_d, T_f)$. This can happen only when the temperature dependence of function F is of the type:

$$F(X, Y) = \frac{\chi(X)}{\chi(Y)}, \text{ where } \chi \text{ is some other function,}$$

So that

$$F(T_f, T_a) = \frac{\chi(T_f)}{\chi(T_a)}, \quad F(T_d, T_f) = \frac{\chi(T_d)}{\chi(T_f)}$$

and $$\qquad F(T_f, T_a)F(T_d, T_f) = \frac{\chi(T_d)}{\chi(T_a)} = F(T_d, T_a)$$ 4.102

Kelvin suggested that the thermodynamic temperature T^{ther} may be defined as,

$$T^{ther} = C\chi(T), \text{ where } C \text{ is a constant}$$ 4.103

This definition leads to

$$\frac{Q_d}{Q_a} = \frac{T_d^{ther}}{T_a^{ther}}$$ 4.104

If a Carnot cycle is run between the thermodynamic temperatures T^{ther} and the triple point of water temperature T_{tri}^{ther}, and if the heat in-flow and out-flow in Carnot cycle are respectively, Q and Q_{tri}, then

$$\frac{Q}{Q_{tri}} = \frac{T^{ther}}{T_{tri}^{ther}} \quad \text{or} \quad T^{ther} = T_{tri}^{ther}\,\frac{Q}{Q_{tri}} \qquad 4.105$$

Substituting the value of the triple point temperature $T_{tri}^{ther} = 273.16\ \text{K}$ as obtained in the case of the gas thermometer, Eq. 4.105 becomes

$$T^{ther} = 273.16\,\frac{Q}{Q_{tri}}\ \text{K} \qquad 4.106$$

Thermodynamic temperature is measured in unit of Kelvin denoted by K. Equation 4.106 may be used to determine the thermodynamic temperature of a given body by running a Carnot cycle between the given body and the triple point temperature of water and to measure the heat in-flow Q and out-flow Q_{tri}.

It is obvious that measurement of thermodynamic temperature is not always practical. However, the significance of thermodynamic temperature lies in the fact that it does not depend on any property of any body or any system. It is, therefore, a natural scale of temperature. Further, Q, that represents the heat in-flow, must be positive, or zero, thermodynamic temperature cannot be negative. The minimum thermodynamic temperature is 0 K (zero Kelvin).

Solved Examples

Note: In numerical problems it is often required to calculate the work done and the heat flow in a process. Often there is confusion between the amount of heat flow and the work done. In the Table S-4.1 below work done, heat flow and change in internal energy for different processes are summarized in a tabular form.

Table S-4.1 Change in internal energy for different processes

Process	Change in system parameter	W-work done by the system	Q-heat flow in the system	ΔU- change in internal energy	Process in P–V plane
Isochoric	$\Delta V = 0$	0	$\mathbb{N}\,C_V\,\Delta T$	$\mathbb{N}\,C_V\,\Delta T$	

Contd.

Contd.

Process	Change in system parameter	W-work done by the system	Q-heat flow in the system	ΔU- change in internal energy	Process in P–V plane
Isobaric	$\Delta P = 0$	$P(V_f - V_i)$	$\mathbb{N}\, C_P \Delta T$	$\mathbb{N}\, C_V\, \Delta T$	
Isothermal	$\Delta T = 0$	$\int_{V_i}^{V_f} PdV$	$\int_{V_i}^{V_f} PdV$	0	
Adiabatic	$Q = 0$	$\mathbb{N}\, C_V \Delta T$	0	$\mathbb{N}\, C_V \Delta T$	
Special case	Both V and P change, but $\Delta V = \Delta P$	$\int_{V_i}^{V_f} PdV$	$\mathbb{N}\left(\dfrac{(C_P + C_v)}{2}\right)\Delta T$	$\mathbb{N}\left(\dfrac{C_v}{2}\right)\Delta T$	

Note: $\mathbb{N}$ is the number of moles (or kilomole) of the gas.

1. Figure S-4.1 shows an enclosure made of adiabatic boundary and divided into two equal halves A and B of volume V_0 by a frictionless and insulated piston. Initially N moles of an ideal gas at pressure P_0 and temperature T_0 K are filled in both parts A and B of the enclosure. The value of γ for the gas is 1.5 and the specific molar thermal capacity of the gas is c_v. Heat is then supplied slowly to part B by passing a current i through the resistance R for time t. As a result

the gas in part B expands and compresses the gas in part A until its pressure becomes $\frac{64}{27} P_0$.

With the provided data, answer the following (a) the processes taking place in parts A and B are reversible or not? In terms of N, c_v and T_0, (b) how much work is done on the gas in part A (c) What is the final temperature of gas in A. (d) What is the final temperature of gas in B (e) How much is the heat flow in part B, (f) what is the relation between c_v, T_0, R, I, and t?

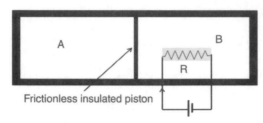

Fig. S-4.1

Solution:

(a) Gas in part B expands but not reversibly. It is because the heat supplied to the gas is by non-reversible dissipative work on the resistor by the current. Therefore, in spite of the fact that the heat is supplied slowly, the total process on side B is non-reversible. On the other hand, compression of the gas in part A is slow, adiabatic and reversible.

(b) The work done on the gas in part A, $W = N c_v (T_1 - T_0)$ S-4.1.1

Here T_1 is the final temperature of the gas on side A. Let the final volume of the gas on side A be V_1. Then,

$$P_1 V_1^\gamma = P_0 V_0^\gamma \quad \text{or} \quad \frac{64}{27} P_0 V_1^{1.5} = P_0 v_0^{1.5} \quad \text{or} \quad V_1 = \frac{9}{16} V_0 \qquad \text{S-4.1.2}$$

Since the gas is ideal $P_0 V_0 = N R T_0$; and $P_1 V_1 = N R T_1$

or $T_1 = \dfrac{P_1 V_1}{P_0 V_0} T_0 = \dfrac{4}{3} T_0$, the final temperature, T_1 of gas on side A = $\dfrac{4}{3} T_0$

Substituting the above value of T_1 in Eq. S-4.1.1, one gets for the work done

$$W = N c_v \left(\frac{4}{3} T_0 - T_0 \right) = \frac{1}{3} N c_v T_0$$

(c) The final temperature of the gas on side A = $T_1 = \dfrac{4}{3} T_0$

(d) Temperature of gas on side B

To calculate the final temperature on side B, we first calculate the final volume V_B. The volume on side A has decreased by the amount $\Delta V = \left(V_0 - \dfrac{9}{16} V_0 \right) = \dfrac{7}{16} V_0$

Final Volume on side B $V_B = V_0 + \Delta V = \dfrac{23}{16} V_0$

If the final pressure and temperature on side B are respectively, P_B and T_B, then,

$$\frac{P_B V_B}{P_0 V_0} = \frac{T_B}{T_0}$$

But final pressure on both sides A and B will be same; so $P_B = 64/27\, P_0$.

So, $T_B = \left(\dfrac{64}{27} \times \dfrac{23}{16}\right) T_0 = \dfrac{92}{27}\, T_0 = 3.407\, T_0$

Temperature of gas on side B $= 3.407\, T_0$

(e) The heat flow on side B

Heat flow on side B $= N\,c_v\,(T_B - T_0) = 2.407\, N\,c_v\,T_0$

(f) Heat flow on side B $= R\,i^2\,t = 2.407\, T_0$,

The desired relation is: $R i^2 t = 2.407\, N\,c_v T_0$

2. Figure S-4.2 shows the steady flow of a fluid through a container. The fluid enters near the base at a height z_1 from the ground with a velocity V_a and leaves at a height z_2 with a velocity V_b. This is achieved by the two frictionless pistons of area of cross section, respectively, S_a and S_b. A stirrer immersed in the fluid and operated externally performs a work W_{sti} per unit time and an amount of heat Q per unit time flows in the system. Derive the energy equation of the system for steady flow and show how Bernoulli's theorem may be obtained from the energy equation.

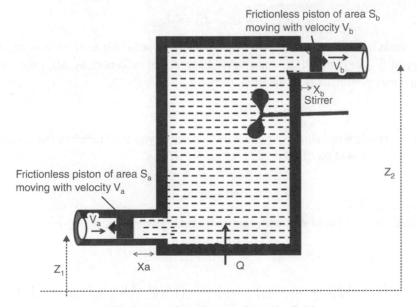

Fig. S-4.2 Steady state flow of a fluid

Solution:

Since the system is in steady state, a mass m of the fluid entering the system (container) per unit time must be equal to the mass m leaving the container per unit time. Let us assume that the pressures on pistons at the entrance and exist are, respectively, P_a and P_b. The force acting on the piston on input $F_a = P_a \times S_a$ and the work W_a done by the input piston in pushing mass m is,

$$W_a = P_a S_a x_a = P_a V_a \qquad \text{S-4.2.1}$$

Here V_a is the volume of the fluid of mass m pushed in

Similarly, work done by the fluid on piston at the output

$$W_b = P_b S_b x_b = P_b V_b \qquad \text{S-4.2.2}$$

Here V_b is the volume of the fluid of mass m pushed out. Therefore, the net work done by pistons is

$$W_{net} = P_b V_b - P_a V_a \qquad \text{S-4.2.3}$$

Also the work done by the stirrer be W_{sti}.

Therefore, the work done by non-conservative forces

$$W_{non} = W_{sti} + (P_b V_b - P_a V_a) \qquad \text{S-4.2.4}$$

The work done W_{cons} by the conservative gravitational force against the lifting of mass m of the fluid from height z_1 to z_2 goes in increasing the potential energy of the mass m of the fluid. Change in the gravitational potential energy of the mass m of the fluid

$$W_{cons} = mg(z_2 - z_1) = \Delta E_{pot} \qquad \text{S-4.2.5}$$

Since some heat Q is flowing in the system and also some stirrer work is done, the internal energy of mass m of the fluid must have changed. Let us denote by ΔU_{int} the change in the internal energy of mass m of the fluid, then

$$\Delta U_{int} = m(u_b - u_a) \qquad \text{S-4.2.6}$$

The fluid enters with velocity V_a and at exit its velocity is V_b. Therefore change in the kinetic energy of mass m of the fluid is ΔE_{kin} which is given by,

$$\Delta E_{kin} = \frac{1}{2} m \left(V_b^2 - V_a^2 \right) \qquad \text{S-4.2.7}$$

Applying the first law of thermodynamics we get,

$$\Delta E_{kin} + \Delta U_{int} + \Delta E_{pot} = Q - \left[W_{non} \right]$$

or $\quad \frac{1}{2} m \left(V_b^2 - V_a^2 \right) + m \left(u_b - u_a \right) + mg \left(z_2 - z_1 \right) = Q - \left[W_{sti} + \left(P_b V_b - P_a V_a \right) \right]$

We divide the above equation by m and substitute $Q = mq$, $W_{sti} = mw_{sti}$, $V_b = mv_b$, and $V_a = mv_a$. Quantities represented by lower case letters are the specific quantities per unit mass.

$$\frac{1}{2}\left(V_b^2 - V_a^2\right) + \left(u_b - u_a\right) + g\left(z_2 - z_1\right) = q - \left[w_{sti} + \left(P_b v_b - P_a v_a\right)\right]$$

Rearranging terms in above equation gives,

$$\left\{u_b + Pv_b + \frac{1}{2}V_b^2 + gz_2\right\} - \left\{u_a + Pv_a + \frac{1}{2}V_a^2 + gz_1\right\} = q - w_{sti}$$

or $$\left(h_b + \frac{1}{2}V_b^2 + gz_2\right) - \left(h_a + \frac{1}{2}V_a^2 + gz_1\right) = q - w_{sti} \qquad \text{S-4.2.8}$$

Equation S-4.2.8 is the thermodynamic energy equation for steady flow of a fluid.

Under following special cases the above equation may be applied to various situations.

(1) If there is no change in the internal energy of the fluid, $h\ (= u + Pv)$ reduces to Pv. Further if there is no stirrer work and no heat flow, $q - w_{sti} = 0$, and Eq. S-4.2.9 reduces to

$$\left(h_b + \frac{1}{2}V_b^2 + gz_2\right) = \left(h_a + \frac{1}{2}V_a^2 + gz_1\right) = \text{constant}$$

On dropping the subscripts, $\left(Pv + \frac{1}{2}V^2 + gz\right) = \text{Constant.}$

or $$P + \frac{1}{2}\frac{1}{v}V^2 + \frac{1}{v}gz \text{ but } \frac{1}{v} = \text{mass per unit volume} = \rho$$

or $$P + \frac{1}{2}\rho V^2 + \rho gz \qquad \text{S-4.2.9}$$

Above equation is nothing but Bernoulli's equation for the steady flow of an uncompressible fluid.

(2) Flow of a fluid through a nozzle

If q and w_{sti} are zero, and the heights of the inlet and out let are same, Eq. S-4.2.8 reduces to,

$$V_b^2 = V_a^2 + 2\left(h_a - h_b\right)$$

Thus the output velocity becomes larger than the input velocity in a nozzle.

Nozzle action $V_b > V_a$

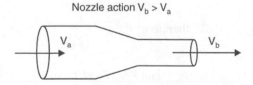

3. A given mass of an ideal gas at temperature T_0, pressure P_0 and volume V_0 undergoes (a) (i) isobaric (ii) isothermal and (iii) adiabatic expansion to the volume $2V_0$. Draw the processes on PV and TV planes and hence show in which of the process minimum work will be done. (b) Repeat the processes in part (a) if the processes are of reduction in volume to a final volume $V_0/2$.

Solution:

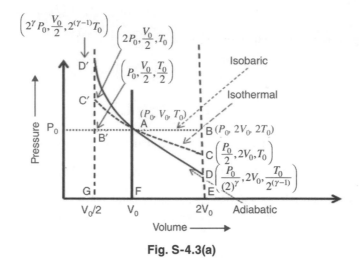

Fig. S-4.3(a)

Figure S-4.3(a) shows in P–V plane the initial position A of the system and final positions B, C and D respectively, after isobaric, isothermal and adiabatic expansion of volume from V_0 to $2V_0$. The coordinates of the points B, C, and D may be obtained using the ideal; gas equation

$$P_0V_0 = NRT_0 \text{ and the conditions:}$$

(1) **Isobaric expansion and compression:** $P_0V_0 = NRT_0$; $P_0 2V_0 = NRT_{final}$, which gives T_{final} = $2T_0$. The coordinates of point B are: $(P_0, 2V_0, 2T_0)$. Similarly for the case of isobaric volume contraction the coordinates of point B' are $\left(P_0, \dfrac{V_0}{2}, \dfrac{T_0}{2} \right)$

(2) **Isothermal expansion and compression:** $P_0V_0 = NRT_0$; $P_{final}2V_0 = NRT_0$ which gives P_{final} $= \dfrac{P_0}{2}$,

The coordinates of C and C' are, therefore, $\left(\dfrac{P_0}{2}, 2V_0, T_0 \right)$ and $\left(2P_0, \dfrac{V_0}{2}, T_0 \right)$

(3) **Adiabatic expansion and compression:** $P_0V_0 = NRT_0$; and $P_0V_0^\gamma = P_{final}V_{final}^\gamma$

So, $P_0V_0^\gamma = P_{final}(2V_0)^\gamma$, therefore, $P_{final} = \dfrac{P_0}{(2)^\gamma}$

Further, $P_{final}(2V_0) = NRT_{final}$ and $P_0V_0 = NRT_0$ so $T_{final} = \dfrac{T_0}{2^{(\gamma-1)}}$

and the coordinates of points D and D' are respectively, $\left(\dfrac{P_0}{(2)^\gamma}, 2V_0, \dfrac{T_0}{2^{(\gamma-1)}} \right)$ and

$\left(2^\gamma P_0, \dfrac{V_0}{2}, 2^{(\gamma-1)}T_0 \right)$

It may be observed that the area ABEFA that corresponds to the work done in isobaric expansion is largest and hence maximum work is done in isobaric expansion. Similarly, the area ADEFA that represents the work done by the system (ideal gas) in adiabatic expansion is least. Work done in isothermal expansion (area ACEFA) is intermediate between the work done in isobaric and adiabatic expansions.

In the case of the reduction of the volume, work is done by the environment on the system and is, therefore, negative. The largest negative work is done in adiabatic compression and the smallest negative work is done in isobaric compression. Since largest negative is least positive, therefore, it may be said that least positive work is done both in adiabatic expansion and compression. Same processes in T–V and P–T planes are shown in Fig. S-4.3 (b) and (c). The coordinates of points A, B, B' etc., are shown in Fig. S-4.3(a).

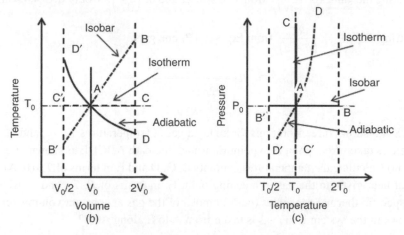

Fig. S-4.3   Isothermal, isobaric and adiabatic processes in (b) *T–V* and (c) *P–T*

4. Derive an expression for the difference of the specific thermal capacities $c_p - c_v$ for Van der Waals gas, given that the specific internal energy for the gas is given by $u = c_v T - \dfrac{a}{v} + u_0$, where u_0 is a constant.

Solution:

According to Eq. 4.32 $C_P - C_V = \left[\left(\dfrac{\partial U}{\partial V} \right)_T + P \right] \left(\dfrac{\partial V}{\partial T} \right)_P$. Writing this equation in terms of the specific quantities one gets,

$$ c_P - c_v = \left[\left(\frac{\partial u}{\partial v} \right)_T + P \right] \left(\frac{\partial v}{\partial T} \right)_P \qquad\qquad \text{S-4.4.1} $$

It is given that $u = c_v T - \dfrac{a}{v} + u_0$, hence $\left(\dfrac{\partial u}{\partial v} \right)_T = \dfrac{a}{v^2}$ S-4.4.2

Also, Van der Waals equation is $P + \dfrac{a}{v^2} = \dfrac{RT}{(v - b)}$ S-4.4.3

In order to find the value of $\left(\dfrac{\partial v}{\partial T}\right)_P$ we differentiate Eq. S-4.4.3 partially with respect to v treating P as constant to get,

$$0 - 2\frac{a}{v^3}\left(\frac{\partial v}{\partial T}\right)_P = -\frac{RT}{(v-b)^2}\left(\frac{\partial v}{\partial T}\right)_P + \frac{R}{(v-b)} \quad \text{or} \quad \left[\frac{RT}{(v-b)^2} - \frac{2a}{v^3}\right]\left(\frac{\partial v}{\partial T}\right)_P = \frac{R}{(v-b)}$$

$$\text{or} \qquad \left(\frac{\partial v}{\partial T}\right)_P = \frac{R}{(v-b)}\frac{(v-b)^2 v^3}{\left[RTv^3 - 2a(v-b)^2\right]} = \frac{Rv^3(v-b)}{\left[RTv^3 - 2a(v-b)^2\right]} \qquad \text{S-4.4.4}$$

Substituting the values of $\left(\dfrac{\partial u}{\partial v}\right)_T$ from Eq. S-4.4.2 and of $\left(\dfrac{\partial v}{\partial T}\right)_P$ from Eq. S-4.4.4 in Eq. S-4.4.1

and putting $P + \dfrac{a}{v^2} = \dfrac{RT}{(v-b)}$ from Eq. S-4.4.3, one gets,

$$c_P - c_v = R\frac{1}{1 - \dfrac{2a(v-b)^2}{RTv^3}}$$

5. Figure S-4.5 shows three isotherms for an ideal gas at temperatures T_1, T_2, and T_3. 1 kilomole of the gas is taken from point A to point E through the path ABCDE as shown by arrows in the figure. (a) Calculate the temperatures at points B, C, D and E in terms of T_1. (b) Also calculate the total heat given to the gas in reaching point E, in terms of R, T_1, and c_v, where c_v is the molar specific thermal capacities (per kilomole) of the gas at constant volume. (c) How much heat flows in the system if the gas is taken from A to E along ACE?

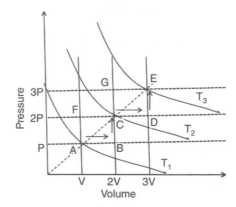

Fig. S-4.5

Solution:

(a) Equation of state for one kilomole of an ideal gas is $PV = RT$. At point A the value of pressure is P and the value of volume is V; hence $PV = RT_1$ where T_1 is the temperature at point A. At point B, pressure is still P but volume is 2V. If T_B is the temperature at point B, then, $P\times(2V) = RT_B$ or $2\,PV = RT_B$. But $PV = RT$, therefore $T_B = 2T_1$. So the temperature of point

B is $2T_1$.Similarly, for point C, $2P \cdot 2V = RT_2 = RT_C$. Therefore the temperature of point C or T_2 is $4T_1$. At point D, $2P \times 3V = RT_D$ Hence $T_D = 6\ T_1$. Further at final point E, $9PV = RT_E = RT_3$. So, $T_3 = T_E = 9\ T_1$.

The temperatures of points A, B, C, D and E are, respectively, T_1, $2\ T_1$, $4\ T_1$, $6\ T_1$ and $9\ T_1$.

(b) Now we calculate the heat supplied in different steps from A to E. There are two types of steps, Isobaric, like AB and, CD. In these steps pressure remains constant and volume increases. Heat flow in isobaric processes $Q = R\ c_p \Delta T$, where c_p is the specific heat capacity at constant pressure and ΔT is the change in temperature.

Heat supplied in going from A to B $= H_1 = c_p\ (2T_1 - T_1) = c_p T_1$.

Heat supplied in going from C to D $= H_2 = c_p\ (6T_1 - 4T_1) = 2\ c_p T_1$.

In parts B to C and D to E volume remain constant and temperature and pressure change. The heat supplied during such processes is given by $c_v\ \Delta T$, where c_v is the specific heat capacity at constant volume and ΔT is the change in temperature at constant volume.

Heat supplied in going from B to C $= c_v(T_c - T_B) = c_v\ (4T_1 - 2T_1) = 2\ c_v\ T_1$.

Heat supplied in going from D TO E $= c_v\ (T_E - T_D) = c_v\ (9T_1 - 6T_1) = 3\ c_v T_1$.

Total heat supplied in going from A to E $= c_p T_1 + 2\ c_p T_1 + 2\ c_v T_1 + 3\ c_v T_1 = 3\ c_p T_1 + 5\ c_v T_1$.

But, $c_p = c_v + R$ for an ideal gas; hence total heat supplied $= 3(c_v + R)\ T_1 + 5\ c_v T_1$
$$= 8\ c_v T_1 + 3\ R\ T_1.$$

If $c_v = 5/2\ R$ (diatomic ideal gas),

Then the total head supplied $= 8\ R \times 5/2 \times T_1 + 3\ R\ T_1 = 23\ R\ T_1$.

On the other hand if the ideal gas is monatomic then $c_v = 3/2\ R$ and

The total heat supplied equals $15\ R\ T_1$.

(c) Heat supplied along the path ACE is

$$\left(\frac{c_V + c_P}{2}\right)\Delta T = \left(\frac{2c_v + R}{2}\right)\Delta T = \left(\frac{2c_v + R}{2}\right)(9T_1 - T_1)$$

$$= \left(\frac{2c_v + R}{2}\right)8T_1 = 4T_1\left(2c_v + R\right)$$

6. A Carnot cycle works with the percent efficiency of 40%. If the temperature of the sink is 300 K, what is the temperature of the heat source? What will be the percentage coefficient of performance of the refrigerator operated between the same temperatures? If Carnot engine draws 2000 Cal from the source, how much work is done by the engine? On the other hand if refrigerator draws 2000 Cal of heat from the sink at 300 K, how much work must be performed by the external motor and how much heat is delivered to the heat source?

Solution:

The efficiency η of a Carnot cycle is given by $\eta = 1 - \dfrac{T_1}{T_2}$ where T_1 and T_2 are the temperatures

of the sink and heat source. It is given that percent efficiency is 40%, which means that efficiency is 0.4. Also it is given that T_1 is 300 K. Therefore,

$$0.4 = 1 - \frac{300}{T_2} \quad \text{or} \quad T_2 = \frac{300}{0.6} = 500 \text{ K}$$

The percentage coefficient of performance of a refrigerator working at temperatures 500 K and 300 K is given by

$$c \times 100 = \left(\frac{T_1}{T_2 - T_1} \right) \times 100 = \frac{300}{200} \times 100 = 150\%$$

If heat engine draws 2000 Cal from the source and rejects Q heat at sink, then, $\dfrac{2000}{Q} = \dfrac{500}{300}$

or $Q = 1200$ Cal. It means the heat rejected at sink is 1200 Cal. The difference $2000 - 1200 = 800$ Cal is converted into work.

In case of the refrigerator if heat drawn from the sink at temperature 300 K is Q_1 and the heat rejected at the source at temperature 500 K is Q_2, then,

$$\frac{Q_1}{Q_2 - Q_1} = \frac{T_1}{T_2 - T_1} = \frac{2000}{Q_2 - 2000} = \frac{300}{500 - 300} = 1.5$$

or $\qquad\qquad Q_2 = 3333.3$ Cal

Hence the external work $= 3333.3 - 2000 = 1333.3$ Cal.

Thus, the external source performs 1333.3 Cal of work in drawing 2000 Cal of heat from the sink and throws out 3333.3 Cal of heat at the higher temperature.

7. A cylinder with frictionless piston contains 0.25 kilomole of a diatomic ideal gas at 2.5×10^5 N m^{-2} pressure and 300 K temperature. The gas first expands isobarically to twice its original volume. It is then compressed isothermally to its original volume and finally brought back isochorically to original pressure. (a) Show the cycle of operation on a P–V diagram. (b) Determine temperature during the isothermal process (c) Determine the maximum pressure. (d) Change in internal energy and the heat flow at each step and in the complete cycle. (e) Work done by the gas and on the gas during the cycle.

Solution:

(a) The P–V diagram of the cycle of operations on the gas is shown in Fig. S-4.7

The gas is at pressure P ($= 2.5 \times 10^5$ N m^{-2}) and volume V, shown by point a, is isobarically allowed to expend to twice of its volume and reaches the point b. It is then isothermally compressed up to point c, and is than isochorically brought back to the initial state at a. The maximum pressure P_{max} is at point c. In part a to b gas performs work, its temperature and internal energy decreases. In part b to c, work is done by the surroundings on the gas in isothermal compression. But the temperature and hence the internal energy of the gas does not change in this part. In part c to a, pressure is reduced at constant volume. Since volume remains constant in this part, no work is done but temperature changes and so the internal energy also changes.

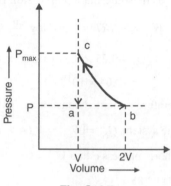

Fig. S-4.7

(b) Let us first calculate the volume V of the gas at point a. Using the ideal gas equation

$$PV = N R T, \text{ where } N \text{ is the number of kilomoles of the gas.}$$

$$2.5 \times 10^5 \times V = 0.25 \times 8.314 \times 10^3 \times 300 \text{ or } V = 2.49 \text{ m}^3$$

So the initial volume of the gas at a is 2.49 m³. At point b the volume of the gas is $2V = 4.98$ m³.

If T_b is the temperature of the gas at b, then $\dfrac{P_a V_a}{T_a} = \dfrac{P_b V_b}{T_b}$ but $P_a = P_b = P$ and $V_b = 2V$ and

$T_a = 300$ K; hence $T_b = 2T_a = 2 \times 300 = 600$ K

Therefore, the isothermal compression of the gas takes place at 600 K.

(c) Maximum pressure P_{max} will be produced at the end of isothermal compression at c. Therefore, at point c:

$$P_{max} V_c = NRT_c \text{ or } P_{max} = \frac{0.25 \times 8.314 \times 10^3 \times 600}{2.49} = 5.0 \times 10^5 \text{ N m}^{-2}$$

During c to a, volume remains constant but pressure changes from 5.0×10^5 to 2.5×10^5 N m⁻² and temperature from 600 to 300 K.

(d) (i) Work done by the gas in going from a to b $W_{ab} = P\Delta V = 2.5 \times 10^5 \times 2.49 = 6.225 \times 10^5$ joule

(ii) Increase in the internal energy of the gas

$$\Delta U_{ab} = Nc_v \Delta T = 0.25 \times \frac{5}{2} R \times (600 - 300)$$

$$\Delta U_{ab} = 15.58 \times 10^5 \text{ J } (c_V \text{ for a diatomic ideal gas is } 5/2 \text{ } R)$$

(iii) Heat flow in the system during a to b,

$$Q_{ab} = Nc_P\Delta T = 0.25 \times \frac{7}{2}R(600 - 300) = -21.82 \times 10^5 \text{ J}$$

(iv) Work done on the gas in isothermal compression from b to c,

$$W_{bc} = -NRT \times 2.303 \log \frac{2v}{v}$$

$$W_{bc} = -0.25 \times 8.314 \times 10^3 \times 600 \times 2.303 \times 0.301$$

$$= -8.64 \times 10^5 \text{ J}$$

(v) $\Delta U_{bc} = 0$, as temperature remained constant.

(vi) Heat flow out of the system $Q_{bc} = |W_{bc}| = 8.64 \times 10^5$ J

(vii) Work done in isochoric process c to a $W_{ca} = 0$

(viii) Change in internal energy

$$\Delta U_{ca} = -Nc_v\Delta T = -0.25 \times \frac{5}{2} R \times (600 - 300) = -15.58 \times 10^5 \text{ J}$$

(ix) Heat flow out of the system in going from c to a

$$Q_{ca} = Nc_V\Delta T = 0.25 \times \frac{5}{2} R \times (600 - 300) = 15.58 \times 10^5 \text{ J}$$

(x) The total work done in the complete cycle $= W_{ab} + W_{bc} + W_{ca} = -2.41 \times 10^5 \text{J}$

(xi) Total change in internal energy= 0

(xii) Total heat flow = 2.41×10^5 J

Since there is no net change in the internal energy of the gas in complete cycle, the net heat flow is equal to net work done.

8. A certain mass of air at 300 K temperature and one atmosphere pressure is adiabatically compressed to 1/15 of its initial volume. Calculate the final temperature if the ratio of specific thermal capacities of air at constant pressure and constant volume is 1.4.

Solution:

Let the initial volume of the gas be V m^3. It is given that the initial pressure P_i was 1 atm = 1.013 $\times 10^5$ N m^{-2} and initial temperature 300 K. Let the final pressure be P_f and the final volume is $V/15$. It is given that $\gamma = 1.4$ for air. Since the process is adiabatic,

$$P_iV^{1.4} = P_f\left(\frac{V}{15}\right)^{1.4}$$

or $\qquad 1.013 \times 10^5 \times V^{1.4} = P_f \times \left(\frac{V}{15}\right)^{1.4}$

$$P_f = 1.013 \times 10^5 \times (15)^{1.4}$$

If T_f is the final temperature, then

$$\frac{P_i V_i}{T_i} = \frac{P_f V_f}{T_f}$$

or $\qquad\qquad T_f = \dfrac{T_i P_f V_f}{P_i V_i} = T_i \times (15)^{0.4} = 300 \times 2.95 = 886.25 \text{ K}$

The final temperature of the gas is 886.25 K

9. Draw the P–V diagram of a Carnot cycle for a two phase working substance, discuss its operation and derive an expression for its efficiency.

Solution:
The P–V diagram for the Carnot cycle of a two phase working substance is shown in Fig. S-4.9. The region where both the liquid and the vapor phases of the working substance may co-exist is out lined by the dome shaped dotted curve. The four steps of the Carnot cycle are depicted by the arms a to b, b to c, c to d and back from d to a.

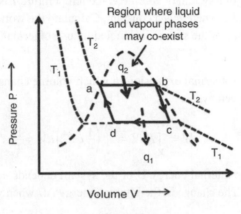

Fig. S-4.9

Step-1: The cycle starts with the state of the working substance shown by point a in the figure. At point a, the working substance is saturated liquid. It undergoes isothermal expansion at temperature T_2 to reach the state b. During the process a to b phase transformation of the working substance takes place and it is converted into vapor phase at point b. During the isothermal expansion heat flow of q_2 per unit mass takes place into the system from the heat reservoir at temperature T_2.

Step-2: Adiabatic expansionof the system takes place in this step from b to c. At point c, generally the liquid and the vapor states co-exist. The temperature of the system falls to a value T_1, lower than T_2. Since the process is adiabatic, no heat flow takes place in this step.

Step-3: This step shown by c to d in the figure is the step of isothermal compression. The working substance remains an equilibrium mixture of liquid and vapor phases but their relative concentrations change from c to d. A heat flow of q_1 per unit mass of the working substance is rejected at the lower temperature (T_1) sink.

Step-4: The working substance undergoes adiabatic compression from d to a in this final step. In the previous isothermal compression step, the mixture was compressed to state d such that the final adiabatic compression brings the system back to the initial state a. No heat flow takes place in this step.

The efficiency η of the cycle is given by,

$$\eta = \frac{q_2 - q_1}{q_2}$$ S-4.9.1

But $q_2 = (h_b - h_a)$ and $q_1 = (h_d - h_c)$, where h_a, h_b, h_c and h_d are, respectively, the specific enthalpies of the working substance at states a, b, c and d. Since the working substance undergoes phase change during the operation of the cycle, heat flows are written in terms of the difference in enthalpies. Putting these values for q_2 and q_1 in Eq. S-4.9.1, the efficiency of the cycle becomes,

$$\eta = \frac{(h_b - h_a) - (h_d - h_c)}{(h_b - h_a)}$$

10. The internal energy U of a system of volume V and temperature T is given as $U = cVT^4$ and the pressure P as $P = \frac{1}{4} cT^4$, where c is a constant. Calculate the work done and heat flow in the system when the volume of the system is doubled in an isothermal process.

Solution:

The work done in an isothermal expansion in which volume changes from an initial value V_i to a final value V_f is given by,

$$W = \int_{V_i}^{V_f} PdV = \int_{V}^{2V} \left(\frac{1}{4}cT^4\right)dV = \frac{1}{4}cT^4V$$ S-4.10.1

In the present case the internal energy U of the system depends not only on temperature but also on the volume V. The change in the internal energy ΔU when volume doubles at constant temperature is,

$$\Delta U = cT^4(2V - V) = cT^4V$$ S-4.10.2

The heat flow Q is equal to the sum of the work done W and the change in the internal energy ΔU,

$$Q = \frac{1}{4}cT^4V + cT^4V = \frac{5}{4}cT^4V$$

Problems

1. Calculate the (a) the work done in phase change of water from liquid to vapor state at $100°C$ and 1 atm pressure, (b) change in the specific internal energy, given that the specific volumes of the liquid water and the vapors are respectively, 1.8 m³ kg⁻¹, and 1×10⁻³ m³ kg⁻¹, and heat of vaporization is 22.6×10⁵ J kg⁻¹.

2. Calculate the heat supplied if in Fig. S-4.5 point E is reached through the path AFCGE.

A fixed amount of an ideal gas is (a) compressed reversibly both isothermally and adiabatically such that the change in pressure is same for both cases, (b) expands isothermally and adiabatically such that the change in volume is same in both cases. Show the processes on a *P–V* diagram and discuss in which process more work is done in the case of compression and in the case of expansion. Support your answer by taking suitable numerical examples.

4. A diatomic ideal gas initially at a pressure of 1.5×10^5 N m^{-2} and volume 0.08 m^3 is compressed reversibly and adiabatically to a volume 0.04 m^3. Determine the (a) final pressure, (b) work done, (c) ratio of the final to initial temperatures. (d) All the above quantities for isothermal compression.

5. One kilomole of a diatomic ideal gas at 1.0 atm pressure and 273 K temperature first undergoes reversible isobaric expansion till its volume increases by 22.4 m^3 and then undergoes reversible isothermal expansion till the pressure becomes 0.5 atm and temperature 546 K. (a) Draw the P–V diagram of the process and calculate (b) the work done, heat flow and change in internal energy during each process.

6. Calculate the latent heat of vaporization of water from the following data; T= 273.2 K, initial specific volume 1 cm^3, final specific volume 1.674 cm^3, $(dP/dT) = 2.71$ cm of Hg K^{-1}.

[Hint: Latent heat, $L = T\left(v_2 - v_1\right)\left(\dfrac{dP}{dT}\right)$]

7. The temperature of 1 kg of water is changed from 50°C to 100°C at constant pressure. Calculate the change in the enthalpy of water, given that the specific molar thermal capacity of water at constant pressure is 75.3×10^3 J K^{-1} kilomole^{-1}.

8. The latent heat of vaporization of a substance is 30 kJ/mol at 60°C and 1.0 atm pressure. Calculate the heat flow Q, change in internal energy ΔU and the work done if 10 mol of the substance vaporizes. Neglect the volume of the substance in liquid state as compared to the volume in vapor state.

9. One kilomole of a diatomic ideal gas at 27°C and 1 atm pressure expands to twice its volume reversibly (a) adiabatically, (b) isothermally, and (c) isobarically. Calculate (i) the work done W, (ii) change in internal energy ΔU, and (iii) heat flow Q in each case.

10. The equation of state and the specific internal energy equation for a gas are respectively, $(P + a)v = RT$ and $u = bT + av + u_0$. Find the value of the specific thermal capacity of the gas at constant volume and of $c_P - c_V$

Short Answer Questions

1. Use thermodynamic relations to show that for an ideal gas
$$c_P - c_V = R$$

2. For a gas that obeys Van der Waals equation of state show that
$$c_P - c_V = R\left(1 + \frac{2a}{RTv}\right) \text{ if } \frac{b}{V} \ll 1$$

3. Draw Carnot cycle in T–P plane and discuss the significance of the area enclosed by the cycle.

4. An ideal gas undergoes reversible adiabatic compression from volume V_1 to V_2. Derive expressions for the work done, heat flow and change in the internal energy of the gas.

5. Write (without derivation) expressions for the efficiency of a heat engine and for the coefficient of performance of an ideal refrigerator.

6. What is the significance of thermodynamic temperature? How can it be measure?

7. Draw the P–V diagram for the Carnot cycle run on a working substance that undergoes phase change and write expression for its efficiency.

8. Define Joule coefficient. What is its value for an ideal gas?

9. What is 'internal pressure' of a gas? Why it is zero for an ideal gas?

10. Write the statement of the first law of thermodynamics.

11. Drive an expression for the work done in a reversible isothermal expansion of an ideal gas at temperature T K from volume V_1 to V_2. How much is the heat flow in this case?

12. Show that in the reversible isobaric expansion of an ideal gas the gas temperature also gets doubled to the initial value if the final volume is twice the original volume.

13. A fixed amount of an ideal gas is enclosed in a diathermic boundary which is surrounded by an adiabatic tank of some fluid. The liquid in the adiabatic tank is very slowly heated by passing electric current through a resistance immersed in the liquid. The temperatures of the liquid as well as of the ideal gas increases slowly. In what respect the temperature rise of the gas is different from that of the liquid?

14. In what respect the adiabatic work between two equilibrium states of a system of same bulk energy is different from the isothermal or isobaric work?

15. How the change in the internal energy of a system may be related to the adiabatic work?

Long Answer Questions

1. State and explain the first law of thermodynamics and hence define the internal energy and the heat flow for the system.

2. Discuss with necessary details the thermodynamic temperature and explain why it is most fundamental or absolute.

3. Describe the working of a Carnot cycle, discuss its significance and derive expressions for the efficiency of an ideal heat engine as well as the performance coefficient of a refrigerator. What are the important characteristics of the Carnot cycle?

4. Starting from the definitions of the thermal heat capacities obtain the relation $C_P - C_V = \left[\left(\frac{\partial U}{\partial V} \right)_T + P \right] \left(\frac{\partial V}{\partial T} \right)_P$

5. Derive the following thermodynamic relations,

   1. $c_P = \left(\frac{\partial h}{\partial T} \right)_P$,

2. $c_P \left(\dfrac{\partial T}{\partial v} \right)_P = \left(\dfrac{\partial h}{\partial v} \right)_P$

3. $c_P \left(\dfrac{\partial P}{\partial v} \right)_T = c_v \left(\dfrac{\partial P}{\partial v} \right)_{adi}$ where symbols have their usual meaning.

6. What was the aim of porous plug experiment? Give a detailed description of the experimental setup and show that enthalpy does not change during the adiabatic expansion of the gas through the porous plug. Also discuss the physical significance of the inversion curve as regards to the liquefaction of gases.

7. What is the difference between the steady state and the state of natural equilibrium? Derive the general equation for the steady flow of a fluid and hence show how Bernoulli's equation may be obtained from it under suitable conditions.

Multiple Choice Questions

Note: *In some of the following questions more than one alternative may be correct. All the correct alternatives must be marked for complete answer.*

1. Which of the following statement (s) about the adiabatic work between two equilibrium states of same bulk energies is/ (are) not true
   (a) It is independent of path.
   (b) It depends on the change in volume.
   (c) No change in the internal energy of the system takes place in adiabatic work.
   (d) Pressure never changes in adiabatic work.

2. When heat flows into a system at constant temperature, its internal energy
   (a) increases
   (b) decreases
   (c) remains constant
   (d) may change depending on the change in other parameters.

3 The work done in a process is proportional to the change in the volume of the system. The process is
   (a) isobaric   (b) isochoric
   (c) isothermal   (d) adiabatic

4 The volume of an ideal gas at temperature T and pressure P is reversibly doubled keeping P constant. The temperature of the gas in final state will become
   (a) $4T$ (b) $2T$
   (c) $T/2$ (d) $T/4$

5. In a reversible process the work done on an ideal gas was found to be proportional to $\log \dfrac{V_f}{V_i}$, where V_f and V_i are, respectively, the final and the initial volumes. The process is
   (a) isobaric   (b) isothermal
   (c) isochoric   (d) none of the above

6. A thermodynamical system consisting of the ideal gas is reversibly and adiabatically taken from point A to a point B on the P–v–T surface. If ΔU, ΔQ and ΔW, respectively, denotes the change in the internal energy, heat flow and the work done in the process, then

(a) $\Delta U = 0$

(b) $\Delta U = \Delta Q = 0$

(c) $\Delta W = 0$

(d) $|\Delta U| = |\Delta W|$

7. Expansion of gas in porous plug experiment is

(a) isothermal

(b) adiabatic and isothermal

(c) adiabatic and isenthalpic

(d) isothermal and isenthalpic

8. If $(h - h_0) = X$, then in case of an ideal gas X is;

(a) $c_P\left(T - T_0\right)$

(b) $c_V\left(T - T_0\right)$

(c) $\dfrac{c_P}{\left(T - T_0\right)}$

(d) $\dfrac{c_V}{\left(T - T_0\right)}$

9. If l_{12}, l_{13} and l_{23} refers, respectively, to the heat of fusion, heat of sublimation and the heat of vaporization of water at its triple point then,

(a) $l_{13} = l_{23} + l_{12}$

(b) $l_{23} = l_{13} + l_{12}$

(c) $l_{12} = l_{13} + l_{23}$

(d) none of the above

10. The absolute temperatures of the heat source and sink of a Carnot cycle of efficiency f are doubled, the new efficiency of the cycle will be,

(a) $4f$

(b) $2f$

(c) f

(d) $f/2$

11. A Carnot cycle works between 200°C and 100°C temperature. What will happen to its efficiency if the temperatures of the source and sink are reduced by half?

(a) Increase by about 80 %

(b) increase by about 8%

(c) decrease by about 80%

(d) decrease by about 8%

12. The ratio of the heat absorbed from the source to the heat rejected at the triple point of water by a Carnot cycle is 0.1. The thermodynamic temperature of the source in Kelvin is

(a) 2.7316

(b) 27.316

(c) 273.16

(d) 2731.6

13. If in a reversible isobaric process the change in temperature and volume of the system containing one mole of an ideal gas are, respectively, ΔT and Δv, then,

(a) $P = \dfrac{R\Delta T}{\Delta v}$,

(b) $R = \dfrac{P\Delta v}{\Delta T}$

(c) $P = \dfrac{\Delta T}{R\Delta v}$

(d) $R = \dfrac{\Delta v}{P\Delta T}$

14. A refrigerator based on the reverse Carnot cycle draws $2Q$ heat from the sink and rejects $3Q$ heat at the high temperature source. The coefficient of performance of the refrigerator is,

(a) 200 %

(b) 100 %

(c) 20 %

(d) 10 %

15. In a reversible isothermal expansion process change in specific internal energy, heat flow per mole and work done per mole are respectively,

(a) $c_p dP, 0, RT \int\limits_{v_{initial}}^{v_{final}} Pdv$

(b) $0, 0, RT \int\limits_{v_{initial}}^{v_{final}} Pdv$

(c) $0, -RT \int\limits_{v_{initial}}^{v_{final}} Pdv, 0$

(d) $0, RT \int\limits_{v_{initial}}^{v_{final}} Pdv, RT \int\limits_{v_{initial}}^{v_{final}} Pdv$

16. In a porous plug experiment a gas is taken from lower pressure to higher pressure across the inversion curve. The temperature of the gas will
    (a) always decrease
    (b) will always increase
    (c) will decrease if the initial temperature is less than the triple point temperature
    (d) will increase if the initial temperature is more than the triple point temperature.

Answers to Numerical and Multiple Choice Questions

Answers to problems

1. (a) 1.7×10^5 J kg^{-1} (b) 20.9×10^5 J kg^{-1}
2. $3c_v T_1 + 5RT_1$
4. (a) 3.96×10^5 N m^{-2}; (b) -9.5×10^3 J; (c) 1.32 (d) 3.0×10^5 N m^{-2}, -8.3×10^3 J, 1.0
5. (b) Step-1: isobaric expansion: Work done = 22.69×10^5 J, Heat flow =79.44×10^5 J, Change in internal energy = 56.74×10^5 J; Step-2: isothermal expansion: Work done = 31.47×105 J, Heat flow = 31.47×105 J, Change in internal energy = 0.
6. 2.26×10^3 J kg^{-1}
7. 13.51×10^5 J
8. $Q = 3.0 \times 10^5$ J; W (done by the system) = 27.6 kJ; $\Delta U = 2.72 \times 10^5$ J
9. (a) $w = 15.11 \times 10^5$ J, $\Delta U = -15.11 \times 10^5$ J, $Q = 0$. (b) $w = 17.29 \times 10^5$ J, $\Delta U = 0$, $Q = 17.29 \times 10^5$ J (c) $w = 24.94 \times 10^5$ J, $\Delta U = 62.35 \times 10^5$ J, $Q = 87.29 \times 10^5$ J.
10. $c_V = b$, $c_p - c_V = R$

Answers to multiple choice questions

1. (c), (d)	2. (c)	3. (a)	4. (b)
5. (b)	6. (d)	7. (c)	8. (a)
9. (a)	10. (c)	11. (b)	12. (b)
13. (a), (b)	14. (a)	15. (d)	16. (b)

Revision

1. The first law of thermodynamics states that "Adiabatic work between two equilibrium states of same bulk potential and kinetic energies is same for all allowed paths"

2. Thermodynamics does not define the absolute value of the internal energy U but the change in the internal energy $dU^{A\to B}$ when system moves from state A to final state B is defined by the following relation,

$$dU^{A\to B} \equiv - dW_{adi}^{A\to B} \qquad 4.1$$

or

$$\int_A^B dU^{A\to B} = -\int_A^B dW_{adi}^{A\to B}$$

or

$$U_B - U_A = -\left(W_{adi}^{A\to B}\right) \quad \text{or} \quad U_A - U_B = W_{adi}^{A\to B} \qquad 4.2$$

3. The adiabatic work is done by a system at the cost of its internal energy.
4. In non-adiabatic process energy in the form of heat flows from the system to the surroundings or from surroundings to the system. The amount of heat flow Q is different both in magnitude and sign for different non-adiabatic paths.
5. The work done in going from an initial state to a final state of equal bulk energies, by a non-adiabatic process $W_{non-adi}^{A\to B}$ is equal to the sum of the adiabatic work $W_{adi}^{A\to B}$ between the same initial and final states and the heat flow $Q_{non-adi}^{A\to B}$ in the non-adiabatic process, i.e.,

$$W_{non-adi}^{A\to B} = W_{adi}^{A\to B} + Q_{non-adi}^{A\to B}$$

or

$$Q_{non-adi}^{A\to B} = W_{non-adi}^{A\to B} - W_{adi}^{A\to B}$$

$$
\underset{\substack{\text{Increase in the}\\\text{internal energy}\\\text{of the system}\\\downarrow}}{(U_B - U_A)} = \underset{\substack{\text{Heat flow}\\\text{into the}\\\text{system}\\\downarrow}}{Q_{non-adi}^{A\to B}} - \underset{\substack{\text{Work done}\\\text{by the}\\\text{system}\\\downarrow}}{W_{non-adi}^{A\to B}}
$$

or

$$dU = dQ - d'W$$

and

$$dU = dQ - PdV \quad \text{For reversible processes}$$

6. **Phase transition:** Three important conclusions about the process of phase transformation may be drawn from Andrew's experiments: (a) there is always a change in volume at constant pressure during the phase transition. (b) there is always a change in the volume during phase transformation and, therefore, configurational work is always done, either by the system or on the system, during phase transition and (c) since temperature remains constant and work is done, therefore, heat flow in to the system or from the system, takes place during the process of phase transformation.

The term $(u + Pv)$ or $(U + PV)$ often appears in thermodynamics and is called the specific enthalpy or enthalpy of the system. Specific enthalpy is denoted by h and enthalpy by the capital letter H.

'Heat flow in any isobaric reversible process is equal to the change in enthalpy'.

$$\ell = h_f - h_i$$

7. In case of reversible process on an ideal gas

$$(\Delta Q)_P = (\Delta Q)_V + \Delta(PV) = (\Delta Q)_V + \Delta(\mathbb{N}\mathcal{R}T)$$

8. Heat capacities at constant pressure C_P and at constant volume C_V.

$$C_P = \frac{(\Delta Q)_P}{dT} = \left(\frac{\partial H}{\partial T}\right)_P \quad \text{and} \quad C_V = \frac{(\Delta Q)_V}{dT} = \left(\frac{\partial U}{\partial T}\right)_V$$

$$C_P - C_V = \left[\left(\frac{\partial U}{\partial V}\right)_T + P\right]\left(\frac{\partial V}{\partial T}\right)_P$$

The term $\left(\dfrac{\partial U}{V}\right)_T$ is called the internal pressure. In thermodynamics this is just a term having dimensions of pressure but according to the kinetic theory the internal pressure originates from the molecular attraction. In case of ideal gas, molecules of which to not exert any force of attraction or repulsion on each other, $\left(\dfrac{\partial u}{\partial v}\right)_T$ must be zero.

9. **Properties of partial derivatives:** If a function $f(x, y, z)$ depends on variables x, y and z and $f(x, y, z) = 0$, then the following two identities hold good.

$$\left(\frac{\partial x}{\partial y}\right)_z = \frac{1}{\left(\dfrac{\partial y}{\partial x}\right)_z}$$

and
$$\left(\frac{\partial x}{\partial y}\right)_z \left(\frac{\partial y}{\partial z}\right)_x \left(\frac{\partial z}{\partial x}\right)_y = -1$$

10. For an isothermal reversible process

$$(dq)_T = \left[P + \left(\frac{\partial u}{\partial v}\right)_T\right]dv = Pdv + \left(\frac{\partial u}{\partial v}\right)_T dv$$

11. For an adiabatic reversible process $dq = 0$

$$(c_v dT)_{adi} = -\left[\left\{P + \left(\frac{\partial u}{\partial v}\right)_T\right\}dv\right]_{adi}$$

12. $\left(\dfrac{\partial u}{\partial P}\right)_v = c_v\left(\dfrac{\partial T}{\partial P}\right)_v$ and $c_v = \left(\dfrac{\partial u}{\partial T}\right)_v$

13. $c_P = \left(\dfrac{\partial h}{\partial T}\right)_P$,

14. $c_P\left(\dfrac{\partial T}{\partial v}\right)_P = \left(\dfrac{\partial h}{\partial v}\right)_P$

15. $c_P\left(\dfrac{\partial P}{\partial v}\right)_T = c_v\left(\dfrac{\partial P}{\partial v}\right)_{adi}$

16. $\left(\dfrac{\partial h}{\partial P}\right)_T = -c_P\left(\dfrac{\partial T}{\partial P}\right)_h$

17. $\left(\dfrac{\partial u}{\partial v}\right)_T = -c_v\left(\dfrac{\partial T}{\partial v}\right)_u$

18. Joule coefficient and is represented by Greek letter η and is defined as,

$$\eta \equiv \left(\frac{\partial T}{\partial v}\right)_u$$

19. Gay-Lussac and Joule carried out experiments to determine η or the internal pressure $\left(\frac{du}{dv}\right)_T$ but failed to record any change in the temperature of the water bath. The experimental setup was defective as the large thermal capacity of the water bath would not have allowed the detection of small change in temperature.

20. The specific internal energy of an ideal gas is given by

$$u = u_0 + c_v\left(T - T_0\right)$$

21. Joule–Thomson or porous plug experiment: In this experiment a gas at high pressure is passed through a porous plug where the molecules of the gas are wire-drawn and the gas expands adiabatically on the other side of the porous plug. Important characteristics of the experimental results are:

   (a) For all real gases it was found that the T_2 is different from T_1, i.e., the temperature of the gas changes when it passes from high pressure side to the low pressure side through the porous plug. At room temperature, for most of the gases, except H_2 and He, there was a fall of temperature. However, the rise or fall of the temperature depends on the values of the temperature T_1, and the pressures P_1 and P_2.

   (b) For a given gas and for fixed values of P_1 and T_1, on changing the pressure P_2 to different values $P_2^a, P_2^b, P_2^c, \ldots$ etc, respectively, different values of temperature $T_2^a, T_2^b, T_2^c, \ldots$ etc., were recorded. When the points (P_2^a, T_2^a), (P_2^b, T_2^b), (P_2^c, T_2^c) etc., were plotted on a P–T graph they fell on a smooth curve that has a maximum, indicated by M in Fig. 4.7.

   (c) In the next step the pressure and temperature on side A were changed to different sets of values (P_1', T_1'); (P_1'', T_1''); (P_1''', T_1'''); etc., and for each set step-2 was repeated to get the corresponding sets of the values of pressure and temperature $(P_2^{a'}, T_2^{a'})$, $(P_2^{b'}, T_2^{b'})$, $(P_2^{c'}, T_2^{c'})$; $(P_2^{a''}, T_2^{a''})$, $(P_2^{b''}, T_2^{b''})$, $(P_2^{c''}, T_2^{c''})$; $(P_2^{a'''}, T_2^{a'''})$, $(P_2^{b'''}, T_2^{b'''})$, $(P_2^{c'''}, T_2^{c'''})$ etc. Family of curves obtained from plotting the above points on a P–T graph are shown in Fig. 4.7.

   (d) The family of curves, one for each setting of (P_1, T_1), represent curves of constant enthalpy. Enthalpy remains constant on each curve but has different values for different curves. Each of these curves shows a point of maximum value denoted by M on the curve. In Fig. 4.7 points of maximum value for different curves are joined by a dotted curve called the inversion curve. On the left hand side of the dotted curve the slope of each isenthalpic curve is positive, while on the right hand side it is negative. Thus the slope of the isenthalpic curve changes sign at the point of maximum value M and, therefore, these points are called the points of inversion and the curve joining the points of inversion (dotted curve) the inversion curve. The physical significance of the inversion curve lies in the fact that cooling will be produced when system crosses the inversion curve from right to left along any isenthalpic curve. For example, in porous plug experiment gas initially at pressure and temperature $\left(P_1'', T_1''\right)$ goes from point A to point B, or A′ to B or A to B′ etc., (in Fig. 4.7) then cooling will be produced, but if it goes from B to A or A to A′ or A′ to A , or B to B′ etc., heating will be produced. This is important from the point of view of liquefying gases.

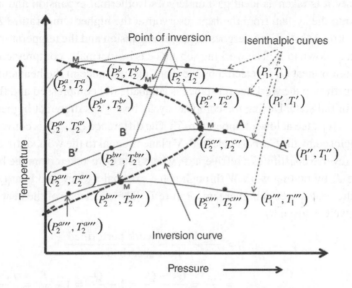

Fig. 4.7   Graphs showing temperature inversion and constant enthalpy curves in porous plug experiment

(e) All isenthalpic curves were found to become flat and horizontal at low pressure and high temperatures.

Joule–Thomson or Joule–Kelvin coefficient: The slope of the isenthalpic curve $\left(\dfrac{\partial T}{\partial P}\right)_H$ is defined as Joule–Thomson (or Joule–Kelvin) coefficient and is represented by Greek letter μ (mu). Therefore,

$$\mu = \left(\frac{\partial T}{\partial P}\right)_H$$

22. Specific enthalpy of an ideal gas is given by : $h = h_0 + c_P\left(T - T_0\right)$ and like specific internal energy is a function of temperature alone

23. Reversible adiabatic process for an ideal gas: Following relations hold good:

$$Pv^\gamma = \text{(constant of integration)}$$

$$Pv^{\gamma-1} = \text{constant}$$

$$Pv^{-(\gamma-1)} = \text{constant}$$

Adiabatic work $w_{adi} = \displaystyle\int_{v_1}^{v_2} Pdv = \int_{v_1}^{v_2} \frac{K}{v^\gamma}\, dv = \frac{1}{(\gamma - 1)}\left(P_1v_1 - P_2v_2\right) = c_v\left(T_1 - T_2\right)$

24. Carnot cycle: The working of a heat engine may be described in terms of four steps of reversible cycle, called the Carnot cycle. The cycle may operate on any working substance and the net result of the cycle is independent of the properties of the cycle. In the first step, the working substance (for simplicity it is taken as ideal gas) undergoes isothermal expansion and an amount Q_2 of heat flows into the system from the heat reservoir at the higher temperature T_2. In the second step, the working substance undergoes adiabatic expansion and the temperature of the working substance goes down to T_1. The working substance is isothermally compressed in the third step and an amount of heat Q_1 is rejected by the working substance in the heat sink at temperature T_1. In the fourth step the working substance is adiabatically compressed and the system comes back to the initial state. During the complete cycle, heat Q_2 is drawn at higher temperature T_2 and heat Q_1 is rejected at lower temperature T_1. The difference $Q_2 - Q_1$ is converted in the work. The area enclosed by the Carnot cycle in P–V plane is equal to the work output. A Carnot cycle operated in reverse constitutes a refrigerator, where heat Q_1 is taken from the heat sink at lower temperature T_1 by putting work W through some external agency and heat $Q_2 = (Q_1 + W)$ is rejected at the heat source at higher temperature T_2. The efficiency of the heat engine based on the Carnot cycle is given by

$$\eta = \frac{\text{mechanical work performed}}{\text{heat absorbed}}$$

$$= \frac{w}{Q_2} = \frac{Q_2 - Q_1}{Q_2} = 1 - \frac{Q_1}{Q_2} = 1 - \frac{T_1}{T_2} = \frac{T_2 - T_1}{T_2}$$

and the coefficient of performance of a refrigerator is denoted by c and defined as,

$$c \equiv \frac{\text{amount of heat removed from the lower temperature}}{\text{Work performed by the external agency or surroundings}} = \frac{Q_1}{w}$$

25. Thermodynamic temperature: It is the natural or absolute temperature because it does not depend on any property of the working substance. This scale of temperature is based on the property of the Carnot cycle. The ratio of the heat absorbed from the heat source and the heat rejected at the heat sink in a Carnot cycle depends only on the ratio of the source to sink temperature and is totally independent of the properties of the working substance. This property of the Carnot cycle is used in defining the absolute or thermodynamic temperature. The only assumption is that it is possible to run a Carnot cycle between any two temperatures and that it is possible to measure the heat absorbed and released by the Carnot cycle. If the thermodynamic temperature at the triple point of water is assigned a value 273.16 K, then any other thermodynamic temperature T^{ther} may be given as,

$$T^{ther} = 273.16 \frac{Q}{Q_{tri}} \text{ K}$$

Where Q and Q_{tri} are, respectively, the heat absorbed from the source at temperature T^{ther} and heat rejected in the sink at the temperature of the triple point.

Second Law of Thermodynamics and some of its Applications

5.0 Need of a System Variable that determines the direction in which an Isolated System will proceed spontaneously

In our daily life as well as in thermodynamics we often find situations where a particular process, though allowed by all known laws of conservation, does not take place. For example, in the initial state we consider a body at temperature T_1 which is placed near a heat reservoir at temperature T_2 ($T_2 > T_1$) and the total system is enclosed in an adiabatic boundary. Slowly the body will gain heat from the source at a higher temperature. The temperature of the body will rise and after some time the temperature of the body will become T_2. No further rise in the temperature of the body will take place and in the final state a thermodynamic equilibrium will get established with both the body and the heat source at the same temperature T_2. During the process of the establishment of the thermodynamic equilibrium the temperature of the heat reservoir or source will not change as by definition a heat source has infinite heat capacity and its temperature does not change when some heat is taken or given to it. The temperature of the body which was initially T_1 will become T_2 in the final state. Since the system is isolated from the surroundings by the adiabatic boundary, no energy enters or leaves the system and the total energy of the system is always same. Further, although the rise of temperature of the body may be very slow, but it cannot be assumed that the process was quasi-static and reversible. It is because the body at temperature T_1 is surrounded by a single source at temperature T_2, not a large number of sources at successively higher temperatures. In nutshell the final equilibrium state of the system is reached through an irreversible process.

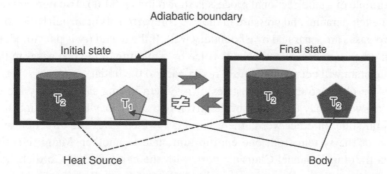

Fig. 5.1 (a) Initial and the final states of a system reached via an irreversible process

The initial and the final states of the system are shown in Fig. 5.1 (a). The problem arises when one looks to the reverse process, i.e., the process in which the body at temperature T_2 loses heat and returns it to the heat source, so that its temperature becomes T_1 and the system reverts back to the initial state. The reverse process, though not disallowed by any conservation law including the law of the conservation of energy, etc., but it never takes place.

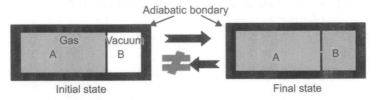

Fig. 5.1   (b) System may go from the initial state to the final state. What stops the system in going from the final to the initial state?

Another example of a unidirectional process is shown in Fig. 5.1 (b). An adiabatic container is divided into two parts A and B, and initially part A is filled with some gas and part B is evacuated. Next a hole is made in the dividing wall to leak the gas from part A to B. This corresponds to the free expansion of the gas in which no work is done by the gas and the temperature of the gas on both sides remains the same. No energy from the outside may enter or leave the system as it is enclosed by an adiabatic boundary. Again, the reverse process, that is the process in which gas from part B moves of its own in to part A and finally all the gas goes back to part A and a vacuum is established in part B, never occurs though no rule or conservation law restricts the reverse process.

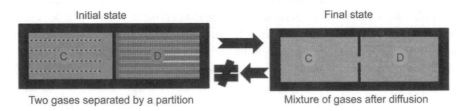

Fig. 5.1   (c) System can go from the initial state to final state but cannot revert back to initial state spontaneously

Another example of a unidirectional process is shown in Fig. 5.1 (c). Two pure but different gases are held at same temperature and pressure in the two compartments of an adiabatic box. Thus in the initial state two gases are separated by a boundary wall. If the boundary is punctured, the two gases mix with each other and the adiabatic box is filled by a mixture of the two gases in the final state. Again, the final state will not spontaneously revert back to the initial state.

It may thus be observed that if two states of a system having same energy are given, we do not have any law which may tell which of the given two states may be the initial state and which may be the final state. In other words there is no law which gives the direction of spontaneous transformation of the system from one equilibrium state to another. Many scientists including Clausius (Rudolf Julius Emanuel Clausius, born with the name Rudolf Gottlieb, was a German physicist and mathematician. He is considered to be one of the central figures in formulating the laws of thermodynamics) considered this problem. Since a state of a system is completely defined by its state variables or parameters, Clausius suggested that *every system has an additional system variable that determines the direction of spontaneous transformation of the system*. He called this property of a system or the system variable as *'Entropy'*. It is said that Clausius selected the name entropy for this state function so that it may sound somewhat like energy which is also a state function and is conserved in all processes.

5.1   Second Law of Thermodynamics in Terms of Entropy

The zeroth law of thermodynamics deals with the state of equilibrium of a system while the first law says that the adiabatic work between the two equilibrium states of a system is path independent. Eventually the first law leads to the equivalence of mechanical work and heat flow and in its analytical form says that the amount of heat flow, dQ, into a system is equal to the sum of the change in the internal energy, dU, of the system and the work, dW, done by the system against the surroundings. The second law of thermodynamics in terms of the entropy of the system may be stated as:

> *'The processes in which the entropy of an isolated system decreases do not take place'*.

Defined in this way, *the entropy of an isolated system may either increase or may remain constant. (Later, it will be shown that the entropy of an isolated system increases in irreversible processes and remains constant if the process is reversible).* It is important to note that the second law of thermodynamics stated above is applicable only to *isolated systems*.

While defining entropy it is assumed (i) that each equilibrium state of a system has a definite value of entropy which is an extensive parameter of the system; (ii) the entropy of a composite system is additive over the constituent subsystems, and that entropy is continuous and differentiable.

5.2   Equilibrium State of an Isolated System and its Entropy

An isolated system, if left to itself, will eventually attain its equilibrium state when no further change in any of its state functions takes place. According to the second law, entropy of an isolated system can only increase; hence the entropy of an isolated system has the maximum value when it is in equilibrium state.

5.3   Entropy of a Non-isolated System

An isolated system may be made up of several components or parts that constantly interact with each other and with their surroundings. Each such component of the isolated system, which is not surrounded by an adiabatic boundary, may be considered to be a non-isolated system. The second law of thermodynamics does not apply to such non-isolated systems or components. As a matter of fact it will be shown that the entropy of such non-isolated components of an isolated system may decrease, remain constant or may increase, when the isolated system undergoes a transition allowed by the first law.

For example, an isolated system enclosed in an adiabatic boundary is shown in Fig. 5.1 (d). The isolated system is made up of three parts or components A, B, and C. Now each part of the system cannot be treated as an isolated system since they are not enclosed in separate adiabatic boundaries and may exchange energies with each other. Suppose the isolated system undergoes some spontaneous transformation so that its entropy changes from the initial value S^i_{sys} to the final value S^f_{sys}.

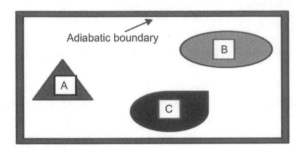

Fig. 5.1   (d) An isolated system is made up of three parts A, B and C. Each part of the system is not an isolated system and the second law of thermodynamics is not applicable to them.

According to the second law of thermodynamics stated above, the entropy of the isolated system should either increase or remain same after the spontaneous transformation. That means,

$$S^f_{sys} \geq S^i_{sys}$$ 5.1

Here S^f_{sys} and S^i_{sys} are respectively the entropies of the system in the initial and the final states.

However, the entropies of non-isolated parts A, B and C of the system are not governed by the second law and may increase, decrease or remain constant when the system undergoes spontaneous transformation, subject to the conditions: $S^i_{sys} = S^i_A + S^i_B + S^i_C$; $S^f_{sys} = S^f_A + S^f_B + S^f_C$; But S^f_A may be smaller or larger than S^i_A, similarly, S^f_A and S^f_B may be larger or smaller than their initial values.

5.4   Quantitative Measure of Entropy

Clausius introduced the concept of entropy as a state function of an isolated thermodynamic system which does not decrease in any possible process. This is only a qualitative definition of entropy. In order to develop a quantitative measure for the state function entropy, we once again consider a Carnot cycle. A Carnot cycle consists of a sequence of four reversible steps: an isothermal expansion at a higher temperature T_2, an adiabatic expansion followed by an isothermal compression at a lower temperature T_1 and finally an adiabatic compression that brings the system back to the initial state. Figures 5.2 (a) and (b) show a Carnot cycle respectively in $(P-v)$ and $T-v$ planes. It may be recalled that in the first step of isothermal expansion a quantity of heat ΔQ_2 at temperature T_2 flows into the system from the reservoir of heat. Since ΔQ_2 is the *heat flow into the system*, it is given a positive (+) sign. No exchange of energy takes place in the second and the fourth steps of the Carnot cycle as both are adiabatic steps. However, in the third step of isothermal compression a quantity of heat,

ΔQ_1 at temperature T_1 $(T_1 < T_2)$, flows out of the system. Since ΔQ_1 is the quantity of *heat that flows out of the system*, it is assigned a negative (–) sign.

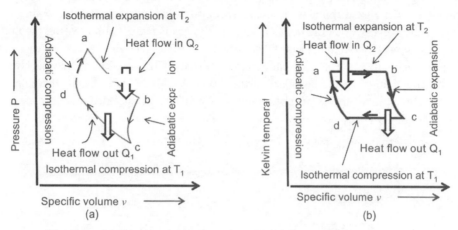

Fig. 5.2   (a) Carnot cycle in $P–v$ plane; (b) Carnot cycle in $T–v$ plane

We know that for a Carnot cycle,

$$\left|\frac{\Delta Q_2}{T_2}\right| = \left|\frac{\Delta Q_1}{T_1}\right|$$

5.2

When the above assigned algebraic signs for ΔQ_2 and ΔQ_1 are used, one gets;

$$\frac{\Delta Q_2}{T_2} = \frac{-\Delta Q_1}{T_1} \text{ or } \frac{\Delta Q_2}{T_2} + \frac{\Delta Q_1}{T_1} = 0 \text{ or } \sum \frac{\Delta Q_T}{T} = 0$$

5.3

Equation 5.3 shows that in a complete operation of a Carnot cycle the algebraic sum of the ratios of heat flow (at constant temperature) to the temperature at different steps is zero. This property of the Carnot cycle will be used to develop a quantitative definition of entropy.

5.5   Representing a Reversible Cyclic Process by a Cascade of Carnot Cycles

A cyclic process consists of a sequence of steps that brings a given thermodynamic system back to its initial state. It may be represented by a closed loop in a two dimensional $(P – v)$ or $T – v$ plot. However, a *reversible cyclic process* is one in which each step of the process is reversible, like the steps of a Carnot cycle. It will be shown here that any reversible cyclic process may be represented by a cascade of Carnot cycles.

A reversible cyclic process is shown in Fig. 5.3 (a) by a closed loop in a $(T – v)$ plot. One may consider that the closed loop is made up of a large number of Carnot cycles, one above the other, covering the area of the closed loop as shown in Fig. 5.3 (a). The horizontal parts of the successive Carnot cycles that represent isothermal expansion in one case and isothermal compression for the next Carnot cycle, are traversed in the opposite directions (one is expansion and the next is compression) and, therefore, get largely cancelled. This leaves the zigzag boundary (made up essentially of un-cancelled horizontal isothermal parts and the adiabatic strokes) of the successive Carnot cycles around the original closed loop representing the cyclic process, as shown in Fig. 5.3 (b). It is simple to visualize that the zigzag boundary of the Carnot cycles will coincide with the boundary of the reversible cyclic process when the number of Carnot cycles is large. On increasing the number of Carnot cycles the adiabatic components of the successive Carnot cycles exactly cancels each other and only the infinitesimally small components coming from the difference of the isothermal components may be left coinciding with the boundary of the reversible closed cycle. It may, therefore, be said that any reversible cyclic process may be treated as equivalent to a succession or cascade or sum of a large number of Carnot cycles, i.e.,

A reversible cyclic process = Σ*Carnot cycles*   5.4

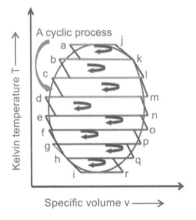
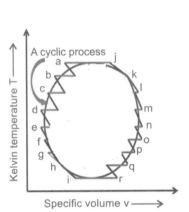

Fig. 5.3   (a) Any closed cyclic process may be represented by succession of Carnot cycles; (b) The horizontal parts of successive Carnot cycles cancel out and only the zigzag boundary is left which smoothes out to the boundary of the reversible cyclic process if the number of Carnot cycles are large.

5.6   Quantitative Definition of Entropy

For a reversible cyclic process, using the results expressed by Eqs. 5.3 and 5.4, one may write

$$\Sigma \frac{\Delta' Q_r}{T} = 0 \qquad\qquad 5.5$$

In Eq. 5.5 $\Delta' Q_r$ represents the heat flow at constant temperature T, the subscript r signifies that the relation holds only for reversible process. It may also be noted that the heat flow is path dependent, and, therefore, one may denote it as $\Delta' Q_r$. The summation, indicated by Σ, is to be carried out on all the Carnot cycles. If the number of Carnot cycles is quite large, the summation sign in Eq. 5.5 may be replaced by the sign of integration, and Δ, denoting a discrete increment by d, denoting continuous differential .Therefore, for a reversible cyclic process,

$$\oint \frac{d' Q_r}{T} = 0 \qquad\qquad 5.6$$

Equation 5.6 tells that the quantity $\dfrac{d'Q_r}{T}$ does not change when an isolated system undergo a reversible cyclic process. It is also known that in a cyclic process (when system comes back to the initial state) the state parameters of the system do not change. It is, therefore, obvious that Eq. 5.6 refers to the change in some state parameter of the system that does not change in the reversible cyclic process. Further, it may also be noted that though the quantity $d'Q_r$ is path dependent but the ratio $\dfrac{d'Q_r}{T}$ is path independent.

Equation 5.6 may be used to define a new state function or parameter denoted by S, called the entropy of the system, such that

$$dS \equiv \frac{d'Q_r}{T} \qquad\qquad 5.7$$

So that in any reversible cyclic process $\oint ds = 0$ 　　　　　　5.8

However, if the system does not undergo a complete cycle and moves through a reversible process from an initial state i to a final state f, the change in the entropy of the system $\int_i^f dS$ may be given as;

$$\int_i^f dS = S_f - S_i \qquad\qquad 5.9$$

It is worth noting that Eq. 5.9 essentially defines the change in entropy of a system and not the absolute value, which is uncertain to the extent of a constant. As is obvious from the above, the classical thermodynamics can only provide the magnitude of the change in the entropy of a system during a process and the absolute value of entropy is uncertain by the amount of a constant of integration. This is a big drawback of classical thermodynamics.

5.6.1 Some important observations about entropy and the change in entropy

(a) Equation 5.9 defines the change in entropy between the two equilibrium states of a system.
(b) Equation 5.9 defines the change in entropy when a system in equilibrium state i is taken to the equilibrium state f through a reversible process.
(c) It may, however, be mentioned that entropy of an equilibrium state of a system is a state function (or parameter) like the volume, temperature, pressure, internal energy of the state, etc., and it does not depend how that state has been reached, i.e., the past history of the state. The entropy difference between two equilibrium states of a system as defined by Eq. 5.9, therefore, does not depend if the state f is reached from state i via a reversible or an irreversible process.
(d) Since $d'Q_r$ is zero for a reversible adiabatic process, the *change in entropy in a reversible adiabatic process is also zero. Change in entropy in an irreversible adiabatic process may not be zero.*
(e) Though Eq. 5.9 defines the change in entropy between two equilibrium state of the system and not the absolute value of the entropy of any equilibrium state, however, for practical purpose the entropy of some convenient equilibrium state of the system may be arbitrarily assigned

an absolute value zero and then the absolute entropies of the other equilibrium states may be calculated using expression 5.9. Mechanical engineers assume the absolute entropy of water (in liquid phase) at 273 K temperature and one atmosphere pressure as zero and develop entropy tables for different equilibrium states of water at different sets of temperatures and pressures.

(f) Entropy is an extensive property of the system and depends on the mass and/or the number of kilomole of the system. One often uses either the specific molal or specific entropy per unit mass of the system which is denoted with lower case letter s such that

$$s = \frac{S}{\mathbb{N}} \text{ or } \frac{S}{m} \qquad\qquad 5.10$$

where $\mathbb{N}$ and m are respectively the number of kilomole/mole and mass of the system.

The MKS unit of entropy S is joule per Kelvin (J/K) while that of specific entropy, s, is $JK^{-1}kg^{-1}$ or JK^{-1} kilomole^{-1}.

5.7   Change in Entropy in Reversible and Irreversible Processes: the Principle of Increase of Entropy of the Universe

All thermodynamic process may be divided into two broad categories: reversible and irreversible. The basic difference in the change of entropy in a reversible and the corresponding irreversible processes is discussed below.

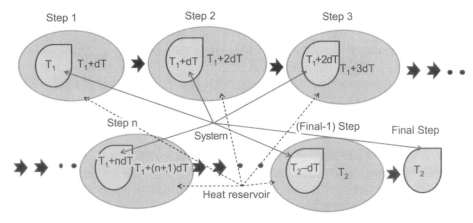

Fig. 5.4   Successive steps of the reversible process by which the temperature of a system may be raised from T_1 to T_2

A typical reversible process of increasing the temperature of a system from a lower value T_1 to a higher value T_2 is shown in Fig. 5.4. As shown in the figure, reversible temperature increase is a multistep process in which the system is enclosed successively by heat reservoirs at an infinitesimally higher temperature and at each step the system and the reservoir are allowed to attain equilibrium. This ensures that the heat transfer at each step remains reversible. The result is that at each step some heat Δq flows into the system at system temperature T and exactly the same amount of heat $(-\Delta q)$ flows out of the reservoir at temperature T, since the temperature of the reservoir and the system are same in the state of equilibrium. The net result is that at each step the entropy ΔS_{sys} of the system increases by the amount $\dfrac{\Delta q}{T}$ while the entropy of the reservoir decreases by exactly the same amount $\left(-\dfrac{\Delta q}{T}\right)$. The total gain in the entropy of the system in going from the initial state to the final state ΔS_{sys} may be given by,

$$\Delta S_{sys} = \sum_{T_1}^{T_2} \frac{\Delta q}{T} \qquad\qquad 5.11$$

Similarly, the total loss of entropy by the reservoirs ΔS_{res} may be written as

$$\Delta S_{res} = -\sum_{T_1}^{T_2} \frac{\Delta q}{T} \qquad\qquad 5.12$$

It may be noticed that in a reversible process the heat reservoir at each step of the process acts as the environment of the system and as such in a reversible process system plus reservoir makes the universe. Further, the net change in the entropy of the universe (system plus the reservoirs) $\Delta S_{sys} = \Delta S_{res}$ is zero. *It may, therefore, be said that in a reversible process there is no change in the entropy of the universe. In a reversible process if system gains (or loses) some entropy, exactly same amount of entropy is lost (or gained) by the environment.*

If the process depicted in Fig. 5.4 is irreversible, i.e., the temperature of the system is changed in a single step and there are no intermediate steps, the change in the entropy of the system ΔS_{sys} is still the same as it was in the reversible case (since the initial and the final states of the system are same) but the change in entropy of the reservoirs ΔS_{res} is not same as in the reversible case. It is because there are no intermediate steps and reservoirs at the intermediate steps are missing in the irreversible process. It is obvious that in irreversible process $|\Delta S_{sys}| \neq |\Delta S_{res}|$ and, therefore, $(\Delta S_{sys} + \Delta S_{res}) > 0$. *It may, therefore, be said that in an irreversible process the entropy of the universe increases. This leads to what is called the principle of the increase of entropy of the universe in all irreversible processes.*

It is interesting to note that *all natural processes are irreversible processes* as they take place under finite differences (and not with infinitesimally small differences) of temperatures, pressures and or volumes, etc. Reversible process is an idealized concept that does not occur in nature. As a result of these naturally occurring irreversible processes the entropy of the universe is increasing with time. The principle of increase of entropy may be looked in contrast to the other universal laws like the law of conservation of energy, the law of conservation of momentum, etc. In a naturally occurring process the *total energy of the universe is conserved but the entropy of the universe increases.* As a consequence of the naturally occurring processes the universe has to bear the burden of ever increasing entropy in contrast to the total energy which is conserved.

5.8   Physical Significance of the Principle of the Increase of Entropy

The increase of entropy or the change of entropy in general may be looked in two different ways. Any process will spontaneously or naturally occur only and only if the entropy of the system plus environment is more than its initial value. This approach is particularly useful in determining whether a particular chemical reaction under given set of parameters like temperature and pressure, will take place or not. The reaction will occur if the entropy increases in the reaction. Since the initial and final state entropies depend on the values of the temperature and pressure, etc., it is possible that the same reaction that was prohibited from the principle of the increase of entropy for one set of the values of the temperature and pressure may occur for another set of these parameters.

The principle of the increase of entropy has an altogether different consequence for mechanical engineers. Mechanical engineers try to convert the internal energy of a system into mechanical work. Temperature of the system is a measure of its internal energy which may be replenished by the flow of heat into the system. The most efficient engine is the Carnot engine that requires two systems at two different temperatures to work. For example if there are two reservoirs of water at temperatures T_2 and T_1 ($T_2 > T_1$) then it is possible to run a Carnot engine between the two water reservoirs and a part of the internal energy of the reservoir at temperature T_2 may be converted into work. On the other hand if water of the two reservoirs is mixed via an irreversible process, the entropy of the mixture will increase but the reservoirs at two different temperatures will become a single reservoir at an intermediate temperature. Now it will not be possible to run a Carnot engine as there is only a single reservoir at one temperature. Thus an irreversible process of mixing the two reservoirs has robbed the systems of the opportunity to convert a part of their internal energy into work. In general it may, therefore, be said that each irreversible process in nature is robbing the nature of an opportunity to convert heat (internal energy) into work. It is not difficult to imagine that irreversible processes taking place in nature in due course of time will remove all temperature differences in the universe and the universe will become a single heat reservoir at a single temperature incapable of producing work. This will lead to what is called the *thermal death* of the universe.

5.9   Other Statements of the Second Law

In the preceding section entropy was defined in terms of the ratio of the heat flow to the temperature in a reversible process. Two other statements, one proposed by Clausius and the other jointly by Kelvin and Planck, may also be taken as the statements of the second law of thermodynamics.

5.9.1   Clausius' statement of the second law

According to the Clausius statement, *'No process is possible whose sole result is a heat flow out of a system at a given temperature and a heat flow of the same magnitude into a second system at a higher temperature'*. Figure 5.5 (a) shows the impossible process referred to in Clausius statement.

5.9.2   Kelvin–Planck's statement of the second law

The Kelvin–Planck statement says, *'No process is possible whose sole result is the abstraction of heat from a single reservoir and the performance of an equivalent amount of work'*. The impossible process referred in this statement is shown in Fig. 5.5 (b).

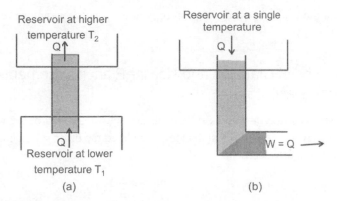

Fig. 5.5 Pictorial representation of impossible processes referred in (a) Clausius and (b) Kelvin-Planck statements

5.9.3 Derivation of Clausius and Kelvin–Planck statements from the principle of the increase of entropy

It can be shown that the two statements mentioned above follow directly from the principle of increase of entropy. In case of the Clausius statement, there will be no violation of the first law if a quantity of heat Q is drawn from the reservoir at lower temperature T_1, no work is done, and the same amount of heat Q is delivered to reservoir at higher temperature T_2. In this process the loss in the internal energy of the reservoir at lower temperature will be exactly equal to the gain in the internal energy of the reservoir at higher temperature and, therefore, the first law will hold good. Further, since no work is done, the process of heat transfer is the *sole* process. When this process is looked from the view point of the change in entropies of the two reservoirs it is found that the decrease in the entropy $-\Delta S_{T_1}$ of the reservoir at lower temperature T_1 is;

$$(\Delta S_{T_1}) = -\frac{Q}{T_1}$$

While the increase (ΔS_{T_2}) in the entropy of the reservoir at higher temperature T_2 is;

$$(\Delta S_{T_2}) = \frac{Q}{T_2}$$

The total change of the entropy in the process ΔS is:

$$\Delta S = \left[(\Delta S_{T_1}) + (\Delta S_{T_2}) \right] = -\frac{Q}{T_1} + \frac{Q}{T_2} = \text{negative as } T_2 > T_1$$

It may be observed that the entropy will decrease in this process and, therefore, the process is not allowed from the principle of the increase of entropy.

In the case of the Kelvin–Planck statement, there is no violation of the first law as the loss in internal energy of a single reservoir at temperature T is exactly equal to the work W. However, the decrease in the entropy of the single reservoir ($-\frac{Q}{T}$) is not balanced by the change of entropy of any other system. Again there is a net decrease of entropy in the process which is prohibited by the law of the increase of entropy.

5.9.4   Equivalence of Clausius and Kelvin–Planck statements of the second law

The fact that Clausius and Kelvin–Planck statements are equivalent may be shown by proving that when one statement, say Kelvin–Planck statement, is violated the other statement, i.e., the Clausius statement is also violated or vice-versa.

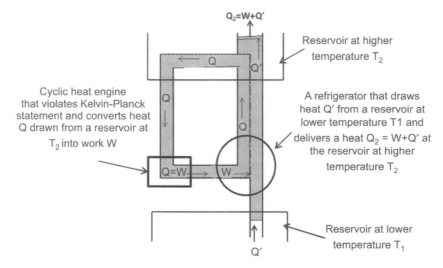

Fig. 5.6   Schematic proof of the equivalence of Kelvin-Planck and Clausius statements

In Fig. 5.6 it is assumed that a cyclic engine (shown by rectangle) violates Kelvin–Planck statement and converts heat Q drawn from a reservoir at higher temperature T_2 totally into work W ($W = Q$). A refrigerator, (shown by circle) absorbs the work W generated by the assumed engine, draws a heat Q' from a reservoir at lower temperature T_1 and rejects heat $Q_2 = (W+Q')$ in the reservoir at higher temperature T_2. Such self-acting system, if possible, will not violate the first law as it neither generates nor destroys energy. However, it will transfer an amount of heat Q' from a reservoir at lower temperature to a reservoir at higher temperature. This will contradict Clausius statement. It may thus be seen that violation of Kelvin–Planck statement leads to the violation of Clausius statement, proving the equivalence of the two statements.

5.9.5   Perpetual motion machine of second kind

The perpetual motion machine of second kind is an imaginary machine that may draw heat from a single reservoir and convert it totally in work. Such a machine will not violate the first law, since neither any energy is created nor is lost in the operation of the machine. A perpetual motion machine if possible will be of great advantage as it will be able to produce infinite amount of work using atmosphere or oceans as the single source of heat. However, such a machine violates Kelvin–Planck statement of the second law and is, therefore, not possible. Sometimes the second law of thermodynamics is also stated as: *A perpetual motion machine of the second kind is impossible.*

5.9.6 Upper limit to the thermal efficiency of a heat engine and to the coefficient of performance of a refrigerator

The Clausius statement of the second law may be used to show that no heat engine working between two reservoirs of given temperatures can have thermal efficiency greater than that of a reversible Carnot engine working between the same reservoirs.

To prove the above statement let us take a reversible Carnot engine that works between two heat reservoirs at temperatures T_2 and T_1, $(T_2 > T_1)$ as shown on the left hand side of Fig. 5.7 (a). The Carnot engine takes heat Q_2 from reservoir T_2, produces the work W and rejects heat Q_1 in reservoir T_1 at lower temperature. Let the efficiency of the Carnot engine be η, so that

$$\eta = \frac{W}{|Q_2|} \qquad 5.13$$

Let us now assume that there is another engine of whatever type which is more efficient than the Carnot engine; it has an efficiency η' that is larger than η. This more efficient engine also works between the same reservoirs and is shown by a rectangular block on the right hand side in Fig. 5.7 (a). The more efficient engine draws a heat Q'_2 from reservoir T_2, produces the same work W and rejects heat Q'_1 in reservoir T_1. The efficiency η' of this more efficient engine is given by;

$$\eta' = \frac{W}{|Q'_2|} \qquad 5.14$$

Since $\eta' > \eta$, it follows from Eqs. 5.13 and 5.14 that,

$$|Q_2'| < |Q_2| \qquad 5.15$$

Also, $W = (Q_2 - Q_1) = (Q_2' - Q_1')$ but $|Q_2'| < |Q_2|$, hence

$$|Q_1'| < Q_1| \qquad 5.16$$

It is thus observed that the more efficient engine absorbs smaller amount of heat from the higher temperature reservoir T_2 and also rejects smaller amount of heat in the lower temperature reservoir T_1 to produce the same work W, as compared to the Carnot engine. This is shown in Fig. 5.7 (a) by taking pipes of smaller diameters for the more efficient engine.

In Fig. 5.7 (b) the Carnot engine is replaced by a Carnot refrigerator. Since Carnot engine is a reversible engine, the magnitudes of $|Q_2|$, $|Q_1|$ and $|W|$ will not change. The Carnot refrigerator will draw a heat Q_1 from reservoir T_1 and will deliver the heat Q_2 at reservoir T_2 when work W is supplied to it. If we now couple the work output of the more efficient engine to the input of the Carnot refrigerator, as shown in Fig. 5.7 (b), the system will become self-working, unaided from any other external agency as the work output of the more efficient engine will make the Carnot refrigerator run. The sole result of the coupled system will be to transfer an amount of heat Q given by

$$Q = [(Q_2 - Q_2') = (Q_1 - Q_1')] \qquad 5.17$$

from the reservoir at lower temperature T_1 to the reservoir at higher temperature T_2. This will violate Clausius statement of the second law and is not possible. Hence the assumption that the other engine has higher efficiency than the Carnot engine is not valid. This proves that the upper limit to the thermal efficiency to any engine is the efficiency of the Carnot engine working between the reservoirs of corresponding temperatures.

It is left as an exercise to show that the upper limit of the coefficient of performance of any refrigerator is the coefficient of performance of a reversible Carnot refrigerator.

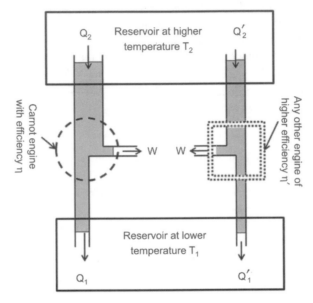

Fig. 5.7 (a) Schematic representation of Carnot engine of efficiency η and any other engine of efficiency $\eta' > \eta$

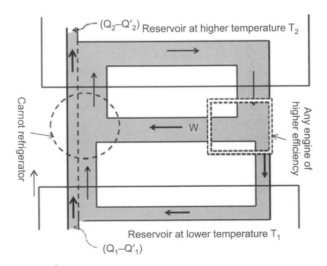

Fig. 5.7 (b) A Carnot refrigerator coupled to the engine of higher efficiency would violate the Clausius statement of the second law

5.9.7 The Clausius inequality

The Clausius inequality is a relationship between the ratios of the heat exchanged to the temperature of an arbitrary number of heat reservoirs when some system consisting of some working substance that interacts with the reservoirs and absorbs or rejects heat to reservoirs, is taken through some arbitrary cyclic process. The process should be cyclic but not necessarily reversible. Since the nature and properties of the working substance do not enter in the final result, the relation is independent of the working substance.

For simplicity, we will derive the desired relationship taking only three reservoirs at temperatures T_3, T_2, and T_1. However, the final result may be generalized to include an arbitrary number of reservoirs. It is further assumed that the working substance of the system absorbs amounts of heat Q_2 and Q_1 respectively from reservoirs at temperatures T_2, and T_1, rejecting an amount of heat Q_3 to the reservoir at temperature T_3. Work W is also produced in the process. In general, there can be no restrictions on the directions of the heat flows and the mechanical work, but they must be consistent with the first and the second laws of thermodynamics. A schematic diagram of the process is shown in Fig. 5.8.

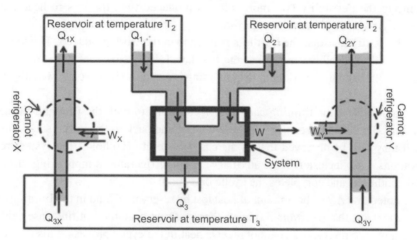

Fig. 5.8 Schematic illustration of Clausius inequality

In order to test whether this confirms to the first and the second laws of thermodynamics, we include two auxiliary Carnot refrigerators X and Y as shown in the figure. Refrigerator X absorbs heat Q_{3X} from reservoir T_3, when some work W_x is performed on it, and rejects heat Q_{1X} at reservoir T_1. Similarly, refrigerator Y absorbs heat Q_{3Y} from reservoir T_3, rejects heat Q_{2Y} to reservoir T_2 on supplying work W_Y. We now put the condition:

$$Q_{1X} = Q_1 \quad \text{and} \quad Q_{2Y} = Q_2 \qquad\qquad 5.18$$

The above condition ensures that there is no change in reservoirs T_1 and T_2 and they remain in their initial state, since there is no net change in their heat contents. Let us now consider the status of reservoir T_3 and the net amount of work involved. The net work *produced* as a result of the operation of Carnot refrigerators and the system is given by,

$$\Delta W = [W - (W_X + W_Y)] \qquad\qquad 5.19$$

The net heat *removed* from reservoir T_3 is:

$$\Delta Q = [(Q_{3X} + Q_{3Y}) - Q_3] \qquad\qquad 5.20$$

Note that the magnitude of ΔQ in Eq. 5.20 is taken positive when heat flows out from the reservoir. There exist the following four possibilities regarding the magnitudes of ΔW and ΔQ;

$$\Delta W = \Delta Q \qquad\qquad (A)$$

$$\Delta W > \Delta Q \qquad\qquad (B)$$

$$\Delta W < \Delta Q \qquad\qquad (C)$$

$$\Delta W = \Delta Q = 0 \qquad\qquad (D)$$

The possibility (A) though conforms to the first law (work produced is equal to the heat lost, total energy is conserved), does not conform to the second law. It contradicts the Kelvin–Planck statement of the second law, because the complete operation amounts to the withdrawal of heat from a single reservoir and conversion of whole of it into work. As such possibility A is not valid.

According to the possibility (B), more work is produced than the loss in heat contents. This contradicts the first law and hence is not valid.

The possibility (C) says that the heat loss is more than the work produced. This possibility is in conformation to the second law as well as the first law. It is possible that a part of the heat loss by reservoir T_3 is converted into work and the remaining is ejected. The first law will also hold in such a situation.

The possibility (D) is the limiting case of (C), conforms to both the first and the second laws. This possibility (D) also ensures that the system has come back to the initial state after the cyclic operation. It may also be observed that in the case possibility (D) holds, the system returns to its initial conditions. Thus inclusion of the auxiliary Carnot refrigerators simply brings the system back to its initial conditions and completes the cyclic operation.

It may be noted that ΔQ is the amount of heat lost by reservoir T_3 and in the limiting case it could be zero, which means that reservoir T_3 must either lose heat or in the limiting case neither lose nor gain heat. In any case reservoir T_3 cannot receive heat in the cyclic process. This may be put in the following analytic form:

$$Q_3 + Q_{3X} + Q_{3Y} \le 0 \qquad\qquad 5.21$$

It is obvious that quantities in Eq. 5.21 have different algebraic signs.

The operations of Carnot Cycles X and Y and conditions of reservoirs 1 and 2 may be represented by the following analytical relations,

$$\frac{Q_{1X}}{T_1} + \frac{Q_{3X}}{T_3} = 0 \qquad\qquad 5.22\,(a)$$

$$\frac{Q_{2Y}}{T_2} + \frac{Q_{3y}}{T_3} = 0 \qquad\qquad 5.22\,(b)$$

$$Q_{1X} + Q_1 = 0 \qquad\qquad 5.22\,(c)$$

$$Q_{2X} + Q_2 = 0 \qquad\qquad 5.22\,(d)$$

Again, quantities appearing in above equations have their own algebraic signs.
From Eqs. 5.22 (a) and 5.22 (c) one may get,

$$Q_{3X} = T_3 \frac{Q_1}{T_1} \qquad\qquad 5.22 \text{ (e)}$$

Similarly Eqs. 5.22 (b) and 5.22 (d) yields:

$$Q_{3Y} = T_3 \frac{Q_2}{T_2} \qquad\qquad 5.22 \text{ (f)}$$

Substitution of the above values of Q_{3X} and Q_{3Y} in Eq. 5.21 gives,

$$Q_3 + T_3 \frac{Q_1}{T_1} + T_3 \frac{Q_2}{T_2} \leq 0$$

or
$$\frac{Q_1}{T_1} + \frac{Q_2}{T_2} + \frac{Q_3}{T_3} \leq 0 \qquad\qquad 5.23$$

Generalizing the result obtained in Eq. 5.23, one may write,

$$\sum \frac{Q}{T} \leq 0 \text{ or } \oint \frac{d'Q}{T} \leq 0 \qquad\qquad 5.24$$

The summation in the above relation is to be carried out over all the reservoirs. An important point to remember is that the *algebraic sign of Q is taken positive if heat flows out* of the reservoir and *negative if it flows into the reservoir*. In case of the continuous variation the integration is to be performed over the path of the transition. Equations 5.24 represent the Clausius inequality. It is a relation between the quantities of heat absorbed or liberated by a number of reservoirs and the temperatures of these reservoirs when a working substance is carried through a cyclic process of any kind, reversible or irreversible. The quantity Q or $d'Q$ has positive sign when heat is given up by a reservoir and negative when it is absorbed by the reservoir. The inequality is independent of the properties of the system (or the working substance) since the temperature, etc., of the system does not enter in the derivation. Further, the equality sign in Eq. 5.24 holds if the cyclic process is reversible, while the inequality sign < in case of the irreversible cyclic process.

$$[\oint \frac{d'Q}{T} = 0]_{rev} \text{ and } [\oint \frac{d'Q}{T} < 0]_{irr}$$

It is easy to realize that Clausius inequality is another way of stating the principle of increase of entropy in irreversible processes and its conservation in reversible processes.

5.10 Calculating the change of Entropy in some processes

As has already been said, the classical thermodynamics does not provide the absolute value of entropy for an equilibrium state of a system. However, it is possible to calculate the change in entropy of a system when it goes from one equilibrium state to another using the tools of classical thermodynamics. Calculation of entropy change for some typical processes is discussed in the following.

5.10.1   Change of entropy in reversible adiabatic process

By definition in reversible adiabatic process there is neither an in-flow nor out-flow of heat from the system. Therefore, $\dfrac{d'Q_r}{T}$ is zero, and hence ΔS, the change in entropy is also zero. As such the reversible adiabatic processes are processes in which entropy of the system remains constant. It is for this reason that reversible adiabatic processes are also called *isentropic* processes and are often specified by putting the subscript s.

5.10.2   Change of entropy in reversible isothermal processes

It may be recalled that the entropy of a system is a state parameter and, therefore, depends only on the state of the system. The entropy of a system in an initial equilibrium state *i* is, say, S_i and in another equilibrium state *f* of the same system be S_f. When the system is taken from state *i* to state *f in whatever way*, reversible or irreversible process, the change in entropy ΔS will always be $(S_f - S_i)$. The magnitude of the change in the entropy of a system from a given initial state to the same final state in processes both reversible and irreversible will be equal. However, what happens to the entropy of the surroundings (or the environment) of the system is quite different in case of the reversible and the irreversible processes. In case of the reversible process, the change in the entropy of the surroundings is exactly equal in amount and opposite in sign to the change of the entropy of the system. However, in case of the irreversible process the change in the entropies of the system and the surroundings are of opposite sign but not equal in amounts.

In an isothermal process the temperature T during the process remains constant and may be taken out of the integral sign. Hence,

$$\Delta(S_f - S_i) = \Delta S_i^f = \frac{1}{T}\int_i^f d'Q_r = \frac{Q_r}{T} \qquad 5.25$$

Equation 5.13 may be used to calculate the change in entropy for the reversible isothermal process.

5.10.3   Change of entropy in a reversible processes where heat flow is accompanied with the change in temperature

Very often we come across processes in which heat flow is accompanied with the change of the system temperature. In such cases the integral $\int_i^f \dfrac{d'Q_r}{T}$ is required to be evaluated for the calculation of the entropy change. Such processes may occur at constant volume or at constant pressure. In the case when the volume of the system remains constant, $d'Q_r = C_V dT$, where C_V is the thermal capacity of the system at constant volume. In the case when C_V may be treated as constant during the process then,

$$\Delta S_i^f = C_V \int_i^f \frac{dT}{T} = C_V \ln \frac{T_f}{T_i} = 2.303\, C_V \log \frac{T_f}{T_i} \qquad 5.26$$

Similarly, for the case where the pressure remains constant during the process and the thermal capacity at constant pressure C_P of the system may be treated as constant during the process, the change in entropy is given by,

$$\Delta S_i^f = C_P \int_i^f \frac{dT}{T} = C_P \ln \frac{T_f}{T_i} = 2.303 \, C_P \, \log \frac{T_f}{T_i} \qquad \qquad 5.27$$

In Eqs. 5.26 and 5.27, T_i and T_f are the system temperatures in the initial and the final states. Further, the change in specific entropies, Δs_i^f, may be obtained by replacing the thermal capacities C_V and C_P by specific thermal capacities c_V and c_P, respectively.

$$\Delta s_i^f = c_V \ln \frac{T_f}{T_i} = 2.303 c_V \log \frac{T_f}{T_i}$$

or
$$\Delta s_i^f = c_P \ln \frac{T_f}{T_i} = 2.303 c_P \log \frac{T_f}{T_i} \qquad \qquad 5.28$$

5.10.4.1   *Entropy change in reversible isothermal expansion of an ideal gas*

As an example, let us calculate the change in entropy in a reversible isothermal expansion of an ideal gas. Let $\mathbb{N}$ moles of an ideal gas be contained in a volume V_i at pressure P in thermal contact with a heat reservoir at temperature T. Let us assume that the gas is in thermal equilibrium with the reservoir. The gas is then allowed to expand in infinitesimal small steps to a volume V_f. The gas performs a work dW in expansion from volume V_i to V_f that is given by,

$$-dW = PdV = \mathbb{N} \frac{RT}{V} dV \qquad \qquad \text{(For an ideal gas } P = \mathbb{N} \frac{RT}{V} \text{)}$$

The work $-dW$ is done by the gas at the cost of its internal energy and as such has negative sign. This loss in the internal energy is replenished by the flow of heat dQ_r from the reservoir to the gas in the reversible isothermal expansion. Therefore,

$$dQ_r = -dW = \mathbb{N} \frac{RT}{V} dV$$

or
$$Q_r = \int_{V_i}^{V_f} dQ_r = \mathbb{N}RT \int_{V_i}^{V_f} \frac{dV}{V} = \mathbb{N}RT \ln \frac{V_f}{V_i}$$

and
$$\Delta S = \frac{Q_r}{T} = \mathbb{N}R \ln \frac{V_f}{V_i} = 2.303 \, \mathbb{N} \, R \log \frac{V_f}{V_i} \qquad \qquad 5.29$$

Since $V_f > V_i$, being expansion, the change in entropy ΔS is positive. The entropy of an ideal gas increases in reversible isothermal expansion.

5.10.4.2   *Change in entropy in reversible change of state of the substance*

Different physical states, solid, liquid, and gaseous of substances are specified, respectively, by subscripts 1, 2, and 3. The specific heats of transformation (also called the specific latent heat) from one physical state to the other, for a given substance have definite values and are denoted by letter ℓ with appropriate subscript. For example, the specific heat of transformation of water from liquid state to the gaseous state ℓ_{23}, at temperature 373 K and one atmospheric pressure has the value 22.6 $\times 10^5$ J kg^{-1}. In general, change of state of substances occurs at almost constant temperature T and pressure P. The change in specific entropy in reversible process of transformation of a substance from one physical state x to the physical state y is given by,

$$s_y - s_x = \frac{\ell_{xy}}{T}$$ 5.30

For example in case of water the change in specific entropy from water to steam

$$\Delta S_{32} = (s_3 - s_2) = \frac{\ell_{23}}{373} = \frac{22.6 \times 10^5}{373} = 6.06 \times 10^3 \text{ J kg}^{-1} \text{ K}^{-1}$$

This means that each kilogram of steam at 373 K (100°C) has 6.06 kilojoule of entropy in excess to that of one kilogram of water at 373 K. Since $\Delta s_{xy} = -\Delta s_{yx}$, the entropy of steam will decrease when its state changes to liquid.

5.10.4.3   *Entropy change in reversible change of temperature without change of state*

Expressions in Eq. 5.28 may be used to calculate the change in specific entropy when the temperature of the material is reversibly changed from T_i to T_f at constant volume and constant pressure, provided the respective specific capacities remain constant in the temperature interval $(T_f - T_i)$.

As an example we consider the case of water. Specific heats of water at constant pressure c_P (at constant pressure of 1 atm) for different physical states are given below.

$$c_P^{ice} = 2.108 \times 10^3 \text{ J kg}^{-1} \text{ K}^{-1},$$

$$c_P^{water} = 4.184 \times 10^3 \text{ J kg}^{-1} \text{ K}^{-1},$$

$$c_P^{steam} = 1.996 \times 10^3 \text{ J kg}^{-1} \text{ K}^{-1}$$

The change in the specific entropy when ice is heated from (-50°C) 223 K to (0°C) 273 K at constant pressure of one atmosphere is given by

$$\Delta s_{223-273}^{ice} = 2.303 \times 2.108 \times 10^3 \times \log\frac{273}{223} = 427 \text{ J kg}^{-1}\text{K}^{-1},$$

It is obvious that c_P^{ice} is assumed to remain constant in the temperature interval 223 K to 273 K.

5.10.5   Entropy changes in irreversible processes

In the previous section, entropy changes in some reversible processes were calculated. Change in entropy between two equilibrium states of a system depends only on the initial and final states and not on how the final state is reached from the initial state. It is, therefore, possible to calculate entropy change in an irreversible process by calculating the entropy change in some reversible process in which the initial and the final states are the same as in the irreversible process. This approach is used in following examples.

5.10.5.1 *Change in entropy of an ideal gas in free expansion*

When an ideal gas undergoes an irreversible free expansion the work done by the gas is zero, as the gas expands against vacuum, i.e., zero pressure. Hence there is no change in the internal energy and the temperature of the gas. The initial and the final states of an ideal gas in free expansion from volume V_i to volume V_f are same as those in the reversible isothermal expansion of the ideal gas from volume V_i to V_f. Change in entropy in free expansion of an ideal gas may, therefore, be calculated using the expression 5.29, developed for the reversible isothermal expansion.

$$\Delta S_{irreversible\ free\ expansion} = 2.303\ \mathbb{N}R\ \log \frac{V_f}{V_i}\ \text{J K}^{-1} \qquad 5.31$$

In free expansion V_f is always greater than V_i and, therefore, entropy increases in free expansion of the gas. This is the reason why free expansion is a spontaneous process and the reverse process of free expansion cannot take place since entropy will decrease in the reverse process.

5.10.5.2 *Change in the entropy of a body in irreversible rise of temperature at constant pressure and constant volume*

The temperature of a body rises irreversibly from the initial value T_i to the final value T_f when it is brought in thermal contact with a single reservoir at temperature T_f. The initial and final states will, however, be same when the temperature is raised by reversible process employing large number of intermediate reservoirs at successively higher temperatures. The expressions contained in Eq. 5.28 may be used for the calculation of change of entropy in case of irreversible rise of temperature.

$$\Delta S_{irre}^{body} = C_P\left(or\,C_V\right)\ln\frac{T_f}{T_i}= 2.303 C_P\left(or\,C_V\right)\log\frac{T_f}{T_i} \qquad 5.32$$

5.10.5.3 *Change in the entropy of the reservoir*

When the temperature of a body is irreversibly changed from an initial value T_i to the final value T_f, the body is surrounded by a heat reservoir at temperature T_f. Heat flows from the reservoir to the body if $T_f > T_i$, and into the reservoir if $T_f < T_i$. This heat-flow is at constant temperature T_f. As a result of the heat- flow the entropy of the reservoir changes. If Q is the magnitude of the heat flow, then the change in the entropy of the reservoir is

$$\Delta S_{irre}^{reservoir} = \mp\frac{Q}{T_f} \qquad 5.33$$

The negative sign in Eq. 5.33 holds for $T_f > T_i$ and the positive for $T_f < T_i$. The heat Q is lost (or gained) by the reservoir at constant temperature T_f but the same amount of heat Q is gained (or lost) by the body at different temperatures varying from T_i to T_f. The amount of heat Q gained (or lost) by the body is given by,

$$Q = C_P (or\ C_V)\ \{T_f \sim T_i\} \qquad 5.34$$

Here C_P (or C_V) is the thermal capacity of the body at constant pressure or volume, as the case may be. Substituting this value of Q in Eq. 5.33, one gets:

$$\Delta S_{irre}^{reservoir} = \mp \frac{C_P\left(or\, C_V\right)\left\{T_f \sim T_i\right\}}{T_f} \qquad 5.35$$

We now take the special case of $T_f > T_i$ and of heat transfer at constant pressure. In this case

$$\Delta S_{irre}^{body} = C_P \ln\frac{T_f}{T_i} \quad \text{and} \quad \Delta S_{irre}^{reservoir} = -\frac{C_P\left\{T_f - T_i\right\}}{T_f}$$

The net change in the entropy of the body plus the reservoir is:

$$\Delta S_{net}^{(body+reservoir)} = C_P\left[\ln\frac{T_f}{T_i} - \frac{\left\{T_f - T_i\right\}}{T_f}\right] = C_P\left[\left(\ln\frac{T_f}{T_i} + \frac{T_i}{T_f}\right) - 1\right]$$

$$C_P\left[\left(\ln x + \frac{1}{x}\right) - 1\right] \qquad 5.36$$

Here $x = \dfrac{T_f}{T_i}$ is the ratio of the final temperature to the initial temperature. In case $T_f > T_i$, $X >$

1 and in the other case of $T_f < T_i$, $X < 1$. The function (In x + 1/x) for positive values of X is plotted against X in Fig. 5.9.

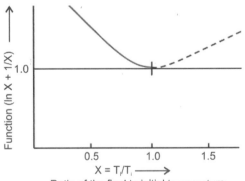

Fig. 5.9 The value of function (In x + 1/x) is positive and
greater than 1 for all positive values of x

As may be observed from the figure, the function has a positive value greater than 1, both for x > 1 (broken line) and X <1 (solid line). In both the cases when heat is lost by the reservoir ($T_f > T_i$, X > 1) or when it is lost by the body ($T_f < T_i$, X < 1), the value of net change in entropy is positive. This shows that in irreversible change of the temperature of a body the entropy of the body plus the reservoir (surroundings) always increases. This is in confirmation to the principle of increase of entropy.

5.10.6   Increase of entropy and degradation of energy

It will be shown here that an increase in entropy is in general accompanied with the degradation of energy.

Suppose there are three heat reservoirs at temperatures T_3, T_2 and T_1 such that $T_3 > T_2 > T_1$. Let a quantity of heat Q be taken from the reservoir at the highest temperature T_3. The maximum work that can be obtained from this quantity Q of heat taken from reservoir at temperature T_3 will be when a reversible Carnot cycle is run between the reservoirs at temperature T_3 and the reservoir at the lowest temperature T_1. It may be recalled that the efficiency of Carnot cycle depends on the difference of the source and sink temperatures, i.e., in the present case on $(T_3 - T_1)$. The maximum work $W_{T_3-T_1}^{max}$

in this case is given by,

$$W_{T_3-T_1}^{max} = Q\left(1 - \frac{T_1}{T_3}\right) \qquad 5.37$$

Now suppose the same quantity of heat Q is transferred from the reservoir at temperature T_3 to the reservoir at lower temperature T_2. This process of heat transfer may be achieved by connecting a metallic rod between the reservoir T_3 and the reservoir T_2 and this process of heat transfer does not violate any law of thermodynamics. The reservoir at lower temperature T_2 now has additional quantity of heat Q. The maximum work that this heat Q can now perform is when a Carnot cycle is run between T_2 and T_1. The maximum work $W_{T_2-T_1}^{max}$ in this case is

$$W_{T_2-T_1}^{max} = Q\left(1 - \frac{T_1}{T_2}\right) \qquad 5.38$$

Since $T_3 > T_2$, $W_{T_3-T_1}^{max} > W_{T_2-T_1}^{max}$. It means that the same quantity of heat Q was capable of producing more work when it was with reservoir at higher temperature T_3 and its capability of performing work has decreased when it is with a reservoir at lower temperature T_2. This is called the energy degradation and is given by:

$$E_{degradation} = W_{T_3-T_1}^{max} - W_{T_2-T_1}^{max} = Q\left(\frac{T_1}{T_2} - \frac{T_1}{T_3}\right) = QT_1\left(\frac{1}{T_2} - \frac{1}{T_3}\right) \qquad 5.39$$

Next let us look to the total change in the entropies of the two reservoirs when heat Q is transferred from reservoir at temperature T_3 to the one at temperature T_2. Reservoir T_3 has lost heat Q at constant temperature T_3 and, therefore, its entropy has decreased by the amount $\Delta S^{T_3} = -\frac{Q}{T_3}$. On the other hand reservoir at temperature T_2 has gained heat Q at constant temperature T_2 and its entropy has increased by the amount $\Delta S^{T_2} = \frac{Q}{T_2}$. Since $T_2 < T_3$, $\Delta S = (\Delta S^{T_2} - \Delta S^{T_3}) = \left(\frac{Q}{T_2} - \frac{Q}{T_3}\right) =$ positive.

Thus the entropy of the total system has increased in the process while the capability of performing work by the same amount of heat has decreased. We see that there is direct correlation between the energy degradation and the increase in entropy,

$$E_{degradation} = QT_1\left(\frac{1}{T_2} - \frac{1}{T_3}\right) = T_1\Delta S \qquad 5.40$$

Thus an overall increase of entropy signifies that energy, while conserved in irreversible processes, becomes less capable of performing work. This statement is known as Kelvin's principle of energy degradation.

5.11   Schematic Representation of Processes in different Two Dimensional Planes

An equilibrium state of a system may be represented by a point and a process by a curve in a two dimensional plane made up of any two state functions of the system. The shape of the curve for the same process to some extent depends on the functional relationship between P, v and T. Carnot cycle for ideal gas as the working substance in four two dimensional planes, (s–T), (s–v), (s–P) and (s–u) are shown respectively in Figs. 5.10 (a), (b), (c), and (d).The processes, isothermal expansion, adiabatic expansion, isothermal compression, and adiabatic compressions are indicated by symbols a, b, c, and d by the side of the curve.

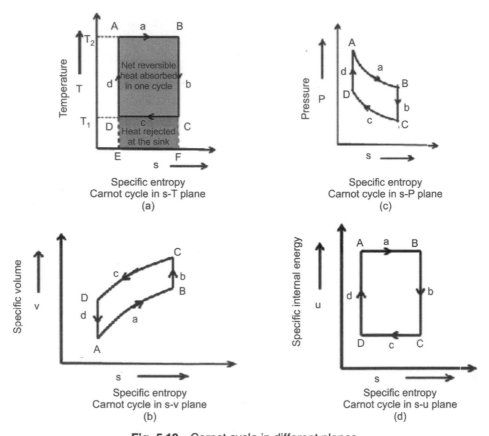

Fig. 5.10   Carnot cycle in different planes

A Carnot cycle draws an amount Q_2 of heat from reservoir at higher temperature T_2, rejects an amount Q_1 at reservoir at lower temperature T_1 and converts the net amount of heat $Q = Q_2 - Q_1$ into work. Figure 5.10 (a) shows the Carnot cycle in s–T plane. The area ABEF shows Q_2, total heat absorbed from reservoir at temperature T_2; area BEFC, Q_1, the heat rejected at the sink at temperature T_1; and area ABCD, the heat Q converted into work. It may be observed that the area enclosed by the Carnot cycle in s–T plane shows the net amount of heat absorbed just like the case of V–P graph for a system the area enclosed shows the work done.

Solved Examples

1. An engine of efficiency of 25% draws 1 kJ from the higher temperature reservoir at 300 K. Calculate the heat rejected by the engine at the sink of temperature 150 K and show that the operation of the engine confirms to Clausius inequality.

   Solution:

   Since the efficiency of the given engine is 25%, it converts 25% of the heat that it draws from the reservoir at higher temperature into work. The heat converted to work is 1000 x 25/100 =250 J. The heat rejected at the sink is 1000-250 =750 J.

   Now we calculate $\Sigma \dfrac{Q}{T}$. There are two reservoirs; the one at higher temperature 300 K loses 1000 J of energy. Since it loses energy Q/T for this is $\dfrac{Q}{T} = \dfrac{1000}{300}$ and for the sink at temperature 150 K $\dfrac{Q}{T} = -\dfrac{750}{150}$. Therefore, $\Sigma \dfrac{Q}{T} = \dfrac{1000}{300} - \dfrac{750}{150} = 3.33 - 5.0 =$ negative or < 0, this shows that the operation of the engine confirms to Clausius inequality.

2. A Carnot engine working between temperatures of 373 K and 273 K draws 1000 J from the high temperature reservoir. Calculate the efficiency of the engine and show that the operation of the engine confirms to Clausius inequality.

   Solution:

   The efficiency of the engine is given by $1 - \dfrac{\text{Temperature of the sink}}{\text{Temperature of the source reservoir}}$. Hence, the efficiency of the engine is $1 - \dfrac{273}{373} = 26.8\%$. It means that 26.8% of 1000 = 268 J of heat is converted into work and the remaining $(1000 - 268 =)$ 732 J of heat is rejected at the sink at temperature 273 K. Now,

   $$\left(\frac{Q}{T}\right)_{\text{source reservoir}} = \frac{1000}{373} = 2.68 \text{ and } \left(\frac{Q}{T}\right)_{\text{sink}} = \frac{-732}{273} = -2.68$$

   Hence, $\Sigma \dfrac{Q}{T} = 0$, which confirms to Clausius inequality

3. 10 kg of ice is heated from 200 K to a temperature 400 K when it is converted in to superheated steam. Calculate the total change of entropy in the whole process. It is given that $C_P^{\text{ice}} = 2.09 \times 10^3$ J kg^{-1} K^{-1}; $c_P^{\text{water}} = 4.18 \times 10^3$ J kg^{-1} K^{-1}; $c_P^{\text{steam}} = c_P^{\text{ice}}$; ℓ_{12} at (273 K) = 3.34×10^5 J kg^{-1}; ℓ_{32} at (373 K) = 22.6×10^5 J kg^{-1}

   Solution:

   The whole process consists of four steps. In the first step, the entropy of ice increases when it is heated from 200 K to its melting point 273 K at constant pressure of one atm. Increase in entropy in first step ΔS_1

$$\Delta S_1 = \int_{T=200K}^{T=273K} \frac{d'Q}{T} = \int_{T=200K}^{T=273K} \frac{C_P dT}{T} = 10 \times c_P^{ice} \boxtimes \int_{T=200K}^{T=273K} \frac{dT}{T} = 10 \times 2.09 \times 10^3 \times \ln\frac{273}{200}$$

or $$\Delta S_1 = 6.50 \; x \; 103 \; J \; K^{-1}$$

In second step ice at 273 K is converted into water at 273 K, and absorbs latent heat of melting at constant temperature $T = 273$ K. Heat absorbed $= 10 \times \ell_{12}$; and hence the rise of entropy in second step ΔS_2

$$\Delta S_2 = \frac{10 \times \ell_{12}}{273} = \frac{10 \times 3.34 \times 10^5}{273} = 12.2 \times 10^3 \; JK^{-1}$$

Third step consists of the heating of water from 273 K to 373 K at constant pressure. The rise in entropy of water in step three ΔS_3 is given by,

$$\Delta S_3 = 10 \times c_P^{water} \int_{273}^{373} \frac{dT}{T} = 10 \times 4.18 \times 10^3 \times \ln\frac{373}{273} = 13.05 \times 10^3 \; JK^{-1}$$

In fourth step water at 373 K is converted into steam at 373 K. Heat absorbed in this phase transition at constant temperature 373 $K = 10 \times \ell_{23}$ and the rise of entropy in this step ΔS_4 is,

$$\Delta S_4 = \frac{10 \times 22.6 \times 10^5}{373} = 60.6 \times 10^3 \; JK^{-1}$$

In the final fifth step steam of temperature 373 K absorbs heat and becomes superheated steam of 400 K. Change in entropy ΔS_5 is

$$\Delta S_5 = 10 \times c_P^{steam} \ln\frac{400}{373} v$$

$$= 10 \times 2.09 \times 10^3 73 \times \text{In } 1.072$$

$$= 1.46 \times 10^3 \; J \; K^{-1}$$

The total increase in entropy in the complete process $\Delta S_{total} = \Delta S_1 + \Delta S_2 + \Delta S_3 + \Delta S_4 + \Delta S_5$

$$\Delta S_{total} = 93.81 \; J \; K^{-1}$$

4. A rheostat of resistance 10 kilo ohm is carrying a current of 10 amp. Cool air of temperature 273 K is continuously blown through the rheostat to keep its temperature constant at 273 K. Calculate the change in the entropy of the rheostat and the surroundings in one minute.

Solution:

An amount of heat $dH = RI^2$ is generated per second when a current of I amp passes through a resistance of R ohm. Heat generated in one minute through the rheostat H is given by,

$$H = 10 \times 10^3 \times 10^2 \times 60 \; J = 6.0 \times 10^7 \; J$$

Now the temperature of the rheostat does not change as the heat produced is taken away by the cool air into the surroundings. Hence the change in the entropy of the rheostat is zero, but the entropy of the surroundings increases by $\Delta S_{surrounding} = \frac{6.0 \times 10^7}{273} = 2.2 \times 10^6 \; JK^{-1}$

Net rise in the entropy of the universe in 1 minute is $2.2 \times 10^6 \; J \; K^{-1}$

5. One kilomole of a monatomic ideal gas at 10×10^5 Nm^{-2} pressure and volume 2 m^3 is taken through a cycle as shown in Fig. S-5.5. The curve AB shows reversible adiabatic expansion, BC isobaric compression and CA isochoric increase of pressure. Calculate; (a) the values of pressure, volume and temperature of the gas at states shown by points A, B and C, (b) heats drawn and rejected by the system, (c) efficiency of the system, (d) change in entropy of the system at each process involved in the cycle.

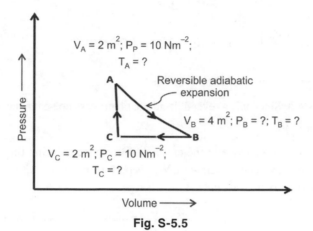

Fig. S-5.5

Solution:

It is given that 1 kilomole of a monatomic ideal gas is taken, hence, the specific thermal capacities of the gas at constant volume and pressure are respectively, $c_V =3/2\ R$, and $c_P = 5/2\ R$. The corresponding thermal capacities

$$C_V = 3/2 \text{x } 8.314 \text{ x } 10^3 \text{x } 1 \text{ kilomole J K}^{-1} = 12.48 \text{ x } 10^3 \text{ J/K}$$

and
$$C_P = 5/2 \text{ x} 8.314 \text{ x} 10^3 \text{ x } 1 \text{ kilomole J K}^{-1} = 20.8 \text{ x} 10^3 \text{ J/K}$$

Also
$$\gamma = \frac{c_P}{c_V} = 1.66$$

(a) Calculations of the parameters at states A, B and C

For the state of the gas at point A:

$$P_A V_A = N R T_A, \text{ hence,}$$

$$T_A = (P_A V_A / N R) = \frac{10 \times 10^5 \times 2}{1 \times 8.314 \times 10^3} = 2.41 \times 10^2 K$$

For the state of the gas at point B; State B is reached via a reversible adiabatic expansion from A, hence

$$P_A V_A^\gamma = P_B V_B^\gamma \text{ and } V_B = 4 \text{ m}^3$$

$$P_B = P_A \left(\left(\frac{V_A}{V_B}\right)^\gamma \right) = 10 \times 10^5 \times \left(\frac{2}{4}\right)^{1.66} = 3.16 \times 10^5 \ N\,m^{-2}$$

Also at point B , $T_B = \dfrac{P_B V_B}{NR} = \dfrac{3.16 \times 10^5 \times 4}{1 \times 8.314 \times 10^3} = 1.52 \times 10^2 \, K$

For the state of the gas at point C,

$$P_C = P_B = 3.16 \times 10^5 \, Nm^2$$

$$V_C = V_A = 2m^3$$

Hence, $T_C = T_B \left(\dfrac{V_B}{V_C} \right) = 1.52 \times 10^2 \times \left(\dfrac{2}{4} \right) = 0.76 \times 10^2 \, K$

(b) Calculation of the heat absorbed or evolved

(1) **Process AB**: Since it is a reversible adiabatic process no heat flow takes place here and so $\Delta Q_{AB} = 0$

Also, there is no change in the entropy of the system during this reversible adiabatic process; $\Delta S_{AB} = 0$. However, work W_{AB} is performed by the gas in adiabatic expansion, at the cost of its internal energy.

$$W_{AB} = C_V (T_A - T_B) = 12.48 \times 10^3 \times (1.52 \times 10^2 - 2.41 \times 10^2)$$

$$= -11.11 \times 10^5 \, J$$

Negative sign in the above equation indicates that the work is performed by the system (gas).

(2) **Process BC:** It is an isobaric compression in which the temperature of the gas drops from 152 K to 76 K. Hence heat is given out in this process,

$$\Delta Q_{BC} = C_P (T_C - T_B) = -20.8 \times 10^3 \, (152 - 76) = -15.8 \times 10^5 \, J$$

That means that 15.8×10^5 J of heat is rejected per cycle in the process represented by BC. To calculate the change in entropy ΔS_{BC}

$$\Delta S_{BC} = \int_{T_B}^{T_C} \frac{C_P dT}{T} = C_P \int_{T_B}^{T_C} \frac{dT}{T} = C_P \ln \frac{T_C}{T_B} = 20.8 \times 10^3 \times \ln \frac{76}{152}$$

$$= -14.42 \times 10^3 \, J \, K^{-1}$$

Negative sign indicates that entropy decreases during B to C.

In process BC work W_{BC} is done on the system in compressing it at constant pressure. This is given by

$$W_{BC} = P_B \times (V_B - V_C) = 3.16 \times 10^3 \, (2.41 \times 10^2 - 1.52 \times 10^2)$$

$$= 6.32 \times 10^5 \, J$$

(3) **Process CA:** This is an isochoric process and the heat flow ΔQ_{CA} in this process is given by,

$$\Delta Q_{CA} = \int_{T_C}^{T_A} C_V dT = C_V \left(T_A - T_C\right) = 12.48 \times 10^3 \left(241 - 76\right)$$

$$= 20.59 \times 10^5 \, J$$

Thus $20.59 \times 10^5 \, J$ of energy is absorbed in the process C to A. Since energy is absorbed the entropy will increase in this process

$$\Delta S_{CA} = \int_{T_C}^{T_A} \frac{C_V dT}{T} = C_V \int_{T_C}^{T_A} \frac{dT}{T} = 12.48 \times 10^3 \, \ln \frac{241}{76}$$

$$= 14.42 \times 10^3 \, J \, K^{-1}$$

Since there is no change in the volume of the gas no mechanical work is performed either by the gas or on the gas in this process.

Efficiency of the cycle: $\eta = \dfrac{(20.59 - 15.80) \times 10^5}{20.59 \times 10^5} \times 100 = 23\%$

Net mechanical work performed in one cycle by the system

$$= 11.11 \times 10^5 - 6.32 \times 10^5 = 4.79 \times 10^5 \, J$$

This is equal to the difference of the heat absorbed and rejected

$$= (20.59 - 15.80) \times 10^5 = 4.79 \times 10^5 \, J$$

Note: As expected the total change in the entropy of the system is zero. In this problem the change of entropy of the surrounding could not be calculated since the temperature of the surroundings is not given.

6. 5 kilomole of an ideal gas for which γ is 1.4 is taken through a reversible Carnot cycle such that the largest volume of the gas becomes 6 times the original volume. If the efficiency of the Carnot engine is 50%, calculate the change in entropy of the gas in the process of isothermal expansion.

Solution:

A Carnot cycle in V–P plane is shown in Fig. S-5.6. We shall denote the system parameters at points A, B, C, and D by the corresponding subscripts. As is indicated in the figure, the maximum value of the volume occurs at point C and it is given that, $V_C = 6V_A$, the number of kilomole of gas $\mathbb{N} = 5$ and $\gamma = 1.4$

The heat Q_{AB} that inflows in the system during the isothermal expansion AB is equal to the work done in this expension and is given by,

$$Q_{AB} = \int_A^B PdV = \int_{V_A}^{V_B} \frac{\mathbb{N}RT_A}{V} dV = \mathbb{N}RT_A \int_{V_A}^{V_B} \frac{dV}{V} = \mathbb{N}RT_A \ln \frac{V_B}{V_A} \quad \text{S-5.1}$$

Change in entropy in isothermal expansion $\Delta S_{AB} = \dfrac{Q_{AB}}{T_A} = \mathbb{N}R \ln \dfrac{V_B}{V_A}$ $\qquad$ S-5.2

$\mathbb{N}$, R and T_A in above equation are respectively the kilomole of the gas, gas constant and the constant temperature of isothermal expansion. V_A and V_B the initial and final volumes respectively at point A and B.

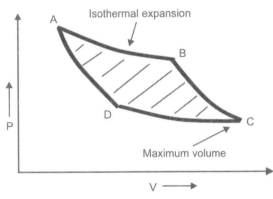

Fig. S-5.6

It is given that the efficiency of the engine is 50%, hence,

$$\eta = 50\,\% = 0.5 = \frac{T_A - T_B}{T_A}\; ; hence \quad T_B = 0.5T_A \qquad \text{S-5.3}$$

Also points B and C lies on the same adiabatic BC, hence

$$T_A V_B{}^{(\gamma-1)} = T_B V_C{}^{(\gamma-1)} \; or \; \left(\frac{V_B}{V_C}\right)^{(\gamma-1)} = \left(\frac{T_B}{T_A}\right) = 0.5 \; or \; \left(\frac{V_B}{V_C}\right) = (0.5)^{\left(\frac{1}{1.4-1}\right)} = 0.177$$

But $V_C = 6.0 \times V_A$, putting this value of V_C in above *Eq.* one gets, $\left(\dfrac{V_B}{V_A}\right) = 1.062$

Substituting the above value of $\left(\dfrac{V_B}{V_C}\right)$ in Eq. S-5.2, one gets:

$$\Delta S_{AB} = \mathbb{N}R \ln \frac{V_B}{V_A} = 5 \times 8.314 \times 10^3 \times \text{In } 1.062 = 2.50 \times 10^3 \; J\,K^{-1}$$

The entropy of the gas increases by $2.50 \times 10^3 \; J\,K^{-1}$ in isothermal expansion.

7. A reversible engine operates between three reservoirs A, B and C respectively at temperatures 400 K, 300 K and 200K. In some complete cycles the engine draws 1200 J of heat from reservoir A and exchange heats Q_B and Q_C with reservoirs B and C respectively. Select from the following

the correct alternative regarding Q_B and Q_C, give reasons for your selection and calculate the change in entropies of the reservoirs and the universe for your choice.

(a) $Q_B = 300\ J$, heat absorbed by reservoir B and $Q_C = 600\ J$ heat absorbed by reservoir C

(b) $Q_B = 600\ J$, heat released by reservoir B and $Q_C = 600\ J$ heat absorbed by reservoir C

(c) $Q_B = 600\ J$, heat released by reservoir B and $Q_C = 600\ J$ heat released by reservoir C

(d) $Q_B = 1200\ J$, heat absorbed by reservoir B and $Q_C = 200\ J$ heat absorbed by reservoir C

(e) $Q_B = 1200\ J$, heat absorbed by reservoir B and $Q_C = 200\ J$ heat given out by reservoir C

Solution:

In the first place we assign algebraic signs to the heat flows. If heat flows out of the reservoir it is taken as positive and when it is received or absorbed the reservoir negative. For example the engine draws 1200 J heat from reservoir A and, therefore, reservoir A loses heat. As such $Q_A = 1200\ J$ is positive. When several reservoirs are involved the heat exchange from reservoirs must follow Clausius inequality, $\sum \frac{Q}{T} \leq 0$. In the present case it is given that the engine is reversible, hence the sign of equality in Clausius inequality equation will hold. The Clausius inequality equation for the present case may be written as,

$$\frac{Q_A}{T_A} + \frac{Q_B}{T_B} + \frac{Q_C}{T_C} = 0$$

Putting $Q_A = +1200\ and\ T_A = 400$ in above equation one gets,

$$\frac{Q_B}{T_B} + \frac{Q_C}{T_C} = -3 \qquad\qquad\qquad \text{S-5.4}$$

Now according to alternative (a) the value of $\frac{Q_B}{T_B} = \frac{-300}{300} = -1\ and\ \frac{Q_C}{T_C} = \frac{-600}{200} = -3$. The

Left hand side (LHS) of Eq. S-5.3 is, therefore, -4. This does not satisfy Eq. S-5.4. Hence alternative (a) is not correct. If same procedure is applied for other alternatives it will be found that the value of the LHS of Eq. S-5.4 comes out to be -1,+5, -6,and -3, respectively for alternatives (b), (c),(d) and (e).It may thus been observed that only alternative (e) satisfies Eq. S-5.4, and, therefore, is the correct alternative. It may also be observed that the total heat absorbed by the engine from reservoir A should be more than the sum of the heats exchanged with the other two reservoirs. It is because a part of the heat absorbed by the engine will be converted into work. According to alternative (e) $Q_B + Q_C = -1200 + 200 = 1000\ J$

This shows that 200 J is converted into work.

It is simple to show that for the correct alternative (e) the change in entropies of reservoirs A, B, and C are respectively, $-3, + 4, -1\ J\ K^{-1}$. Since the process is reversible, there is no change in the entropy of the universe.

8. Two identical metal blocks A and B having thermal capacities C_p at constant pressure are initially at temperatures T_A and T_B, $(T_A > T_B)$. A Carnot engine is operated between the blocks that draws heat from A in infinitesimal reversible cycles and rejects a part of it at block B. The complete operation takes place at constant pressure and without the change of phase of any component. Determine the final temperature of the blocks and the work done by the engine.

Solution:

As the engine will draw heat from block A and rejects a part of the heat to block B, the temperature of block A will fall and that of B will rise. Thus with each cycle the temperature difference between A and B will go on reducing and will ultimately become zero. Let the final temperature of both the blocks be T_f.

Let dQ_A denote the heat lost by reservoir A per cycle when the temperature falls by dT. Then,

$$dQ_A = C_p dT \text{ and the total heat loss by A is given by } Q_A = C_P \int_{T_A}^{T_f} dT$$

or
$$Q_A = C_P(T_f - T_A) = -C_P(T_A - T_f) \qquad \text{S-5.5}$$

Since the blocks are not of infinite thermal capacity, their temperature changes with the transfer of heat from them or to them. Let dS_A denote the change of entropy of block S at each step then,

$$\Delta S_A = \int_{T_A}^{T_f} dS_A = \int_{T_A}^{T_f} \frac{dQ_A}{T} = C_P \int_{T_A}^{T_f} \frac{dT}{T} = C_P \ln\frac{T_f}{T_A} = -C_P \ln\frac{T_A}{T_f} \qquad \text{S-5.6}$$

In a similar way the amount of heat transferred to block B and the change in the entropy of block B may be calculated to,

$$Q_B = C_P(T_F - T_B) \text{ and } \Delta S_B = C_P \ln\frac{T_f}{T_B} \qquad \text{S-5.7}$$

Now the work done by the engine $\quad W = |Q_A| - |Q_B| = C_P(T_A + T_B - 2T_f)$ \qquad S-5.8

Since the engine is a reversible engine the entropy lost by block A is equal to entropy gained by B

So,
$$|\Delta S_A| = |\Delta S_B| \text{ or } \ln\frac{T_A}{T_f} = \ln\frac{T_f}{T_B} \text{ or } T_f = \sqrt{T_A T_B} \qquad \text{S-5.9}$$

Substituting the value of T_f from Eq. S-5.9 in Eq. S-5.8,

$$W = C_P\left(T_A + T_B - 2\sqrt{T_A T_B}\right)$$

The final temperature and the work done are given, respectively, by Eqs. S-5.9 and S-5.8.

9. Calculate the change of entropy of the universe when two equal masses weighing 10 kg each of the same liquid at two different temperatures 400 K and 500 K are mixed and left to establish thermal equilibrium with the surroundings at 300 K. Assume that the specific thermal capacity at constant pressure C_p of the liquid is 3.00×10^3 J/kg/K at 1 atmosphere pressure at which the mixing takes place. There is no change of phase during the process.

Solution:

On mixing the two equal masses of the same liquid the temperature of the mixture will become $T_f = \dfrac{400 + 500}{2} = 450\,K$. The total process of mixing and equilibration of the liquid with its surroundings may be divided into three steps. Step-1: The mass of the liquid at higher temperature loses heat and entropy when its temperature changes from 500 K to 450 K. Step-II: the mass of the liquid at lower temperature gains heat and entropy when its temperature changes from 400 K to 450 K. Step-III: the total mass 20 kg of the liquid at temperature 450 K loses heat and entropy when it comes in thermal equilibrium with surroundings at 300 K. Step wise detailed calculations are done here.

Step-I: Heat lost by 10 kg of liquid, $\Delta Q_1 = m\,c_p\,(500 - 450) = 10 \times 3.0 \times 10^3 \times 50 = 15.0 \times 10^5\,J$

Change in the entropy in this step, $\Delta S_1 = mc_P \displaystyle\int_{500}^{450} \frac{dT}{T} = 10 \times 3.0 \times x10^3\ \ln\dfrac{450}{500}$

$$= -3.16 \times 10^3\,J\,/\,K \qquad\qquad\qquad\qquad\text{S-5.10}$$

The negative sign in the above equation shows that the entropy of this mass of the liquid decreases.

Step-II: Heat gained by 10 kg of liquid, $\Delta Q_2 = m\,c_P\,(450 - 400) = 10 \times 3.0 \times 10^3 \times 50$

$$= 15.0 \times 10^5\,J$$

Change in the entropy in this step, $\Delta S_2 = mc_P \displaystyle\int_{400}^{450} \frac{dT}{T} = 10 \times 3.0 \times 10^3\ \ln\dfrac{450}{400}$

$$= 3.53 \times 10^3\,J/K \qquad\qquad\qquad\qquad\text{S-5.11}$$

Step-III: Heat lost to the surroundings by 20 kg of liquid at 450 K temperature in coming to equilibrium with surroundings at temperature 300 K, $\Delta Q_3 = 20 \times 3.0 \times 10^3 \times (450 - 300)$

$$= 90 \times 10^5\,J$$

Heat ΔQ_3 is gained by the surroundings at constant temperature of 300 K, therefore, the rise in the entropy of the surroundings, $\Delta S_{surr} = \dfrac{\Delta Q_3}{300} = \dfrac{90 \times 10^5}{300} = 30 \times 10^3\,J\,/\,K$

Loss in the entropy of the liquid $\Delta S_{total\ liq} = 20 \times 3.0 \times 10^3 \times \ln\dfrac{300}{450} = -24.33 \times 10^3\,J/K$

Total change in the entropy of the universe (liquid + surroundings)

$$= (-3.16 + 3.53 + 30.0 - 24.33) \times 10^3\,J/K$$

$$= 6.04 \times 10^3\,J/K$$

10. Show that the ratio $\left|\dfrac{Q_2}{Q_1}\right|$ must be same for all Carnot cycles operating between the same pair of reservoirs.

Solution:

Figure S-5.10 (a) shows two Carnot engines, the one on the left draws an amount of heat Q_2 per cycle from reservoir T_2 and rejects per cycle heat Q_1 at reservoir T_1. The engine on the right draws heat Q_2' from reservoir at temperature T_2 and rejects heat Q_1' in reservoir T_1 in each cycle.

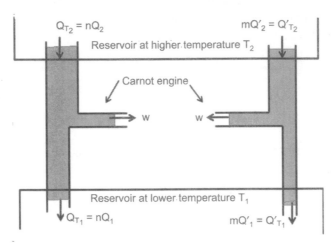

$Q_{T_2} = nQ_2$ $mQ_2' = Q_{T_2}'$

Reservoir at higher temperature T_2

Carnot engine

W   W

Reservoir at lower temperature T_1

$Q_{T_1} = nQ_1$ $mQ_1' = Q_{T_1}'$

Fig. S-5.10(a)

Let the engine on the left in n complete cycles absorbs total heat, $Q_{T_2} = (nQ_1)$, rejects the total amount of heat $Q_{T_1} = (nQ_1)$ at reservoir of temperature T_1 and produces total work W. The other engine on the left, absorbs, say, a total heat $Q_{T_2}' = (mQ_2')$ and rejects a total heat $Q_{T_1}' = (mQ_1')$ in m complete cycles to produce the same work W.

In general, the magnitudes of Q_{T_2} and Q_{T_2}' may be different so also the magnitudes of Q_{T_1} and Q_{T_1}'. Let us assume that $Q_{T_2} > Q_{T_2}'$. Since both the engines accumulate the same amount of work W, then $Q_{T_1} > Q_{T_1}'$. Since Carnot engine is reversible at each stage, the magnitudes of heats absorbed or released do not change if the engine is replaced by a refrigerator. We replace the engine on the right by a corresponding refrigerator and couple the work port of the refrigerator to the work port of engine on the right, as is shown in Fig. S-5.10 (b). It can be seen in Fig. S-5.10 (b) that the Carnot engine on the right will produce exactly the same amount of work as is required to run the refrigerator on right. The combination of the refrigerator with the engine will make a self-acting device un-added from outside which will remove heat $\left(Q_{T_1} - Q_{T_1}'\right) =$ $\left(Q_{T_2} - Q_{T_2}'\right)$ from the reservoir at lower temperature T_1 and will transport it to the reservoir at higher temperature T_2. This is shown as a fine pipe of heat on the extreme left in Fig. S-5.10 (b).This will contradict the Clausius statement of the second law. Therefore, for the validity of the Clausius statement, it is necessary that

$$\left(Q_{T_2} = Q_{T_2}'\right) \text{ and } \left(Q_{T_1} - Q_{T_1}'\right)$$

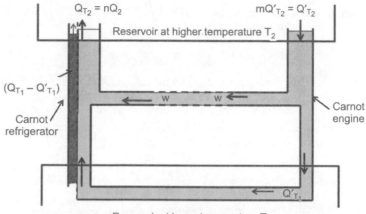

Fig. S-5.10(b)

This means that, $\qquad Q_{T_2} = nQ_2 = mQ_2'$ and $Q_{T_1} = nQ_1 = mQ_1'$

and $\qquad\qquad\qquad \dfrac{Q_{T_2}}{Q_{T_1}} = \dfrac{Q_2}{Q_1} = \dfrac{Q_2'}{Q_1'}$

This proves that the ratio of heat absorbed to heat rejected per cycle for all Carnot engines working between the same reservoirs is the same.

11. The temperature of a house is to be kept at 25°C when the temperature of the outside drops to 0-10°C. The house is estimated to be losing heat at the rate of 120×10^6 J/h . Calculate the minimum power required to run a heat pump for maintaining the temperature of the house and calculate the change in the entropy of the universe per second.

Solution:

A heat pump is just like a refrigerator that draws heat from the outside atmosphere (at lower temperature) and dumps this heat plus a part of the work input to the refrigerator into the house. To run such a refrigerator some work input is required. The work-input is a minimum when the refrigerator works with a reversible Carnot cycle.

It is given that the inside temperature of the house T_{in} is 25°C = 298 K and the outside temperature T_{out} is –10°C = 263 K. Further, it is given that the estimated heat loss H_{loss} of the house is 120 $\times$ 10^6 J/h. we convert this heat loss in joule per second, $H_{loss} = \dfrac{120 x 10^6}{3600} = 33.3 \text{ kJs}^{-1}$

= 33.3 kW. The required refrigerator should pump 33.3 kW of power to the interior of the house.

The required Carnot refrigerator should work between T_{in} and T_{out} and will have the coefficient of performance $COP = \dfrac{1}{1 - \dfrac{T_{out}}{T_{in}}} = \dfrac{1}{1 - \dfrac{263}{298}} = 8.51$.

If Q is the heat drawn from outside, then $COP = \dfrac{H_{loss} - Q}{Q} = 8.51$, hence

$$Q = \frac{H_{loss}}{9.51} = \frac{33.3 \ kW}{9.51} = 3.50 \ kW$$

Therefore, the minimum work required to run the refrigerator = 33.3 – 3.5 = 29.8 kW.

It may be seen that the refrigerator when supplied a work of 29.8 kJ per second draws 3.5 kJ of heat per second from the outside (at –10°C) and deposits a total of 33.3 kJ (29.8+3.5) of heat per second to compensate for the loss of 33.3 kJ per second.

Change in the entropy of the universe $= \dfrac{33.3 \, kW}{298 \, K} - \dfrac{3.5 \, kW}{263 \, K} = 98.4 \, WK^{-1} = 98.4 \, Js^{-1} \, K^{-1}$

The entropy of the universe is increasing by 98.4 J/K per second.

Problems

1. 10 kilomole of a monatomic ideal gas at 241 K occupies a volume of 2 m³. It undergoes adiabatic expansion so that the volume becomes twice the original volume. Calculate the initial pressure, final pressure, final temperature and the change in entropy of the gas. What would have been the change in the entropy if the expansion was isothermal at constant pressure and temperature of 241 K? Also draw curves for the process in V-S and T-S planes.

2. Calculate the change in the specific entropy of water in the form of ice when it is heated at constant pressure of one atmosphere from 200 K to 400 K (superheated steam). Following data may be used for these calculations.
$c_P^{ice} = 2.09 \times 10^3$ J kg⁻¹ K⁻¹; $c_P^{water} = 4.18 \times 10^3$ J kg⁻¹ K⁻¹; $c_P^{steam} = c_P^{ice}$; ℓ_{12} at (273 K) = 3.34 × 10⁵ Jkg^{-1}; ℓ_{32} at (373 K) = 22.6 × 10⁵ Jkg⁻¹

3. 10 kg of water at 100°C is mixed with an equal amount of water at 0°C in an adiabatic container. Neglecting the small variation of specific heat capacities, calculate the change in entropy of water. Use the data provided in problem 2.

4. 10 kg of some substance, the specific thermal capacity of which is given as $c_P = 1.2 + 0.03T$, is heated at constant pressure from 300 K to 500 K. Compute the change in entropy of the substance.

5. Equal amounts of a liquid at two different temperatures of A and B Kelvin are mixed together in an adiabatic container. Calculate the change in the specific entropy of the universe. Given that the specific thermal capacity at constant pressure C_P of the liquid remains constant during the mixing.

6. Three engines, A, B and C draw the same amount of heat 420 J from a reservoir at temperature 600 K. They, however, reject respectively, 110 J, 210 J, and 310 J of heat at a reservoir at temperature 300 K. Using Clausius inequality, determine which of the engine is reversible and which is not possible.

7. 50 kg of water at 20°C is kept in a perfectly insulating container. A block of iron weighing 100 kg and at temperature 100°C is dropped in water and the system is kept insulated from

the surroundings. The specific thermal capacities at constant pressure c_p of water and iron are respectively, 4.18 kJ K^{-1} and 0.45 kJ K^{-1}. Calculate the change in entropy of the water plus iron system when they reach thermal equilibrium.

8. An inventor claims that he has developed an engine that draws 1000 kJ energy from a reservoir at temperature 500 K and rejects 475 kJ at the sink at 300 K. Will you believe the inventor? Give reasons for your answer.

9. An adiabatic air compressor converts 2 kg of air of initial pressure 1 bar and temperature 310 K into air of pressure 7 bar and temperature 560 K, per second. Calculate the change in the specific entropy of air per second, given that the specific thermal capacity at constant pressure for air is 1 kJ kg^{-1} K^{-1} and the gas constant for air is 0.287 kJ kg^{-1} K^{-1}. No phase change occurs and c_p may be assumed to remain constant during the process. Comment about the reversibility of the process.

10. One kilomole of a monatomic ideal gas undergoes the process AB represented in Fig. P-5.10. In this process the entropy of the gas increases by 14.406 x 10³ JK^{-1}. Identify the process and find the ratio of the final and initial volumes of the gas.

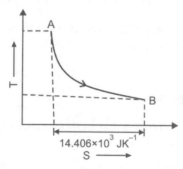

Fig. P-5.10

Short Answer Questions

1. Show that Clausius statement of the second law is equivalent to Planck-Kelvin statement.
2. State and explain the physical significance of Clausius inequality.
3. Prove that no refrigerator can have coefficient of performance better than that of a Carnot refrigerator working between the same pair of reservoirs.
4. Define entropy and obtain an expression for the change of entropy in isothermal expansion of an ideal gas.
5. Change in entropy of a system between two given states does not depend on whether the final state is reached by a reversible or an irreversible process. Explain.
6. Prove that the ratio of the heat drawn to the heat rejected by different Carnot engines between the same pair of reservoirs is same.
7. Every natural process robs the universe of the opportunity of producing work. Explain.

8. A reversible engine draws Q_1 and Q_3 amounts of heat respectively from reservoirs at temperatures T_1 and T_3 and rejects Q_2 and Q_4 amounts of heat at reservoirs respectively at temperatures T_2 and T_4. Show that $\dfrac{Q_1}{T_1} + \dfrac{Q_3}{T_3} = \dfrac{Q_2}{T_2} + \dfrac{Q_4}{T_4}$

9. State the principle of increase of entropy.

10. Energy and entropy are both state parameters but their behavior are very different. Explain.

11. Calculate the change in entropy when exactly 1 mol of solid iodine, I_2, at a temperature of 360 K is heated at constant pressure to produce liquid iodine at a temperature of 410 K. The constant pressure molar heat capacity of solid iodine is 54.44 J K^{-1} mol^{-1} and of liquid iodine is 80.67 J K^{-1} mol^{-1}. The melting temperature of iodine is 387 K, and the molar enthalpy of fusion of iodine is 7.87 kJ mol^{-1}.

12. At low temperatures, the heat capacity of solid chlorine, Cl_2, follows the Debye T^3 law, $C_{p,m} = a\,T^3$, with $a = 1.24 \times 10^{-3}$ J K^{-4} mol^{-1}. Estimate the molar entropy of solid chlorine at 10 K. [Hint: At low temperatures for substances that obey Debye law the molar entropy is given by $S^{molar}(T) = \frac{1}{3} c_P^m(T)$, where $c_P^m(T)$ the molar thermal capacity at constant pressure.]

13. Show that the change in the entropy of a system in an isobaric process is always more than that in a corresponding isothermal process.

Long Answer Questions

1. Show that any reversible cyclic process may be represented by a cascade of Carnot cycles, and hence define the change in entropy of a system.

2. State and proof Clausius inequality.

3. State and proof Carnot theorem.

4. Derive expressions for the change in entropy of a given mass of an ideal gas for (a) Isothermal expansion at constant pressure (b) Isochoric compression (c) free expansion.

5. Taking a suitable example show that the entropy of a system increases in an irreversible process.

6. Write a detailed note on entropy.

Multiple Choice Questions

Note: Some of the following questions have more than one correct alternative. All correct alternatives must be ticked in such cases for getting full marks.

1. The entropy of a non-isolated system in any natural process
   (a) must increase
   (b) must decrease
   (c) remains unchanged
   (d) may increase, decrease or remain unchanged.

2. Tick all the correct statements.
   (a) Increasing the temperature difference between the hot and the cold reservoirs increases the efficiency of the Carnot engine.

(b) Decreasing the temperature difference between the hot and the cold reservoirs increases the coefficient of performance of a refrigerator.

(c) Entropy is conserved in natural processes

(d) Entropy is generated in natural processes.

3. A reversible engine draws equal amounts of heat Q from there different reservoirs at identical temperatures T_0 and rejects heat of the same amount Q in a reservoir at temperature T_4. The temperature T_4 is,

    (a) $=3\,T_0$ (b) $=T_0$

    (c) $=T_0/3$ (d) $< T_0/3$

4. A reversible engine draws equal amounts of heat Q from there different reservoirs at identical temperatures T_0 and rejects heat of the same amount Q in a reservoir at temperature T_4. The efficiency of the engine is,

    (a) 66.6 %   (b) 50 %

    (c) 33.3 %   (d) 16.6 %

5. By increasing the temperature difference between the cold and the hot reservoir by a factor of four and keeping the temperature of the hot reservoir constant, the efficiency of Carnot engine may be

    (a) Increased by a factor of 2   (b) increased by a factor of 4

    (c) decreased by a factor of 2   (d) decreased by a factor of 4.

6. If ΔS_{system} and $\Delta S_{surroundings}$ denote, respectively, the change in the entropies of the system and the surroundings, then for a spontaneous reaction;

    (a) $\Delta S_{system} - \Delta S_{surroundings} < 0$ (b) $\Delta S_{system} \times \Delta S_{surroundings} = 1$

    (c) $\Delta S_{system} + \Delta S_{surroundings} > 0$ (d) $\Delta S_{system} + \Delta S_{surroundings} < 0$

7. The entropy of a system has the largest value when the system is

    (a) undergoing isothermal change   (b) in adiabatic state

    (c) undergoing isochoric change   (d) in equilibrium.

8. Two isolated systems in their initial states have entropies S_1^i and S_2^i. The system were then mixed in an adiabatic container so that their final entropies become S_1^f and S_2^f. Which of the following statement (s) is (are) correct?

    (a) $\left(S_1^i + S_2^i\right) < \left(S_1^f + S_2^f\right)$ (b) $\left(S_1^i + S_2^i\right) > \left(S_1^f + S_2^f\right)$

    (c) $\left(S_1^i + S_2^i\right) = \left(S_1^f + S_2^f\right)$ (d) $\left(S_1^i\right) = \left(S_1^f\right) and \left(S_2^i\right) = \left(S_2^f\right)$

9. $\oint \dfrac{d'Q}{T} = 0$ for

    (a) any closed cycle   (b) only for adiabatic closed cycle

    (c) only for isothermal closed cycle   (d) not for any closed cycle.

10. A system is taken from a given initial state to a final state through an isothermal process in which the volume changes by a factor of X. Next the same system is also taken from the same initial state via an isobaric process in which the volume change is again by the factor X. If S_{ther} and S_{baric} respectively, represents the change in the entropy of the system in the isothermal and isobaric processes then;

(a) $S_{baric} = S_{ther}$

(b) $S_{baric} > S_{ther}$

(c) $S_{baric} < S_{ther}$

(d) $S_{baric} \leq S_{ther}$

11. Two bodies A and B are enclosed in a adiabatic box which is kept in some surroundings. Some natural interactions take place between A and B and the system contained in the box reaches equilibrium. If in the following relations, S denotes the entropy of some object (specified by the superscript) in the given state (indicated by the subscript). Tick all those alternatives that always hold.

(a) $S_A^{final} > S_A^{initial}$

(b) $S_B^{final} > S_B^{initial}$

(c) $\left(S_A^{final} - S_A^{initial} \right) + \left(S_B^{final} - S_B^{initial} \right) > 0$

(d) $S_{surrounding}^{initial} = S_{surrounding}^{final}$

12. For a given substance the specific latent heat for a phase transition at absolute temperature T is given by ℓ. The change in the specific entropy of the substance in the phase change is given by;

(a) $T\ell$;

(b) ℓ/T;

(c) T/ℓ

(d) $T + \ell$

Answers to Numerical and Multiple Choice Questions

Answers to problems

1. 10×10^5 N m^{-2}; 3.16×10^5 N m^{-2}; 152 K; Zero, -398 J K^{-1}
2. 9.4×10^3 J kg^{-1} K^{-1}
3. 1.01×10^3 J K^{-1}
4. 66.12 J K^{-1}
5. $2c_P \ln\left(\dfrac{A + B}{2\sqrt{AB}} \right)$
6. Engine A is not possible and B is reversible. Engine C is irreversible and possible
7. 1.24×10^3 J K^{-1}
8. No.
9. 33.0 J kg^{-1} K^{-1} s^{-1}, irreversible
10. Isobaric expansion, 1:2
11. 28.9 J K^{-1} mol^{-1}

Answers to multiple choice questions

1. (d)	2. (a), (b), (d)	3. (c)	4. (a)
5. (b)	6. (c)	7. (d)	8. (c)
9. (a)	10. (b)	11. (c) and (d)	12. (b)

Revision

1. Clausius suggested that *every system has an additional system variable that determines the direction of spontaneous transformation of the system.* He called this property of a system or the system variable as *'Entropy'*.

Second Law of thermodynamics in terms of Entropy: *"The processes in which the entropy of an isolated system decreases do not take place".*

2. The entropy of an isolated system has the maximum value when it is in equilibrium state.
3. *Any reversible cyclic process* may be represented by a cascade of Carnot cycles.
4. Change in entropy $dS \equiv \dfrac{d'Q_r}{T}$, where d'Q is the path dependent small amount of heat flow at constant temperature T.
5. *In a reversible process there is no change in the entropy of the universe. In a reversible process if system gains (or loses) some entropy, exactly same amount of entropy is lost (or gained) by the environment.*
6. In an irreversible process the entropy of the universe increases. This leads to what is called the principle of the increase of entropy of the universe in all irreversible processes.
7. All natural or spontaneous processes are irreversible
8. Irreversible processes in nature in due course of time will remove all temperature differences in the universe and the universe will become a single heat reservoir at a single temperature incapable of producing work. This will lead to what is called the *thermal death* of the universe.
9. Clausius statement of the second law: *"No process is possible whose sole result is a heat flow out of a system at a given temperature and a heat flow of the same magnitude into a second system at a higher temperature.*
10. Kelvin–Planck statement of the second law: *"No process is possible whose sole result is the abstraction of heat from a single reservoir and the performance of an equivalent amount of work".*
11. It is possible to show that Clausius and Planck-Kelvin statements of the second law are equivalent.
12. Perpetual motion machine of second kind: The perpetual motion machine of second kind is an imaginary machine that may draw heat from a single reservoir and convert it totally in work. Such a machine will not violate the first law, since neither any energy is created nor is lost in the operation of the machine. However, it will violate the second law of thermodynamics and hence it is not possible to have it.
13. The efficiency of a Carnot engine and the coefficient of performance (COP) of a Carnot refrigerator puts upper limit respectively to the efficiency of any engine and COP of any refrigerator.
14. The Clausius inequality: The Clausius inequality is a relationship between the ratios of the heat exchanged to the temperature of an arbitrary number of heat reservoirs when some system consisting of some working substance that interacts with the reservoirs and absorbs or rejects heat to reservoirs, is taken through some arbitrary cyclic process. The process should be cyclic but not necessarily reversible. Since the nature and properties of the working substance do not enter in the final result, the relation is independent of the working substance. It may be put in the following algebraic form;

$$\Sigma \frac{Q}{T} \le 0 \ \text{ or } \ \oint \frac{d'Q}{T} \le 0$$

Here summation or integration is to be carried out over all the reservoirs involved in heat transfer. Q is taken positive if heat flows out of the reservoir and negative if it flows into the reservoir.

15. Change of entropy in reversible adiabatic process is zero.

16. Change of entropy in a reversible processes where heat flow is accompanied with the change in temperature:

(a) $\Delta S_i^f = C_V \int_i^f \dfrac{dT}{T} = C_V \ln \dfrac{T_f}{T_i} = 2.303 \, C_V \, \log \dfrac{T_f}{T_i}$

(b) $\Delta S_i^f = C_P \int_i^f \dfrac{dT}{T} = C_P \ln \dfrac{T_f}{T_i} = 2.303 \, C_P \, \log \dfrac{T_f}{T_i}$

17. Entropy change in reversible isothermal expansion of an ideal gas:

$$\Delta S = \dfrac{Q_r}{T} = N_m R \ln \dfrac{V_f}{V_i} = 2.303 \; N_m R \log \dfrac{V_f}{V_i}$$

18. Change in entropy in reversible change of state of the substance: The change in the specific entropy when a given substance undergo phase change from physical state X to the physical state y at constant temperature T.

$$s_y - s_x = \dfrac{\ell_{xy}}{T}$$

19. Change in entropy of an ideal gas in free expansion:

$$\Delta S_{irreversible} \, free \; expansion = 2.303 \; N_m R \log \dfrac{V_f}{V_i} \; \text{J K}^{-1}$$

20. Change in the entropy of a body in irreversible rise of temperature at constant pressure and constant volume:

$$\Delta S_{irre}^{body} = C_P \left(or \, C_V \right) \ln \dfrac{T_f}{T_i} = 2.303 C_P \left(or \, C_V \right) \log \dfrac{T_f}{T_i}$$

An equilibrium state of a system may be represented by a point on a two dimensional plane made up of any two system parameters and any thermodynamic process in this two dimensional plane may be represented by a curve. As such it is possible to draw a schematic figure of the various processes that a given system has undergone in a two dimensional plane by joining different curves each representing a process. The shape of the curves representing different processes depends on the selection of the state parameters for defining the two dimensional plane.

21. An overall increase of entropy signifies that energy, while conserved in irreversible processes, becomes less capable of performing work. This statement is known as Kelvin's principle of energy degradation.

Tds Equations and their Applications

6.0   Introduction

The first and the second laws of thermodynamics may be written, respectively, in the following differential forms;

$$d'Q = dU + d'W \qquad\qquad 6.1$$

and
$$d'Q_r = T\,dS \qquad\qquad 6.2$$

In Eq. 6.1, that represents the first law of thermodynamics, $d'Q$ represents the path dependent heat flow in the system, dU the increase in the internal energy, and $d'W$ the path dependent work done by the system. Equation 6.2, which is the mathematical representation of the second law, is *applicable only for reversible processes* and tells that (for reversible processes) heat flow to the system at constant temperature $d'Q_r$, which is path dependent, is equal to the multiplication of the constant temperature T and the change in the entropy of the system dS. Substituting the value of $d'Q$ from Eq. 6.2 in Eq. 6.1, one gets for reversible processes,

$$T\,dS = dU + d'W \qquad\qquad 6.3$$

However, for reversible processes $d'W = P\,dV$ and so Eq. 6.3 reduces to

$$T\,dS = dU + P\,dV \qquad\qquad 6.4$$

Equation 6.4 that contains both the first and the second laws is in principle applicable for reversible processes. However, a closer look to this equation reveals that no quantity that depends on the process or the path is involved in the equation. The equation relates the small variations of state parameters, S, U and V. As a matter of fact Eq. 6.4 gives a relation between the state functions of two nearby equilibrium states of the system. Since state functions do not depend on path, Eq. 6.4 is applicable for any two nearby equilibrium states of a system, irrespective of whether the other state is reached from the first state by an irreversible or a reversible process. Sometimes it may also happen that the final state with entropy $(S + dS)$, internal energy $(U + dU)$ and volume $(V + dV)$ may not be reached from the initial state of entropy S, internal energy U and volume V, by any process and vice-versa. However, Eq. 6.4 will still hold as it gives relationship between the state parameters of two nearby equilibrium states of a system.

Let us look more closely to Eq. 6.1. This Eq. gives a relation between the heat flow, change in internal energy and the path dependent work done. The relationship holds good for both the reversible and irreversible processes. If the process is reversible then $d'W$ may be written as $(P\,dV)$, but if the process is irreversible then $d'W$ cannot be written as $(P\,dV)$. As such only the value of $d'W$ changes with the type of the process, it may even be zero for an isochoric process.

It may, therefore, be said that both Eq. 6.1 and Eq. 6.4 are of very general nature, the former can be applied to any process, reversible and irreversible and the later to any two nearby equilibrium states of a system, irrespective of the process through which one of the state is reached from the other or even in those cases when states are not reachable from each other.

6.1 The Tds Equations

A state of a thermodynamic system is characterized by its state parameters U, T, V, P, S etc. However, only three state parameters are required to completely define the state. Moreover, out of the chosen three parameters, any two may be taken as independent variables and the third is then related to the other two by the equation of state. Since pressure, volume and temperature are the most frequently measured state functions, any pair of them is generally taken as independent variables and the other state parameters may then be written as functions of the two parameters chosen as independent. This leads to three possibilities: (i) T and P independent, (ii) T and v independent, and (iii) P and v independent. Some useful relations involving partial derivatives of P, T, V, u, h, etc., based on these possibilities have already been obtained in chapter-4. Similar procedure may also be followed for the state parameter entropy S introduced in chapter-5.

(i) Pressure P and specific volume v as independent variables

As an example let us take pressure P and specific volume v as independent variables. Now, specific entropy s may be written as a function of P and v so that,

$$s = s\,(P, v) \text{ and } ds = \left(\frac{\partial s}{\partial v}\right)_P dv + \left(\frac{\partial s}{\partial P}\right)_v dP \qquad 6.5$$

But from Eq. 6.4,

$$ds = \frac{1}{T}\left[du + Pdv\right] \qquad 6.6$$

Also,

$$u = u\,(P, v) \text{ and } du = \left(\frac{\partial u}{\partial v}\right)_P dv + \left(\frac{\partial u}{\partial P}\right)_v dP \qquad 6.7$$

Substituting the value of du from Eq. 6.7 in Eq. 6.6, one gets;

$$ds = \frac{1}{T}\left[\left(\frac{\partial u}{\partial v}\right)_P dv + \left(\frac{\partial u}{\partial P}\right)_v dP + Pdv\right] = \frac{1}{T}\left[\left(\frac{\partial u}{\partial P}\right)_v dP + \left(\left(\frac{\partial u}{\partial v}\right)_P + P\right)dv\right] \qquad 6.8$$

Equating Eq. 6.5 and 6.8,

$$\left(\frac{\partial s}{\partial P}\right)_v = \frac{1}{T}\left(\frac{\partial u}{\partial P}\right)_v \qquad 6.9$$

and

$$\left(\frac{\partial s}{\partial v}\right)_P = \frac{1}{T}\left[\left(\left(\frac{\partial u}{\partial v}\right)_P + P\right)\right] \qquad 6.10$$

Now from Eq. 4.47 (chapter-4)

$$\left(\frac{\partial u}{\partial P}\right)_v = c_v\left(\frac{\partial T}{\partial P}\right)_v = c_v\left(\frac{\partial T}{\partial v}\right)_P\left(\frac{\partial v}{\partial P}\right)_T = c_v\frac{\kappa}{\beta} \qquad 6.11$$

Equations 3.35 and 3.37 which define $|\kappa| \equiv \dfrac{1}{v}\left(\dfrac{\partial v}{\partial P}\right)_T$ and $\beta \equiv \dfrac{\left(\dfrac{\partial v}{\partial T}\right)_P}{v}$ have been used in driving

Eq. 6.11.

From Eqs. 6.11 and 6. 9 , it follows that;

$$\left(\frac{\partial s}{\partial P}\right)_v = \frac{1}{T}\left(\frac{\partial u}{\partial P}\right)_v = \frac{c_v}{T}\frac{\kappa}{\beta} \qquad 6.12$$

Also, from Eq. 6.10, $\qquad \left(\frac{\partial s}{\partial v}\right)_P = \frac{1}{T}\left[\left(\left(\frac{\partial u}{\partial v}\right)_P + P\right)\right]$

But, $\qquad\qquad u = h - Pv \quad$ and $\quad \left(\frac{\partial u}{\partial v}\right)_P = \left[\left(\frac{\partial h}{\partial v}\right)_P - P\right] \qquad 6.13$

Putting the value of $\left(\frac{\partial u}{\partial v}\right)_P$ from Eq. 6.13 in Eq. 6.10, one gets,

$$\left(\frac{\partial s}{\partial v}\right)_P = \frac{1}{T}\left[\left(\left(\frac{\partial h}{\partial v}\right)_P\right)\right] = \frac{1}{T}\left[\left(\frac{\partial h}{\partial T}\right)_P\left(\frac{\partial T}{\partial v}\right)_P\right]$$

But from Eq. 4.52, $\left(\frac{\partial h}{\partial T}\right)_P = c_P$ and, therefore,

$$\left(\frac{\partial s}{\partial v}\right)_P = \frac{c_P}{T}\left(\frac{\partial T}{\partial v}\right)_P = \frac{c_P}{Tv\beta} \qquad 6.14$$

On substituting the values of $\left(\frac{\partial s}{\partial v}\right)_P$ and $\left(\frac{\partial s}{\partial P}\right)_v$ respectively, from Eq. 6.14 and Eq. 6.12 in Eq.

6.5, one gets:

$$Tds = c_P\left(\frac{\partial T}{\partial v}\right)_P dv + c_v\left(\frac{\partial T}{\partial P}\right)_v dP \qquad 6.15$$

(ii) Temperature *T* and molar volume *v* independent variables

If T and v are taken as independent variables then specific entropy s may be written as a function of T and v,

$$s = s\,(T,\,v),$$

and $\qquad\qquad\qquad ds = \left(\frac{\partial s}{\partial T}\right)_v dT + \left(\frac{\partial s}{\partial v}\right)_T dv \qquad 6.16$

Again writing the specific internal energy u as a function of T and v, it is easy to show that

$$\left(\frac{\partial s}{\partial T}\right)_v = \frac{1}{T}\left(\frac{\partial u}{\partial T}\right)_v = \frac{c_v}{T} \qquad 6.17$$

and
$$\left(\frac{\partial s}{\partial v}\right)_T = \frac{1}{T}\left[\left(\frac{\partial u}{\partial v}\right)_T + P\right] \qquad 6.18$$

Differentiating Eq. 6.17 partially with respect to ∂v and Eq. 6.18 with respect to ∂T one may get,

$$\frac{\partial^2 s}{\partial v \partial T} = \frac{1}{T}\frac{\partial^2 u}{\partial v \partial T} \qquad 6.19$$

and
$$\frac{\partial^2 s}{\partial T \partial v} = \frac{1}{T}\left[\frac{\partial^2 u}{\partial T \partial v} + \left(\frac{\partial P}{\partial T}\right)_T\right] - \frac{1}{T^2}\left[\left(\frac{\partial u}{\partial v}\right)_T + P\right] \qquad 6.20$$

Since $\dfrac{\partial^2 s}{\partial v \partial T} = \dfrac{\partial^2 s}{\partial T \partial v}$, equating Eq. 6.19 and 6.20, and simplifying one gets;

$$\left(\frac{\partial u}{\partial v}\right)_T = T\left(\frac{\partial P}{\partial T}\right)_v - P = \frac{T\beta}{\kappa} - P \qquad 6.21$$

Also,
$$u = u\,(T,\,v) \text{ and } du = \left(\frac{\partial u}{\partial v}\right)_T dv + \left(\frac{\partial u}{\partial T}\right)_v dT \qquad 6.22$$

But $\left(\dfrac{\partial u}{\partial T}\right)_v = c_v$ and from Eq. 6.21 $\left(\dfrac{\partial u}{\partial v}\right)_T = T\left(\dfrac{\partial P}{\partial T}\right)_v - P$; putting these values in Eq. 6.22, one

gets:

$$du = c_v dT + \left[T\left(\frac{\partial P}{\partial T}\right)_v - P\right]dv \qquad 6.23$$

In Eq. 4.32 of chapter-4, it was derived that,

$$c_P - c_v = \left[\left(\frac{\partial u}{\partial v}\right)_T + P\right]\left(\frac{\partial v}{\partial T}\right)_P$$

Substituting the value of $\left(\dfrac{\partial u}{\partial v}\right)_T$ from Eq. 6.21 in the above equation one gets,

$$c_P - c_v = T\left(\frac{\partial v}{\partial T}\right)_P\left(\frac{\partial P}{\partial T}\right)_v = \frac{Tv\beta^2}{\kappa} \qquad 6.24\text{ (a)}$$

$$= -T\frac{\left[\left(\frac{\partial P}{\partial T}\right)_v\right]^2}{\left(\frac{\partial P}{\partial v}\right)_T} \qquad 6.24\text{ (b)}$$

Again from Eq. 6.17,

$$\left(\frac{\partial s}{\partial T}\right)_v = \frac{1}{T}\left(\frac{\partial u}{\partial T}\right)_v = \frac{c_v}{T} \qquad 6.25$$

and from Eq. 6.18,

$$\left(\frac{\partial s}{\partial v}\right)_T = \frac{1}{T}\left[\left(\frac{\partial u}{\partial v}\right)_T + P\right] = \frac{1}{T}\left[\left\{T\left(\frac{dP}{dT}\right)_v - P\right\} + P\right] = \left(\frac{dP}{dT}\right)_v \qquad 6.26$$

When values of $\left(\dfrac{\partial s}{\partial T}\right)_v$ and $\left(\dfrac{\partial s}{\partial v}\right)_T$, respectively, from Eq. 6.25 and 6.26 are substituted in Eq. 6.16, one gets,

$$Tds = c_v dT + T\left(\frac{dP}{dT}\right)_v dv \qquad 6.27$$

(iii) Pressure P and temperature T as independent variables

Finally, following relations may be obtained when P and T are taken as independent variables.

We have $\quad ds = \dfrac{1}{T}\left[du + Pdv\right] = \dfrac{1}{T}\left[d(h - Pv) + Pdv\right] = \dfrac{1}{T}\left[dh - vdP\right]$ \qquad 6.28

Also, $\qquad\qquad\qquad h = h(P,T)$ and $dh = \left(\dfrac{\partial h}{\partial P}\right)_T dP + \left(\dfrac{\partial h}{\partial T}\right)_P dT$ \qquad 6.29

Similarly, $\qquad\qquad s = s\,(P,\,T)$ and $ds = \left(\dfrac{\partial s}{\partial P}\right)_T dP + \left(\dfrac{\partial s}{\partial T}\right)_P dT$ \qquad 6.30

Substituting the value of dh from Eq. 6.29 into Eq. 6.28, one gets,

$$ds = \frac{1}{T}\left[\left(\left(\frac{\partial h}{\partial P}\right)_T dP + \left(\frac{\partial h}{\partial T}\right)_P dT\right) - vdP\right] = \frac{1}{T}\left[\left(\frac{\partial h}{\partial P}\right)_T - v\right]dP + \frac{1}{T}\left(\frac{\partial h}{\partial T}\right)_P dT \qquad 6.31$$

Comparing the coefficients of dP and dT in Eq. 6.31 and Eq. 6.30, one obtains

$$\left(\frac{\partial s}{\partial P}\right)_T = \frac{1}{T}\left[\left(\frac{\partial h}{\partial P}\right)_T - v\right] \qquad 6.32$$

and $\qquad\qquad\qquad \left(\dfrac{\partial s}{\partial T}\right)_P = \dfrac{1}{T}\left(\dfrac{\partial h}{\partial T}\right)_P = \dfrac{c_P}{T}$ \qquad 6.33

When Eq. 6.32 is partially differentiated with respect to T and Eq. 6.33 with respect to P, the second order partial derivatives of $\left(\dfrac{\partial^2 s}{\partial T \partial P}\right)$ and $\left(\dfrac{\partial^2 s}{\partial P \partial T}\right)$ will be obtained, that may be equated to get,

$$\left(\frac{\partial h}{\partial P}\right)_T = -T\left(\frac{\partial v}{\partial T}\right)_P + v = -\beta vT + v \qquad 6.34$$

Also, $\left(\dfrac{\partial h}{\partial T}\right)_P = c_P$, Eq. 6.29 gives,

$$dh = c_P dT - \left[T\left(\frac{\partial v}{\partial T}\right)_P - v\right]dP \qquad 6.35$$

The partial derivatives of s, with the help of Eq. 6.32, Eq. 6.33 and Eq. 6.34 may be given as,

$$\left(\frac{\partial s}{\partial T}\right)_P = \frac{c_P}{T} \quad \text{and} \quad \left(\frac{\partial s}{\partial P}\right)_T = -\left(\frac{\partial v}{\partial T}\right)_P \qquad 6.36$$

Therefore,

$$Tds = c_P dT - T\left(\frac{\partial v}{\partial T}\right)_P dP \qquad 6.37$$

Equations 6.15, 6.27 and 6.37 are called Tds equations and are important as one can calculate the heat flow $d'Q_{rev}$ in reversible processes for each pair of independent variables. Further, when divided by T, these equations enable one to calculate the change in entropy ds in reversible processes. The three Tds equations are collected below

$$Tds = c_P\left(\frac{\partial T}{\partial v}\right)_P dv + c_v\left(\frac{\partial T}{\partial P}\right)_v dP \qquad 6.38$$

$$Tds = c_v dT + T\left(\frac{dP}{dT}\right)_v dv \qquad 6.39$$

$$Tds = c_P dT - T\left(\frac{\partial v}{\partial T}\right)_P dP \qquad 6.40$$

Since, $\left(\frac{\partial T}{\partial v}\right)_P = \frac{1}{\beta v}; \left(\frac{\partial T}{\partial P}\right)_v = \frac{\kappa}{\beta}$; the three Tds equations may also be written as;

$$Tds = \frac{c_P}{\beta v} dv + \frac{\kappa c_v}{\beta} dP \qquad 6.38\ (a)$$

$$Tds = c_v dT + \frac{\beta T}{\kappa} dv \qquad 6.39\ (a)$$

$$Tds = c_P dT - T\beta v dP \qquad 6.40\ (a)$$

The values of the first partial derivatives of specific entropy and of the specific thermal capacity at constant pressure, derived earlier, are collected below for ready reference, as they are frequently required for calculations.

$$\left(\frac{\partial s}{\partial T}\right)_P = \frac{c_P}{T} \qquad 6.41\ (a)$$

$$\left(\frac{\partial s}{\partial v}\right)_P = \frac{c_P}{\beta v T} \qquad 6.41\ (b)$$

$$\left(\frac{\partial s}{\partial P}\right)_v = \frac{c_v \kappa}{\beta T} \qquad 6.41\ (c)$$

$$\left(\frac{\partial s}{\partial P}\right)_T = \beta v \qquad 6.41\ (d)$$

$$\left(\frac{\partial c_P}{\partial P}\right)_T = -T\left(\frac{\partial^2 v}{\partial T^2}\right)_P \qquad 6.41\ (e)$$

6.2 Application of Tds Equations

The three Tds equations provide three different but equivalent methods for calculating the change in entropy of a system. If it is assumed that the system in some state that has specific entropy s_0, specific volume v_0, temperature T_0 and pressure P_0. Then it follows from Eqs. 6.38 (a), 6.39 (a), and 6.40 (a)

$$ds = \frac{c_P}{\beta v_0 T_0} dv + \frac{\kappa c_v}{\beta T_0} dP \qquad 6.42\ (a)$$

$$ds = c_v \frac{dT}{T_0} + \frac{\beta}{\kappa} dv \qquad\qquad \text{6.42 (b)}$$

$$ds = c_P \frac{dT}{T_0} - \beta v_0 dP \qquad\qquad \text{6.42 (c)}$$

6.2.1 Ideal gas

In the case of an Ideal gas c_P and c_v, the specific thermal capacities are functions of temperature alone, and $\beta = \frac{1}{T}$, and $\kappa = \frac{1}{P}$. Therefore, Eqs. 6.42 (a)–(c) reduce to:

$$ds = c_P \frac{dv}{v} + c_v \frac{dP}{P} ; \ ds = c_v \frac{dT}{T} + \frac{P}{T} dv = c_v \frac{dT}{T} + R \frac{dv}{v} ; \ ds = c_P \frac{dT}{T} - R \frac{dP}{P} . \qquad \text{6.42 (d)}$$

The above equations on integration give,

$$s = s_0 + \int_{v_0}^{v_1} c_P \frac{dv}{v} + \int_{P_0}^{P_1} c_v \frac{dP}{P} \qquad\qquad \text{6.43 (a)}$$

$$s = s_0 + \int_{T_0}^{T_1} c_v \frac{dT}{T} + R \int_{v_0}^{v_1} \frac{dv}{v} \qquad\qquad \text{6.43 (b)}$$

$$s = s_0 + \int_{T_0}^{T_1} c_P \frac{dT}{T} - R \int_{P_0}^{P_1} \frac{dP}{P} \qquad\qquad \text{6.43 (c)}$$

These equations may be further simplified if it is assumed that c_P & and c_v are constant over the ranges of the changes in temperature, pressure and volume.

$$s = s_0 + c_P \ln \frac{v_1}{v_0} + c_v \ln \frac{P_1}{P_0} \qquad\qquad \text{6.44 (a)}$$

$$s = s_0 + c_v \ln \frac{T_1}{T_0} + R \ln \frac{v_1}{v_0} \qquad\qquad \text{6.44 (b)}$$

$$s = s_0 + c_P \ln \frac{T_1}{T_0} - R \ln \frac{P_1}{P_0} \qquad\qquad \text{6.44 (c)}$$

It is simple to show that each of these equations can be obtained from the other with the help of the equation of state $Pv = RT$ for an ideal gas. For example, if one starts from the last equation 6.44 (c) and put $P_0 = R\frac{T_0}{v_0}$ and $P_1 = R\frac{T_1}{v_1}$ to get $= s_0 + c_P \ln \frac{T_1}{T_0} - R \ln \frac{T_1 v_0}{T_0 v_1} = s = s_0 + (c_P - R) \ln$

$\frac{T_1}{T_0} + R \ln \frac{v_1}{v_0} = s = s_0 + c_v \ln \frac{T_1}{T_0} + R \ln \frac{v_1}{v_0}$, that is Eq. 6.44 (b). Further insight to these equations

may be obtained from Fig. 6.1 in which two nearby equilibrium states of an ideal gas are represented by points A_0 (T_0, P_0, v_0) and A_1 (T_1, P_1, v_1). Three equations, 6.44 (a), 6.44 (b) and 6.44 (c) give the magnitude of the entropy of the state shown by point A_1 in terms of the entropy of state at A_0 and the changes in P, V, and T. As shown in the figure state A_1 may be reached from state A_0 via three different paths, $A_0 a A_1$;

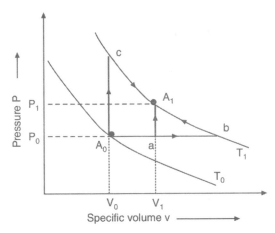

Fig. 6.1 Three different integration paths

A_0bA_1 and A_0cA_1. In Eq. 6.44 (a), the term $(c_P \ln \frac{v_1}{v_0})$ represents the change in entropy in the

process represented by A_0a, and the term $(c_v \ln \frac{P_1}{P_0})$ gives the change in entropy during the process

aA_1. Similarly, in Eq. 6.44 (b), the term $(c_v \ln \frac{T_1}{T_0})$ gives the change in entropy when the initial system

moves along A_0c at constant volume, while the term $\left(R \ln \frac{v_1}{v_0} \right)$ represents the change in the entropy

in the isothermal transition cA_1. It is now easy to identify that the third Eq. 6.44 (c) is represented in Fig.6.1 by the path A_0bA_1. The figure also illustrates that the change in state parameters, like the change in entropy, does not depend on the path through which the system has moved from the initial state to the final state.

6.2.2   Van der Waals gas

The equation of state of the Van der Waals gas is $\left(P + \frac{a}{v^2} \right) (v - b) = RT$, and for Van der Waals gas,

$$\beta \equiv \frac{1}{v}\left(\frac{\partial v}{\partial T}\right)_P = \frac{Rv^2(v-b)}{RTv^3 - 2a(v-b)^2} \text{ , while } \kappa \equiv -\frac{1}{v}\left(-\frac{\partial v}{\partial P}\right)_T = \frac{v^2(v-b)^2}{RTv^3 - 2a(v-b)^2} \qquad 6.45$$

Further, in this case it is simpler if one starts from Eq. 6.43 (b),

$$s = s_0 + \int_{T_0}^{T_1} c_v \frac{dT}{T} + R \int_{v_0}^{v_1} \frac{dv}{v}$$

$$s = s_0 + c_v \ln\frac{T_1}{T_0} + R \ln\left(\frac{v_1 - b}{v_0 - b}\right) \qquad 6.46$$

Now from the equation of state of the Van der Waals gas $\left(P + \dfrac{a}{v^2}\right)(v - b) = RT$ one gets,

$$\frac{(v_1 - b)}{(v_0 - b)} = \frac{RT_1}{\left(P_1 + \dfrac{a}{v_1^2}\right)} \frac{\left(P_0 + \dfrac{a}{v_0^2}\right)}{RT_0} = \frac{T_1}{T_0} \frac{\left(P_0 + \dfrac{a}{v_0^2}\right)}{\left(P_1 + \dfrac{a}{v_1^2}\right)}$$

Substituting this value in Eq. 6.46 one obtains,

$$s = s_0 + c_v \ln\frac{T_1}{T_0} + R \ln\frac{T_1}{T_0} \frac{\left(P_0 + \dfrac{a}{v_0^2}\right)}{\left(P_1 + \dfrac{a}{v_1^2}\right)} = s_0 + (c_v + R)\ln\frac{T_1}{T_0} - R \ln\frac{\left(P_1 + \dfrac{a}{v_1^2}\right)}{\left(P_0 + \dfrac{a}{v_0^2}\right)}$$

or

$$s = s_0 + c_P \ln\frac{T_1}{T_0} - R \ln\frac{\left(P_1 + \dfrac{a}{v_1^2}\right)}{\left(P_0 + \dfrac{a}{v_0^2}\right)} \tag{6.47}$$

Equation 6.47 is nothing but the equivalent of Eq. 6.44 (c) for the ideal gas case. It is left as an exercise to obtain the equation equivalent to Eq. 6.44 (a) for Van der Waals case.

6.2.2.1 *Expression for the specific internal energy of Van der Waals gas*

It is shown by Eq. 6.23 that the change in internal energy is given by,

$$du = c_v dT + \left[T\left(\frac{\partial P}{\partial T}\right)_v - P\right]dv \tag{6.48}$$

One can find the value of $\left[T\left(\dfrac{\partial P}{\partial T}\right)_v - P\right]$ for Van der Waals gas from the equation of state in the following manner.

$$\left(P + \frac{a}{v^2}\right)(v - b) = RT$$

Differentiating this partially with respect to T at constant volume gives,

$$\left(\frac{\partial P}{\partial T}\right)_v (v - b) = R; \quad \text{and} \quad T\left(\frac{\partial P}{\partial T}\right)_v = \frac{RT}{(v - b)}, \text{ therefore,}$$

$$\left[T\left(\frac{\partial P}{\partial T}\right)_v - P\right] = \frac{a}{v^2}$$

When the above value of $\left[T\left(\dfrac{\partial P}{\partial T}\right)_v - P\right]$ is put in Eq. 6.48 it gives,

$$du = c_v dT + \frac{a}{v^2} dv$$ 6.49

Equation 6.49 on integration yields,

$$u = u_0 + c_v (T_1 - T_0) - a\left(\frac{1}{v_1} - \frac{1}{v_0}\right)$$ 6.50

u_0 in the above equation is the specific internal energy of the initial state. The following important conclusions regarding the internal energy of Van der Waal gas may be drawn from Eq. 6.50:

1. Internal energy of Van der Waals gas depends both on specific volume and temperature.
2. It depends on the value of constant a, but is independent of the value of constant b. This is expected since the constant a is related to the molecular attraction, and when specific volume changes, the intra molecules distance also changes in turn changing the potential energy between the molecules. Constant b is a measure of the volume occupied by the molecules and does not affect the potential or the kinetic energy of the molecule.

6.2.2.2 *Difference between c_P and c_V for Van der Waals gas*

Equation 6.24 gives the difference in the two specific heats of a system as,

$$c_P - c_v = T\left(\frac{\partial v}{\partial T}\right)_P \left(\frac{\partial P}{\partial T}\right)_v = \frac{Tv\beta^2}{\kappa}$$

Putting the values of β and κ from Eq. 6.45, it is easy to show that for Van der Waals gas,

$$c_P - c_v = R\frac{1}{1 - \left\{\dfrac{2a(v-b)^2}{RTv^3}\right\}}$$ 6.51

The term $\left\{\dfrac{2a(v-b)^2}{RTv^3}\right\}$ in the denominator may be looked as a correction term. Further, in this

correction term $(v - b)$ may be approximated by v where $v = \dfrac{RT}{P}$, the value for the ideal gas. Under

these approximations, Eq. 6.51 may be written as,

$$c_P - c_v = R\frac{1}{1 - \left\{\dfrac{2a(v-b)^2}{RTv^3}\right\}} = R\frac{1}{1 - \left\{\dfrac{2a}{RTv}\right\}} = R\left[\left(1 - \frac{2aP}{R^2T^2}\right)^{-1}\right] \approx R\left(1 + \frac{2aP}{R^2T^2}\right)$$ 6.52

The factor $\dfrac{2aP}{R^2T^2}$ for gases like carbon dioxide at room temperature is of the order of 10^{-2} that

shows that for Van der Waals gas $c_P - c_v$ is almost equal to R, the value for ideal gas.

6.3   Temperature Entropy Diagram

Like any other state function, specific entropy (or entropy) of a substance may be plotted in a three-dimensional coordinate system consisting of any other pair of state functions as independent variables. The specific entropy surface for an ideal gas as a function of pressure and temperature is shown in Fig. 6.2. As shown in the figure, the specific entropy of an ideal gas decreases logarithmically with pressure and increases logarithmically with temperature. Straight lines ABC and BD, respectively shows the partial derivatives $\left(\dfrac{\partial s}{\partial T}\right)_P$ and $\left(\dfrac{\partial s}{\partial P}\right)_T$ of specific entropy with respect to temperature at constant pressure and with respect to pressure at constant temperature.

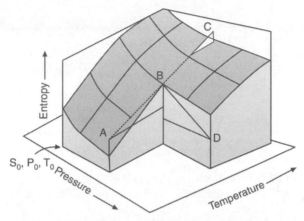

Fig. 6.2   Entropy of an ideal gas as a function of temperature and pressure

6.4   Analysis of Joule and Joule–Thomson Experiments

A preliminary analysis of Joule and Joule–Thomson experiments based on the first law of thermodynamics has been presented in chapter-4. More insight in these experiments may be obtained by analyzing them under both the first and the second laws. Two factors, the slope of the temperature verses specific volume curve at constant specific internal energy $\left(\dfrac{\partial T}{\partial v}\right)_u \equiv \eta$, denoted by η, called the *Joule coefficient,* and the slope of the temperature verses pressure curve at constant specific enthalpy $\left(\dfrac{\partial T}{\partial P}\right)_h \equiv \mu$, denoted by μ, called the *Joule–Thomson* or *Joule–Kelvin coefficient,* play important roles in liquefaction of gases. According to Eq. 4.61,

$$\mu = \left(\frac{\partial T}{\partial P}\right)_h = -\frac{1}{c_P}\left(\frac{\partial h}{\partial P}\right)_T \qquad\qquad 6.53$$

and from Eq. 4.62,
$$\eta = \left(\frac{\partial T}{\partial v}\right)_u = -\frac{1}{c_v}\left(\frac{\partial u}{\partial v}\right)_T \qquad\qquad 6.54$$

The values of $\left(\dfrac{\partial h}{\partial P}\right)_T$ and $\left(\dfrac{\partial u}{\partial v}\right)_T$ obtained using both the first and the second laws are given by Eq. 6.34 and Eq. 6.21;

$$\left(\frac{\partial h}{\partial P}\right)_T = -T\left(\frac{\partial v}{\partial T}\right)_P + v \qquad 6.55$$

$$\left(\frac{\partial u}{\partial v}\right)_T = T\left(\frac{\partial P}{\partial T}\right)_v - P \qquad 6.56$$

Equations 6.55 and 6.56 tell that these partial derivatives may be obtained from the equation of state of the substance. For example the equation of state for an ideal gas is $Pv = RT$ or $v = \dfrac{RT}{P}$ and

$\left(\dfrac{\partial v}{\partial T}\right)_P^{ideal} = \dfrac{R}{P}$. Therefore, $[-T\left(\dfrac{\partial v}{\partial T}\right)_v^{ideal} + v] = 0$ and hence from Eq. 6.55, $\left(\dfrac{\partial h}{\partial P}\right)_T^{ideal} = 0$. Similarly

for an ideal gas $P = \dfrac{RT}{v}, \left(\dfrac{\partial P}{\partial T}\right)_v^{ideal} = \dfrac{R}{v}$, and $\left[T\left(\dfrac{\partial P}{\partial T}\right)_v^{ideal} - P\right] = 0$, which means that for an ideal

gas, $\left(\dfrac{\partial u}{\partial v}\right)_T^{ideal} = 0$. Since for an ideal gas c_V and c_p both are finite, it follows from Eqs. 6.53 and

6.54, that both the Joule coefficient η and Joule–Thomson coefficient μ are zero for an ideal gas.

Next let us calculate $\left(\dfrac{\partial h}{\partial P}\right)_T^{Van}$ and $\left(\dfrac{\partial u}{\partial v}\right)_T^{Van}$ for Van der Waals gas, the equation of state for which

is $\left(P + \dfrac{a}{v^2}\right)(v - b) = RT$; Let us first calculate $\left(\dfrac{\partial P}{\partial T}\right)_v^{Van}$ for Van der Waals gas. For that we write,

$P = \dfrac{RT}{(v - b)} - \dfrac{a}{v^2}$, and differentiate P partially with respect to T keeping v constant. This gives,

$$\left(\frac{\partial P}{\partial T}\right)_v^{Van} = \frac{R}{(v - b)}.$$

Hence, $$\left(\frac{\partial u}{\partial v}\right)_T^{Van} = \left[T\left(\frac{\partial P}{\partial T}\right)_v^{Van} - P\right] = \frac{RT}{(v-b)} - \left\{\frac{RT}{(v-b)} - \frac{a}{v^2}\right\} = \frac{a}{v^2} \quad 6.57$$

Putting this value of $\left(\dfrac{\partial u}{\partial v}\right)_T^{Van}$ in Eq. 6.54, the Joule coefficient η for Van der Waals gas turns out to be,

$$\eta^{Van} = -\frac{a}{c_v v^2} \qquad 6.58$$

In order to calculate the Joule–Thomson coefficient for Van der Waals gas, μ^{Van} we need to calculate the value of $\left(\dfrac{\partial v}{\partial T}\right)_P^{Van}$. It is, however, easier to calculate $\left(\dfrac{\partial T}{\partial v}\right)_P^{Van}$ and then take the reciprocal of it. It may be shown that

$$\left(\frac{\partial h}{\partial P}\right)_T^{Van} = \frac{RTbv^3 - 2av(v-b)^2}{RTv^3 - 2a(v-b)^2} \qquad 6.59$$

and using Eq. 6.53, one gets

$$\mu^{Van} = -\frac{1}{c_P} \frac{RTbv^3 - 2a(v-b)^2}{\left\{RTv^3 - 2a(v-b)^2\right\}} \qquad 6.60$$

In chapter-4, Fig. 4.7 (reproduced below for ready reference) shows the results of Joule–Thomson experiment. The dotted curve in this figure, called the inversion curve, is the locus of the points for which $\left(\frac{\partial h}{\partial P}\right)_T = 0$. The temperature of inversion for the Van der Waals gas may be obtained by setting $\left(\frac{\partial h}{\partial P}\right)_T^{Van} = 0$ in Eq. 6.59. This gives the temperature of inversion for Van der Waals gas,

$$T_{inv}^{Van} = \frac{2a(v-b)^2}{Rbv^2} \qquad 6.61$$

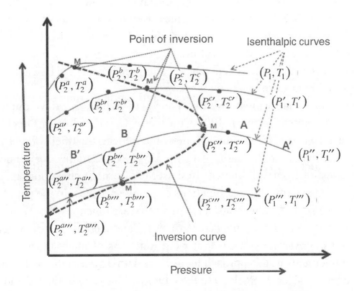

Fig. 4.7 Graph showing temperature inversion and constant enthalpy curves in porous plug experiment

Temperature of inversion has got special significance in the context of liquefaction of gases. A gas must be cooled below its temperature of inversion before reducing the pressure to liquefy it. At the temperature of inversion, pressure is low and volume of the gas is large so that the constant b in Eq. 6.61 may be neglected in comparison to v, and the expression for the temperature of inversion reduces to the approximate value,

$$T_{inv}^{Van} \approx \frac{2a}{Rb} \qquad 6.62$$

Equation 6.62 tells that the temperature for inversion of Van der Waals gas depends both on the constants a and b. Since the magnitude of b is almost same for most of the gases, inversion temperature, according to Eq. 6.62, is nearly linearly proportional to the value of constant a. Moreover, there is remarkable agreement between the values of $\dfrac{2a}{Rb}$ and the observed temperature of inversion for some real gases like CO_2, H_2, He, etc.

6.5 Axiomatic Thermodynamics: Caratheodory Principle

The laws of classical thermodynamics discussed so far are empirical in nature and need little imagination as they are largely based on some observations. For example, the zeroth law follows from the observed fact that all macroscopic systems if left to themselves, will attain a state of equilibrium with their surroundings when no further change in their state parameters takes place. The first law that establishes equivalence of work and energy and the conservation of their sum follows from Joule's observation of conversion of work into heat in boring of gunmetal barrels of guns. The second law of thermodynamics is the outcome of the fact that no engine can have efficiency greater than that of a Carnot engine working between the heat reservoirs having the same temperatures. It has already been described how both the Clausius and the Planck–Kelvin statements of the second law may be shown equivalent to each other by using Carnot engines. Further, the working of an ideal Carnot cycle does not need complicated and abstract imagination. It is precisely because of these reasons of practical observations and simplicity that the preceding treatment of classical thermodynamics is popular and is retained in modern books on the subject.

It is, however, a fact that almost twenty five years before the development of classical thermodynamics in its present form, in 1909 Constantin Caratheodory, a Greek mathematician, working in Germany, developed an axiomatic (purely theoretical) treatment of thermodynamics and published his research in the journal, Maths. Ann. 67, 355 (1909). Caratheodory under took this piece of research on a suggestion from well-known physicist Max Born. Caratheodory in his original work considered very general systems that are functions of many variables. In the conventional classical thermodynamics worked out so far, we consider systems that are functions of only three variables, P, v, T, and out of them only two may be independent variables. Caratheodory, in a more general theoretical approach considered multivariate systems that were functions of many independent variables. In order to appreciate the multivariate approach let us start with a system that is a function of not two but three independent variables. In analogy to the a two variant system, the function corresponding to entropy S in a three variant system may be written as,

$$S = S(T, x_1, x_2) \qquad\qquad 6.63$$

where, T is an intensive independent variable, like the temperature in a two variant system while x_1 and x_2 are the other two independent extensive variables of the system. As a matter of fact, it may be shown, using differential calculus, that for a system of any number of independent variables, it is always possible to find a denominator T such that the path dependent energy transfer Q_r' in a reversible process, when divided by T becomes path independent state function analogous to entropy.

The state function S defined by Eq. 6.63 for a three-variant system is an analog to the entropy in a two-variant system and may be represented by a two-dimensional surface in a three-dimensional coordinate system consisting of T, x_1, and x_2 as shown in Fig. 6.3. In this figure three isentropic planes with constant values of entropies S_1, S_2 and S_3 are shown and if $S_3 > S_2 > S_1$, then according to the second law the system may move only in the direction of the increase of entropy. Thus these isentropic planes divide the space into regions that may be accessible to the system in future and the region through which the system had dwelled in the past. Extending the analogy further, when the system is a function of m independent variables, the isentropic surfaces are planes of $(m\text{-}1)$ dimensions in m-dimensional hyperspace and the hyperspace with respect to a given system may again be decomposed into past and future. It may be emphasized that the division of the space into two components, one representing the past of the system and the other representing the future, has been possible by the principle of increases of entropy that is imbedded in the second law of conventional classical thermodynamics.

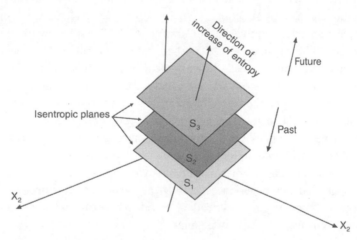

Fig. 6.3   Isentropic planes of constant entropy S_1, S_2, and S_3 in the three dimensional space made up of variants x_1, x_2 and T

Caratheodory on the other hand, developed an equivalent thermodynamics, purely theoretical in nature that is based on one axiom or assumption. His axiom, also called the Caratheodory Principle, was not based on any observation or experience. The axiom Caratheodory made is,

"In the neighborhood (however close) of an equilibrium state of a system of any number of independent coordinates, there exist other equilibrium states that are inaccessible by means of reversible adiabatic processes".

In the first step of his work, Caratheodory proved the theorem related to the integrability of a special type of differential equations called the Pfaffian equations. The theorem may be stated as:

Given an expression of the type Pdx + Qdy + Rdz + ..., where P, Q, R, ... are functions of x, y, z, ... If, in the neighborhood of a point, there are other points which cannot be reached along the solution curves of the equation Pdx + Qdy + Rdz + ... = 0, then their exist functions Δ and Σ of x, y, z, ... such that Pdx + Qdy + Rdz + ... = ΔdΣ.

Using this theorem in conjunction with the Caratheodory axiom, it is possible to show that the small amount of heat dQ_r absorbed reversibly by any system at temperature t (on any scale) may be written as:

$$dQ_r = \varphi(t) f(\sigma) d\sigma \qquad\qquad 6.64$$

Here σ is an undetermined function of the independent coordinates of the system. The Kelvin temperature T may then be identified with $\varphi(t)$ and the change in entropy dS with $f(\sigma) d\sigma$.

Caratheodory' s proof of the above mentioned theorem involves complicated mathematical operations, and therefore, axiomatic thermodynamics is generally not given in books. Several attempts have been made to simplify the axiomatic approach, particularly by M. Born, S. Chandrasekhar and others, but still the mathematics of the subject remains complicated. In further attempts to simplify the Caratheodory axiom approach, L. A. Turner showed how the Caratheodory axiom may lead to the concept of reversible adiabatic surfaces by invoking a simple geometrical argument, thereby discarding the Caratheodory theorem completely. Once the set of non-intersecting adiabatic surfaces is established, Turner showed that simple mathematical tools may be used to infer the concepts of absolute temperature and entropy function. F. W. Sears further elucidated and simplified Turner's work, and later, gave an axiom equivalent to Caratheodory's axiom but involving adiabatic work instead of the inaccessibility of states. P. T. Lundsberg has shown that the Caratheodory axiom may be derived from the conventional Planck– Kelvin statement of the second law.

In a way the second law of thermodynamics includes the first law of energy conservation along with the concepts of the past and the future of the system. In conventional thermodynamics this is achieved via the principle of the increase of entropy in any irreversible process that is essentially based on the results of various experiences. On the other hand, in the axiomatic approach of Caratheodory the same has been achieved by the axiom of inaccessible states in the neighborhood of any point in hyperspace. The conventional statements of thermodynamics are simple, easy to understand and since they follow from common experiences, are logical and elegant. It is because of these reasons that the pure theoretical approach of Caratheodory, based on his principle of inaccessible states, and involving complicated mathematics, is not popular.

Solved Examples

1. Given that the coefficient of volume expansion, compressibility and the specific molar volume for copper are respectively, 4.9×10^{-5} K^{-1}, 7.7×10^{-12} m^2 N^{-1}, and 7.15×10^{-3} mole^{-1}, calculate the difference between the specific molar thermal capacities for copper at 300 K. Also compare the value with that of an ideal gas.

Solution:

Equation 6.24 gives, $c_P - c_v = T \left(\dfrac{\partial v}{\partial T}\right)_P \left(\dfrac{\partial P}{\partial T}\right)_v = \dfrac{Tv\beta^2}{\kappa}$, Substituting the given values,

$$c_P - c_v = \frac{300 \times 7.15 \times 10^{-3} \times \left(4.9 \times 10^{-5}\right)^2}{7.7 \times 10^{-12}} = 667 \text{ J mol}^{-1} \text{K}^{-1}$$

The corresponding value of $c_P - c_v = R = 8.314 \times 10^3$ *Jmol*$^{-1}$ *K*$^{-1}$, for an ideal gas.

2. In an experiment on ^{4}He at 6.0 K and 19.7 atmospheric pressure and for specific molar volume of 2.64×10^{-3} m^3 kilomole^{-1}, Hill and Lounasmaa measured the isothermal compressibility to be 9.42×10^{-8} m^2 N^{-1} and expansivity β to be 5.35×10^{-2} K^{-1}. They determined the value of

the molal thermal capacity at constant volume for ^{4}He to be 9950 J kilomole^{-1} K^{-1}. Using the data calculate the values of the following for ^{4}He at 6.0 K temperature and 19.7 atm. pressure;

$$\left(\frac{\partial u}{\partial v}\right)_T, (c_P - c_v), \left(\frac{\partial s}{\partial T}\right)_v, \left(\frac{\partial s}{\partial v}\right)_T, \left(\frac{\partial s}{\partial P}\right)_v, \text{ and } \left(\frac{\partial s}{\partial v}\right)_P$$

Solution:

From Eq. 6.21 one has, $\left(\dfrac{\partial u}{\partial v}\right)_T = T\left(\dfrac{\partial P}{\partial T}\right)_v - P = \dfrac{T\beta}{\kappa} - P$; therefore,

$$\left(\frac{\partial u}{\partial v}\right)_T = \frac{6.0 \times 5.35 \times 10^{-2}}{9.42 \times 10^{-8}} - (19.7 \times 1.013 \times 10^5)$$

$$= 3.407 \times 10^6 - 1.995 \times 10^6$$

$$= 1.412 \times 10^6 \text{ Jm}^{-3}.$$

Also from Eq. 6.24, $\qquad c_P - c_v = T\left(\dfrac{\partial v}{\partial T}\right)_P \left(\dfrac{\partial P}{\partial T}\right)_v = \dfrac{T v \beta^2}{\kappa}$

$$= \frac{6.0 \times 2.64 \times 10^{-3} \times \left(5.35 \times 10^{-2}\right)^2}{9.42 \times 10^{-8}}$$

$$= 4813 \text{ J kilomole}^{-1} \text{ K}^{-1}$$

Since the value of c_v is given to be 9950 J kilomole^{-1} K^{-1},

The value of $c_P = 9950 + 4813 = 14763$ J kilomole^{-1} K^{-1}

From Eq. 6.17 $\qquad \left(\dfrac{\partial s}{\partial T}\right)_v = \dfrac{1}{T}\left(\dfrac{\partial u}{\partial T}\right)_v = \dfrac{c_v}{T}$

$$= \frac{9950}{6.0} = 1.66 \times 10^3 \text{ J kilomole}^{-1} \text{ K}^{-2}$$

and from Eq. 6.26 $\qquad \left(\dfrac{\partial s}{\partial v}\right)_T = \left(\dfrac{dP}{dT}\right)_v = \dfrac{\beta}{\kappa}$

$$= \frac{5.35 \times 10^{-2}}{9.42 \times 10^{-8}} = 5.67 \times 10^5 \text{ Jm}^{-3} \text{ K}^{-1}$$

Also, from Eq. 6.12, $\qquad \left(\dfrac{\partial s}{\partial P}\right)_v = \dfrac{1}{T}\left(\dfrac{\partial u}{\partial P}\right)_v = \dfrac{c_v}{T}\dfrac{\kappa}{\beta}$

$$= \frac{9950 \times 9.42 \times 10^{-8}}{6.0 \times 5.35 \times 10^{-2}} = 2.919 \times 10^{-3} \text{ m}^3 \text{ kilomole}^{-1} \text{ K}^{-1}$$

and from Eq. 6.14 $\qquad \left(\dfrac{\partial s}{\partial v}\right)_P = \dfrac{c_P}{T}\left(\dfrac{\partial T}{\partial v}\right)_P = \dfrac{c_P}{T v \beta}$

$$= \frac{14763}{6.0 \times 2.64 \times 10^{-3} \times 5.35 \times 10^{-2}} = 1.742 \times 10^6 \text{ Jm}^{-3} \text{ K}^{-1}$$

3. Derive an expression for the mechanical work done in reversible isothermal expansion of Van der Waals gas from initial specific molar volume v_i to a final specific molar volume v_f

Solution:

In a reversible isothermal change in volume of a gas, a quantity of heat dQ is either absorbed (if there is expansion) or rejected by the gas (if there is reduction in the volume). At the same time some work dw is performed either by the gas or on the gas in the change of the volume. Further, there may be some change du in the internal energy of the gas. From the first law,

$$dQ = du + dw, \text{ and } dw = dQ - du$$

Therefore, to find the work done we need to find the values of dQ and du.

Equation 6.39 gives, $Tds = c_v dT + T\left(\dfrac{dP}{dT}\right)_v dv$

However, $Tds = dQ$ for a reversible process and in case of isothermal reversible expansion dT is zero. The above equation then reduces to,

$$Tds = dQ = T\left(\frac{dP}{dT}\right)_v dv \qquad\qquad \text{S-6.1}$$

But for Van der Waals gas $P = \dfrac{RT}{(v-b)} - \dfrac{a}{v^2}$; therefore, $\left(\dfrac{dP}{dT}\right)_v = \dfrac{R}{(v-b)}$

Hence, $\qquad\qquad dQ = \dfrac{RT}{(v-b)} dv \qquad\qquad\qquad\qquad$ S-6.2

Also, it follows from Eq. 6.49 for Van der Waals gas $du = c_v dT + \dfrac{a}{v^2} dv$, which in the case of the isothermal process $(dT = 0)$ reduces to

$$du = \frac{a}{v^2} dv \qquad\qquad \text{S-6.3}$$

From Eq. S-6.2 and Eq. S-6.3, one gets,

$$dw = dQ - du = \frac{RT}{(v-b)} dv - \frac{a}{v^2} dv \qquad\qquad \text{S-6.4}$$

and $\quad w = \displaystyle\int_{v_i}^{v_f} dw = \int_{v_i}^{v_f} \frac{RT}{(v-b)} dv - \int_{v_i}^{v_f} \frac{a}{v^2} dv = RT\ln\frac{(v_f - b)}{(v_i - b)} - a\left(\frac{1}{v_i} - \frac{1}{v_f}\right)$

The required expression for the work done is $\quad RT\ln\dfrac{(v_f - b)}{(v_i - b)} - a\left(\dfrac{1}{v_i} - \dfrac{1}{v_f}\right)$

4. Assuming that the specific compressibility and the specific expansivity of liquids and solids do not change with pressure, temperature and volume, obtain approximate expressions for the specific volume v, change in specific entropy Δs, and the change in the specific enthalpy Δh of the solid or liquid when its pressure, volume and temperature are changed from P_i, v_i, T_i to the value P_f, v_f, T_f.

Solution:

To derive the desired relations we have to start with relations that are written in terms of β and κ, since they do not change much and may be considered as constant with the change of the parameters of the system that is liquid or the solid. We start with the expression for the specific volume.

Taking $v = v(T, P)$, $dv = \left(\dfrac{\partial v}{\partial T}\right)_P dT + \left(\dfrac{\partial v}{\partial P}\right)_T dP = v\left[\dfrac{1}{v}\left(\dfrac{\partial v}{\partial T}\right)_P\right]dT + v\left[-\dfrac{1}{v}\left(\dfrac{\partial v}{\partial P}\right)_T\right]dP$

or $dv = v\beta dT - v\kappa dP$ and $v = \displaystyle\int_{T_i}^{T_f} v\beta dT - \int_{P_i}^{P_f} v\kappa dP + \text{constant of integration}$

Assuming that there is not much change in the volume of solids and liquids when temperature and pressures are changed, the specific volume may also be taken out of the sign of integration. If the initial volume is taken as v_i which is assumed to be constant and since β and κ are also assumed to be constant, one gets the approximate relation,

$$v_f = v_i\beta(T_f - T_i) - v_i\kappa(P_f - P_i) + v_i \qquad\qquad \text{S-6.5}$$

To determine the change in specific entropy, one has to choose the *Tds* relation which involves T and P, i.e., Eq. 6.40, which gives,

$$Tds = c_p dT - T\left(\dfrac{\partial v}{\partial T}\right)_P dP$$

or
$$ds = c_P \dfrac{dT}{T} - \left(\dfrac{\partial v}{\partial T}\right)_P dP = c_P \dfrac{dT}{T} - v\beta\, dP$$

Assuming that c_p does not change much with the change of temperature and specific volume v does not change with pressure, the above equation may be written approximately as,

$$s_f = \int ds = \int_{T_i}^{T_f} c_P \dfrac{dT}{T} - \int_{P_i}^{P_f} v\beta\, dP = s_i + c_P \ln\dfrac{T_f}{T_i} - v_i\beta\left(P_f - P_i\right)$$

or
$$\Delta s = s_f - s_i = c_P \ln\dfrac{T_f}{T_i} - v_i\beta\left(P_f - P_i\right)$$

To calculate the change in the enthalpy with the change of temperature and pressure we use Eq. 6.35,

$$dh = c_p dT - \left[T\left(\dfrac{\partial v}{\partial T}\right)_P - v\right]dP$$

In the above equation $\left(\dfrac{\partial v}{\partial T}\right)_P$ may be replaced by $(v_i\beta)$ to get,

$$\Delta h = h_f - h_i = c_P(T_f - T_i) - [T_i v_i\beta - v_i]\,(P_f - P_i)$$

5. Let κ_{iso} and κ_{adi} respectively denote the isothermal and adiabatic compressibility of a material. Derive an expression for the difference $\kappa_{iso} - \kappa_{adi}$ in terms of expansivity, temperature, specific volume and specific thermal capacity at constant pressure for the substance.

Solution:

The isothermal and the adiabatic compressibility are defined respectively by the following equations:

$$\kappa_{iso} = \kappa \equiv -\frac{1}{v}\left(\frac{\partial v}{\partial P}\right)_T \text{ and } \kappa_{adi} \equiv -\frac{1}{v}\left(\frac{\partial v}{\partial P}\right)_{adi}$$

The isothermal compressibility is one which is generally used and is denoted by κ. The value of κ_{adi} may be obtained from Tds equation given by Eq. 6.38 (a)

$Tds = \frac{c_P}{\beta v}dv + \frac{\kappa c_v}{\beta}dP$, which in case of adiabatic transformation ($ds = 0$) reduces to,

$$0 = \frac{c_P}{\beta v}(dv)_{adi} + \frac{\kappa c_v}{\beta}(dP)_{adi} \quad or \quad \left(\frac{\partial v}{\partial P}\right)_{adi} = -\frac{\kappa v c_v}{c_P}$$

Hence

$$\kappa_{adi} \equiv -\frac{1}{v}\left(\frac{\partial v}{\partial P}\right)_{adi} = \frac{\kappa c_v}{c_P}$$

Now,

$$(\kappa_{iso} - \kappa_{adi}) = \left(\kappa - \frac{\kappa c_v}{c_P}\right) = \frac{\kappa\left(c_P - c_v\right)}{c_P} \quad \text{S-6.6}$$

But from Eq. 6.24,

$$c_P - c_v = T\left(\frac{\partial v}{\partial T}\right)_P\left(\frac{\partial P}{\partial T}\right)_v = \frac{Tv\beta^2}{\kappa}$$

Substituting the value of $(c_P - c_v)$ from the above Eq. in Eq. S-6.6, one gets the desired expression

$$(\kappa_{iso} - \kappa_{adi}) = \frac{Tv\beta^2}{c_P}$$

6. Derive the equation of state of a substance for which $\left(\frac{\partial u}{\partial v}\right)_T = 0 = \left(\frac{\partial h}{\partial P}\right)_T$.

Solution:

From Eq. 6.21 we have,

$\left(\frac{\partial u}{\partial v}\right)_T = T\left(\frac{\partial P}{\partial T}\right)_v - P$, Putting it equal to zero gives $T\left(\frac{\partial P}{\partial T}\right)_v = P$ S-6.7

or $\left[\frac{1}{P}dP = \frac{1}{T}dT\right]_v$ which means that $\frac{T}{P} = $ constant which may a function of $v = A(v)$

or $T = A(v)P$ S-6.8

Similarly from Eq. 6.34, $\left(\frac{\partial h}{\partial P}\right)_T = -T\left(\frac{\partial v}{\partial T}\right)_P + v = 0$

or
$$T\left(\frac{\partial v}{\partial T}\right)_P = v$$

or
$$\frac{T}{v} = \text{a constant that may be a function of } P = B(P)$$

or
$$T = B(P)v \qquad\qquad\qquad \text{S-6.9}$$

A comparison of Eq. S-6.8 and Eq. S-6.9 tells that the constant $A(v)$, that is a function of (v) can only be *const* $\times v$, i.e., Kv, where K is a constant. Putting this value of $A(v)$ in Eq. (S-6.8) gives the desired expression

$$T = KvP$$

7. The state of a system is determined by the state variables y and z. The energy of the state is indicated by $U(y, z)$ and $U = \sin(y + z)\, dy + \sin(y + z)\, dz$. The element of work done is given by $dW = \sin y \cos z\, dy$. Show that U is a state function and W is not a state function. Find $U(y,z)$ and dQ, the element of heat flow.

Solution:

A state function by definition is an exact differential. In the present case we test both dU and dW to see whether they are exact differentials or not. If $\dfrac{\partial^2 U}{\partial y \partial z} = \dfrac{\partial^2 U}{\partial z \partial y}$, then U is a state function, otherwise it is not. Similarly, if $\dfrac{\partial^2 W}{\partial y \partial z} = \dfrac{\partial^2 W}{\partial z \partial y}$, then W is a state function, otherwise not.

Now $\dfrac{\partial^2 U}{\partial y \partial z} = \dfrac{\partial^2 U}{\partial z \partial y} = \sin(y + z) + \sin(y + z)$, hence U is a state function.

In the case of W, the function is not symmetrical with respect to dy and dz and, therefore,

$\dfrac{\partial^2 W}{\partial y \partial z} \neq \dfrac{\partial^2 W}{\partial z \partial y}$, hence it is not an exact differential and is not a state function.

However, $dQ = dU + dW = [(\sin(y + z)) + \sin y \sin z)\, dy + \sin(y + z)\, dz]$

8. A system is completely described by the four state variables $T, S, P,$ and V.

   1. Show that the general form of the energy equation is $U = V\, f(S/V)$
   2. If $U = S^a\, V^{(1-a)}$ find the equations of state $P(S,V)$ and $T(S,V)$. Also solve for $T(S,V)$
   3. Calculate the efficiency of a Carnot engine using the material with energy equation
      $$U = S^a\, V^{(1-a)}$$

Solution:

   1. Since U is extensive variable it can always be written as the multiplication of V with some specific function that depends on two independent variables out of the remaining three. It is because out of the four variables there are only two independent variables. As such it is always possible to write

$$U = V f(S, V). \qquad \text{S-6.10}$$

Suppose we multiply each extensive variable by some scaling factor say, z, then:

$$zU = zV f(zS, zV) \qquad \text{S-6.11}$$

Equation S-6.11 gives, $U = V f(zS, zV) \qquad \text{S-6.12}$

Comparing Eq. S-6.10 and Eq. S-6.12, one gets,

$$f(zS, zV) = f(S, V)$$

This is possible only if the function is of the form, $f\left(\dfrac{S}{V}\right)$, because in that case $f(zS, zV)$

$$= f\left(\frac{zS}{zV}\right) = f\left(\frac{S}{V}\right)$$

Hence the general form of U will be $U = V f(S/V)$

2. It is given that $U = S^a V^{(1-a)} \qquad \text{S-6.13}$

We know $dQ = dU + PdV$, but $dQ = T\,dS$,

Hence $dU = TdS - PdV$, it follows from here that,

$$T = \left(\frac{\partial U}{\partial S}\right)_V, \text{ and } P = -\left(\frac{\partial U}{\partial V}\right)_S \qquad \text{S-6.14}$$

On differentiating partially Eq. S-6.12 once with respect to S and then with respect to V,

$$\left(\frac{\partial U}{\partial S}\right)_V = a\, S^{(a-1)} V^{(1-a)} = a\frac{U}{S}$$

and
$$\left(\frac{\partial U}{\partial V}\right)_S = (1-a) S^a V^{(1-a-1)} = -(a-1)\frac{U}{V}$$

Substituting the values of $\left(\dfrac{\partial U}{\partial S}\right)_V$ and $\left(\dfrac{\partial U}{\partial V}\right)_S$ obtained in Eq. S-6.14, one gets:

$$T = a\frac{U}{S} = a\frac{S^a V^{(1-a)}}{S} = a\left(\frac{S}{V}\right)^{(a-1)} \qquad \text{S-6.15}$$

and
$$P = (a-1)\frac{U}{V} = (a-1)\frac{S^a V^{(1-a)}}{V} = (a-1)\left(\frac{S}{V}\right)^a \qquad \text{S-6.16}$$

Equations S-6.15 and S-6.16 give the required equations of state.

Also, from Eq. S-6.15, $\left(\dfrac{S}{V}\right) = \left(\dfrac{T}{a}\right)^{\left(\frac{1}{a-1}\right)}$ put this value in Eq. S-6.16 to get,

$$P = (a-1)\frac{U}{V} = (a-1)\frac{S^a V^{(1-a)}}{V} = (a-1)\left(\frac{S}{V}\right)^a = (a-1)\left(\frac{T}{a}\right)^{\left(\frac{a}{a-1}\right)}$$

The desired equations of state are: $P = (a-1)\left(\dfrac{T}{a}\right)^{\left(\frac{a}{a-1}\right)}$, $P = (a-1)\left(\dfrac{S}{V}\right)^{a}$ and $T = a\left(\dfrac{S}{V}\right)^{(a-1)}$

It is important to note that for the system that has energy equation $U = S^a V^{(1-a)}$, Pressure P is not a function of V. It means that if temperature remains constant and the volume is changed from V_1 to V_2 pressure does not change. Also, for such a substance $U = \dfrac{TS}{a} = \dfrac{PV}{(a-1)}$

and $PV = \dfrac{a-1}{a} TS$.

The Carnot engine: It consists of four steps. Step-1 is of Isothermal expansion of the substance from volume V_1 to V_2 at temperature T_2. For the present substance pressure does not change in isothermal expansion. Let the pressure in the I-step be P_1.

Work done in first step $W_I = P_1 (V_2 - V_1) = \left(\dfrac{a-1}{a}\right) T_2 (S_1 - S_2)$

Change in the internal energy in first step $= U_2 - U_1 = \dfrac{P_1(V_2 - V_1)}{(a-1)}$

Heat absorbed in first step $Q_2 = W_I + (U_2 - U_1) = \dfrac{a P_1(V_2 - V_1)}{(a-1)} = T_2(S_2 - S_1)$ S-6.17

Step-II: In this step substance undergoes adiabatic expansion. No heat is exchanged but the temperature falls to T_1. Also, during reversible adiabatic process entropy does not change. Let S_2 denote the entropy at the end of step-I

Work done in this step $W_{II} = U_2 - U_3 = \dfrac{(T_2 - T_1)S_2}{a}$

Step-III: Isothermal compression: just like step-I, but it is of compression so heat is rejected at constant temperature T_1 which is lower than T_2. Let the constant pressure in this step be P_3

Work done in step-III $W_{III} = P_3 (V_4 - V_3) = \left(\dfrac{a-1}{a}\right) T_1 (S_2 - S_1)$

and heat rejected $Q_1 = \dfrac{a P_3 (V_4 - V_3)}{(a-1)}$

Step-IV: Adiabatic compression; No heat is exchanged, entropy remains constant. Volume decreases from V_4 to V_1.

Work done in step-IV $W_{IV} = U_4 - U_1 = \dfrac{(T_2 - T_1)S_1}{a}$

The total work in the complete cycle is

$$W_{Total} = \sum_{I}^{IV} W_i = \left(\dfrac{a-1}{a}\right)T_2(S_1 - S_2) + \left(\dfrac{a-1}{a}\right)T_1(S_2 - S_1) + \dfrac{(S_2 - S_1)(T_2 - T_1)}{a}$$

$$= (T_2 - T_1)(S_2 - S_1)$$

The efficiency of Carnot engine $\eta = \dfrac{W_{total}}{Q_2} = \dfrac{(T_2 - T_1)(S_2 - S_1)}{T_2(S_2 - S_1)} = \dfrac{(T_2 - T_1)}{T_2}$

9. Compute the difference between the molar specific heats $(c_P - c_v)$, the isothermal and adiabatic compressibility's κ_T, κ_S and expansivity β for Van der Waals gas assuming it to be a monatomic gas.

Solution:

The equation of state for Van der Waals gas is,

$$P = \frac{RT}{(v - b)} - \frac{a}{v^2}$$

Therefore, $\left(\dfrac{\partial P}{\partial T}\right)_v = \dfrac{R}{(v - b)}$ and $\left(\dfrac{\partial P}{\partial v}\right)_T = \dfrac{2a(v - b)^2 - RTv^3}{(v - b)^2 v^3}$ S-6.18

Now, from Eq. 6.24 b, $c_P - c_v = -T\dfrac{\left[\left(\dfrac{\partial P}{\partial T}\right)_v\right]^2}{\left(\dfrac{\partial P}{\partial v}\right)_T} = \dfrac{R}{1 - 2a(v - b)^2 \ / \ RTv^3}$

Also, $\kappa_T = -\dfrac{1}{v}\left(\dfrac{\partial v}{\partial P}\right)_T = -\dfrac{1}{v}\left(\dfrac{\partial P}{\partial v}\right)_T^{-1} = \dfrac{(v - b)^2 v^2}{RTv^3 - 2a(v - b)^2}$

But $\kappa_s = \dfrac{\kappa_T}{\gamma}$ and $\gamma = 1.66$ for a monatomic gas

Hence, $\kappa_s = 0.6\,\kappa_T = 0.6\dfrac{(v - b)^2 v^2}{RTv^3 - 2a(v - b)^2}$

Similarly, $\beta = \dfrac{1}{v}\left(\dfrac{\partial v}{\partial T}\right)_v = \dfrac{Rv^2(v - b)}{RTv^3 - 2a(v - b)^2}$

10. A material has thermal expansivity $\beta = \dfrac{1}{v}\left(\dfrac{R}{P} + \dfrac{a}{RT^2}\right)$ and isothermal compressibility

$\kappa_T = \dfrac{1}{v}\left(Tf(P) + \dfrac{b}{P}\right)$. Find the value of function $f(P)$ and the equation of state of the substance.

Also discuss the condition for the stability of the substance.

Solution:

By definition $\beta = \dfrac{1}{v}\left(\dfrac{\partial v}{\partial T}\right)_P = \dfrac{1}{v}\left(\dfrac{R}{P} + \dfrac{a}{RT^2}\right),$

Hence, $\left(\dfrac{\partial v}{\partial T}\right)_P = \left(\dfrac{R}{P} + \dfrac{a}{RT^2}\right)$ S-6.19

Similarly,
$$\kappa_T = -\frac{1}{v}\left(\frac{\partial v}{\partial P}\right)_T = \frac{1}{v}\left(T f(P) + \frac{b}{P}\right)$$

Therefore,
$$\left(\frac{\partial v}{\partial P}\right)_T = -\left(T f(P) + \frac{b}{P}\right) \qquad \text{S-6.20}$$

Specific molar volume v is a state function and so it must be an exact differential. Differentiating partially Eq. S-6.19 with respect to P, at constant T, and Eq. S-6.20 with respect to T at constant P, and equating $\left(\dfrac{\partial^2 v}{\partial P \partial T}\right) = \left(\dfrac{\partial^2 v}{\partial T \partial P}\right)$, one gets:

$$f(P) = \frac{R}{P^2}$$

Putting this value of $f(P)$ in Eq. S-6.20, one gets:

$$\left(\frac{\partial v}{\partial P}\right)_T = -\left(\frac{RT}{P^2} + \frac{b}{P}\right)$$

$$\int_{v_0}^{v} dv = (v - v_0)$$

$$= \int_{P_0}^{P}\left(\frac{\partial v}{\partial P}\right) dP = \frac{RT}{P - P_0} - b \ln\left(P - P_0\right) + C(T) \qquad \text{S-6.21}$$

where $C(T)$ is a constant of integration that may be a function of T. Partially differentiating Eq. S-6.21 with respect T at constant pressure P and equating it to Eq. S-6.21, we get:

$$\left(\frac{\partial v}{\partial T}\right)_P = \frac{R}{P} + \left(\frac{\partial C(T)}{\partial T}\right) = \left(\frac{R}{P} + \frac{a}{RT^2}\right)$$

Hence,
$$\left(\frac{\partial C(T)}{\partial T}\right) = \frac{a}{RT^2} \text{ and } C(T) = -\frac{a}{RT} \qquad \text{S-6.22}$$

Substituting the value of $C(T)$ in Eq. S-6.21, one gets the desired equation of state of the substance,

$$(v - v_0) = \frac{RT}{P - P_0} - b \ln\frac{P}{P_0} - \frac{a}{RT}$$

Since the compressibility of a stable substance must be positive,

$$kt = \frac{1}{v}\left(T f(P) + \frac{b}{P}\right) = \frac{1}{v}\left(\frac{RT}{P^2} + \frac{b}{P}\right) > 0 \text{ or } \frac{RT}{P^2} + \frac{b}{P} > 0$$

This reduces to: $P/T < R/b$, which is the desired condition of stability.

Problems

1. The differential of some function u that depends on the state parameters x and y, has the following form for three different systems A, B and C. Find out for which of the three systems, u may be a state function. Also obtain the value of u when it is a state function.

For system A: $u = \left(2y^2 - 3x\right)dx - 4xy\,dy$

For system B: $u = \dfrac{x\,dy - y\,dx}{\left(x^2 + y^2\right)}$

For system C: $u = \left(y - x^2\right)dx + \left(x + y^2\right)dy$

2. Compute the molal specific volume of a monatomic ideal gas for which at 10 K, the thermal expansivity and isothermal compressibility are respectively, 5×10^{-5} K^{-1} and 10×10^{-12} $m^2\,N^{-1}$.

3. In an experiment 10 m^3 kilomole^{-1} volume of a monatomic gas was passed through a throttle value to expand to a volume of 20 m^3 kilomole^{-1}. Assuming the gas to be a monatomic Van der Waals gas and taking the values of constant a $=400$ J m^3 kilomole^{-2} and b $= 0.05$ m^3 kilomole^{-1}, calculate the fall in temperature of the gas.

4. Assuming oxygen to be a Van der Waals gas with constants a $= 138 \times 10^3$ J m^3 kilomole^{-2}, b $= 0.0318$ m^3 kilomole^{-1}, compute the maximum temperature up to which the gas can be liquefied by the Joule–Thomson expansion.

5. Define Joule and Joule–Kelvin coefficients and calculate the approximate values for these coefficients for one kilomole of CO_2, at 1 K, assuming it a Van der Waals gas with $a = 366 \times 10^3$ J m^3 kilomole^{-2}, $b = 0.0429$ m^3 kilomole^{-1}, and $\gamma\left(= \dfrac{c_P}{c_V}\right) = 1.29$. Assume that $(c_p - c_V = R)$ for the Van der Waals gas.

6. Calculate the amount of mechanical work done in isothermal expansion of molar specific volume of 5 m^3 of CO_2 to a value of 10 m^3 at a temperature of 1.0 K. Assume CO_2 as an Van der Waals gas. Use the values of constants $a = 366 \times 10^3$ J m^3 kilomole^{-2}, $b = 0.0429$ m^3 kilomole^{-1}, and $\gamma\left(= \dfrac{c_P}{c_V}\right) = 1.29$ for CO_2.

[*Hint*: Mechanical work done in isothermal expansion of Van der Waals gas is given by,

$$W = RT \ln \frac{\left(v_{final} - b\right)}{\left(v_{initial} - b\right)} + a\left(\frac{1}{v_{final}} - \frac{1}{v_{initial}}\right)$$

7. The temperature of a fixed volume of a Van der Waals gas is changed from 5 K to 10 K. Compute the ratio of the change in specific internal energy per Kelvin to the change in the specific entropy of the gas, given that the constants of the gas are, a $= 366 \times 10^3$ J m^3 kilomole^{-2}, b $= 0.0429$ m^3 kilomole^{-1}, and $\gamma\left(= \dfrac{c_P}{c_V}\right) = 1.29$.

8. The rate of change of internal energy with volume at constant temperature of 10 K and pressure of 20×10^5 Nm^{-2}, for a substance is 1.5×10^6 Jm^{-3}. If the isothermal compressibility of the substance is 1.0×10^{-7} m^2 N^{-1}, compute the expansibility of the substance.

9. With the following data compute the rate of change specific entropy of a substance at temperature 10 K (i) with temperature at constant volume (ii) with volume at constant temperature. Also give the units for the two calculated quantities. $c_V = 10 \times 10^3$ J kilomole^{-1}., isothermal compressibility $k_{iso} = 10.0 \times 10^{-8}$ m^2 N^{-1}, expansivity $\beta = 5.0 \times 10^{-2}$ K^{-1}, $c_P = 15 \times 10^3$ J kilomole^{-1}

10. Compute the change in the temperature when 1×10^{-5} m^3 kilomole^{-1} volume of a liquid is adiabatically compressed at 10 K (i) the change in volume is 10 % (ii) the change in pressure is 10 %. Discuss the sign of the temperature change. The required data of the substance is given below and it may be assumed that the values of the parameters do not change during the adiabatic process. $c_V = 10 \times 10^3$ J kilomole^{-1}., isothermal compressibility $k_{iso} = 10.0 \times 10^{-8}$ m^2 N^{-1}, expansivity $\beta = 5.0 \times 10^{-2}$ K^{-1}, $c_P = 15 \times 10^3$ J kilomole^{-1}

Short Answer Questions

1. Starting from the appropriate Tds equation drive an expression for the change in the temperature of a substance in adiabatic change in its volume in terms of the isothermal compressibility, expansivity, etc.

2. Drive an expression for the isothermal rate of change in specific enthalpy with pressure for a substance.

3. Without derivation write an expression for $(c_P - c_V)$ in terms of β, T, v and κ.

4. Chose the appropriate Tds equation to derive an expression for the molar entropy of Van der Waals gas and show how it may be reduced to the corresponding expression for an idcal gas.

5. Write expressions for isothermal compressibility, expansivity, specific internal energy and specific entropy for Van der Waals gas and compare them with those for an ideal gas.

6. With the appropriate Tds equation drive a relation between T and v for Van der Waals gas under reversible adiabatic process.

7. Show that the heat absorbed in a reversible isothermal process by a Van der Waals gas is

$$d'Q = RT \frac{dv}{(v-b)} .$$

8. In equation $Tds + T\left(\dfrac{\partial v}{\partial T}\right)_v dP = X$, identify the value of X and the independent variables of

the system.

9. Show that $\left(\dfrac{\partial^2 P}{\partial T^2}\right)_v = \dfrac{1}{T}\left(\dfrac{\partial c_v}{\partial v}\right)_T$

10. Obtain the relation $Tds = dU + PdV$

Long Answer Questions

1. Write the three Tds equations and derive one of them.

2. Derive the relation $c_{P_1} = c_P + T\displaystyle\int_{P_1}^{P_2}\left(\dfrac{d^2v}{\partial T^2}\right)_P dP$ and discuss its significance.

3. With the help of a suitable example show how the second law of thermodynamics divides the multi-dimensional hyperspace into past and future in case of multi-variant thermodynamic systems.

4. Write a detailed note on axiomatic thermodynamics developed by Caratheodory; state the Caratheodory principle and state why this treatment has not gained popularity.

5. Discuss Joule and Joule–Thomson experiments and derive expressions for the Joule and Joule–Kelvin coefficients for a Van der Waals gas.

6. Define the maximum temperature of inversion and obtain an expression for it in case of a Van der Waals gas. Discuss why the maximum temperature of inversion depends on constant a but not on constant b in case of the Van der Waals gas.

7. Obtain expressions for specific internal energy, specific entropy and specific enthalpy for a gas the equation of state for which is,

$$(P + b)v = RT$$

8. Starting from the three Tds equations, derive expressions for the specific entropy of an ideal gas and show that the three different expressions refer to three different paths of reaching the final state from a given initial state.

9. Obtain an expression for the work done in an isothermal process by a Van der Waals gas when it expands.

10. Assuming that the compressibility and expansivity of solids and liquids remain constant, show that their specific volume may be given as,

$$v_{final} = v_{ini} [1 + b(T_{final} - T_{ini}) - \kappa(P_{final} - P_{ini})]$$

Here, subscript *final* and *ini*, respectively, means final and initial value of the referred parameter.

Multiple Choice Questions

Note: Some of the multiple choice questions may have more than one correct alternative. All correct alternatives must be marked for complete answer in such cases.

1 The MKS units of Van der Waals constants a and b are respectively,
   (a) J kilomole^{-1} m^3; kilomole^{-1} m^3
   (b) J kilomole^{-1} m^3; kilomole^{-1} m^3 K^{-1}
   (c) J kilomole^{-1} K^{-1}; kilomole^{-1} m^3
   (d) J^{-1} kilomole m^3; kilomole^{-1} m^3 K^{-1}

2. Tick the correct relation (s)
   (a) $\left(\dfrac{\partial U}{\partial V}\right)_T = T\left(\dfrac{\partial P}{\partial T}\right)_T - P$
   (b) $\left(\dfrac{\partial U}{\partial V}\right)_T = T\left(\dfrac{\partial P}{\partial T}\right)_s - P$
   (c) $\left(\dfrac{\partial U}{\partial V}\right)_T = T\left(\dfrac{\partial P}{\partial T}\right)_V - P$
   (d) $\left(\dfrac{\partial S}{\partial V}\right)_T = T\left(\dfrac{\partial P}{\partial T}\right)_T - P$

3. The maximum value of the temperature of inversion for Van der Waals gas is
   (a) $\dfrac{2a^2(v-b)}{Rv^2 b}$
   (b) $\dfrac{2a(v-b)}{Rv^3}$
   (c) $\dfrac{2a(v-b)}{Rb}$
   (d) $\dfrac{2a(v-b)}{Rv^2 b}$

4. The difference $(c_P - c_V)$ for Van der Waals gas depends on
   (a) Pressure P of the gas
   (b) Temperature T of the gas
   (c) constant a
   (d) constant b

5. The change in entropy of Van der Waals gas depends on
   (a) Change in the volume of the gas
   (b) change in the temperature of the gas
   (b) the value of constant a
   (d) the value of constant b.

6 For an ideal gas, specific enthalpy h is
   (a) $c_P (T - T_0) + h_0$
   (b) $c_V (T - T_0) + h_0$
   (c) $c_P (V - V_0) + h_0$
   (d) $c_P (P - P_0) + h_0$

7 The coefficient of cubical expansion for an ideal gas is
   (a) T
   (b) V
   (c) P
   (d) S
   Here symbols have their usual meaning.

8 Chose the correct relation(s)
   (a) $\left(\dfrac{\partial s}{\partial T}\right)_P = \left(\dfrac{c_P}{T}\right)$
   (b) $\left(\dfrac{\partial s}{\partial T}\right)_P = \left(\dfrac{c_V}{T}\right)$
   (c) $\left(\dfrac{\partial c_P}{\partial P}\right)_T = -T\left(\dfrac{\partial^2 v}{\partial T^2}\right)_P$
   (d) $\left(\dfrac{\partial c_P}{\partial P}\right)_T = v\left(\dfrac{\partial^2 v}{\partial T^2}\right)_P$

9. The temperature of water under adiabatic compression decreases if the temperature of water lies between,
   (a) $-20°C$ and $-4°C$
   (b) $-4°C - 0°C$
   (c) $0°C - +4°C$
   (d) above $+4°C$

10. The ratio of the thermal capacities $\dfrac{c_P}{c_V}$ for a substance is:
   (a) $> \dfrac{\text{isothermal compressibility}}{\text{adiabatic comressibility}}$
   (b) $= \dfrac{\text{isothermal compressibility}}{\text{adiabatic comressibility}}$
   (c) $= \dfrac{\text{adiabatic comressibility}}{\text{isothermal compressibility}}$
   (d) $< \dfrac{\text{isothermal compressibility}}{\text{adiabatic comressibility}}$

Answers to Numerical and Multiple Choice Questions

Answers to problems

1 u is a state function for systems B and C; $u_B = -\arctan(x/y)$, $u_C = yx + (y^3 - x^3)/3$
2 3.33 m^3 kilomole^{-1}
3 1.6 K
4 1044 K
5 $\mu = 2.26 \times 10^{-5}$ and $\eta = 12.77$
6 30.8 kJ

7 1.44

8 3.5×10^{-2} K^{-1}

9 (i) 1.0×10^3 J kilomole^{-1} K^{-2}, (ii) 5.0×10^5 J K^{-1} m^{-3}.

10 (i) 5.0×10^{-4} K; (ii) 0.33×10^{-4} K.

Answers to short answer questions

1. $dT_s = -\dfrac{\beta T}{\kappa c_V}(dv_s)$

2. $\left(\dfrac{\partial h}{\partial P}\right)_T = -\beta vT + v$

3. $(c_P - c_V) = \beta^2 \kappa_{iso}^{-1} T v$

4. Start with $Tds = c_v dT + T\left(\dfrac{\partial P}{\partial T}\right)_v dv$ and obtain $s = c_V \ln\dfrac{T_{final}}{T_{ini}} + R\ln\left(\dfrac{v_{final} - b}{v_{ini} - b}\right) + s_0$

6. $T(v - b)^{\frac{R}{c_v}} = $ constant .

8. $X = c_P dT$; temperature T and Pressure P are the independent variables.

Answers to multiple choice questions

1. (a)   2. (c)   3. (d)   4. (a), (b), (c)
5. (a), (b), (d)   6. (a)   7. (a)   8. (a), (c)
9. (c)   10. (b)

Revision

1. For any process, reversible or irreversible, $TdS = dU + PdV$
2. Following Tds equations and expressions for some important derivatives be remembered by heart

$$TdS = c_P\left(\dfrac{\partial T}{\partial v}\right)_P dv + c_v\left(\dfrac{\partial T}{\partial P}\right)_v dP = \dfrac{c_P}{\beta v}dv + \dfrac{\kappa c_v}{\beta}dP$$

$$TdS = c_v dT + T\left(\dfrac{dP}{dT}\right)_v dv = c_v dT + \dfrac{\beta T}{\kappa}dv$$

$$TdS = c_P dT - T\left(\dfrac{\partial v}{\partial T}\right)_P dP = c_P dT - T\beta vdP$$

$$\left(\dfrac{\partial s}{\partial T}\right)_P = \dfrac{c_P}{T}$$

$$\left(\dfrac{\partial s}{\partial v}\right)_P = \dfrac{c_P}{\beta vT}$$

$$\left(\frac{\partial s}{\partial P}\right)_v = \frac{c_v \kappa}{\beta T}$$

$$\left(\frac{\partial s}{\partial P}\right)_T = \beta v$$

$$\left(\frac{\partial c_P}{\partial P}\right)_T = -T\left(\frac{\partial^2 v}{\partial T^2}\right)_P$$

3. For an ideal gas

$$s = s_0 + c_P \ln\frac{v_1}{v_0} + c_v \ln\frac{P_1}{P_0}$$

$$s = s_0 + c_v \ln\frac{T_1}{T_0} + R \ln\frac{v_1}{v_0}$$

$$s = s_0 + c_P \ln\frac{T_1}{T_0} - R \ln\frac{P_1}{P_0}$$

4. For Van der Waals gas

$$\beta \equiv \frac{1}{v}\left(\frac{\partial v}{\partial T}\right)_P = \frac{Rv^2(v-b)}{RTv^3 - 2a(v-b)^2} \text{ and } \kappa \equiv -\frac{1}{v}\left(-\frac{\partial v}{\partial P}\right)_T = \frac{v^2(v-b)^2}{RTv^3 - 2a(v-b)^2}$$

$$s = s_0 + \int_{T_0}^{T_1} c_v \frac{dT}{T} + R \int_{v_0}^{v_1} \frac{dv}{v}$$

$$s = s_0 + c_v \ln\frac{T_1}{T_0} + R \ln\left(\frac{v_1 - b}{v_0 - b}\right)$$

$$s = s_0 + c_P \ln\frac{T_1}{T_0} - R \ln\frac{\left(P_1 + \dfrac{a}{v_1^2}\right)}{\left(P_0 + \dfrac{a}{v_0^2}\right)}$$

$$u = c_v dT + \frac{a}{v^2} dv$$

$$u = u_0 + c_v(T_1 - T_0) - a\left(\frac{1}{v_1} - \frac{1}{v_0}\right)$$

5. Internal energy of Van der Waals gas depends both on specific volume and temperature.
6. It depends on the value of constant a, but is independent of the value of constant b. This is expected since the constant a is related to the molecular attraction, and when specific volume changes, the intra molecules distance also changes in turn changing the potential energy between the molecules. Constant b is a measure of the volume occupied by the molecules and does not affect the energy, potential or kinetic, of the molecule.

$$c_P - c_v = R \dfrac{1}{1 - \left\{\dfrac{2a(v-b)^2}{RTv^3}\right\}} = R \dfrac{1}{1 - \left\{\dfrac{2a}{RTv}\right\}}$$

$$= R\left[\left(1 - \dfrac{2aP}{R^2 T^2}\right)^{-1}\right] \approx R\left(1 + \dfrac{2aP}{R^2 T^2}\right)$$

$$\mu = \left(\dfrac{\partial T}{\partial P}\right)_h = -\dfrac{1}{c_P}\left(\dfrac{\partial h}{\partial P}\right)_T$$

$$\eta = \left(\dfrac{\partial T}{\partial v}\right)_u = -\dfrac{1}{c_v}\left(\dfrac{\partial u}{\partial v}\right)_T$$

$$\eta^{Van} = -\dfrac{a}{c_v v^2}$$

$$\mu^{Van} = -\dfrac{1}{c_P}\dfrac{RTbv^3 - 2a(v-b)^2}{\left\{RTv^3 - 2a(v-b)^2\right\}}$$

$$T_{inv}^{Van} = \dfrac{2a(v-b)^2}{Rbv^2}$$

(Maximum temperature of inversion for Van der Waals gas)

7. Caratheodory principle: '*In the neighborhood (however close) of an equilibrium state of a system of any number of independent coordinates, there exist other equilibrium states that are inaccessible by means of reversible adiabatic processes*'.

Thermodynamic Functions, Potentials, Maxwell's Equations, the Third Law and Equilibrium

7.0 Introduction

State variables or state functions are parameters of a system, which depend on the specific equilibrium state of the system, and are independent of the history of how that state has been reached. Further, all the state variables are in general not independent and some of them are related through the equation of state. A combination of the state functions of an equilibrium state is, obviously, also another state function. Some combinations of state functions, either because of their frequent use in thermodynamic calculations or because they are associated with some special property of the system, are given specific name like *Enthalpy*. Enthalpy, denoted by H (the specific value is denoted by h), Helmholtz function denoted by F (specific value by f), and Gibb's function denoted by G (specific value by g) are three such functions. Enthalpy and its importance in phase transitions have already been discussed earlier. Helmholtz and Gibb's functions are discussed below.

7.1 The Helmholtz Function

Hermann Ludwig Ferdinand von Helmholtz (August 31, 1821–September 8, 1894), a German mathematician, physicist and philosopher introduced a new combination of state functions, called the Helmholtz function which is denoted by F, and is defined as,

$$F = U - TS.$$

And the specific Helmholtz function $f = u - Ts$ 7.1

7.1.1 Significance of Helmholtz function

In order to appreciate the physical significance of Helmholtz function let us go back to the first law that says that the work W done by a system in a process is equal to the reduction in the internal energy of the system $(U_1 - U_2)$ plus the amount of heat flow into the system Q. This may be written as,

$$W = (U_1 - U_2) + Q$$ 7.2

Here U_1 and U_2 are the initial and final values of the internal energy of the system. One point that may be clearly seen from Eq. 7.2 is that, in general, both the reduction in internal energy and the amount of heat inflow contribute to the work that the system may perform. Now there may be two extreme situations: (a) the process is adiabatic when Q is zero, and in that case the work will be done totally at the cost of the internal energy; (b) there is no change in the internal energy of the system, or if there is any it is replenished by some external agency, then the work will be done by the heat flow. In general, however, both the internal energy and heat supplied to the system will jointly contribute to the work that the system may perform.

Now, if the heat flow to the system is at constant temperature T, that will happen if the system is thermally connected to a heat reservoir at temperature T, the amount of heat flow to the system may be denoted by Q_T and the work done by W_T, then Eq. 7.2 becomes,

$$W_T = (U_1 - U_2) + Q_T \qquad \qquad 7.3$$

Figure 7.1 (a) is a schematic presentation of Eq. 7.3.

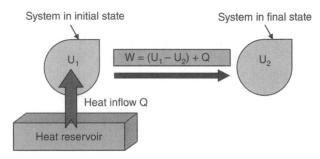

Fig. 7.1 (a) Schematic representation of Equation 7.3

We now wish to find an upper limit for W_T that is the maximum amount of work that a system in contact with a constant temperature reservoir may perform. In order to find the maximum value of W_T, we make use of the second law which says that in any process the sum of the entropies of all components must either increase or may remain constant, if the process is reversible. In the present case of the heat transfer from reservoir at temperature T to the system, the increase in the entropy of the system equals $(S_2 - S_1)$, where S_1 is the entropy of the system before the heat transfer and S_2 after the heat transfer. The decrease in the entropy of the reservoir is $\left(-\frac{Q_T}{T}\right)$. Hence, from the principle of increase of entropy,

$$\left(S_2 - S_1\right) + \left(-\frac{Q_T}{T}\right) \geq 0 \qquad \qquad 7.4$$

And if T is not zero, $\qquad T\,(S_2 - S_1) \geq Q_T \qquad \qquad 7.5$

From Eq. 7.5 it is obvious that the maximum value of Q_T may be $T(S_2 - S_1)$, otherwise Q_T will always be less than $T(S_2 - S_1)$. Therefore, from Eq. 7.3, W_T will have maximum value when $Q_T = T(S_2 - S_1)$. Equation 7.3 may now be rewritten as:

$$W_T \leq (U_1 - U_2) + T\,(S_2 - S_1) \qquad \qquad 7.6$$

Or $\qquad\qquad\qquad\qquad W_T \leq (U_1 - TS_1) - (U_2 - TS_2) \qquad \qquad 7.7$

Or
$$W_T \le (F_1 - F_2)$$
7.8

Here, $F_1 = (U_1 - TS_1)$ and $F_2 = (U_2 - TS_2)$ are, respectively, the values of the Helmholtz functions in the initial and final states of the system. A schematic view of Eq. 7.8 is given in Fig. 7.1 (b).

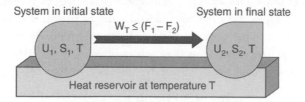

System in initial state   System in final state

$W_T \le (F_1 - F_2)$

U_1, S_1, T U_2, S_2, T

Heat reservoir at temperature T

Fig. 7.1 (b) Schematic representation of Equation 7.8

Equation 7.8 tells that *the decrease in the Helmholtz function puts the upper limit to the amount of work that may be obtained in any process between the two equilibrium states of a system at same temperature, during which there is a heat flow into the system from a single reservoir.* This statement provides the physical significance of the Helmholtz function. The maximum value of the work W_T^{Max} will occur for a reversible process, in which case,

$$W_T^{Max} = (F_1 - F_2)$$
7.9

In all other irreversible processes W_T is less than W_T^{Max}. It may once again be emphasized that Eqs. 7.8 and 7.9 give the upper limit and the maximum value of the work that may be obtained, but the energy for this work comes only partly from the internal energy of the system and the remaining part of energy for the work is provided by the heat reservoir.

The specific Helmholtz function $f = u - Ts$ may be obtained using the appropriate expressions for u and s. For example in case of an ideal gas the change in specific Helmholtz function in going from state of temperature T_i and specific volume v_i to a state of temperature T_f and specific volume v_f is given by,

$$f = f_f - f_i = (u_f - u_i) - [T_f s_f - T_i s_i] = \left\{ c_v \left(T_f - T_i \right) \right\} - T_f \left[c_v \ln \frac{T_f}{T_i} + R \ln \frac{v_f}{v_i} \right] + s_i \left(T_f - T_i \right) \text{ 7.10 (a)}$$

Other expressions for f may be readily obtained by using (P, v) and (P, T) dependent expressions of entropy. In case of a Van der Waals gas, it is easy to get the following expression for the change in the specific Helmholtz function, using the expressions for the specific internal energy u and specific entropy s obtained in chaper-6.

$$f^{Van} = c_v \left(T_f - T_i - T_f \ln \frac{T_f}{T_i} \right) - RT_f \ln \left(\frac{v_f - b}{v_i - b} \right) - a \left(\frac{1}{v_f} - \frac{1}{v_i} \right) - s_i \left(T_f - T_i \right) \qquad \text{7.10 (b)}$$

In his original work Helmholtz called $\Delta F = (F_1 - F_2)$ *free energy*, as it is the amount of the maximum energy that may be released in an isothermal process and is available for work. The term *Helmholtz free energy* has been used in older literature for ΔF. However, to avoid any confusion with another similar quantity, the Gibb's free energy, we shall not use the term free energy and will refer them as the change in the values of Helmholtz and Gibb's functions.

7.1.2 Configurational work W_T

The work W in Eq. 7.1 is the sum of all configurational works. Configurational work, in general, may be written as the multiplication of an intrinsic function and the differential of the corresponding extrinsic function, for example PdV, or HdM (in magnetic systems), etc. Depending on the nature of the system many different types of configurational works may be performed by the system. Since PdV type of configurational work is performed in almost all systems, it may be treated separately and all other works may be lumped together. Thus W may be written as $W = P\, dV + A$, where A represents all other non-PdV configurational works. When the process is isothermal, the work

$$W_T = (PdV)_T + A_T \qquad\qquad 7.11$$

Substituting this value of W_T in Eq. 7.8, one gets

$$\{(PdV)_T + A_T\} \leq (F_1 - F_2) \qquad\qquad 7.12$$

If the process is both isothermal and isochoric then dV is zero and Eq. 7.12 reduces to

$$A_{T,v} \leq (F_1 - F_2) \qquad\qquad 7.13$$

According to Eq. 7.13, decrease in Helmholtz function in a process at constant volume and at constant temperature puts an upper limit to the non-PdV work.

Finally, if no configurational work, including PdV is performed in an isothermal process, i.e., $A_T = 0$ and $PdV = 0$ then

$$0 \leq (F_1 - F_2)$$

This means that in processes at constant temperature and volume when A is zero, $F_2 \leq F_1$, that is, in such processes the Helmholtz function either decreases or remains constant (if the process is reversible). One may also say that such processes are possible only when the Helmholtz function of the system decreases.

7.2 The Gibbs Function

In the preceding sections we considered the isochoric and isothermal processes. We now consider isothermal and isobaric processes, i.e., the processes in which the temperature and the external pressure remain constant. In this case then from Eq. 7.12

$$\{(PdV)_{T,P} + A_T\} \leq (F_1 - F_2)$$

Or $\qquad\qquad [\{P(V_2 - V_1)\}_{T,P} + A_{T,P}] \leq (F_1 - F_2) \leq [\{U_1 - TS_1\} - \{U_2 - TS_2\}]$

Or $\qquad\qquad A_{T,P} \leq [\{U_1 - TS_1\} - \{U_2 - TS_2\}] - \{P\,(V_2 - V_1)\}_{T,P}$

Or $\qquad\qquad A_{T,P} \leq [\{U_1 - TS_1 + PV_1\} - \{U_2 - TS_2 + PV_2\} \qquad\qquad 7.14$

A function G, called the Gibb's function, is defined as,

$$G \equiv (U - TS + PV) = (F + PV) = (H - TS) \qquad\qquad 7.15$$

So that Eq. 7.14 becomes

Or
$$A_{T,P} \leq [G_1 - G_2] \qquad\qquad 7.16$$

where
$$G_1 \equiv \{U_1 - TS_1 + PV_1\} \text{ and } G_2 \equiv \{U_2 - TS_2 + PV_2\} \qquad 7.17$$

According to Eq. 7.16, the decrease in Gibbs function sets an upper limit to the non-Pdv work that may be obtained in any process between the two states of a system at same temperature and pressure. If the process is reversible, the work is equal to the decrease in the value of the Gibbs function. Figure 7.1 (c) shows how the situation of Eq. 7.16 may be represented in a figure.

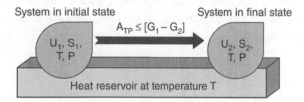

Fig. 7.1   (c) Schematic representation of Equation 7.16

In the case when $A_{T,P} = 0$, that is, all works except of *Pdv* work are absent then

$$[G_1 - G_2] \geq 0 \qquad\qquad 7.18$$

It means that in the case when only Pdv work is performed by any process between the two equilibrium states of a system at same temperature and pressure the Gibbs function either decreases or remains same. There is no change in the value of Gibbs function when the process is reversible. Conversely, it may be said that it is not possible without decreasing the Gibbs function to reach an equilibrium state of a system that has same temperature and pressure as the initial state by a process in which only *PdV* work is done.

It is left as an exercise to show that in case of an ideal gas the change in specific Gibb's function $g (= u - Ts + Pv = h - Ts)$ in going from an equilibrium state of temperature T_i, and pressure P_i to a state of final values T_f, P_f is given by,

$$\Delta g = RT (\ln P_f + \omega) \qquad\qquad 7.19$$

where
$$\omega = c_P T_f \left(1 - \ln \frac{T_f}{T_i}\right) - RT_f \ln P_i - c_P T_i - s_i \left(T_f - T_i\right)$$

It may be noted that ω is a function of temperatures only.

7.3   The Characteristic Variables

In systems that have three variables and two of which are independent, each state function has two preferred variables, called "pair of natural or characteristic variables". Characteristic variables are also called canonical variables. The pairs of canonical variables of four important functions are given here within bracket after each function: H (S, P); U (S, V); F (T, V); G (T, P). The significance of characteristic variables lies in the fact that if the function is known in terms of its characteristic variables and the derivatives of the function with respect to the characteristic variables are also known then all thermodynamic properties of the system can be determined or it may be said that the system is completely known. It may be pointed out that in general knowledge of the equation of state and the energy equation of the system is required for the complete knowledge of the system.

7.4   Thermodynamic Potentials

Let us consider the two nearby equilibrium states of a closed system. Since the system is closed, matter can neither enter nor escape from the system and if the two states are infinitesimally close to each other, the change in the magnitude of system parameters may be represented by their differentials. Therefore,

$$dU = TdS - PdV \tag{7.20}$$

$$dH = dU + PdV + VdP \tag{7.21}$$

$$dF = dU - TdS - SdT \tag{7.22}$$

And $\qquad dG = dU - TdS - SdT + PdV + VdP \tag{7.23}$

On substituting the value of dU from Eq. 7.20 in Eqs. 7.21–7.23, one gets:

$$dH = TdS + VdP \tag{7.24 (a)}$$

$$dF = -PdV - SdT \tag{7.25 (a)}$$

$$dG = VdP - SdT \tag{7.26 (a)}$$

Assuming H, F, G and U as functions of their characteristic variables, (S, P), (V, T), (P, T) and (S, V) respectively, one may write:

$$dH = \left(\frac{\partial H}{\partial S}\right)_P dS + \left(\frac{\partial H}{\partial P}\right)_S dP \tag{7.24 (b)}$$

$$dF = \left(\frac{\partial F}{\partial V}\right)_T dV + \left(\frac{\partial F}{\partial T}\right)_V dT \tag{7.25 (b)}$$

$$dG = \left(\frac{\partial G}{\partial P}\right)_T dP + \left(\frac{\partial G}{\partial T}\right)_P dT \tag{7.26 (b)}$$

$$dU = \left(\frac{\partial U}{\partial S}\right)_V dS + \left(\frac{\partial U}{\partial V}\right)_S dV \tag{7.27}$$

On comparing the coefficients of the corresponding terms in Eq. 7.24 (a) with Eq. 7.24 (b), Eq. 7.25 (a) and Eq. 7.25 (b), Eq. 7.26 (a) with Eq. 7.26 (b) and Eq. 7.20 with Eq. 7.27 one gets,

$$\left(\frac{\partial H}{\partial S}\right)_P = T; \qquad \left(\frac{\partial H}{\partial P}\right)_S = V \tag{7.28}$$

$$\left(\frac{\partial F}{\partial V}\right)_T = -P; \qquad \left(\frac{\partial F}{\partial T}\right)_V = -S \tag{7.29}$$

$$\left(\frac{\partial G}{\partial P}\right)_T = V; \qquad \left(\frac{\partial G}{\partial T}\right)_P = -S \qquad\qquad 7.30$$

And
$$\left(\frac{\partial U}{\partial S}\right)_V = T; \qquad \left(\frac{\partial U}{\partial V}\right)_S = -P \qquad\qquad 7.31$$

Since Eqs. 7.28–7.31 resemble the equations for the components of electrostatic field intensities derived from a scalar potential $\varphi\left[\varepsilon_x = \left(\frac{\partial \varphi}{\partial x}\right)\right]$, H, F, G and U are also called thermodynamic potentials in analogy to the electrical case. However, it is always better to call them functions rather than free energies or potentials to avoid any unnecessary confusion. Relations represented by Eqs. 7.28–7.31 are very important and need to be remembered by heart.

7.4.1 Remembering expressions for the derivatives of thermodynamic potentials

It is desired that expressions for thermodynamic potentials U, H, F, G and for their derivatives dU, dH, dF and dG contained in Eqs. 7.20, 7.24(a), 7.25(a) and 7.26(a) may be readily recalled when required for calculations. One way of remembering these relations is to construct a cube, called the Koenig cube, like the one shown in Fig. 7.2 (d). The cube has four sides at the center of each side is one of the four potentials U, H, G and F. The characteristic variables of each of the four potentials are at the two ends of the arm associated with the potential. For example, the characteristic variables V and S are at the two ends of the arm which contains potential U at the center. The two arrows within the cube show the direction in which the multiplication of the corner quantities is taken positive. The multiplication of corner quantities is taken negative if traversed in the direction opposite to the direction of the arrows. It is easy to remember the arrangement of letters, by remembering the phrase "**V**ery **F**irst **T**ime, **U**nder **G**raduate, **S**tudents **H**ave **P**roblem "in which first letter of each word indicates either some variable or some function.

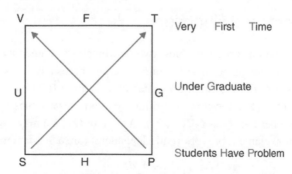

Fig. 7.1   (d) Remembering Thermodynamic relations

The working of the cube may be explained by taking the example of potential U. As is clear from the cube the characteristic variables of U are S and V that are at the ends of the side that has U in the middle. Now the relations of U and dU with V, S, T, and P may be easily written as,

$$U = TS - PV \quad \text{and} \quad dU = TdS - PdV$$

It may be noticed that in going diagonally from V to P motion is in a direction opposite to the arrow and hence the negative sign while in going from S to T diagonal motion is in the direction of the arrow, hence the positive sign. Also, when writing the differential dU, the differentials dS and dV of the characteristic variables of U will appear in the relation and not the differentials of T and P.

7.5   Gibbs–Helmholtz Relations

While discussing the importance of characteristic variables it was mentioned that a system is completely known if some system function (in terms of its characteristic variables) along with the function derivatives with respect to characteristic variables are known. As an example let us take the case of the Gibbs function G. If G as a function of (T, P) is known, i.e., the expression for G in terms of T and P is known and the magnitudes of partial derivatives $\left(\dfrac{\partial G}{\partial P}\right)_T = V$ and $\left(\dfrac{\partial G}{\partial T}\right)_P = -S$

are also known, then all other thermodynamic parameters of the system may be obtained using relations given by Eqs. 7.24 (a)–7.26 (a) and 7.20. For example, the Helmholtz function H for the system will be,

$$H = G + TS = G\,(T,P) - T\left(\frac{\partial G}{\partial T}\right)_P \qquad\qquad 7.32$$

Similarly, if F is known as a function of T and V, then the equation of state of the system, i.e., a relation between P, V and T may be obtained from the relation $\left(\dfrac{\partial F}{\partial V}\right)_T = -P$. Also since $\left(\dfrac{\partial F}{\partial T}\right)_V = -S$,

and therefore, $$U = F + TS = F - T\left(\frac{\partial F}{\partial T}\right)_V \qquad\qquad 7.33$$

Equations 7.32 and 7.33 are called the Gibbs–Helmholtz equations.

7.6   The Generalized Functions

So far systems that depend on three variables, two of which are independent variables have been discussed. However, in many physical situations we come across systems that are functions of many variables. Let us consider a system the states of which are functions of temperature T, two extensive variables X_1 and X_2, and the corresponding two intensive variables Y_1 and Y_2. In such a case temperature T and any other pair (X_1, X_2), (Y_1, Y_2), (X_1, Y_2) or (X_2, Y_1) may be taken as independent variables. The differential element of work (path dependent) for such a system is given by,

$$d'W = Y_1 dX_1 + Y_2 dX_2 \qquad\qquad 7.34$$

And the differential element of internal energy dU by

$$dU = TdS - d'W = TdS - (Y_1 dX_1 + Y_2 dX_2) \qquad\qquad 7.35$$

Let T, X_1 and X_2 be the independent variables.

Now, $$F = U - TS \text{ and } dF = dU - TdS - SdT$$

Or
$$dF = \{TdS - (Y_1 dX_1 + Y_2 dX_2)\} - TdS - SdT$$

Or
$$dF = -Y_1 dX_1 - Y_2 dX_2 - SdT \qquad 7.36$$

Since there are three independent variables, T, X_1 and X_2, the Helmholtz function F is also a function of these three independent variables. That means $F = F(T, X_1, X_2)$ and,

$$dF = \left(\frac{\partial F}{\partial T}\right)_{X_1, X_2} dT + \left(\frac{\partial F}{\partial X_1}\right)_{T, X_2} dX_1 + \left(\frac{\partial F}{\partial X_2}\right)_{T, X_1} dX_2 \qquad 7.37$$

Comparing the coefficients of dT, dX_1, dX_2 in Eqs. 7.36 and 7.37 one gets,

$$\left(\frac{\partial F}{\partial T}\right)_{X_1, X_2} = -S, \left(\frac{\partial F}{\partial X_1}\right)_{T, X_2} = -Y_1, and \left(\frac{\partial F}{\partial X_2}\right)_{T, X_1} = -Y_2 \qquad 7.38$$

Similarly, the Gibbs function for the system may be defined as $G = U - TS + X_1 Y_1 + X_2 Y_2$

And
$$dG = -SdT + X_1 dY_1 + X_2 dY_2 \qquad 7.39$$

Also
$$dG = \left(\frac{\partial G}{\partial T}\right)_{X_1, X_2} dT + \left(\frac{\partial G}{\partial Y_1}\right)_{T, Y_2} dY_1 + \left(\frac{\partial G}{\partial Y_2}\right)_{T, Y_1} dY_2 \qquad 7.40$$

Comparing the coefficients of dT, dY_1, and dY_2 in Eqs. 7.39 and 7.40, one gets,

$$\left(\frac{\partial G}{\partial T}\right)_{X_1, X_2} = -S, \left(\frac{\partial G}{\partial Y_1}\right)_{T, Y_2} = X_1, \left(\frac{\partial G}{\partial Y_2}\right)_{T, Y_1} = X_2 \qquad 7.41$$

This treatment may be extended for larger number of independent variables and in each case the partial derivatives of U, H or G may give the required values of system parameters.

When the system in consideration involves gravitational, electrical or magnetic fields, that is some conservative field, the total energy of the system apart from the internal energy U also contains a component that is due to the potential energy of the field. Since potential energy has the form $X_2 Y_2$, the total energy $E = U + X_2 Y_2$. In such cases one uses the generalized Helmholtz function, denoted by F^*, defined as,

$$F^* \equiv E - TS = U - TS + X_2 Y_2 \qquad 7.42$$

7.7 Maxwell Relations

James Clerk Maxwell (13 June 1831–5 November 1879, a Scottish mathematical physicist) using the property that the derivatives of state functions are exact differentials, derived relationships between the partial derivatives of state variables P, T, V and S. These relations are called the Maxwell relations.

Let $\psi(a, b)$ be a state function that depends on variables a and b and its derivative $d\psi$ is given as

$$d\psi(a, b) = M(a, b)\, da + N(a, b)\, db \qquad 7.43$$

Since $\psi(a, b)$ is a state function $d\psi(a, b)$ should be an exact differential and hence,

$$\left(\frac{\partial M}{\partial b}\right)_a = \left(\frac{\partial N}{\partial a}\right)_b \qquad\qquad 7.44$$

Now, comparing Eq. 7.44 with Eq. 7.20, $dU = TdS - PdV$, one may write

$$\psi (a,b) = U\ (S,V);\ a = S;\ b = V;\ M\ (a,\ b) = T\ (S,\ V)\ \text{and}\ N\ (a,\ b) = -P\ (S.V)$$

Hence, Eq. 7.44 gives; $\qquad \left(\frac{\partial T}{\partial V}\right)_S = \left(\frac{\partial(-P)}{\partial S}\right)_V = -\left(\frac{\partial P}{\partial S}\right)_V$

Using the same method, one may obtain the following Maxwell relations from Eqs. 7.24 (a)–7.26 (a)

$$dU = TdS - PdV \qquad \Rightarrow \quad \left(\frac{\partial T}{\partial V}\right)_S = -\left(\frac{\partial P}{\partial S}\right)_V \qquad 7.45$$

$$dF = -PdV - SdT \qquad \Rightarrow \quad \left(\frac{\partial P}{\partial T}\right)_V = \left(\frac{\partial S}{\partial V}\right)_T \qquad 7.46$$

$$dG = VdP - SdT \qquad \Rightarrow \quad \left(\frac{\partial V}{\partial T}\right)_P = -\left(\frac{\partial S}{\partial P}\right)_T \qquad 7.47$$

$$dH = TdS + VdP \qquad \Rightarrow \quad \left(\frac{\partial T}{\partial P}\right)_S = \left(\frac{\partial V}{\partial S}\right)_P \qquad 7.48$$

These equations are called the Maxwell relations and are important as they provide means to calculate change in entropy as a function of P, T, or V.

7.8   The Third Law of Thermodynamics

(a) Background

The third law of thermodynamics emerged from the study of chemical reactions at low temperatures. Towards the end of the eighteen century Chemists were trying to know the property of the system that drives chemical reactions at constant temperature and pressure, particularly at low temperatures. In most of the chemical reactions studied at that time it was found that heat was emitted in the reaction and that the reaction rate was proportional to the amount of heat evolved. This led to the wrong conclusion that the decrease in the enthalpy of the system during the chemical reaction is the driving force for the reaction. However, the fact that in some reactions heat was absorbed instead of being emitted, led to the correct conclusion that in spontaneous reactions at constant temperature and pressure, it is the change in the Gibbs function, and not the change in enthalpy, that decides whether heat will be emitted or absorbed in the process.

(b) Nernst observation

According to the Gibbs–Helmholtz equation, Eq. 7.32, $H = G(T,P) - T\left(\frac{\partial G}{\partial T}\right)_P$, the change in enthalpy of a system at constant temperature and pressure is given by

$$(H_2 - H_1) = (G_2 - G_1) - T\left[\left(\frac{\partial G_2}{\partial T}\right) - \left(\frac{\partial G_1}{\partial T}\right)\right]_P \qquad 7.49$$

Or

$$\Delta H = \Delta G - T\left(\frac{\partial \Delta G}{\Delta T}\right)_P \qquad 7.50$$

Equation 7.49 tells that ΔH will approch ΔG only when $T\left(\frac{\partial \Delta G}{\Delta T}\right)_P$ approaches zero. Walther

Hermann Nernst (June 25, 1864–November 18, 1941) a German physicist on the basis of some very precise experiments made a very general observation that ΔH approaches ΔG and that their rates of change with temperature $\left(\frac{\partial \Delta H}{\Delta T}\right)_P$ and $\left(\frac{\partial \Delta G}{\Delta T}\right)_P$ also approach zero as temperature approaches absolute zero. Thus in the limit temperature approaching zero, i.e., T $\to$ 0, $\Delta H = \Delta G$ and $\lim_{T \to 0}\left[\left(\frac{\partial G_2}{\partial T}\right) - \left(\frac{\partial G_1}{\partial T}\right)\right]_P$ also approaches zero. But $\left(\frac{\partial G}{\partial T}\right)_P = -S$; therefore, in the limit T $\to$ 0,

$$\lim_{T \to 0}\left[\left(\frac{\partial G_2}{\partial T}\right) - \left(\frac{\partial G_1}{\partial T}\right)\right]_P = 0$$

Or

$$\lim_{T \to 0}\left(S_1 - S_2\right) = 0 \qquad 7.51$$

Equation 7.51 contains one of the few different statements of the third law of thermodynamics and is generally called the Nernst heat theorem. According to the Nernst heat theorem:

'In the vicinity of absolute zero of temperature reactions in liquid and solid systems in the state of internal equilibrium take place with no change in entropy'.

(c) Planck's statement of the third law

Planck formulated the third law of thermodynamics as follows:

'When temperature falls to absolute zero, the entropy of any pure crystalline substance tends to a universal constant, which can be taken to be zero'.

In mathematical notations, the Planck statement may be put as,

$$S \to 0, \text{ as } T \to 0 \qquad 7.52$$

The entropy scale, in which entropy S is taken as zero at absolute zero of temperature, is called absolute entropy scale and the value of entropy is referred as absolute entropy. If entropy S is a function of some system parameter x, that may be volume or some similar other quantity, then in Planck statement, Eq. 7.52, it is inherently assumed that x remains finite at $T = 0$.

The Planck statement of the third law has limited jurisdiction as it is applicable only to crystalline substances and is, therefore, not universal. It is not consistent with the spirit of the laws of thermodynamics which claim universal applications. However, if the Planck statement is read in conjunction with the Nernst heat theorem and the Einstein statement, given below, it becomes a universal law.

(d) Einstein statement

'As temperature falls to absolute zero, the entropy of any substance remains finite'

Mathematically, the Einstein statement may be put as,

$$\lim_{T \to 0} S(T,x) = S_0(x), |S_0(x)| < \infty, as T \to 0, \text{ and } |x| < 0 \qquad 7.53$$

It is an assumption that the system property x remains finite at absolute zero.

An important conclusion that follows from Einstein statement is that thermal capacities both at constant volume and at constant pressure for all substances vanish at absolute zero. It is because in heating of a substance at constant volume the entropy changes by

$$S = \int_0^T \frac{C_V dT}{T}$$

And if S has to be finite at $T = 0$, then C_V must vanish at absolute zero, otherwise the above integral will diverge. Similarly, considering the heating of the substance at constant pressure, it may be shown that at absolute zero of temperature the thermal capacity at constant pressure must also vanish. Large number of experiments involving the study of properties of substances at very low temperatures were carried out at the University of Berlin, Germany, by the research group of Nernst and confirmed the findings of Einstein. It may be noted that thermodynamic systems become quantum systems at low temperatures and, therefore, confirmation of Einstein statement is also a confirmation of the quantum nature of the matter.

(e) Nernst principle of un-attainability of absolute zero

The third law of thermodynamics may also be stated in terms of the principle of un-attainability of absolute zero put forward by Nernst and that says,

'Any thermodynamic process cannot reach the temperature of absolute zero in finite number of steps and in finite time'

The most effective way of producing low temperatures is to take a system and cover it with an adiabatic boundary. If this isolated system is now made to do some adiabatic work at the cost of its internal energy, the temperature of the system will fall. Adiabatic demagnetization of paramagnetic salts is an example of such a process that is often used to produce low temperatures. In an adiabatic process the entropy of the system does not change but some other system parameter x (that may be the state of magnetization, or may be the volume in the case of adiabatic expansion) changes in the process of adiabatic work. The principle of un-attainability may be put into the following mathematical form,

$$S(T, x) - S(0, x + \Delta x) > 0 \text{ when } T > 0 \text{ and } |x| < \infty, |\Delta x| < \infty \qquad 7.54$$

If the principle of un-attainability, represented by Eq. 7.54, is not true i.e., $S(T, x) = S(0, x + \Delta x)$, then it will be possible to reach a state at temperature absolute zero starting from a state of finite temperature. According to the present understanding of the principle of un-attainability in case it is possible that $S(T, x) = S(0, x + \Delta x)$, then the adiabatic process connecting (T, x) state with $(0, x + \Delta x)$ state must be hindered by some other physical factor, for example the process may take infinite time.

Another way of understanding the principle of un-attainability is through Figs. 7.2 (a) and (b) in which the behavior of a system in the neighborhood of absolute zero for two values of the parameter x is shown by solid lines marked X and X + ΔX. For example if the system is some paramagnetic salt, X + ΔX may represent the state of magnetization of the salt while X may refer to the state when the salt is demagnetized. Vertical dotted lines represent adiabatic work performed by the system, adiabatic demagnetization in case of a paramagnetic salt, in which temperature of the system decreases without any change in the entropy. The horizontal arrows represent the isothermal process of bringing the system back to the initial state at a lower temperature.

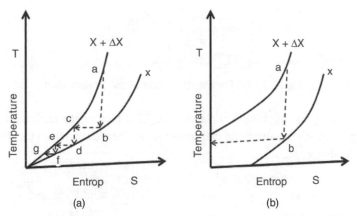

Fig. 7.2   (a) Systems follow Einstein statement; (b) Systems do not follow Einstein statement

The states X and X+ΔX, in Fig. 7.2(a), follow Einstein statement so that their entropies approach zero at absolute zero. As is evident from the ladder like vertical and horizontal arrows in Fig. 7.2(a) showing the adiabatic and isothermal recycling processes, the system will not be able to attain absolute zero in a finite number of steps. In Fig. 7.2 (b), on the other hand, the system does not follow Einstein statement, entropies of the states do not approach a zero value at absolute zero temperature and, therefore, the temperature of absolute zero may be attained in finite number of steps.

7.9   The Concept of Perpetual Motion

The fundamental implications of the laws of thermodynamics are related to engines—a device that may convert heat into work. Total work can be converted into heat but the reverse is not true, that is, the total heat cannot be converted into work. The zeroth, first, and second laws of thermodynamics put some restrictions that prohibit certain types of engines—these are conventionally called perpetual motions of zeroth, first and second kinds, depending upon which law of thermodynamics the engine violates.

Perpetual motion of the zeroth kind is shown in Fig. 7.3(a). The Carnot engine draws heat Q_H from the reservoir at high temperature, rejects heat Q_L to the reservoir at lower temperature and produces work $W = (Q_H - Q_L)$. The zeroth law is violated in the transfer of heat from a low temperature reservoir to a reservoir at higher temperature. Though not explicit, the principle of heat flow from higher to lower temperature is embedded in the zeroth law. No doubt that the second law is also violated in Fig. 7.3(a).

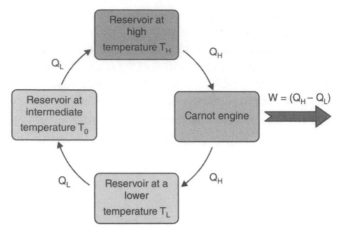

Fig. 7.3   (a) Perpetual motion of the Zeroth kind

Perpetual motion of the first kind is shown in Fig. 7.3 (b) where the engine produces work from nowhere and Fig. 7.3 (c) shows an engine that converts total heat drawn from the reservoir into work, violating the second law.

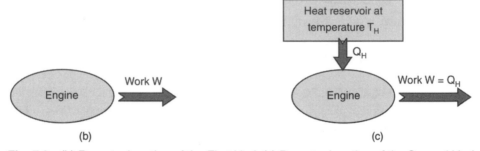

(b)   (c)

Fig. 7.3   (b) Perpetual motion of the First kind; (c) Perpetual motion of the Second kind

Perpetual motion of the third kind is shown in Fig. 7.3(d). The engine violates the principle of un-attainability of absolute zero, dumps all the excess entropy drawn from the high temperature reservoir at an entropy sink at absolute zero and converts total heat drawn into work.

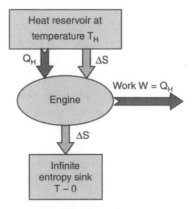

Fig. 7.3   (d) Perpetual motion of the Third kind

7.10   Thermodynamics of Open Systems

A system is said to be a closed one when matter (or mass) can neither enter nor leave the system. The formulations developed so far are applicable only to closed systems. The laws of thermodynamics with suitable modifications are, however, applicable to open systems where matter or molecules may leave or enter the system. Thermodynamics of open systems will be developed in the following.

In general, matter of same kind and / or different kinds and with the same or different values of the intrinsic parameters may enter or leave a given system if it is open. In an open system at a given point of time there may be several different types of particles, each with a given concentration. One essential consequence of the openness of the system is that for an open system the thermodynamic parameters are required to be given as functions of the concentration of different types of particles in the system, in addition to all other parameters that are required for a closed system.

Let us start with a simpler case of an open system that has only one kind of particles or molecules. The number of moles $\mathbb{N}$ (or number of particles as the case may be) in the system will change with time in an open system and, therefore, for any extrinsic property like the internal energy U of the system, a term that depends on the number of moles $\mathbb{N}$ (*or the number*) of molecules will have to be included. For a closed system $U = U(S,V)$ but for an open system

$$U = U(S,V,\mathbb{N}) \quad \text{and} \quad dU = \left(\frac{\partial U}{\partial S}\right)_{V,\mathbb{N}} dS + \left(\frac{\partial U}{\partial V}\right)_{S,\mathbb{N}} dV + \left(\frac{\partial U}{\partial \mathbb{N}}\right)_{V,S} d\mathbb{N} \qquad 7.55$$

In case of an open system, it is required to include a new parameter called the *chemical potential* denoted by μ and defined as,

$$\mu \equiv \left(\frac{\partial U}{\partial \mathbb{N}}\right)_{V,S} d\mathbb{N} \qquad 7.56$$

The chemical potential μ is equal to the change in internal energy per mole (or per particle) of the system at constant entropy and volume. Equation 7.55 must reduce to $dU = \left(\frac{\partial U}{\partial S}\right)_V dS + \left(\frac{\partial U}{\partial S}\right)_S dV$
$= TdS - PdV$ if the system is closed and $d\mathbb{N}$ is zero, the Eq. 7.55 may be written as

$$dU = \left(\frac{\partial U}{\partial S}\right)_{V,\mathbb{N}} dS + \left(\frac{\partial U}{\partial V}\right)_{S,\mathbb{N}} dV + \left(\frac{\partial U}{\partial \mathbb{N}}\right)_{V,S} d\mathbb{N} = TdS - PdV + \mu d\mathbb{N} \qquad 7.57$$

Equation 7.57 is the expression for the change in internal energy for an open system. Equation 7.57 may also be written as

$$dS = \frac{1}{T} dU + \frac{P}{T} dV - \frac{\mu}{T} d\mathbb{N} \qquad 7.58$$

Treating S as a function of $(U, V,$ and $\mathbb{N})$, one may write

$$dS = \left(\frac{\partial S}{\partial U}\right)_{V,\mathbb{N}} dU + \left(\frac{\partial S}{\partial V}\right)_{U,\mathbb{N}} dV + \left(\frac{\partial S}{\partial \mathbb{N}}\right)_{V,U} d\mathbb{N} \qquad 7.59$$

Comparing the coefficients of dU, dV and d$\mathbb{N}$ in Eqs. 7.58 and 7.59 one gets,

$$\left(\frac{\partial S}{\partial U}\right)_{V,\mathbb{N}} = \frac{1}{T} ; \left(\frac{\partial S}{\partial V}\right)_{S,\mathbb{N}} = \frac{P}{T} ; \mu = -T\left(\frac{\partial S}{\partial \mathbb{N}}\right)_{V,U} \qquad 7.60$$

Since $$dF = dU - Tds - SdT$$

The difference in the value of F of two nearby equilibrium states of an open system may be given by

$$dF = -PdV - SdT + \mu d\mathbb{N} \qquad 7.61$$

Also $$dF = \left(\frac{\partial F}{\partial V}\right)_{T,\mathbb{N}} dV + \left(\frac{\partial F}{\partial T}\right)_{V,\mathbb{N}} dT + \left(\frac{\partial F}{\partial \mathbb{N}}\right)_{V,T} d\mathbb{N} \qquad 7.62$$

Comparison of Eqs. 7.61 and 7.62 gives,

$$\mu = \left(\frac{\partial F}{\partial \mathbb{N}}\right)_{V,T} \qquad 7.63$$

Similarly, the difference in the Gibbs function dG in the two nearby states of an open system may be given by the relation

$$dG = VdP - SdT + \mu d\mathbb{N} \qquad 7.64$$

And comparing it with equation $dG = \left(\frac{\partial G}{\partial P}\right)_{T,\mathbb{N}} dP + \left(\frac{\partial G}{\partial T}\right)_{P,\mathbb{N}} dT + \left(\frac{\partial G}{\partial \mathbb{N}}\right)_{T,P} d\mathbb{N}$ one may get,

$$\mu = \left(\frac{\partial G}{\partial \mathbb{N}}\right)_{P,T} \qquad 7.65$$

Collecting values of the chemical potential from Eqs. 7.60, 7.63 and 7.65

$$\mu = -T\left(\frac{\partial S}{\partial \mathbb{N}}\right)_{V,U} = \left(\frac{\partial F}{\partial \mathbb{N}}\right)_{V,T} = \left(\frac{\partial G}{\partial \mathbb{N}}\right)_{P,T} \qquad 7.66$$

Equation 7.66 may be further generalized for any pair of corresponding intensive parameter Y (say, P) and the extensive parameter X (say V).

$$\mu = -T\left(\frac{\partial S}{\partial \mathbb{N}}\right)_{X,U} = \left(\frac{\partial F}{\partial \mathbb{N}}\right)_{X,T} = \left(\frac{\partial G}{\partial \mathbb{N}}\right)_{Y,T} \qquad 7.67$$

7.10.1 General case of a system with different kinds of entities

In the preceding section an open system with $\mathbb{N}$ moles (or number of particles) of only one kind of particles (molecules) was considered. However, it is possible that the system is made up of $\mathbb{N}_1$, $\mathbb{N}_2$, $\mathbb{N}_3$, ... $\mathbb{N}_\gamma$ moles (or particles) of different types of entities contained in volume V. In that case the internal energy U and the entropy S of the system will be functions of the number of moles of different kinds of entities, i.e.,

$$U = U(S, V, \mathbb{N}_1, \mathbb{N}_2, \mathbb{N}_3 ... \mathbb{N}_j ... \mathbb{N}_\gamma) \qquad 7.68$$

And $$S = S(U, V, \mathbb{N}_1, \mathbb{N}_2, \mathbb{N}_3 ... \mathbb{N}_j ... \mathbb{N}_\gamma) \qquad 7.69$$

The difference in the internal energy dU between the two nearby equilibrium states of such a system

$$dU = \left(\frac{\partial U}{\partial S}\right)_{V,\mathbb{N}} dS + \left(\frac{\partial U}{\partial V}\right)_{S,\mathbb{N}} dV + \sum_1^\gamma \left(\frac{\partial U}{\partial \mathbb{N}_j}\right)_{S,V,\mathbb{N}_{i\neq j}} \qquad 7.70$$

where,
$$\left(\frac{\partial U}{\partial S}\right)_{V,\mathbb{N}} = T; \left(\frac{\partial U}{\partial V}\right)_{S,\mathbb{N}} = -P; \left(\frac{\partial U}{\partial \mathbb{N}_j}\right)_{S,V,\mathbb{N}_{i \neq j}} = \mu_j \qquad 7.71$$

And Eq. 7.70 becomes,
$$dU = TdS - PdV + \sum_{1}^{\gamma} \mu_j \qquad 7.72$$

7.11 The Equilibrium

In the following we discuss conditions for different types of equilibriums.

(a) Thermal equilibrium

Let us consider a system that consists of two subsystems A and B in thermal contact. Let each subsystem has a fixed volume and fixed number of particles or moles. The entropy of system is the sum of the entropies of the two subsystems.

$$S = S^A(U^A, V^A, \mathbb{N}^A) + S^B (U^B, V^B, \mathbb{N}^B) \qquad 7.73$$

Then in equilibrium entropy of the system must be maximum and $dS = 0$, i.e.,

$$dS = \left(\frac{\partial S^A}{\partial U^A}\right)_{V^A,\mathbb{N}^A} dU^A + \left(\frac{\partial S^B}{\partial U^B}\right)_{V^B,\mathbb{N}^B} dU^B = 0 \qquad 7.74$$

But from Eq. 7.60 one has $\left(\frac{\partial S^A}{\partial U^A}\right)_{V^A,\mathbb{N}^A} = \frac{1}{T^A}$ and $\left(\frac{\partial S^B}{\partial U^B}\right)_{V^B,\mathbb{N}^B} = \frac{1}{T^B}$,

Here T^A and T^B are the temperatures of the two subsystems. Also, $dU^A = -dU^B$ as the total energy $U = U^A + U^B$ is conserved. With these substitutions Eq. 7.74 reduces to

$$\left(\frac{1}{T^A} - \frac{1}{T^B}\right) dU^A = 0 \text{ and hence, } T^A = T^B \qquad 7.75$$

Therefore, for thermal equilibrium of two systems in contact their temperatures must be equal.

It may be observed that this condition of thermal equilibrium also leads to the fact that energy (heat) must flow from a system at higher temperature to the system at lower temperature to attain thermal equilibrium. To show this let us assume that $T^A > T^B$ and the system is not in thermal equilibrium. In that case $dS > 0$ that means $\left(\frac{1}{T^A} - \frac{1}{T^B}\right) dU^A > 0$. But $\left(\frac{1}{T^A} - \frac{1}{T^B}\right) < 0$ if $T^A > T^B$.

Therefore, U^A must also be less than zero to make $dS > 0$. That means energy flows from system A to system B.

(b) Mechanical equilibrium

Suppose that the two subsystems are separated by a moveable diathermic wall so that the volumes V^A and V^B of the two subsystems may change to attain mechanical equilibrium subject to the condition that the total volume $V = (V^A + V^B)$, total energy $U = (U^A + U^B)$ and the total number of moles of particles of each kind remain same.

Again, from Eq. 7.73, for equilibrium

$$dS = \left(\frac{\partial S^A}{\partial U^A}\right)_{V^A, N_i^A} dU^A + \left(\frac{\partial S^A}{\partial V^A}\right)_{U^A, N_i^A} dV^A + \left(\frac{\partial S^B}{\partial U^B}\right)_{V^B, N_i^B} dU^B + \left(\frac{\partial S^B}{\partial V^B}\right)_{U^B, N_i^B} dV^B = 0$$

Or
$$dS = \frac{1}{T^A} dU^A - \frac{P^A}{T^A} dV^A + \frac{1}{T^B} dU^B - \frac{P^B}{T^B} dV^B = 0 \qquad 7.76$$

But because of the conservation of energy and constant total volume $dU^A = -dU^B$ and $dV^A = -dV^B$

$$dS = \left(\frac{1}{T^A} - \frac{1}{T^B}\right) dU^A - \left(\frac{P^A}{T^A} - \frac{P^B}{T^B}\right) dV^A = 0 \qquad 7.77$$

Since pressure and temperature are independent, Eq. 7.77 can only be satisfied if,

$$T^A = T^B \text{ and } P^A = P^B \qquad 7.78$$

The necessary conditions for mechanical equilibrium are specified by Eq. 7.78.

(c) Equilibrium with respect to mass flow

We now consider that the two subsystems are separated by a diathermic wall that is permeable to only i^{th}-kind of entities and to no other. We are now looking for equilibrium conditions with respect to the temperature and the chemical potential. Therefore, for equilibrium

$$dS = \frac{1}{T^A} dU^A - \frac{\mu^A}{T^A} d\mathbb{N}^A + \frac{1}{T^B} dU^B - \frac{\mu^B}{T^B} d\mathbb{N}^B = 0$$

That gives,
$$T^A = T^B \text{ and } \mu^A = \mu^B \qquad 7.79$$

This shows that as temperature difference is responsible for heat flow, pressure difference for change in volume, the difference in the chemical potential may be looked as the source for the flow of matter. Later it will be shown that chemical potential also plays as deriving force in phase transitions and chemical reactions.

(d) Phase equilibrium

It is known that pure substances tend to stay in three distinct states: solid, liquid and gas. If we take the example of water, when ice is heated it spontaneously converts into water at fixed values of temperature and pressure. On heating further, again at another set of fixed temperature and pressure water is spontaneously converted into vapors. These processes are discontinuous, i.e., they occur at specific state conditions—particular combinations of temperature and pressure. At exactly those conditions, the system may exist simultaneously in two or more phases in equilibrium.

A phase is homogeneous region of matter in which there is no spatial variation in the average density, energy, composition or other macroscopic properties of the matter. Phases may be different in their molecular structure, for example water has multiple ice phases that differ in their crystallographic structures. A phase may be considered as a distinct subsystem with boundaries that are interfaces with container walls or other phases. When one talks of phase equilibrium it means that the system consists of at least two sub systems of different phases separated with phase boundary. These subsystems exist spontaneously, without the use of partitions or other external interventions and since they are in equilibrium they may exchange energy, volume and particles.

In the last section on 'equilibrium with respect to the mass flow' it has been shown that the condition of entropy maximization requires that the temperature, the pressure and the chemical potential of the subsystems must be equal. The subsystems that may exchange particles, volume and energy designated by A and B in the last section, may well be the two phases of a single component system like water. A single component system is the one which is made up of only one kind of particles (molecules) but their arrangement, density or any other physical property may be different in different phases.

Thus in a single component system, the two different phases denoted by subscript 1 and 2, if in equilibrium then,

$$T_1 = T_2; P_1 = P_2; \text{ and } \mu_1 = \mu_2 \qquad\qquad 7.80$$

The three equalities contained in Eq. 7.80 ensure thermal, mechanical and thermodynamic equilibrium in the system. Invoking the P and T dependence of the chemical potential, the three conditions of equilibrium may be reduced to a single condition

$$\mu_1 (P,T) = \mu_2 (P,T) \qquad\qquad 7.81$$

Equation 7.81 showing the equality of two chemical potentials, have two unknown variables T and P, and if one of them is specified, Eq. 7.81 can be solved to get the value or values of for the other. Further, it may be noted that the unknown variables in Eq. 7.81 P and T are the characteristic variables of the Gibbs function.

Figure 7.4 shows the behavior of chemical potential for ice and liquid water as a function of Kelvin temperature. At atmospheric pressure, the chemical potentials of the ice and liquid water phases are equal at temperature 273.15 K where the two phases may co-exist in equilibrium; at temperatures higher than 273.15 K the chemical potential of liquid water is less than that for ice and therefore only liquid phase of water exists. At temperatures lower than 273.15 K, the chemical potential of ice phase is lower and therefore the only phase that may exist at temperatures lower than 273.15 K is the ice phase.

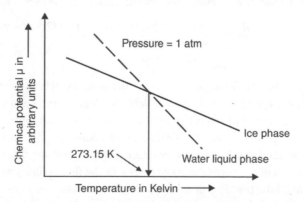

Fig. 7.4   Chemical potential vs temperature for water at 1 atm pressure

More than one phase of a single component system may co-exist in equilibrium at fixed values of temperature and pressure and from Eq. 7.65, at constant temperature and pressure Gibbs function per mole (or the specific Gibbs function), $\left(\dfrac{\partial G}{\partial N}\right)_{T,P} = g = \mu$. Therefore, while considering phases in equilibrium, the specific Gibbs functions for different phase must be equal to each other.

We now move to a more general case. Let there be a multi-component system that has γ different components (number of different types of molecules or entities is γ). Let N_γ be the number of moles of the component γ and the total number of moles be $N = \sum_{k=1}^{k=\gamma} N_k$. The concentration of each component in the system is represented by its molar fraction $x_k = \dfrac{N_k}{N}$. Now $\sum_{k=1}^{k=\gamma} x_k = 1$. Hence, there are $(\gamma - 1)$ independent values of x_k and not γ, because any constrain like the sum of x_k must be equal to one, reduces the degrees of freedom. Let us further assume that α number of different phases out of γ components co-exist in equilibrium at given values of P and T. The essential conditions for the co-existence of these different phases of different components are that the chemical potentials (or specific Gibbs functions) of all the co-existing phases at the given pressure P and temperature T are equal to each other. This equilibrium condition may be written as:

$$\mu_{component\ k,\ phase\ j}\left(P,T,\{x\}_{phase\ j}\right) = \mu_{component\ k,\ phase(j+1)}\left(P,T,\{x\}_{phase(j+1)}\right) \qquad 7.82$$

Equation 7.82 contains many equations, one each for possible value of k and j. The variable k in Eq. 7.82 may take γ different values and the variable j may take $(\alpha - 1)$ different values. Thus the total number of equations for the equality of the chemical potentials contained in Eq. 7.82 is $\gamma(\alpha - 1)$. Next let us calculate the total number of unknown quantities in these equations. We may do it in the following way. For each component (given value of γ) there will be α equations corresponding to different phases, each of which will have one unknown x. Out of the γ equations, the number of independent equations will be $(\gamma-1)$. So the total number of unknown x becomes $\{\alpha(\gamma-1)\}$. Apart from that the temperature T and P which are same in all the above equations are also unknown. Hence the total number of unknown quantities is $[2 + \alpha(Y - 1)]$. The degrees of freedom $d_{freedom}$ or variance of a system is equal to the total number of unknowns minus the number of equations giving the relations between the unknowns (called the constraints). In the present case,

$$d_{freedom} = [2 + \alpha(\gamma - 1)] - \gamma(\alpha - 1) = (2 - \alpha + \gamma)]$$

Or $$d_{freedom} = (2 - \alpha + \gamma) \qquad 7.83$$

Equation 7.83 represents the *Gibbs phase rule*. If the number of phases in equilibrium and the number of components are given, one may calculate the degrees of freedom for the system. For example, in case of water, the system has single component, $\gamma = 1$, and if we are looking for the co-existence of all the three phases, $\alpha = 3$, the degrees of freedom $d_{freedom}$ comes out to be zero. This means that only at one fixed value of the temperature and pressure the three phases of water may co-exist in equilibrium. In other words the co-existence of the three phases will be represented by a point. This point is called the triple point of water. For the co-existence of two phases of water, $\alpha = 2$, $\gamma = 1$ and degrees of freedom $d_{freedom} = 1$. It means that for a fixed value of either P or T the other parameter may be varied along a line in the P-T plane. The phase diagram of water in P-T plane is shown in Fig. 7.5.

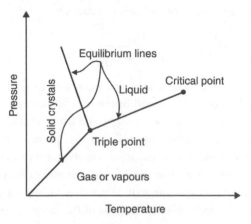

Fig. 7.5   Phase diagram of water

It has been shown that different phases of a multiple component system are in equilibrium when their chemical potentials or specific Gibbs functions are equal. If suppose the phases are not in equilibrium then matter from the phase that has larger value of specific Gibbs function g will flow into the phase of lower value of specific Gibbs function, till the specific Gibbs functions become equal.

(e) The Clausius–Clapeyron equation

Let there be two phases of a single component system that are in equilibrium at temperature T and pressure P. The two phases may be denoted by a and b. According to the phase rule, in case of a single component system if two phases are in equilibrium, the equilibrium extends over a certain range of pressure and temperature values and the specific Gibbs functions for the two phases are equal to each other over this range.

$$g^a = g^b \qquad\qquad 7.84$$

If pressure is changed within this range, it will be accompanied with the change of temperature, the molar concentrations and the specific Gibbs functions of the two phases. Let dg^a and dg^b be the changes in the specific Gibbs functions when pressure and temperature varies within the limits of equilibrium. The specific Gibbs functions for the two phases must also be equal for the new set of temperature and pressure, i.e.,

$$g^a + dg^a = g^b + dg^b$$

Or $$dg^a = dg^b \qquad\qquad 7.85$$

But, according to Eq. 7.26 (a), $dg = vdP - sdT$ and, therefore,

$$v^a dP - s^a\, dT = v^b\, dP - s^b\, dT$$

Or $$(v^b - v^a)\, dP = (s^b - s^a)\, dT$$

But in a single component system $\left(s^b - s^a\right) = \dfrac{\ell_{a,b}}{T}$, where $\ell_{a,b}$ is the latent heat of transformation of phase a to phase b. Therefore, the above equation reduces to

$$\left(\frac{\partial P}{\partial T}\right)_{a,b} = \frac{\ell_{a,b}}{T\left(v^b - v^a\right)} \qquad\qquad 7.86$$

Equation 7.86 is called the Clausius–Clapeyron equation. In geometric terms the Clausius–Clapeyron equation gives the slope of the equilibrium line between the two phases a and b. In Eq. 7.86, latent heat of transformation for liquid to vapor and from solid to liquid, i.e., $\ell_{2,3}$ and $\ell_{1,3}$ for most of the cases are positive, and temperature T is always positive. The sign of the slope of the equilibrium line between the liquid–vapor and solid–vapor equilibrium lines is decided by the sign of the factor $(v^b - v^a)$. If the specific volume in final phase is more than in the initial phase, $(v^b - v^a)$ is positive, the slope of the equilibrium line will be positive. Otherwise it will be negative. In case of ice, the volume decreases on melting and the slope of the ice-water equilibrium line is negative as shown in Fig. 7.5.

(f) Order of phase transition

Paul Ehrenfest (Jan 18,1880–Sep 25, 1933, an Austrian theoretical physicist) gave a scheme for classifying phase transitions. His classification was based on the continuity of the specific Gibbs function and its derivatives across the phase boundary. When in a single component system one moves from one phase to the other, generally the specific Gibbs function g remains continuous but the derivative $\left(\dfrac{\partial g}{\partial T}\right)_P = -s$ has discontinuity at the boundary as specific entropies of different phases

are different. Such phase transitions are called the first order phase transitions. Most of the solid–liquid, solid–vapor and liquid–vapor phase transitions are of the first order. The second order phase transitions are those in which both the specific Gibbs function g and its first derivatives remain continuous through the phases but the second derivative of the specific Gibbs function $\left(\dfrac{\partial^2 g}{\partial T^2}\right)_P = -\dfrac{c_P}{T}$

shows discontinuity at the phase boundary. Liquid–vapor transitions at critical point and super conducting state to non-super conducting state phase transitions are examples of second order phase transitions. In Ehrenfest scheme, in third order phase transitions the specific Gibbs function, its first derivatives and its second derivatives are all continuous but the third derivative of g shows discontinuity at the boundary. In this way still higher orders of phase transitions may be defined.

However, a typical phase transition in which the variation of the specific thermal capacity at constant pressure c_P with temperature has the shape of Greek letter lambda is commonly called the lambda transition. It is found in case of liquid helium when it goes from a liquid to a super fluid phase.

In statistical mechanics phase transitions are classified in terms of molecular scale correlations which are beyond the scope of present discussion.

Solved Examples

1. A system contained in volume V at temperature T has internal energy U given by $U = aVT^4$ and pressure given by $P = 1/3\ aT^4$, where a is a constant. Obtain the Helmholtz potential, entropy and chemical potential for the system.

Solution:

Using the relations $P = -\left(\dfrac{\partial F}{\partial V}\right)_T$ and $S = -\left(\dfrac{\partial F}{\partial T}\right)_V$ one gets,

$[dF = -PdV]_T$ and $F = -\left[\dfrac{1}{3}aT^4\right]V$ Substituting for U in above equation,

$$F = -\left[\frac{1}{3}aT^4\right]V = -\frac{U}{3} \qquad \text{S-7.1.1}$$

Also,

$$S = -\left(\frac{\partial F}{\partial T}\right)_V = \frac{4}{3}aT^3V = \frac{4U}{T} \qquad \text{S-7.1.2}$$

Since F is not a function of $\mathbb{N}$, the number of moles, the chemical potential $\mu = \left(\dfrac{\partial F}{\partial \mathbb{N}}\right)_{T,V}$ is zero.

2. A container of adiabatic walls has a volume V. It is divided into two parts of volumes V_1 and V_2 by a diaphragm. Two different but both ideal gases are filled in the two compartments at the same temperature T and pressure P. How much work can be derived from the system if partition between the two parts is removed?

Solution:

On removing the partition the two gases will diffuse into each other. As a result of this diffusion the total pressure P and temperature of the container will not change as the gases are ideal. However, the Gibbs function of the mixture will decrease. The decrease in Gibbs function will be equal to the work that may be derived from the system. Essentially it is required to calculate the initial and the final Gibbs functions of the system.

Let $\mathbb{N}_1$ and $\mathbb{N}_2$ be the number of moles of the gases in the two compartments and g_1^i and g_2^i their initial specific Gibbs functions. The initial value of the total Gibbs function G_{ini} is:

$$G_{ini} = \mathbb{N}_1 g_1^i + \mathbb{N}_2 g_2^i \qquad \text{S-7.2.1}$$

And from Eq. 7.19, the specific Gibbs function for an ideal gas is given by $g = RT(\ln P_f + \omega)$

where ω is a function of temperature alone. Therefore, for the two ideal gases the values of ω_1 and ω_2 will remain same both in the initial and the final states. Specific Gibbs functions for the two ideal gasses will have different values, g_1^f, g_2^f in the final state because in the final state each gas will experience its own partial pressures P_1^f, P_2^f .

Let $\mathbb{N} = \mathbb{N}_1 + \mathbb{N}_2$ so that the molar fractions of the two gases in the mixture in final state are

$$x_1 = \frac{N_1}{N} \text{ and } x_2 = \frac{N_2}{N} \qquad \text{S-7.2.2}$$

In the final state both the gases are contained in volume $V = (V_1 + V_2)$ at the same temperature T but they have different partial pressures.

$$P_1^f V = N_1 RT \text{ and } P_2^f V = N_2 RT, \text{ also } PV = \mathbb{N} RT$$

Therefore,
$$P_1^f = \frac{N_1 RT}{V} = \frac{N_1 RT}{\dfrac{N_1 RT}{P}} = Px_1 \text{ and similarly } P_2^f = Px_2$$

And $g_1^i = RT \left(\ln P + \omega_1 \right); g_1^f = RT \left(\ln P_1^f + \omega_1 \right) = RT \left(\ln Px_1 + \omega_1 \right) = RT \left(\ln P + \ln x_1 + \omega_1 \right)$

$g_2^i = RT \left(\ln P + \omega_2 \right); g_2^f = RT \left(\ln P_2^f + \omega_2 \right) = RT \left(\ln Px_2 + \omega_2 \right) = RT \left(\ln P + \ln x_2 + \omega_2 \right)$

Now change in the Gibbs function $\Delta G = G_{fin} - G_{ini}$ is,

$$\Delta G = N_1 \left(g_1^f - g_1^i \right) + N_2 \left(g_2^f - g_2^i \right) = RT \left[N_1 \ln x_1 + N_2 \ln x_2 \right] \qquad \text{S-7.2.3}$$

Equation S-7.2.3 gives the amount of the work that may be derived from the system. It may be noted that both $\ln x_1$ and $\ln x_2$ are negative quantities since x_1 and x_2 are fractions less than one. Therefore, the change in Gibbs function is negative, as expected since Gibbs function decreases in any irreversible process.

3. Obtain specific Helmholtz function for Van der Waals gas assuming that the specific thermal capacity c_V for the gas is (3/2)R.

Solution:

The equation of state for Van der Waals gas is $\left(P + \dfrac{a}{v^2} \right)(v - b) = RT$

Hence

$$P = \frac{RT}{(v - b)} - \frac{a}{v^2}$$

But
$$\left(\frac{\partial F}{\partial v} \right)_T = -P \text{ and } \left(\frac{\partial F}{\partial T} \right)_T = -s$$

Hence
$$F = \int P dv + C(T) = -\left[\frac{RT}{(v - b)} - \frac{a}{v^2} \right] dv + C(T)$$

where, $C(T)$ is a constant which may be a function of T.

$$F = -[RT \ln (v - b) + \frac{a}{v}] + C(T) \qquad \text{S-7.3.0}$$

Now
$$\left(\frac{\partial F}{\partial T} \right)_T = -s, \quad \text{so, } -s = -[R \ln (v - b)] + \frac{dC(T)}{dT} \qquad \text{S-7.3.1}$$

Also, $s = \int_{T_i}^{T_f} \dfrac{c_V dT}{T} = c_V \ln \dfrac{T_f}{T_i} = \dfrac{3}{2} R \ln \dfrac{T_f}{T_i}$ (given that $c_V = \dfrac{3}{2} R$)   S-7.3.2

T_f and T_i are the final and initial temperatures.

Equating S-7.3.1 and S-7.3.2

$$\frac{3}{2} R \ln \frac{T_f}{T_i} = [R \ln (v - b)] - \frac{dC(T)}{dT}$$

Or $$C(T) = \left[\frac{3}{2} RT - \frac{3}{2} RT \ln \frac{T_f}{T_i} \right] + \text{constant that does not depend on } T$$

Substituting this value of C(T) in Eq. S-7.3.0 one gets,

$$F = -[RT \ln (v - b) + \frac{a}{v}] + \left[\frac{3}{2} RT - \frac{3}{2} RT \ln \frac{T_f}{T_i} \right] + \text{constant}$$

4. The equation of state of a paramagnetic salt is given by $m = \left(\dfrac{D}{T} \right) H$ and the molar heat capacity

at constant magnetization by c_m, that may be taken as constant. Here m is molar magnetization, H magnetizing field, T temperature and D a constant. Obtain thermodynamic functions for the system.

Solution:

We first calculate the molar internal energy $u\,(m,T)$ that will be the sum of two components, one magnetic, denoted by u_m and the other thermal, denoted by u_T. These components are given by,

$$u_m = \int_0^m H dm = \int_0^m \frac{T}{D} m\, dm = \frac{T}{2D} m^2 \text{ and } u_T = c_m T$$

Therefore,

$$u\,(m,T) = u_m + u_T = T \left(c_m + \frac{m^2}{2D} \right) \text{and } T = \frac{u(m,T)}{\left(c_m + \dfrac{m^2}{2D} \right)}   \text{S-7.4.1}$$

Now we know $$\left(\frac{\partial s}{\partial u} \right)_m = \frac{1}{T} = \frac{\left(c_m + \dfrac{m^2}{2D} \right)}{u(m,T)};$$

therefore, $$s = \left(c_m + \frac{m^2}{2D} \right) \ln \frac{u(m,T)}{u_0} \text{ or } u(m,T) = u_0 e^{\left(\frac{s}{c_m + \frac{m^2}{2D}} \right)}   \text{S-7.4.2}$$

where u_0 is the value of the molar internal energy at $m = 0$, and is a constant of integration

Now, the specific Helmholtz function $f(m, T)$

$$= u(m,T) - Ts = T\left(c_m + \frac{m^2}{2D}\right) - T\left(c_m + \frac{m^2}{2D}\right)\ln\frac{u(m,T)}{u_0}$$

Or

$$f(m, T) = T\left(c_m + \frac{m^2}{2D}\right)\left[1 - \ln\frac{T\left(c_m + \frac{m^2}{2D}\right)}{u_0}\right] \qquad \text{S-7.4.3}$$

And molar enthalpy $h(H,T) = u - Hm = T\left(c_m + \frac{DH^2}{2T^2}\right) - \left(\frac{D}{T}\right)H^2 = T\left(c_m - \frac{DH^2}{2T^2}\right)$ S-7.4.4

Similarly, $\quad g = F - Hm = T\left(c_m - \frac{DH^2}{2T^2}\right) - T\left(c_m + \frac{DH^2}{2T^2}\right)\ln\frac{T\left(c_m + \frac{DH^2}{2T^2}\right)}{u_0}$ S-7.4.5

5. Given that molar Helmholtz function f is known as a function of V and T. Drive expressions for molar enthalpy h and molar Gibbs function g in terms of F, V and T.

Solution:

We know that $\qquad\qquad g = f + Pv \quad$ and $\quad h = g + Ts$ \qquad\qquad S-7.5.1

Also, $\qquad\qquad \left(\frac{\partial f}{\partial v}\right)_T = -P \quad$ and $\quad \left(\frac{\partial f}{\partial T}\right)_v = -s$ \qquad\qquad S-7.5.2

Substituting the values of P and s from Eq. S-7.5.2 into Eq. S-7.5.1, one gets:

$$g = f + Pv = f - v\left(\frac{\partial f}{\partial v}\right)_T \text{ and } h = g + Ts = f - v\left(\frac{\partial f}{\partial v}\right)_T - T\left(\frac{\partial f}{\partial T}\right)_v$$

6. A container of volume V is filled with N moles of an ideal gas. The container is divided by a moveable frictionless diaphragm into two parts A and B of volumes V_1 and $(V - V_1)$. The gas in the part of volume V_1 contains N_1 moles and the gas molecules cannot pass through the diaphragm. The gas in parts A and B is kept at the fixed temperature T. Assuming that the diaphragm moves very slowly changing the volume V_1, identify which of the thermodynamic potential should be minimized with respect to V_1 to obtain the condition of equilibrium for the gas in part A and B.

Solution:

Since the temperature and the number of moles of the gas on the two sides is kept constant and their volumes are allowed to change, it should be the total Helmholtz function of parts A plus part B that must be minimized with respect to V_1 for obtaining the condition of equilibrium. If F_{tot}, F_A and F_B are, respectively, the Helmholtz's functions of the total system and parts A and B, then,

$$F_{tot} = F_A + F_B = F\ (T,\ N_1,V_1) + F\ (T,(N - N_1),(V - V_1)) \qquad \text{S-7.6.1}$$

For equilibrium $\left(\dfrac{\partial F_{tot}}{\partial V_1}\right)_{T,N} = 0$, i.e.,

$$\left(\frac{\partial F\left(T,N_1,V_1\right)}{\partial V_1}\right)_{T,N} + \left(\frac{\partial F\left(T,\left(N-N_1\right),\left(V-V_1\right)\right)}{\partial V_1}\right)_{T,N}\left(\frac{\partial\left(V-V_1\right)}{\partial V_1}\right) = 0 \qquad \text{S-7.6.2}$$

But $\left(\dfrac{\partial F\left(T,N_1,V_1\right)}{\partial V_1}\right)_{T,N} = -P\,(T,\,N_1,\,V_1)$ and

$$\left(\frac{\partial F\left(T,\left(N-N_1\right),\left(V-V_1\right)\right)}{\partial V_1}\right)_{T,N} = -P\,((T,\,(N-N_1),\,(V-V_1)))$$

Also $\left(\dfrac{\partial\left(V-V_1\right)}{\partial V_1}\right) = -1$

With these substitutions in Eq. S-7.6.2 one gets,

$$-P(T,\,N_1,\,V_1) - [(P(T,(N-N_1)\,(V-V_1))\,(-1)]$$

That is, $\qquad P(T,\,N_1,\,V_1) = [(P(T,(N-N_1)\,(V-V_1))]$

This means that the pressures on the two sides must be equal for equilibrium for all values of V_1.

7. The internal energy of a system is given by $U(S, V) = (S^2/V)$. Derive the equation of state and determine C_P, C_V, κ_s, κ_T and β for the system.

Solution: It is given that $U\,(S,\,V) = \dfrac{S^2}{V}$ but $\left(\dfrac{\partial U}{\partial S}\right)_V = T$ and $\left(\dfrac{\partial U}{\partial V}\right)_S = -P$

Therefore, $\qquad T = \dfrac{2S}{V}$ and $-P = -\dfrac{S^2}{V^2}$ $\qquad$ S-7.7.1

Eliminating S from the two equations above, the equation of state $T^2 = 4P$ is obtained.

Also from Eq. S-7.7.1, $\qquad S = \dfrac{1}{2}TV$ and so, $U\,(T,\,V) = \dfrac{\left(\frac{1}{2}TV\right)^2}{V} = \dfrac{1}{4}T^2V$ $\qquad$ S-7.7.2

Now $\qquad C_V = \left(\dfrac{\partial U}{\partial T}\right)_V = \dfrac{1}{2}TV$. It is clear that $C_P = \left(\dfrac{\partial U}{\partial T}\right)_P = 0$ $\qquad$ S-7.7.3

From Eq. S-7.7.1 $\qquad V = \dfrac{S}{P^{\frac{1}{2}}}$ and $\kappa_s \equiv -\dfrac{1}{V}\left(\dfrac{\partial V}{\partial P}\right)_s = \dfrac{P^{\frac{1}{2}}}{S}\left\{\dfrac{S}{2P^{\frac{3}{2}}}\right\} = \dfrac{1}{2P}$ $\qquad$ S-7.7.4

Also, $\qquad \kappa_T \equiv -\dfrac{1}{V}\left(\dfrac{\partial V}{\partial P}\right)_T$ and $\beta = \dfrac{1}{V}\left(\dfrac{\partial V}{\partial T}\right)_P$

are undefined as volume cannot be changed by pressure at constant T and V cannot be changed by temperature T at constant Pressure P.

8. The equation of state of a non-ideal gas is PV = NRT. The volume of the gas at constant N and constant temperature T expands by a small amount ΔV. Calculate (i) $\left(\dfrac{\partial S}{\partial V}\right)_{N,T}$, (ii) the work done in this expansion of volume by ΔV (iii) heat transferred in the expansion process (iv) change in internal energy in the process (v) obtain the parameters on which U depends.

Solution:

The equation of state of the gas is given as $PV = NRT$. So $\left(\dfrac{\partial P}{\partial T}\right)_{V,N} = \dfrac{NR}{V}$, But

(i) $\left(\dfrac{\partial s}{\partial V}\right)_{T,N} = \left(\dfrac{\partial P}{\partial T}\right)_{V,N} = \dfrac{NR}{V}$ S-7.8.1

(ii) Work done in expansion $\Delta V = W = P\Delta V = \dfrac{NRT}{V}\Delta V$

(iii) Heat transferred in expansion $\Delta V = Q = T\left(\dfrac{\partial s}{\partial V}\right)_{T,N}\Delta V = T\dfrac{NR}{V}\Delta V$

(iv) Change in internal energy $\Delta U = W - Q = \dfrac{NRT}{V}\Delta V - T\dfrac{NR}{V}\Delta V = 0$. This means that $\left(\dfrac{\partial U}{\partial V}\right)_{,N} = 0$.

(v)   This is possible if $U = U(T, N)$. Since U is an extensive property $U = Nu$ and $u = u(T)$, here u is the internal energy per mole and is a function of temperature alone.

9. (a) N moles of a Van der Waals gas in equilibrium at pressure, volume and temperature P_1, V_1, and T_1 respectively, is contained in a box which is connected to another box that has vacuum. The two boxes are initially cut off from each other by a stopcock as shown in part (A) of Fig.(Sa-7.9.1). In the final state (part B of the figure) the stopcock is removed and the gas comes to an equilibrium, respectively, at pressure, volume and temperature P_2, V_2, and T_2, where V_2 is the total volume of the gas in final state. Calculate the final temperature T_2 of the gas given that the molar thermal capacity at constant volume of the Van der Waals gas (c_V) is (3/2)R.

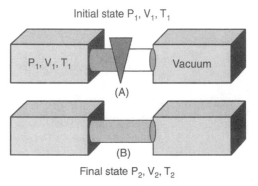

Initial state P_1, V_1, T_1

P_1, V_1, T_1 Vacuum

(A)

(B)

Final state P_2, V_2, T_2

Fig. S-7.9.1

(b) What will be the temperature if the gas is an ideal gas?

Solution:

(a) In the present case the gas undergoes free expansion. It is important to remember that in the free expansion by a gas no mechanical work is done by the gas against the external pressure and the internal energy U of the gas remains constant. However, in case of the Van der Waals gas the temperature of the gas may change from T_1 to T_2, in such a way that internal energy U remains constant, when volume changes from V_1 to V_2. If $\left(\dfrac{\partial T}{\partial V}\right)_{N,U}$ denotes the rate of change in temperature with volume then the change in temperature $(T_2 - T_1)$ is given by,

$$(T_2 - T_1) = \int_{V_1}^{V_2} \left(\frac{\partial T}{\partial V}\right)_{N,U} dV \qquad \text{S-7.9.1}$$

Here $\left(\dfrac{\partial T}{\partial V}\right)_{N,U}$ gives the rate of change of temperature with volume when the internal energy and the total number of particles are kept constant. To get $\left(\dfrac{\partial T}{\partial V}\right)_{N,U}$, we use the relation

$$\left(\frac{\partial T}{\partial V}\right)_{N,U} = -\left(\frac{\partial T}{\partial U}\right)_{N,V}\left(\frac{\partial U}{\partial V}\right)_{N,T} = -\frac{1}{C_V}\left(\frac{\partial U}{\partial V}\right)_{N,T} = -\frac{1}{Nc_V}\left(\frac{\partial U}{\partial V}\right)_{N,T} \qquad \text{S-7.9.2}$$

We used the relation $\left(\dfrac{\partial U}{\partial T}\right)_{N,V} = C_V = Nc_V$. Now to calculate $\left(\dfrac{\partial U}{\partial V}\right)_{N,T}$ following expressions may be used,

$$\left(\frac{\partial U}{\partial V}\right)_{N,T} = \left(\frac{\partial U}{\partial V}\right)_{N,S} + \left(\frac{\partial U}{\partial S}\right)_{N,V}\left(\frac{\partial S}{\partial V}\right)_{N,T} = -P + T\left(\frac{\partial S}{\partial V}\right)_{N,T} \qquad \text{S-7.9.3}$$

Since $\left(\dfrac{\partial U}{\partial V}\right)_{N,S} = -P$ and $\left(\dfrac{\partial U}{\partial S}\right)_{N,V} = T$

Also, $\left(\dfrac{\partial S}{\partial V}\right)_{N,T} = \left(\dfrac{\partial P}{\partial T}\right)_{N,V} = \dfrac{NR}{(V - Nb)}$ for Van der Waals gas, as $P = \dfrac{NRT}{(V - Nb)} - \dfrac{N^2 a}{V^2}$

Substituting the above values in Eq. S-7.9.2, $\left(\dfrac{\partial U}{\partial V}\right)_{N,T} = \dfrac{N^2 a}{V^2}$ putting this value in Eq. S-7.9.2 gives,

$$\left(\frac{\partial T}{\partial V}\right)_{N,U} = -\frac{1}{Nc_V}\left(\frac{\partial U}{\partial V}\right)_{N,T} = -\frac{1}{Nc_V}\left(\frac{N^2 a}{V^2}\right) = -\frac{Na}{c_V}\frac{1}{V^2},$$

Putting this value in Eq. S -7.9.1 gives

$$(T_2 - T_1) = \int_{V_1}^{V_2} \left(\frac{\partial T}{\partial V}\right)_{N,U} dV = -\frac{Na}{c_V}\int_{V_1}^{V_2}\frac{dV}{V^2} = \frac{Na}{c_V}\left[\frac{1}{V_2} - \frac{1}{V_1}\right]$$

Since it is given that $c_V = \frac{3}{2} R$, the above equation gives,

$$(T_2 - T_1) = \frac{2Na}{3R}\left[\frac{1}{V_2} - \frac{1}{V_1}\right]$$ S-7.9.3

Since $V_2 > V_1$, $(T_2 - T_1)$ is negative and temperature falls in the free expansion of Van der Waals gas.

(b) In case of an ideal gas, the internal energy is a function of temperature alone and since no work is performed in free expansion, the internal energy and hence the temperature does not change in case of an ideal gas.

10. A Carnot refrigerator is used to cool down a body of thermal capacity $C = KT^3$, K being a constant, from room temperature T_1 to absolute zero. The refrigerator works between the body and the room. Compute (a) amount of work required to be done in complete cooling of the body (b) change in the entropy and internal energy of the body (c) change in the entropy and internal energy of the room.

Solution:

(a) Let Q_B, Q_R, T_R and W, respectively, denote, the heat drawn by the refrigerator from the body at temperature T, out of which Q_R is rejected in the room at temperature T_R and a work W is performed on the refrigerator. Now,

$$Q_R = Q_B + W \text{ and } Q_B = Q_R - W$$ S-7.10.1

And the Coefficient of performance of the refrigerator, (COF) = $\frac{W}{Q_R} = \frac{(T_R - T)}{T_R}$

or $$Q_R = \frac{WT_R}{(T_R - T)}$$

Substituting this value of Q_R in Eq. S-7.10.1 gives:

$$Q_B = Q_R - W = \frac{WT_R}{(T_R - T)} - W = W\left(\frac{T}{(T_R - T)}\right)$$

or $$W = \left(\frac{(T_R - T)}{T}\right)Q_B$$ S-7.10.2

Equation (S-7.10.2) shows that in order to draw a quantity of heat Q_B from the body at temperature T the amount of work required to be done is W which is equal to $\left(\frac{(T_R - T)}{T}\right)Q_B$.

Since the body has thermal capacity $C = K T^3$, the amount of heat ΔQ_B that must be withdrawn from the body to reduce its temperature from T to $(T - \Delta T)$, i.e., by ΔT is $\Delta Q_B = C \Delta T = K T^3 \Delta T$. Therefore, the work ΔW required to be done to reduce the temperature of the body by ΔT, using Eq. S-7.10.2 is:

$$\Delta W = \left(\frac{(T_R - T)}{T}\right)\Delta Q_B = KT^2(T_R - T)\Delta T$$

Replacing ΔW and ΔT by dW and dT, one gets:

$$dW = KT^2 (T_R - T)\, dT$$

or $$W = K \int_{T_R}^{0} T^2 \left(T_R - T\right) dT = -\frac{K}{12} T_R^4 \qquad \text{S-7.10.3}$$

The negative sign of the work indicates that the work is to be done on the system, the refrigerator.

(b) The thermal capacity (at constant volume) is defined as $C_V \equiv \left(\dfrac{\partial U}{\partial T}\right)_V$, which in the present

case is KT^3. Using the relation $\left(\dfrac{\partial S}{\partial T}\right)_V = \dfrac{1}{T}\left(\dfrac{\partial U}{\partial T}\right)_V = \dfrac{KT^3}{T} = KT^2$, and hence,

$$\Delta S = \int_{T_R}^{0} KT^2 dT = -\frac{K}{3} T_R^3$$

Negative sign indicates that this is the decrease in the entropy.

Change in internal energy of the body may be calculated using the relation

$$dU = CdT \quad \text{or} \quad \Delta U = \int_{T_R}^{0} KT^3 dT = -\frac{K}{4} T_R^4$$

(c) The increase in the internal energy of the room = |decrease in internal energy of body| + |Work W|

$$\frac{K}{4} T_R^4 + \frac{K}{12} T_R^4 = \frac{1}{3} T_R^4$$

Increase in the entropy of the room = loss in the entropy of the body = $\dfrac{K}{3} T_R^3$

11. Show that in adiabatic compression of a gas through a porous plug the enthalpy of the system remains unaltered. Further, derive an expression for the Joule–Thomson coefficient μ.

Solution: A schematic diagram of a porous plug experiment is shown in Fig. S-7.11

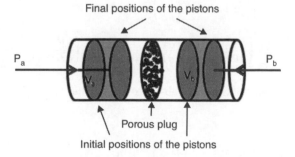

Final positions of the pistons

P_a P_b

Porous plug

Initial positions of the pistons

Fig. S-7.11

As shown in this figure, a piston on the left hand side of the porous plug pushes the gas through it with a pressure P_a and a volume V_a of the gas is pushed through in some given time. The work done by the piston on left W_a is equal to $P_a V_a$. On the other side the gas is held by another piston on which a pressure P_b is applied. Since $P_b < P_a$, the piston on the right hand side moves further to the right. If the increase in volume of the gas on the right hand side is V_b, then the work done by the gas on the piston W_b is equal to $P_b V_b$. Since the process is adiabatic, there is no flow of heat to or from the system and the net work done by the gas is equal to the fall in its internal energy. If U_a and U_b, respectively, denote the initial internal energy of the gas on the right hand side and the final internal energy on the left of the porous plug then,

$$U_a - U_b = P_b V_b - P_a V_a \text{ or } U_a + P_a V_a = U_b + P_b V_b \text{ or } H_a = H_b$$

Thus enthalpy is conserved in the porous plug experiment.

The Joule–Thomson coefficient is defined as

$$\eta = \left(\frac{\partial T}{\partial P}\right)_{H,N} \hspace{4cm} \text{S-7.11.1}$$

But

$$\left(\frac{\partial T}{\partial P}\right)_{H,N} = -\left(\frac{\partial T}{\partial H}\right)_{P,N}\left(\frac{\partial H}{\partial P}\right)_{T,N} = -\left\{\left(\frac{\partial H}{\partial T}\right)_{P,N}\right\}^{-1}\left(\frac{\partial H}{\partial P}\right)_{T,N} \hspace{1cm} \text{S-7.11.2}$$

Also,

$dH = TdS + VdP$ and if P is constant, $dP = 0$, then at constant temperature $\left(\frac{\partial H}{\partial T}\right)_{P,N}$ is given,

$$\left(\frac{\partial H}{\partial T}\right)_{PN} = T\left(\frac{\partial S}{\partial T}\right)_{P,N} = C_P$$

Substituting this value in Eq. S-7.11.2, one gets

$$\eta = \left(\frac{\partial T}{\partial P}\right)_{H,N} = -\left\{\left(\frac{\partial H}{\partial T}\right)_{P,N}\right\}^{-1}\left(\frac{\partial H}{\partial P}\right)_{T,N} = -\frac{1}{C_P}\left(\frac{\partial H}{\partial P}\right)_{T,N} \hspace{1cm} \text{S-7.11.3}$$

Again from the relation $dH = TdS + VdP$ one gets

$$\left(\frac{\partial H}{\partial P}\right)_{T,N} = T\left(\frac{\partial S}{\partial P}\right)_{T,N} + V \hspace{4cm} \text{S-7.11.4}$$

But

$$\left(\frac{\partial S}{\partial P}\right)_{T,N} = -\left(\frac{\partial V}{\partial T}\right)_{P,N} = -V\left[\frac{1}{V}\left(\frac{\partial V}{\partial T}\right)_{P,N}\right] = -V\alpha \hspace{1cm} \text{S-7.11.5}$$

Here α is the expansivity of the gas. Substituting the value of $\left(\frac{\partial S}{\partial P}\right)_{T,N}$ from above in Eq. (S-7.11.4)

$$\left(\frac{\partial H}{\partial P}\right)_{T,N} = T\left(\frac{\partial S}{\partial P}\right)_{T,N} + V = V(1 - \alpha T) \hspace{3cm} \text{S-7.11.5}$$

And from Eq. S-7.11.3

$$\eta = \left(\frac{\partial T}{\partial P}\right)_{H,N} = -\frac{1}{C_P}\left(\frac{\partial H}{\partial P}\right)_{T,N} = -\frac{1}{C_P}V(1-\alpha T) \qquad \text{S-7.11.6}$$

Equation (S-7.11.6) gives the desired relation. Further, it is left to show that $\alpha T = 1$ for an ideal gas and therefore, the Joule–Thomson coefficient is zero for an ideal gas. The reason is that in ideal gas there is no attraction between the molecules of the gas and hence no work is done in pulling the molecules apart. In case of a real gas work is done against the force of molecular attraction at the cost of the internal energy of the gas.

12. A system consists of a rubber band of fixed mass and of original length L_0 stretched to a length L under a tension T, at temperature T. The following relations are found to hold for the system,

$$\left(\frac{\partial T}{\partial T}\right)_L = \frac{aL}{L_0}\left[1-\left(\frac{L_0}{L}\right)^3\right] \text{ and } \left(\frac{\partial T}{\partial L}\right)_T = \frac{aT}{L_0}\left[1+2\left(\frac{L_0}{L}\right)^3\right]$$

Here a is a constant. Compute the form of the equation of state of the system and determine $\left(\frac{\partial L}{\partial T}\right)_T$ and discuss its physical significance. Also determine the work required to be done to adiabatically and reversible stretch the wire to double of its original length. What will be the change in the temperature of the wire?

Solution:

The three variables of the system are the length, the tension and the temperature. A relationship between these three variables will give the equation of state. The form of the equation of state, therefore, may be found by integrating any one of the given two relations. If the first relation is integrated one gets,

$$T = \frac{aL}{L_0}\left[1-\left(\frac{L_0}{L}\right)^3\right]T + constant \qquad \text{S-7.12.1}$$

While the integration of the second relation gives,

$$T = \frac{aT}{L_0}\left[1-\left(\frac{L_0}{L}\right)^3\right]L + constant \qquad \text{S-7.12.2}$$

Thus leaving the constants of integrations, the form of the equation of state of the system is,

$$T = \frac{aTL}{L_0}\left[1-\left(\frac{L_0}{L}\right)^3\right] \qquad \text{S-7.12.3}$$

To calculate at fixed tension the rate of change of length with temperature $\left(\frac{\partial L}{\partial T}\right)_T$, the following procedure may be followed,

$$\left(\frac{\partial L}{\partial T}\right)_T \left(\frac{\partial T}{\partial T}\right)_L \left(\frac{\partial T}{\partial L}\right)_T = -1$$

And, therefore,

$$\left(\frac{\partial L}{\partial T}\right)_T = -\frac{1}{\left[\left(\frac{\partial T}{\partial T}\right)_L\left(\frac{\partial T}{\partial L}\right)_T\right]} = -\frac{\left(\frac{\partial T}{\partial T}\right)_L}{\left(\frac{\partial T}{\partial L}\right)_T} = -\frac{\dfrac{aL}{L_0}\left[1-\left(\dfrac{L_0}{L}\right)^3\right]}{\dfrac{aT}{L_0}\left[1+2\left(\dfrac{L_0}{L}\right)^3\right]} = \frac{L\left[1+2\left(\dfrac{L_0}{L}\right)^3\right]}{T\left[1+2\left(\dfrac{L_0}{L}\right)^3\right]} \quad \text{S-7.12.4}$$

The above equation gives the desired relation between given parameters. The physical significance of $\left(\dfrac{\partial L}{\partial T}\right)_T$ lies in the fact that $\{\dfrac{1}{L}\left(\dfrac{\partial L}{\partial T}\right)_T\}$ gives the linear expansivity of the rubber band corresponding to the volume expansivity $\alpha = \{\dfrac{1}{V}\left(\dfrac{\partial V}{\partial T}\right)_V\}$ of other systems.

In adiabatic expansion of the rubber band $dQ = 0$, and $dU = C_L\,dT$, where C_L is the thermal capacity of the band at constant length and may be taken as constant, while the work done against the tension is $dW = TdL$. Hence from the first law,

$$CLdT + TdL = 0$$

Or

$$\frac{dT}{dL} = -\frac{T(T,L)}{C_L} = -\frac{1}{C_L}\frac{aTL}{L_0}\left[1-\left(\frac{L_0}{L}\right)^3\right]$$

Or

$$\frac{dT}{T} = -\frac{1}{C_L}\frac{aL}{L_0}\left[1-\left(\frac{L_0}{L}\right)^3\right]dL = -\frac{a}{C_L L_0}\left[L-L_0^3 L^{-2}\right]dL \quad \text{S-7.12.5}$$

Equation S-7.12.5 contains the variables T and L on the two sides and may be integrated. If T_i, T_f, L_0 and L_f represent, respectively, the initial temperature at which the length of the rubber band was L_0 and T_f as the final temperature with band length L_f then,

$$\int_{T_i}^{T_f}\frac{dT}{T} = -\frac{a}{C_L L_0}\int_{L_0}^{L_f}\left[L-L_0^3 L^{-2}\right]dL$$

Or

$$\ln\frac{T_f}{T_i} = -\frac{aL_0}{2C_L}\left(\frac{L_f}{L_0}\right)^2\left[1+2\left(\frac{L_0}{L_f}\right)^3\right] \quad \text{S-7.12.6}$$

Equation S-7.12.6 provides a relation between the length of the wire and the temperature. The mechanical work in stretching the band from L_o to L_f as a result of which the temperature of the band rises from T_i to T_f may be calculated using the relation,

$$W = C_L\,(T_f - T_i)$$

Problems

1. Two vessels, insulated from the outside world, one of volume V_1 and the other of volume V_2, contain equal numbers N molecules of the same ideal gas. The gas in each vessel is originally

at temperature T_i. The vessels are then connected and allowed to reach equilibrium in such a way that the combined vessel is also insulated from the outside world. The final volume is V $=V_1+V_2$.

What is the maximum work that can be obtained by connecting these insulated vessels?

Express your answer in terms of T_i, V_1, V_2, and N.

2. Starting with the definition of expansivity $\alpha = \dfrac{1}{V}\left(\dfrac{\partial V}{\partial T}\right)_P$ and $C_P = \left(\dfrac{\partial S}{\partial T}\right)_P$ derive the relation

$Tds = C_P dT - \alpha TV\, dP$, if the amount of the material is kept constant. Further show that in an adiabatic process one can never obtain the temperature of 0K by a finite change in pressure.

3. Discuss the change in the internal energy of a system in an isochoric and isentropic spontaneous process.

4. A system containing a fixed amount of the material performs work $W = NRT_i \ln \dfrac{V_f}{V_i}$ when its

volume expands from V_i at temperature T_i to the value V_f. The entropy of the system at temperature

T is given by $S = NR\left(\dfrac{V_0}{V}\right)\left(\dfrac{T}{T_0}\right)^2$. Obtain the Helmholtz function and the equation of state

of the system.

5. The Gibbs function for a system containing N moles of the material is given by

$$G = NRT \ln \dfrac{P}{P_0} - N\,PB(T)$$

Here B(T) is a function of temperature alone. Derive expressions for molar internal energy, molar Helmholtz function, molar entropy and molar enthalpy of the system.

6. A non-ideal gas the equation of state for which is $P = \dfrac{NRT}{V}\left[1 + \dfrac{N}{V}A(T)\right]$ under goes small

isothermal expansion ΔV when a quantity of heat ΔQ is supplied to the gas at constant N. Establish a relation between ΔQ and ΔV retaining only first order terms, evaluate the partial derivatives involved in the relation and discuss the significance of the terms appearing in the final result.

7. Compute the heat evolved Q and the work done W in isothermally compressing a given mass M of a fluid of density ρ from initial pressure P_i to a final pressure P_f. Put your result in terms of the isothermal compressibility κ_T and the constant pressure expansivity β.

8. The equation of state and the energy equation for an ideal gas are respectively, PV= NRT and U = (3/2) NRT. Use thermodynamic relations to obtain the values of C_V, C_P, H, and [S (T,V,N) − S(T_0, V, N)] .

9. The expansivity β and the isothermal compressibility κ_T for solids are very small as well they may be taken as constant without committing much error. Under these assumptions show that the approximate equation of state for solids may be given as:

$$v = v_0\,[1 + \beta\,(T - T_0) - k_T\,(P - P_0)]$$

Here v_0 and v are the specific volumes of the solid, respectively, at temperature T_0, pressure P_0 and at T and P.

10. Assuming that the equation of state for a solid is given as,

$$v = v_0 \left[1 + \beta (T - T_0) - \kappa_T (P - P_0) \right]$$

Here v_0 and v are the specific volumes of the solid, respectively, at temperature T_0, pressure P_0 and at T and P. And that the expansivity β and compressibility κ_T are small and constant, derive expressions for the specific entropy s, the specific enthalpy h, specific internal energy u and the difference $(c_P - c_V)$.

11. Choose a suitable *Tds* equation to show that the Clausius–Clapeyron equation, in case of a second order phase transition in which the initial and the final molar entropies are equal, gives that

The rate of change of pressure with temperature =

$$\frac{1}{Tv} \left(\frac{\text{difference between the initial and final values of molar heat capacity at constant pressure}}{\text{difference between the initial and final values of expansivity}} \right)$$

12. The internal energy U of a system is given as $U = KVT^2$, where K is a constant. Obtain expressions for entropy S (T,V) and the equation of state for the system assuming entropy to be zero at absolute zero.

Short Answer Questions

1. Obtain conditions for thermal and mechanical equilibriums in a system.
2. Describe the physical significance of chemical potential.
3. State and explain Gibbs phase rule.
4. Explain why the three phases of water can co-exist in equilibrium only at fixed values of temperature and pressure.
5. Write the three *Tds* equations.
6. Without derivation give the Maxwell's relations.
7. Write expressions for the change in Helmholtz function and Gibbs function between the two nearby equilibrium states of an open system.
8. Define an open system.
9. With the help of a schematic diagram explain the concepts of the second and third types of perpetual motions.
10. State the principle of un-attainability of absolute zero.
11. Give statement of the Nernst heat theorem
12. State Einstein's statement of third law of thermodynamics and give its limitations.
13. Why must the specific heat capacities vanish at absolute zero?
14. State the significance of characteristic variables of a thermodynamic function.

15. The decrement in the values of Gibbs and Helmholtz functions in going from one to another nearby equilibrium state of a system put upper limits to what types of work and under what conditions?

Long Answer Questions

1. Derive an expression for the difference in the value of Helmholtz function between two nearby states of a system which are connected to a single heat reservoir. Discuss the physical significance of this difference.

2. Differentiate between Gibbs and Helmholtz functions and discuss the conditions when these functions assume significance.

3. What required the formulation of the third law of thermodynamics? Discuss the various statements of the third law and some consequences of the law.

4. Discuss the establishment of equilibrium in a closed and in an open system.

5. What is chemical potential? Establish its relation with parameters G, H, U and F of a system.

6. Derive expressions for specific Helmholtz and specific Gibbs functions for ideal gas.

7. Derive expressions for specific Helmholtz and specific Gibbs functions for Van der Waals gas.

Multiple Choice Questions

Note: Some of the multiple choice questions may have more than one correct alternative. All correct alternatives must be marked for complete answer in such cases.

1. The only variables of a system are T, S, P and V; the expression $U - TS + PV$ represents
   (a) Helmholtz function
   (b) Enthalpy
   (c) Gibbs function
   (d) Isothermal compressibility.

2. $-\left(\dfrac{\partial S}{\partial P}\right)_N$ is equal to

   (a) $\left(\dfrac{\partial^2 G}{\partial P \partial T}\right)_N$
   (b) βv

   (c) $\left(\dfrac{\partial^2 F}{\partial P \partial T}\right)_N$
   (d) $v \kappa_T$

3. Which of the following partial derivative represents the pressure of the system?
   (a) $-\left(\dfrac{\partial S}{\partial P}\right)_N$
   (b) $-\left(\dfrac{\partial F}{\partial V}\right)_T$

   (c) $-\left(\dfrac{\partial F}{\partial T}\right)_V$
   (d) $-\left(\dfrac{\partial U}{\partial V}\right)_S$

4. The correct Gibbs–Helmholtz equations are
   (a) $U = F - T\left(\dfrac{\partial F}{\partial T}\right)_V$
   (b) $F = U - T\left(\dfrac{\partial U}{\partial T}\right)_V$

   (c) $H = G - T\left(\dfrac{\partial G}{\partial T}\right)_P$
   (d) $G = H - T\left(\dfrac{\partial H}{\partial T}\right)_V$

5. Which of the following must be positive for the stability of a system?

   (a) κ_s

   (b) $\left(\dfrac{\partial \mu}{\partial N}\right)_{V,S}$

   (c) $\left(\dfrac{\partial F}{\partial T}\right)_V$

   (d) $\left(\dfrac{\partial G}{\partial T}\right)_P$

6. For a monatomic ideal gas the ratio $\left(\dfrac{\kappa_T}{\kappa_s}\right)$ is

   (a) 2/3   (b) 3/2
   (c) 3/5   (d) 5/3

7. In the state of equilibrium of a system, which pair of its properties has minimum and maximum value?

   (a) S, U   (b) U, S
   (c) G, F   (d) S, G

8. Which principle or law prohibits the attainment of absolute zero of temperature in finite number of thermodynamic processes?

   (a) the first law   (b) the second law
   (c) Caratheodory principle   (d) Nernst principle

9. A two-component system has three phases in equilibrium. The degrees of freedom are

   (a) 4   (b) 3
   (c) 2   (d) 1

10. In a multi-component system three phases co-exist in equilibrium with three degrees of freedom, the components in the system are

    (a) 4   (b) 3
    (c) 2   (d) 1

11. Two equilibrium states A and B of a system are connected to a single heat reservoir at temperature T. The state parameters of the two states are represented by putting subscripts A and B to the appropriate symbol. The maximum reversible work that can be drawn from the combination of two states is

    (a) $F_A - F_B$ (b) $G_A - G_B$

    (c) $U_A - U_B$ (d) $H_A - H_B$

12. Which thermodynamic parameter/ function puts an upper limit to the non-PdV work between the two equilibrium states of a system having same temperature and pressure?

    (a) Entropy   (b) Enthalpy
    (c) Gibbs function   (d) Helmholtz function

13. In the second order phase transitions 'X' shows discontinuity with respect to temperature. The 'X' is

    (a) $\left(\dfrac{\partial s}{\partial T}\right)_P$

    (b) $\left(\dfrac{\partial g}{\partial T}\right)_P$

    (c) $\left(\dfrac{\partial f}{\partial T}\right)_P$

    (d) $\left(\dfrac{\partial^2 s}{\partial T^2}\right)_P$

14. For a single component system the chemical potential μ is equal in magnitude to the value of the specific
   (a) Entropy
   (b) Helmholtz function
   (c) Enthalpy
   (d) Gibbs function

15. Two gases A and B are held in two parts of an adiabatic box separated by a partition and diffuse into each other when partition is removed. The diffusion process will stop when
   (a) the partial pressures of the gases become equal
   (b) the temperatures on both sides become equal
   (c) the entropies of the two gases become equal
   (d) the chemical potentials of the two gases become equal

16. A multi-component system is in mechanical equilibrium only when
   (a) temperature in every part of the system is same
   (b) pressure in every part of the system is same
   (c) the Gibbs function of the system has minimum value
   (d) the chemical potentials for each component are equal

17. Superconductor to conductor phase transitions are generally transitions of
   (a) zeroth order
   (b) first order
   (c) second order
   (d) lambda type

18. Which of the following relations is not true?
   (a) $\left(\dfrac{\partial G}{\partial P}\right)_T = \left(\dfrac{\partial H}{\partial P}\right)_S$
   (b) $\left(\dfrac{\partial U}{\partial V}\right)_S = \left(\dfrac{\partial F}{\partial V}\right)_T$
   (c) $\left(\dfrac{\partial G}{\partial P}\right)_T = \left(\dfrac{\partial H}{\partial S}\right)_P$
   (d) $\left(\dfrac{\partial U}{\partial S}\right)_V = \left(\dfrac{\partial H}{\partial S}\right)_P$

19. Tick the correct relations.
   (a) $\left(\dfrac{\partial T}{\partial V}\right)_S = -\left(\dfrac{\partial P}{\partial S}\right)_V$
   (b) $\left(\dfrac{\partial T}{\partial P}\right)_S = \left(\dfrac{\partial V}{\partial P}\right)_S$
   (c) $\left(\dfrac{\partial S}{\partial V}\right)_T = \left(\dfrac{\partial P}{\partial T}\right)_V$
   (d) $\left(\dfrac{\partial S}{\partial P}\right)_T = \left(\dfrac{\partial T}{\partial P}\right)_S$

20. The maximum inversion temperature for Van der Waals gas is
   (a) $\dfrac{2b}{Ra}$
   (b) $\dfrac{ab}{2R}$
   (c) $\dfrac{2a}{Rb}$
   (d) $\dfrac{Rb}{2a}$

Answers to Numerical and Multiple Choice Questions

Answer to problems

1. $NkTi \ln \dfrac{(V_1 + V_2)^2}{4V_1 V_2}$

3. The internal energy will decrease

4. $F(T,V) = F(T_0,V_0) - NRT_0 \left[\ln \dfrac{V}{V_0} - \dfrac{1}{3}\left(\dfrac{V_0}{V}\right)\left\{\left(\dfrac{T}{T_0}\right)^3 - 1\right\}\right]$

$P = \dfrac{NRT_0}{V} - \dfrac{1}{3}\dfrac{NRT_0 V_0}{V^2}\left\{\left(\dfrac{T}{T_0}\right)^3 - 1\right\}$

5. $u = T\left[P\dfrac{dB}{dT} - R\right], f = RT\left[\ln\dfrac{P}{P_0} - 1\right], s = -R\ln\dfrac{P}{P_0} + P\dfrac{dB}{dT}, h = P\left[T\dfrac{dB}{dT} - B\right]$

6. $\Delta Q = T\left(\dfrac{\partial S}{\partial V}\right)_{T,N} \Delta V = P\Delta V + \dfrac{N^2}{V^2}RT^2\dfrac{dA}{dT}$, the term $P\Delta V$ gives the work done by the system

against the atmosphere and the second term $\dfrac{N^2}{V^2}RT^2\dfrac{dA}{dT}$ gives the change in the internal energy.

7. $Q = -\dfrac{M}{\rho}T\beta\left(P_f - P_i\right)\left[1 - \dfrac{\kappa_T}{2}\left(P_f - P_i\right)\right]$, $W = \dfrac{M}{2\rho}\kappa_T\left(P_f^2 - P_i^2\right)$

8. $C_V = \dfrac{3}{2}NR, C_P = \dfrac{5}{2}NR, H = \dfrac{5}{2}NRT, \left[S(T,V,N) - S(T_0,V,N)\right] = \dfrac{3}{2}NR\ln\dfrac{T}{T_0}$

10. $s = c_P\ln\dfrac{T}{T_0} - \beta v_0\left[P - P_0\right] + s_{0;}$

$h = c_P\left(T - T_0\right) + v_0\left(P - P_0\right)\left[1 - \beta T_0 - \dfrac{\kappa_T}{2}\left(P - P_0\right)\right] + h_0$

$u = c_v\left(T - T_0\right) + \left(v - v_0\right)\left[\dfrac{1}{2\kappa_T}\left(2\beta T_0 + \dfrac{v}{v_0} - 1\right)\right] + u_0 \; ; \left(c_P - c_V\right) = \dfrac{Tv\beta^2}{\kappa_T}$

12. $S(T,V) = 2KVT; PV = KVT^2$

Answer to multiple choice questions

1. (c)	2. (a), (b)	3. (b), (d)	4. (a), (c)
5. (a), (b)	6. (d)	7. (b)	8. (d)
9. (d)	10. (a)	11. (a)	12. (c)
13. (a)	14. (d)	15. (d)	16. (a), (b)
17. (c)	18. (c)	19. (a), (c)	20. (c)

Revision

1. Helmholtz function

$$F \equiv U - TS \text{ and } f \equiv u - Ts$$

2. The decrease in the Helmholtz function puts the upper limit to the amount of work that may be obtained in any process between the two equilibrium states of a system at same temperature, during which there is a heat flow into the system from a single reservoir.

3. For an ideal gas

$$f = f_f - f_i = \left(u_f - u_i\right) - T_f\left[s_f - s_i\right] = \left\{c_v\left(T_f - T_i\right)\right\} - T_f\left[c_v \ln\frac{T_f}{T_i} + R \ln\frac{v_f}{v_i}\right] + s_i\left(T_f - T_i\right)$$

4. For Van der Waals gas

$$f^{van} = c_v\left(T_f - T_i - T_f \ln\frac{T_f}{T_i}\right) - RT_f \ln\left(\frac{v_f - b}{v_i - b}\right) - a\left(\frac{1}{v_f} - \frac{1}{v_i}\right) - s_i\left(T_f - T_i\right)$$

5. Work W_T may be divided into two parts: one the PdV work and the other the sum of all other types of works that maybe denoted by A_T. $W_T = (PdV)_T + A_T$. If the process is both isothermal and isochoric then dV is zero and, $A_{T,v} \leq (F_1 - F_2)$
   which means that decrease in the Helmholtz function in a process at constant volume and temperature puts an upper limit to the non-PdV work.
   If no configurational work, including PdV is performed in an isothermal process, i.e., $A_T = 0$ and $PdV = 0$, then $1 \leq (F_1 - F_2)$
   This means that in processes at constant temperature and volume when A is zero, $F_2 \leq F_1$, that is in such processes Helmholtz function decreases or remains constant (if the process is reversible). One may also say that such processes are possible only when the Helmholtz function of the system decreases.

6. In a process in which the temperature and the external pressure both remain constant

$$A_{T,P} \leq [\{U_1 - TS_1 + PV_1\} - \{U_2 - TS_2 + PV_2\}]$$

$$A_{T,P} \leq [G_1 - G_2]$$

Where
$$G_1 \equiv \{U_1 - TS_1 + PV_1\} \text{ and } G_2 \equiv \{U_2 - TS_2 + PV_2\}$$

7. When only Pdv work is performed by any process between the two equilibrium states of a system at same temperature and pressure the Gibbs function either decreases or remains same. $[G_1 - G_2] \geq 0$

8. In case of an ideal gas the change in specific entropy $g(= u - Ts + Pv = h - Ts)$ in going from an equilibrium state of temperature T_i, and pressure P_i to a state of final values T_f, P_f is given by,

$$g = RT (\ln P_f + \omega)$$

where
$$\omega = c_P T_f\left(1 - \ln\frac{T_f}{T_i}\right) - RT_f \ln P_i - c_P T_i - s_i\left(T_f - T_i\right)$$

It may be noted that ω is a function of temperature only.

9. The characteristic variables
   The significance of characteristic variables lies in the fact that if the function is known in terms of its characteristic variables and the derivatives of the function with respect to the characteristic variables are also known then all thermodynamic properties of the system can be determined or it may be said that the system is completely known.

10. *U*, *H*, *F* and *G* are also called thermodynamic potentials since their derivatives resembles to the components of a scalar potential.

$$\left(\frac{\partial H}{\partial S}\right)_P = T; \qquad \left(\frac{\partial H}{\partial P}\right)_S = V$$

$$\left(\frac{\partial F}{\partial V}\right)_T = -P; \qquad \left(\frac{\partial F}{\partial T}\right)_V = -S$$

$$\left(\frac{\partial G}{\partial P}\right)_T = V; \qquad \left(\frac{\partial G}{\partial T}\right)_P = -S$$

And
$$\left(\frac{\partial U}{\partial S}\right)_V = T; \qquad \left(\frac{\partial U}{\partial V}\right)_S = -P$$

11. Remembering thermodynamic potentials and their derivatives

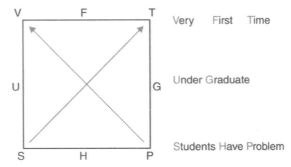

Very   First   Time

Under Graduate

Students Have Problem

12. Gibbs–Helmholtz relations

$$H = G + TS = G(T,P) - T\left(\frac{\partial G}{\partial T}\right)_P$$

$$U = F + TS = F - T\left(\frac{\partial F}{\partial T}\right)_V$$

13. Generalized functions

Let us consider a system the states of which are functions of temperature T, two extensive variables X_1, X_2 and the corresponding two intensive variables Y_1 and Y_2. In such a case the temperature, T, and any other pair (X_1, X_2), (Y_1, Y_2), (X_1, Y_2) or (X_2, Y_1) may be taken as independent variables. The differential element of reversible work (path dependent) for such a system is given by,

$$dW = d'W = Y_1 dX_1 + Y_2 dX_2$$

And the differential element of internal energy dU by

$$dU = TdS - d'W = TdS - (Y_1 dX_1 + T_2 dX_2)$$

Let T, X_1 and X_2 be the independent variables.

$$F = U - TS \text{ and } dF = dU - TdS - SdT$$

$$dF = -Y_1 dX_1 - Y_2 dX_2 - SdT$$

$$\left(\frac{\partial F}{\partial T}\right)_{X_1 X_2} = -S, \left(\frac{\partial F}{\partial X_1}\right)_{T,X_2} = -Y_1, \text{ and } \left(\frac{\partial F}{\partial X_2}\right)_{T,X_1} = -Y_2$$

$$dG = -SdT + X_1 dY_1 + X_2 dY_2$$

$$\left(\frac{\partial G}{\partial T}\right)_{X_1 X_2} = -S, \left(\frac{\partial G}{\partial Y_1}\right)_{T,Y_2} = X_1, \left(\frac{\partial G}{\partial Y_2}\right)_{T,Y_1} = X_2$$

14. Maxwell relations

$$dU = TdS - PdV \qquad \Rightarrow \qquad \left(\frac{\partial T}{\partial V}\right)_S = -\left(\frac{\partial P}{\partial S}\right)_V$$

$$dF = -PdV - SdT \qquad \Rightarrow \qquad \left(\frac{\partial P}{\partial T}\right)_V = \left(\frac{\partial S}{\partial V}\right)_T$$

$$dG = VdP - SdT \qquad \Rightarrow \qquad \left(\frac{\partial V}{\partial T}\right)_P = -\left(\frac{\partial S}{\partial P}\right)_T$$

$$dH = TdS + VdP \qquad \Rightarrow \qquad \left(\frac{\partial T}{\partial P}\right)_S = \left(\frac{\partial V}{\partial S}\right)_P$$

15. The third law of thermodynamics
    Nernst heat theorem:

> *'In the vicinity of absolute zero of temperature reactions in liquid and solid systems in the state of internal equilibrium take place with no change in entropy'.*

Planck statement of the third law: Planck formulated the third law of thermodynamics as follows.

> *'When temperature falls to absolute zero, the entropy of any pure crystalline substance tends to a universal constant, which can be taken to be zero'.*

In mathematical terms, $S \to 0$, as $T \to 0$
Einstein statement:

> *'As temperature falls to absolute zero, the entropy of any substance remains finite'.*

An important conclusion that follows from Einstein statement is that thermal capacities both at constant volume and at constant pressure for all substances vanish at absolute zero.
Nernst principle of un-attainability of absolute zero: The third law of thermodynamics may also be stated in terms of the principle of un-attainability of absolute zero put forward by Nernst and that says,

> *'Any thermodynamic process cannot reach the temperature of absolute zero in finite number of steps and in finite time'.*

16. The concept of perpetual motion

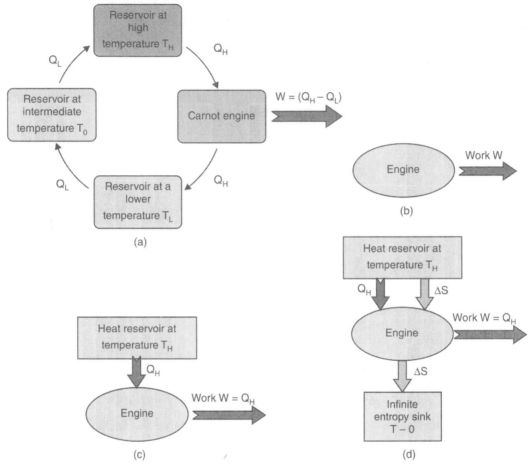

Fig. 7.3   (a) Perpetual motion of the Zeroth kind; (b) Perpetual motion of the First kind;
(c) Perpetual motion of the Second kind; (d) Perpetual motion of the Third kind

17. Thermodynamics of open systems:
For an open system the system parameters are also functions of the concentrations of different components apart with other parameters like temperature, pressure etc.
For a single component open system

$$U = U(S, V, N) \text{ and } dU = \left(\frac{\partial U}{\partial S}\right)_{V,N} dS + \left(\frac{\partial U}{\partial S}\right)_{S,N} dV + \left(\frac{\partial U}{\partial N}\right)_{V,S} dN$$

$$TdS - PdV + \mu dN$$

where the chemical potential $\mu \equiv \left(\frac{\partial U}{\partial N}\right)_{V,S} dN$

Also,

$$\mu = -T\left(\frac{\partial S}{\partial N}\right)_{V,U} = \left(\frac{\partial F}{\partial N}\right)_{V,T} = \left(\frac{\partial G}{\partial N}\right)_{P,T}$$

General case of a system with different kinds of entities

$$dU = \left(\frac{\partial U}{\partial S}\right)_{V,N} dS + \left(\frac{\partial U}{\partial V}\right)_{S,N} dV + \sum_{1}^{\gamma} \left(\frac{\partial U}{\partial N_j}\right)_{S,V,N_{i \neq j}}$$

$$\left(\frac{\partial U}{\partial S}\right)_{V,N} = T; \left(\frac{\partial U}{\partial V}\right)_{S,N} = \boxtimes P; \left(\frac{\partial U}{\partial N_j}\right)_{S,V,N_{i \neq j}} = \mu_j$$

18. The equilibrium

Thermal equilibrium: For thermal equilibrium of two systems in contact their temperatures must be equal

Mechanical equilibrium: For mechanical equilibrium in two systems in contact the temperatures and pressures of the two systems must be equal.

Equilibrium with respect to mass flow: Temperature difference is responsible for heat flow, pressure difference for change in volume, the difference in the chemical potential may be looked as the source for the flow of matter. Later it will be shown that chemical potential also plays as deriving force in phase transitions and chemical reactions.

Phase equilibrium: A phase is homogeneous region of matter in which there is no spatial variation in the average density, energy, composition or other macroscopic properties of the matter. Phases may be different in their molecular structure, for example water has multiple ice phases that differ in their crystallographic structures. A phase may be considered as a distinct subsystem with boundaries that are interfaces with container walls or other phases. When one talks of phase equilibrium it means that the system consists of at least two sub systems of different phases separated with phase boundary. These subsystems exist spontaneously, without the use of partitions or other external interventions and since they are in equilibrium they may exchange energy, volume and particles.

The conditions for the co-existence of different phases numbering α in equilibrium with each other in a system consisting of γ components are given by the set of equations

$$\mu_{component, phase\ j} (P, T, \{x\}_{phase\ j}) = \mu_{component\ k, phase\ (j+1)} (P, T, \{x\}_{phase\ (j+1)})$$

The degrees of freedom or the number of variants are given by the following equation called Gibbs phase rule

$$d_{freedom} \text{ or number of variants of the system} = (2 - \alpha + \gamma)$$

19. The Clausius–Clapeyron equation:

In case of a single component system, the two phases a and b will be in equilibrium when the specific Gibbs functions and small variations in them are equal.

$$g^a = g^b, \text{ and } g^a + dg^a = g^b + dg^b$$

Or $$dg^a = dg^b$$

This leads to $(v^b - v^a)dP = (s^b - s^a)dT$

And $\left(\frac{\partial P}{\partial T}\right)_{a,b} = \dfrac{\ell_{a,b}}{T\left(v^b - v^a\right)}$ where $\ell_{a,b}$ is the latent heat of transformation of phase a to phase b

Order of phase transition: Paul Ehrenfest (Jan 18, 1880–Sept.25, 1933, an Austrian theoretical physicist) gave a scheme for classifying phase transitions. His classification was based on the continuity of the specific Gibbs function and its derivatives across the phase boundary. When in a single component system one moves from one phase to the other, generally the specific Gibbs function g remains continuous but the derivative $\left(\dfrac{\partial g}{\partial T}\right)_P = -s$ has discontinuity at the boundary as specific entropies of different phases are different. Such phase transitions are called First Order phase transitions. Most of the solid-liquid, solid-vapor and liquid- vapor phase transitions are of the first order. The Second Order phase transitions are those in which both the specific Gibbs function g and its first derivatives remain continuous through the phases but the second derivative of the specific Gibbs function $\left(\dfrac{\partial^2 g}{\partial T^2}\right)_P = -\dfrac{c_P}{T}$ show discontinuity at the phase boundary. Liquid-vapor transitions at critical point and super conducting state to non-super conducting state phase transitions are examples of second order phase transitions. In Ehrenfest scheme, the specific Gibbs function, its first derivatives and its second derivatives are all continuous but the third derivatives of g show discontinuity at the boundary. In this way still higher orders of phase transitions may be defined.

A typical phase transition in which the variation of the specific thermal capacity at constant pressure c_P with temperature has the shape of Greek letter lambda is commonly called the lambda transition. It is found in case of liquid helium when it goes from a liquid to a super fluid phase.

Some Applications of Thermodynamics to Problems of Physics and Engineering

8.0   Introduction

The four laws of thermodynamics and some of their implications have been discussed in the previous chapters. In this chapter we shall study the applications of the principles of thermodynamics to some problems of physics and engineering. Thermodynamics is frequently used for the analysis of chemical reactions. However, the sign convention used in chemistry for defining work and other thermodynamic functions is different than that used in physics, and, therefore, the application of thermodynamics to chemical reactions, also called thermochemistry, will be dealt separately to avoid any confusion.

Pedagogic development of thermodynamics suggested that it can be applied only to gases, liquids and to some extent to solids. However, thermodynamics has a very wide scope of applications and can be applied to any macroscopic system, irrespective of it being a solid, liquid, gas, radiations, magnetic, or electric system etc. The following examples are chosen to illustrate the methodology of applying thermodynamics to other systems.

8.1   The Blackbody Radiation

Blackbody radiation is an idealized concept of a bundle of radiations consisting of electromagnetic waves of all frequencies from zero to infinity with a well-defined strength of each frequency component at a given temperature. Such radiation is assumed to be emitted when a so-called blackbody is heated. A blackbody is a body that absorbs all electromagnetic radiations that fall on it and emits them all on heating. Many attempts, some highly technical and sophisticated have been made to build a blackbody, but a simple one can be made by taking a spherical hollow ball like body with a small hole, the interior of which is painted black. A conical projection opposite to the hole scatters the radiations entering the ball through the hole which are absorbed by the blackened walls. Thus essentially all radiations falling at the hole of the body are absorbed by the inside walls and none is able to come out through the opening making the ball a blackbody for incident radiations. If the ball is made of a material that may withstand high temperature, the interior of the ball (also called the cavity) will be filled with electromagnetic radiations characteristic of blackbody radiation at that temperature to which the body is heated.

A simple blackbody

Let us consider a blackbody the cavity of which is initially evacuated and then get filled with blackbody radiations in equilibrium at some temperature T. It is assumed that some radiations that leak through the small hole do not disturb the equilibrium existing in the cavity. Actual experiments carried out on the out-coming blackbody radiations have reveled two important facts: (i) The energy density u, the energy contained per unit volume of the radiations, is constant at a given temperature and is a function only of the temperature T of the cavity, i.e., $u = u(T)$. (ii) From some earlier experiments it was known that radiations falling on any surface exert a pressure on the surface. In case of the blackbody radiations it was experimentally found that the pressure P exerted by blackbody radiations is equal to one third of the energy density u, i.e., $P = \frac{1}{3}u(T)$. These two experimental

facts may be used to develop a thermodynamic description of the blackbody radiation in equilibrium inside the cavity.

8.1.1 Equation of state and thermodynamic functions for blackbody radiation

The cavity filled with blackbody radiations in equilibrium at temperature T, may be considered to be a thermodynamic system that is in equilibrium at temperature T and is exerting a pressure P on its boundary walls. If V is the volume of the cavity than the total energy of the thermodynamic system is

$$U(T) = Vu(T) \text{ and the pressure } P = \frac{1}{3}u \qquad 8.1$$

From Eq. 8.1, it follows that

$$\left(\frac{\partial U}{\partial V}\right)_T = u(T) \, and \, \left(\frac{\partial P}{\partial T}\right)_V = \frac{1}{3}\left(\frac{\partial u(T)}{\partial T}\right)_V = \frac{1}{3}\frac{du}{dT} \qquad 8.2$$

But we have the thermodynamic relation

$$\left(\frac{\partial U}{\partial V}\right)_T = T\left(\frac{\partial P}{\partial T}\right)_V - P \qquad 8.3$$

Substituting the values of $\left(\frac{\partial U}{\partial V}\right)_T$ and $\left(\frac{\partial P}{\partial T}\right)_V$ in Eq. 8.3 from Eq. 8.2 and of P from Eq. 8.1, one gets:

$$u(T) = T\frac{1}{3}\frac{du(T)}{dT} - \frac{1}{3}u(T) \text{ or } \frac{4}{3}u(T) = \frac{1}{3}T\frac{du(T)}{dT}$$

Or
$$4\frac{dT}{T} = \frac{du(T)}{u} \text{ or In u(T) = 4 In T + constant}$$

$$u(T) = \sigma T^4 \qquad\qquad 8.4$$

Here σ is a constant. The energy density u may be identified with the specific internal energy of the thermodynamic system of blackbody radiation inside the cavity and T as the thermodynamic temperature of this system. Using the thermodynamic relation of Eq. 8.3 and the two experimental facts regarding the blackbody radiations, we arrived at the relation Eq. 8.4, which tells that the specific internal energy of the system is proportional to the fourth power of thermodynamic temperature. The same result was mathematically derived by Jozef Stefan (24 March 1835–7 Jan 1893, a Slovenian origin Austrian mathematician and physicist and a Slovenian language poet) from the experimental work of French scientists Dulong (Pierre-Louis Dulong, 12Feb, 1785–19 July, 1838) and Petit (Alexis Therese Petit, 2 Oct. 1791–21 June, 1820). He determined the value of the constant σ, called Stefan's constant as $7.56 \times 10^{-16}\ JK^{-4}\ m^{-3}$.

The equation of state of the thermodynamic system of blackbody radiations may be easily obtained from relations given by Eq. 8.1 and Eq. 8.4,

$$P = \frac{1}{3}u \text{ and } u = \sigma T^4,$$

And, therefore,
$$P = \frac{\sigma}{3}T^4 \qquad\qquad 8.5 \text{ (a)}$$

In Eq. 8.5 (a) it is important to note that the pressure and temperature of the system are interdependent and one of them cannot be changed without changing the other.

The total internal energy of the thermodynamic system of blackbody radiations

$$U = Vu = \sigma V T^4 \qquad\qquad 8.5 \text{ (b)}$$

The heat capacity at constant volume

$$C_V = \left(\frac{\partial U}{\partial T}\right)_V = 4\sigma V T^3 \qquad\qquad 8.6$$

And the entropy
$$S = \int_0^T \frac{1}{T}C_V dT = \int_0^T \frac{1}{T}\left(4\sigma V T^3\right)dT = \frac{4}{3}\sigma V T^3 \qquad\qquad 8.7$$

Thermodynamic relations $H = U + PV$, $F = U - TS$ and $G = F + PV$ may now be used to get the following values for

The enthalpy
$$H = \frac{4}{3}\sigma V T^4, \qquad\qquad 8.8$$

Helmholtz function $\qquad\qquad F = -\frac{1}{3}\sigma V T^4$ $\qquad\qquad\qquad$ 8.9

And Gibbs function $\qquad\qquad G = 0$ $\qquad\qquad\qquad\qquad$ 8.10

It may be noted that G and hence specific Gibbs function g are identically zero for blackbody radiation. This is because P and T cannot be varied independently and, therefore, g is undefined for the system. Since for a single component system the chemical potential μ is equal to g, chemical potential is also undefined in the case of the blackbody radiation. A more appropriate interpretation for $\mu = 0$ may be that the chemical potential is defined with respect to a constant number of particles N in the system. However in case of blackbody radiation there is continuous absorption and re-emission of photons, therefore, the concept of chemical potential is not valid in this case.

It is easy to write thermodynamic functions for blackbody radiation as functions of pairs of parameters like *PV, PT, TS*, etc. These forms are tabulated below in Table 8.1.

Table 8.1 Thermodynamic functions for blackbody radiation

P	U	S	H	F	G	μ
$\frac{1}{3}\sigma T^4$	$\sigma T^4 V$	$\frac{4}{3}\sigma V T^3$	$\frac{4}{3}\sigma V T^4$	$-\frac{1}{3}\sigma V T^4$	0	0
$\frac{1}{3}\frac{U}{V}$	$3PV$	$\frac{4}{3}\frac{U}{T}$	$\frac{4}{3}U$	$-PV$		
	$\left(\frac{3S}{4}\right)^{\frac{4}{3}}(\sigma V)^{-\frac{1}{3}}$	$4\left(\frac{\sigma}{3}\right)^{\frac{1}{4}} V P^{\frac{3}{4}}$	TS	$-\frac{1}{4}TS$		
		$\frac{4}{3}(\sigma V)^{\frac{1}{4}} U^{\frac{3}{4}}$	$4PV$	$-\frac{1}{3}U$		
			$S\left(\frac{3P}{\sigma}\right)^{\frac{1}{4}}$			

8.1.2 C_P, κ_T and β for blackbody radiation

The heat capacity at constant pressure C_P, the isothermal compressibility κ_T and expansivity β for blackbody radiation are all zero or undefined. The reason is again the interdependence of T and P. C_P is defined as the amount of heat requited to raise the temperature of a given volume of the system by 1 K keeping pressure constant. However, in case of blackbody radiation it is not possible to raise the temperature without changing the pressure. Similar arguments may be given about the isothermal compressibility and the expansivity. C_V on the other hand is finite as it is equal to the amount of heat required to raise the temperature of a given amount of the system by 1 K keeping the volume constant.

8.1.3   The third law of thermodynamics and blackbody radiation

The third law of thermodynamics appears to be valid for blackbody radiation because entropy, given by Eq. 8.7 as $S = = \frac{4}{3}\sigma V T^3$ approaches zero as T goes to zero.

8.1.4   Reversible isothermal expansion or compression of blackbody radiation

Let us first consider reversible isothermal expansion of a volume V of blackbody radiation by a small amount, ΔV, at constant temperature T. The amount of heat $(\Delta Q)_T$ supplied to maintain the temperature can be easily calculated from

$$(\Delta Q)_T = T\left(\frac{\partial S}{\partial V}\right)_T \Delta V = T\left(\frac{4}{3}\sigma T^3\right)\Delta V = \frac{4}{3}\sigma T^4 \Delta V \qquad 8.11$$

With the change of volume the internal energy also changes and the change in internal energy ΔU is

$$\Delta U = \sigma T^4 \Delta V \qquad 8.12$$

The work done in expansion in volume expansion at constant temperature is

$$\Delta W = P\Delta V = \frac{1}{3}\sigma T^4 \Delta V \qquad 8.13$$

Equations 8.11, 8.12 and 8.13 show that

$$\Delta Q = \Delta U + \Delta W \qquad 8.14$$

That shows that the system follows the first law of thermodynamics. In case of reversible isothermal compression, the change in volume ΔV, the work ΔW and the change in internal energy ΔU all will have negative sign. As such ultimately equation Eq. 8.14 will remain unaltered.

8.1.5   Reversible adiabatic expansion

In reversible adiabatic process entropy S remains constant. The entropy of blackbody radiation is given by $S = \frac{4}{3}\sigma VT^3$. This implies that the multiplication VT^3 is constant. Now,

$$T = \left(\frac{3S}{4\sigma V}\right)^{-\frac{1}{3}} \text{ and } P = \frac{1}{3}\sigma\left(\frac{3S}{4\sigma}\right)^{\frac{4}{3}} V^{-\frac{4}{3}} \text{ Or } PV^{\frac{4}{3}} = \text{constant} \qquad 8.15$$

It may thus be noted that in an adiabatic process taking place in blackbody radiation the quantity $PV^{\frac{4}{3}}$ remains constant. This may be compared to the relation PV^γ in case of an ideal gas. However, it is important to realize the term $\frac{4}{3}$ is not the ratio of the two heat capacities C_P/C_V in case of blackbody radiation. It is now simple to calculate the work done in reversible adiabatic expansion of blackbody radiation from an initial state P_i and V_i to the final state P_f and V_f

$$W_{ad} = \int_{P_i,V_i}^{P_f,V_f} PdV = 3\left(P_f V_f - P_i V_i\right) = -3(P_i V_i - P_f V_f) \qquad 8.16$$

In adiabatic expansion the work is done by the system (blackbody radiations) at the cost of its internal energy as no heat may enter or leave the system and, therefore, $W_{ad} = U_i - U_f$. Since $U = 3PV$, the magnitude of W_{ad} comes out to be $3(P_i V_i - P_f V_f)$ from internal energy considerations also.

8.1.6   Free expansion of blackbody radiation

Imaginary experiment of free expansion of blackbody radiation may be done by assuming that the cavity is divided into two compartments: one completely empty at temperature T = 0 with pressure also zero (P = $\frac{1}{3}\sigma T^4$) and the other filled with blackbody radiation at temperature *T*. The dividing partition is made of perfectly black screen that absorbs all radiations falling on it. The blackbody radiation undergoes free expansion when the partition is removed. Since radiation expands against zero pressure no work is done by radiations and its internal energy U (= $\sigma T^4 V$) remains constant while volume increases. This results in the drop of the temperature of blackbody radiation to compensate for the increase of volume on free expansion. It is in contrast of the free expansion of ideal gas the temperature of which does not decrease on free expansion. The pressure P of blackbody radiation decreases on free expansion as it is proportional to the fourth power of temperature but the entropy increases as it is equal to (4/3)(U/T) and T decreases while U remains constant.

8.1.7   Joule–Thomson cooling of blackbody radiation

In a throttling process, blackbody radiations are passed from high pressure to a low pressure region through a porous plug. Enthalpy of blackbody radiation remains unaltered in the process. Since enthalpy of blackbody radiation is (4/3) U, the internal energy of the system also remains unaltered but temperature which is proportion to pressure, drops on the lower pressure side. Thus for all temperatures the temperature of blackbody radiation falls on throttling. This may also be verified by calculating the Joule-Thomson coefficient $\mu = \left(\dfrac{\partial T}{\partial P}\right)_H = \dfrac{3}{4\sigma T^3}$ for blackbody radiation and is always positive (meaning that that cooling will be produced or (T_i-T_f) remains positive for all values of T_i. This is in contrast to both the ideal gas for which μ is identically zero at all temperatures and the real gas for which μ is positive only if the gas is pre-cooled below the maximum inversion temperature.

This illustrates how the concepts of thermodynamic may be successfully applied to a system of blackbody radiation in equilibrium in a cavity.

8.2   Thermodynamics of Paramagnetic System

In this section, we will see how the general framework of thermodynamics may be readily applied to magnetic systems. This will provide yet another illustration of the very general applicability of thermodynamic concepts. We shall confine our studies only to paramagnetic materials. Paramagnetism is a relatively weak form of magnetism characterized by the fact that the material has magnetization only when it is subjected to an applied magnetic field; turning the external magnetic field off removes all traces of the magnetic activities of the material. The external magnetic field is generally denoted by $\mathcal{H}$ and is assumed to be highly uniform. In practice field $\mathcal{H}$ may be provided by a very long solenoid currying a current I. As shown in Fig. 8.2, if L is the length of the circuit and p the number of turns per unit length of the solenoid the magnitude of the external magnetization field $\mathcal{H}$ may be given as,

$$\mathcal{H} = \mu_0 p I \qquad 8.17$$

Here μ_0 is the permeability of vacuum and is a constant. A sample made of some paramagnetic material is now inserted in the solenoid. For simplicity it is assumed that the sample is of long cylindrical shape that fits perfectly in the solenoid. Since the material is of magnetic nature, the magnetic field inside the sample is different from $\mathcal{H}$, is called magnetic induction and is denoted by **B.** At a given location inside the solenoid, in absence of the sample the magnitude of magnetic field is $\mathcal{H}$ while in presence of the magnetic material it becomes B. A quantity '*magnetization*', denoted by M, is defined in terms of the difference between B and $\mathcal{H}$, in the following way,

$$\mathbf{M} \equiv \frac{(B - \mathcal{H})}{\mu_0} \quad or \quad B = \mathcal{H} + \mu_0 M \qquad 8.18$$

The fields B and $\mathcal{H}$ are measured in Tesla (τ) and the units of magnetization is Joule per meter cube per Tesla ($J\ m^{-3}\ \tau^{-1}$). Further, $M = 0$ if $\mathcal{H} = 0$ and for paramagnetic substance M is in the direction of $\mathcal{H}$. Therefore, the magnetic induction **B** is always larger than the magnetizing field $\mathcal{H}$.

Magnetization M originates from the magnetic dipole moment of molecules of the paramagnetic substance. Each molecule of the paramagnetic material, because of the associated magnetic dipole moment μ, may be considered as a tiny bar magnet. In absence of any external magnetic field $\mathcal{H}$, these molecular tiny bar magnets are randomly oriented on account of the thermal motion of molecules due to the finite temperature of the sample material. In absence of any preferred direction the magnetic moments of tiny molecular magnets are randomly oriented and add incoherently giving a net zero magnetization. However, when the external field $\mathcal{H}$ is switched on, the magnetic moments of molecular magnets have a tendency to align themselves in the direction of $\mathcal{H}$, adding coherently to $\mathcal{H}$ making B larger than $\mathcal{H}$.

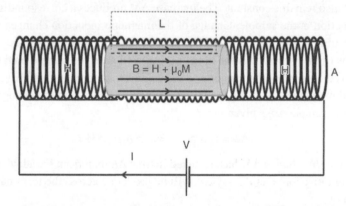

Fig. 8.1 Paramagnetic sample in uniform magnetic field H

The vectorial sum of all the magnetic moments of molecular magnets in the sample defines the total magnetic moment of the sample, the magnitude of which is denoted by $\mathcal{M}$. The total magnetic moment per unit volume of the sample $\left(\dfrac{\mathcal{M}}{V}\right)$ is equal to the magnetization M.

$$\mathcal{M} = MV$$

8.2.1   Equation of state of a paramagnetic system

The thermodynamic state parameters of a paramagnetic sample are the magnetizing field $\mathcal{H}$, the total magnetic moment $\mathcal{M}$ or magnetization M and the temperature T. The magnitude of the total magnetic moment $\mathcal{M}$ is directly proportional to the magnetizing field $\mathcal{H}$ and is inversely proportional to the temperature T of the sample. The equation of state, a relation between state parameters, for paramagnetic sample is given by,

$$\mathcal{M} = \alpha \; \mathbb{N} R' \frac{\mathcal{H}}{T} \qquad\qquad 8.19$$

Here $\mathbb{N}$ is the number of moles in the sample and R′ a constant having the value 10.4 $J^2 T^{-2}$. Alpha (α) in Eq. 8.19 is a dimensionless constant, the value of which depends on the sample, for example α is 63 for gadolinium sulfate and 35 for ion ammonium alum.

8.2.2   Work done in changing the magnitude of magnetic induction B in a paramagnetic sample

The volume of paramagnetic material does not change appreciably with the increase or decrease of magnetic field across it. Therefore, the work (PdV) is not considered in case of paramagnetic substances. However, work is performed when the magnetization M in a sample undergoes a change. There are two possible ways in which M in a sample may change: (i) by the change of $\mathcal{H}$ and (ii) by the change of temperature T. We calculate the work required to change M by a small amount ΔM keeping $\mathcal{H}$ constant.

Let there be an increase ΔM in the magnitude of the magnetization M, while the current I in the circuit is kept constant. The increase ΔM may be produced by reducing the temperature. Since current I is constant so $\mathcal{H}$ also remains constant. The increase ΔM produces a corresponding increase ΔB in the magnetic induction of the sample. Change of the magnetic induction changes the magnitude of the magnetic flux linked with the sample. The change in the magnetic flux lined with each turn of the solenoid coil is ($A\Delta B$), where A is the area of cross section of the sample (and the solenoid coil, see Fig. 8.2). If L is the length of the sample and p the number of turns per unit length of the solenoid coil, then the total magnetic flux linked with the sample is $LpA\Delta B$ and the change in the magnetic flux linked with the sample $\Delta\phi$ is given by,

$$\Delta\phi = L \, p \, A \; \Delta B = L \, p \, A \; \mu_0 \; \Delta M \qquad\qquad 8.20$$

Now suppose that this change ΔM has occurred in time Δt, then from Faraday's law of induction, an induced electromotive force (emf), say $\Delta\varepsilon$, will be produced across the terminals of the solenoid that will be given by,

$$\Delta\varepsilon = \frac{\Delta\phi}{\Delta t} = L \, p \, A \mu_0 \, \frac{\Delta M}{\Delta t} = p V \mu_0 \, \frac{\Delta M}{\Delta t} \qquad\qquad 8.21$$

Here $V = L \, A$ is the volume of the sample.

The potential $\Delta\varepsilon$ induced across the terminals of the coil opposes the already present potential of the battery. As it is required to keep the current I constant, the battery must supply additional power to maintain the current at its value I. This additional power ΔP required to be supplied by the battery is given by

$$\Delta P = I\Delta\mathcal{E} = I\,pV\,\mu_0\,\frac{\Delta M}{\Delta t}$$

Since power is the rate of doing work, the extra work $\Delta W'$ performed by the battery equals $\Delta P\,\Delta t$,

Hence
$$\Delta W' = I\,p\,V\,\mu_0\,\Delta M = \mathcal{H}V\,\Delta M \qquad\qquad 8.22$$

Here the relation $\mathcal{H} = I\,p\,\mu_0$ given by Eq. 8.17 has been used.

It may, therefore, be said that when the magnitude of magnetization in a paramagnetic system is increased by dM, a work $\Delta W'$ is performed on the system. Since work is performed on the system we assign a negative sign to the work. Hence,

$$dW' = -\mathcal{H}VdM = -\mathcal{H}d\mathcal{M} \qquad\qquad 8.23$$

Here we have used the relation $VdM = d\mathcal{M}$

If dM is decrease in the magnetization, then the work is done by the paramagnetic system and dW' is positive.

8.2.3 Potential energy and the internal energy

Paramagnetic sample acquires a potential energy Ep when it is put in an external magnetic field $\mathcal{H}$. The magnitude of the potential energy is given by,

$$E_p = -\mathcal{H}M \qquad\qquad 8.24$$

The potential energy is assigned a negative sign as it is the outcome of the work done by the external magnetic field $\mathcal{H}$ in aligning the magnetic moments of molecules in the sample. If U denotes the thermodynamic internal energy of the sample then the total energy E of the system is,

$$E = U + E_p = U - \mathcal{H}M \qquad\qquad 8.25$$

and
$$dE = dU - \mathcal{H}d\mathcal{M} - \mathcal{M}d\mathcal{H} \qquad\qquad 8.26\ (a)$$

or
$$dE + \mathcal{M}d\mathcal{H} = dU - \mathcal{H}d\mathcal{M} \qquad\qquad 8.26\ (b)$$

Now from the first and second laws of thermodynamics,

$$Tds = dU + dW'$$

In the case of a paramagnetic system the above equation becomes,

$$Tds = dU - \mathcal{H}d\mathcal{M} \qquad\qquad 8.27$$

Equation 8.27 may also be written in terms of the total energy E as follows,

$$Tds = dU - \mathcal{H}d\mathcal{M} = dE + \mathcal{M}d\mathcal{H} \qquad\qquad 8.28$$

It may be recalled that in case of the (P, T, V) system used for gases we had

$$Tds = dQ = dU + PdV = dH - VdP$$

or
$$Tds = dH - VdP \qquad\qquad 8.29$$

To some extent Eq. 8.28 may be compared to Eq. 8.29. The role played by H (enthalpy) in (P, V, T) system appears to be played by the total energy E in paramagnetic system, while $\mathcal{H}$ plays the role of intrinsic parameter P. Total energy E is sometimes called the enthalpy of the magnetic system and is denoted by H^*. However, the analogy is superfluous since enthalpy $H = U + PV$, while $E = U - \mathcal{H}M$. It may be noted that P and V appearing in enthalpy H are both the properties of the (P, V, T) system, while M is a property of the magnetic system but $\mathcal{H}$ is external to it.

8.2.4   The heat capacities of a paramagnetic system

The two heat capacities for paramagnetic system may be defined as follows:

$$C_M = \left(\frac{\partial Q}{\partial T} \right)_M = \left(\frac{\partial U}{\partial T} \right)_M \qquad \text{8.30 (a)}$$

$$C_{\mathcal{H}} = \left(\frac{\partial Q}{\partial T} \right)_{\mathcal{H}} = \left(\frac{\partial E}{\partial T} \right)_{\mathcal{H}} \qquad \text{8.30 (b)}$$

Experiments have shown that the magnitude of C_M for paramagnetic substances may be given by the following empirical relation,

$$C_M = \frac{\beta \, \mathbb{N} R}{T^2} \qquad \text{8.31}$$

Here $\mathbb{N}$ is the number of moles of the substance, R the gas constant, T the temperature and β is a constant that has different value for different substances in unit of K^2. For example, the magnitude of β for gadolinium sulfate is 0.35 K^2 and that for iron ammonium alum 0.013 K^2.

8.2.5   Thermodynamic functions for paramagnetic system

The thermodynamic functions, namely, the magnetic internal energy U, magnetic enthalpy H^*, Helmholtz function F, magnetic Helmholtz function F^*, Gibbs function G and magnetic Gibbs function G^* and derivatives of these functions are defined according to the following Table 8.2.

Table 8.2   State functions for paramagnetic system

Function	Derivative of the function	Characteristic parameters
U	$dU = TdS - \mathcal{H}dM$	$U(S, M)$
$H^* = E = U + E_p$	$dE = TdS - Md\mathcal{H}$	$E(S, \mathcal{H})$
$H^* = U - \mathcal{H}M$	$dH^* = TdS - Md\mathcal{H}$	$H^*(S, \mathcal{H})$
$F = U - TS$	$dF = -SdT + \mathcal{H}dM$	$F(T, M)$
$F^* = E - TS$	$dF^* = -SdT - Md\mathcal{H}$	$F^*(T, \mathcal{H})$
$G = U - TS + \mathcal{H}M$	$dG = -SdT - Md\mathcal{H}$	$G(T, \mathcal{H})$
$G^* = E - TS + \mathcal{H}M$	$dG^* = -SdT - \mathcal{H}dM$	$G^*(T, M)$

It is simple to obtain the following relations from the function derivatives tabulated above.

$$\left(\frac{\partial H^*}{\partial S}\right)_{\mathcal{H}} = T \; ; \qquad \left(\frac{\partial H^*}{\partial \mathcal{H}}\right)_{S} = -M \qquad\qquad \text{8.32 (a)}$$

$$\left(\frac{\partial F^*}{\partial T}\right)_{\mathcal{H}} = -S \; ; \qquad \left(\frac{\partial H^*}{\partial \mathcal{H}}\right)_{T} = -M \qquad\qquad \text{8.32 (b)}$$

$$\left(\frac{\partial G^*}{\partial T}\right)_{M} = S \; ; \qquad \left(\frac{\partial G^*}{\partial M}\right)_{T} = \mathcal{H} \qquad\qquad \text{8.32 (c)}$$

The property of the state functions that they are perfect differentials and using the following relations for the perfect differential $\psi\,(a,b)$

If $$d\psi\,(a,\,b) = X\,(a,\,b)\,da + Y\,(a,\,b)\,db$$

Then, $$\left(\frac{\partial X}{\partial b}\right)_{a} = \left(\frac{\partial Y}{\partial a}\right)_{b}$$

One may obtain following Maxwell relations for paramagnetic system:

$$\left(\frac{\partial T}{\partial M}\right)_{S} = -\left(\frac{\partial \mathcal{H}}{\partial S}\right)_{M} \quad ; \quad \left(\frac{\partial T}{\partial \mathcal{H}}\right)_{S} = -\left(\frac{\partial M}{\partial S}\right)_{\mathcal{H}} \qquad \text{8.33 (a)}$$

$$\left(\frac{\partial S}{\partial M}\right)_{T} = -\left(\frac{\partial \mathcal{H}}{\partial T}\right)_{M} \quad ; \quad \left(\frac{\partial S}{\partial \mathcal{H}}\right)_{T} = \left(\frac{\partial M}{\partial T}\right)_{\mathcal{H}} \qquad \text{8.33 (b)}$$

8.2.6   *TdS* equations for paramagnetic system

The two *TdS* equations for a magnetic system are:

$$TdS = C_{M}dT - T\left(\frac{\partial \mathcal{H}}{\partial T}\right)_{M} dM \qquad\qquad \text{8.34}$$

$$TdS = C_{\mathcal{H}}dT + T\left(\frac{\partial M}{\partial T}\right)_{\mathcal{H}} d\mathcal{H} \qquad\qquad \text{8.35}$$

Using the above two equations it is possible to derive important relations between the two heat capacities of a magnetic system. Equating Eq. 8.34 and Eq. 8.35 one gets,

$$(C_{M} - C_{\mathcal{H}})\,dT = T\left[\left(\frac{\partial \mathcal{H}}{\partial T}\right)_{M} dM + \left(\frac{\partial M}{\partial T}\right)_{\mathcal{H}} d\mathcal{H}\right]$$

Or $$dT = \frac{T\left(\frac{\partial \mathcal{H}}{\partial T}\right)_{M} dM}{(C_{M} - C_{\mathcal{H}})} + \frac{T\left(\frac{\partial M}{\partial T}\right)_{\mathcal{H}} d\mathcal{H}}{(C_{M} - C_{\mathcal{H}})}$$

The above equation gives $$\left(\frac{\partial T}{\partial \mathcal{H}}\right)_{M} = \frac{T\left(\frac{\partial M}{\partial T}\right)_{\mathcal{H}}}{(C_{M} - C_{\mathcal{H}})} \qquad\qquad \text{8.36}$$

And $$\left(\frac{\partial T}{\partial M}\right)_{\mathcal{H}} = \frac{T\left(\frac{\partial \mathcal{H}}{\partial T}\right)_{M}}{(C_{M} - C_{\mathcal{H}})} \qquad\qquad \text{8.37}$$

From any of the above equation one may obtain,

$$(C_M - C_{\mathcal{H}}) = T\left(\frac{\partial \mathcal{H}}{\partial T}\right)_M \left(\frac{\partial M}{\partial T}\right)_{\mathcal{H}} \tag{8.38}$$

But

$$\left(\frac{\partial \mathcal{H}}{\partial T}\right)_M = -\left(\frac{\partial \mathcal{H}}{\partial M}\right)_T \left(\frac{\partial M}{\partial T}\right)_{\mathcal{H}} \tag{8.39}$$

Substituting the value of $\left(\frac{\partial \mathcal{H}}{\partial T}\right)_M$ from Eq. 8.39 in Eq. 8.38, one gets,

$$(C_{\mathcal{H}} - C_M) = T\left[\left(\frac{\partial M}{\partial T}\right)_{\mathcal{H}}\right]^2 \left(\frac{\partial \mathcal{H}}{\partial M}\right)_T \tag{8.40}$$

The quantity $\left(\frac{\partial \mathcal{H}}{\partial M}\right)_T$ is called the *magnetic susceptibility*, and may have both positive and negative

signs for magnetic materials. For example diamagnetic substances have negative values of susceptibilities and for ferromagnetic and paramagnetic substances susceptibilities are positive. The

difference of the heat capacities $(C_{\mathcal{H}} - C_M)$, therefore, goes with the sign of the magnetic susceptibility.

From the equation of state of the paramagnetic substance, $M = \alpha \, \mathbb{N}R' \frac{\mathcal{H}}{T}$ (as in Eq. 8.19), it is

possible to calculate

$$\left(\frac{\partial M}{\partial T}\right)_{\mathcal{H}} = -MT \text{ and } \left(\frac{\partial M}{\partial \mathcal{H}}\right)_T = M/\mathcal{H}$$

These values of $\left(\frac{\partial M}{\partial T}\right)_{\mathcal{H}}$ and $\left(\frac{\partial M}{\partial \mathcal{H}}\right)_T$ may now be put in Eq. 8.40 to get,

$$(C_{\mathcal{H}} - C_M) = \frac{\mathcal{H}M}{T} \tag{8.41}$$

Since susceptibility for paramagnetic substances is positive, $C_{\mathcal{H}} > C_M$ for paramagnetic substances.

8.2.7 Magneto–caloric effect: low temperatures by adiabatic demagnetization

One of the most important applications of the thermodynamics of paramagnetic systems for practical use is the magneto–caloric effect, that adiabatic change in magnetization of a paramagnetic material changes the temperature of the specimen. This may be understood using the *Tds* equation (Eq. 8.34)

$$TdS = C_M dT - T\left(\frac{\partial \mathcal{H}}{\partial T}\right)_M dM$$

For an isentropic change ($dS = 0$) the above equation reduces to,

$$C_M\left(\frac{\partial T}{\partial M}\right)_S = T\left(\frac{\partial \mathcal{H}}{\partial T}\right)_M \tag{8.42}$$

Or

$$\left(\frac{\partial T}{\partial M}\right)_S = \frac{T}{C_M}\left(\frac{\partial \mathcal{H}}{\partial T}\right)_M = \frac{TM}{C_M \alpha \mathbb{N}R'} \tag{8.43}$$

Here we have used the relation $\left(\dfrac{\partial \mathcal{H}}{\partial T}\right)_{\mathcal{M}} = \dfrac{\mathcal{M}}{\alpha \mathbb{N} R'}$

The right hand side of Eq. 8.43 is always positive and hence $\left(\dfrac{\partial T}{\partial \mathcal{M}}\right)_S$ is always positive. Which

means that the temperature will decrease if the magnetization of a paramagnetic sample is adiabatically (S- constant) reduced. This method of adiabatic demagnetization of paramagnetic material is frequently used for producing low temperatures of less than 1 K required for the liquefaction of gases.

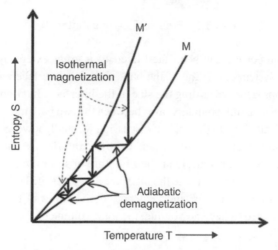

Fig. 8.2   Production of low temperature by adiabatic demagnetization
of a paramagnetic sample

Figure 8.3 shows how cycles of isothermal magnetization and adiabatic demagnetization of a paramagnetic sample can be used to reach temperatures lower than 1K.

8.3   Thermodynamics of Interface or Surface Films: Surface Tension

It is a common observation that a layer, few molecules thick, at the interface of two different materials behaves differently from the bulk material. This thin layer is called the interface or surface film. When a substance is placed in vacuum (or in air), the surface in contact with vacuum (or air) is called the free surface. However, there are three distinct subsystems at the free surface: (i) bulk material (ii) the interface or surface layer and (iii) the bulk vapors, as shown in Fig. 8.4. A few distinctions of the atoms in the surface films are

- The environment of surface atoms is very different from the atoms in the bulk
- Because of the fewer neighbor atoms, the chemical environment of surface atoms is anisotropic and different from the atoms in the bulk
- The balance of energies at the surface is the reason why small drops are curved and why liquid interfaces rise when put in a capillary.

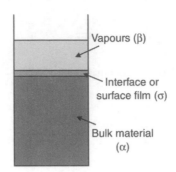

Fig. 8.3   Free surface of a material

Study of interface properties has been facilitated to a great extent by applying the laws of thermodynamics to the system consisting of the bulk material, interface and the (bulk) vapors. In general there are two approaches regarding the size of the interface. In Gibbs approach, interface is assumed to be an infinitesimal thin boundary layer between the two bulk materials while Guggenheim assumes an interface of finite size. In the present discussion we shall, however, follow Gibbs approach for simplicity. The two bulk materials may be denoted by α and β while the interface is denoted by σ. In a very general case the bulk material may be a mixture of several materials denoted by i which may go from 1 to some value m. In such case the interface as well as the bulk vapors will also contain molecules of different materials. Let V^α and V^β denote the volumes of the two bulk materials and since the interface is assumed to have no volume, the total volume of the system V is,

$$V = V^\alpha + V^\beta \qquad\qquad 8.44$$

Now all other extensive quantities, internal energy U, number of particles N, and the entropy of the complete system S may be written as the sum of the individual components as,

$$U = U^\alpha + U^\beta + U^\sigma \qquad\qquad 8.45\ (a)$$

$$S = S^\alpha + S^\beta + S^\sigma \qquad\qquad 8.45\ (b)$$

$$N_i = N_i^\alpha + N_i^\beta + N_i^\sigma \qquad\qquad 8.45\ (c)$$

If the specific energies in α and β phases are respectively u^α and u^β then the internal energy of the interface may be given as:

$$U^\sigma = U - V^\alpha u^\alpha - V^\beta u^\beta \qquad\qquad 8.46$$

If d_i^α and d_i^β, respectively, give the concentration (number per unit volume) of the ith material in phase α and β, then the number of particles of material 'i' in the interface will be given by,

$$N_i^\sigma = N_i - d_i^\alpha V^\alpha - d_i^\beta V^\beta \qquad\qquad 8.47$$

The concentration of particles of material 'i' in the interface, called the *interfacial excess*, is defined as,

$$\Gamma_i = \frac{N_i^\sigma}{A} \qquad\qquad 8.48$$

Here A denotes the area of the interface. The unit of interfacial excess is number of particles per unit area or mole per unit area.

One problem with Gibbs approach of infinitesimal thick interface is where to put it, In the middle of the beta and alpha bulk phases or nearer to alpha or to beta bulk phases? In order to make further calculations invariant with respect to the placing of the interface, a quantity called 'the relative adsorption of component ι with respect to component 1 and denoted by $\Gamma_\iota^{(1)}$,' is defined by the following relation:

$$\Gamma_i^{(1)} = \Gamma_i^\sigma - \Gamma_1^\sigma \frac{\left(d_i^\alpha - d_i^\beta\right)}{\left(d_1^\alpha - d_1^\beta\right)} \qquad 8.49$$

8.3.1 Thermodynamic functions of the system

(a) *Internal energy*

When some non-PdV work dW is done in the system consisting of the bulk phases α, β and the interface σ, the internal energy of the system changes by dU. This variation in the internal energy from the first and the second laws may be given as,

$$dU = TdS - PdV + \Sigma_i \mu_i dN_i + dW \qquad 8.50$$

If γ is the energy required to generate 1unit area of the interface, then the work done in adding an area dA will be γdA, which will be equal to the work dW. γ is called the surface tension of the material.

The variation in internal energy of the system may be split into the change in internal energies of the alpha, beta phases and the σ- interface,

$$dU = dU^\alpha + dU^\beta + dU^\sigma$$

$$= TdS^\alpha - P^\alpha dV^\alpha + \Sigma_i \mu_i^\alpha dN_i^\alpha + TdS^\beta - P^\beta dV^\beta + \Sigma_i \mu_i^\beta dN_i^\beta$$

$$+ TdS^\sigma + \Sigma_i \mu_i^\sigma dN_i^\sigma + \gamma dA \qquad 8.51$$

Each TdS term in Eq. 8.51 corresponds to the change in respective internal energy due to the change in entropy or the heat flow. The μdN term gives the variation in the internal energy due to the change of the composition of the phase. The PdV term gives the change in internal energy because of the work done in changing the volume at constant pressure. The interface on account of its vanishing volume does not contribute to the PdV work.

Since $\qquad\qquad dV = dV^\alpha + dV^\beta \,;\, dV^\alpha = V - dV^\beta \qquad 8.52$

The three TdS terms in Eq. 8.51 can be lumped together and substituting the value of dV^α from Eq. 8.52, in Eq. 8.51, one gets,

$$dU = TdS - P^\alpha dV - \left(P^\beta - P^\alpha\right)dV^\beta + \Sigma_i \mu_i^\alpha dN_i^\alpha + \Sigma_i \mu_i^\beta dN_i^\beta + \Sigma_i \mu_i^\sigma dN_i^\sigma + \gamma dA \qquad 8.53$$

(b) Helmholtz function

Similarly, the variation in Helmholtz function $dF = dF^\alpha + dF^\beta + dF^\sigma$

So, $dF = -SdT - P^\alpha dV - \left(P^\beta - P^\alpha\right) dV^\beta + \sum_i \mu_i^\alpha dN_i^\alpha + \sum_i \mu_i^\beta dN_i^\beta + \sum_i \mu_i^\sigma dN_i^\sigma + \gamma dA$ 8.54

In Eq. 8.54, terms having dT and dV become zero respectively for isothermal and isochoric processes. Further, if the system is closed, i.e., $dN_i = 0$,

$$dN_i^\sigma = -\left(dN_i^\alpha + dN_i^\beta\right).$$ 8.54

When the bulk phases α and β are in equilibrium with the interface σ, the chemical potentials of all the three are equal,

$$\mu_i^\alpha = \mu_i^\beta = \mu_i^\sigma$$ 8.55

Therefore, the change in Helmholtz function for a closed system $(dF)_{N,V,T,equi}$ in a process at constant temperature T and volume V at equilibrium (between the two bulk phases and the interface) will be,

$$(dF)_{N,V,T,equi} = -\left(P^\beta - P^\alpha\right) dV^\beta + \gamma dA$$ 8.56

Also $$\left(\frac{\partial F}{\partial V^\beta}\right) = -\left(P^\beta - P^\alpha\right)$$ 8.57

Equation 8.56 may be used to give the thermodynamic definition of surface tension as,

$$\gamma \equiv \left(\frac{\partial F}{\partial A}\right)_{N,V,T,equi} = \left(\frac{dF}{dA}\right)_{V^\beta,N,V,T,equi}$$ 8.58

The surface tension tells us how the Helmholtz function of the system changes when surface area is increased keeping the total volume, total number of particles, temperature and the volume of phase β constant. The surface tension γ is the energy required to increase the area of the interface by 1 unit and has the unit of Joule/meter^2 (J/m^2), or Newton/meter (N/m).

(c) Young–Laplace equation

It is important to note that the volume V^β and the area of the interface A are not independent. If the area A of the interface changes then the volume V^β also changes. From differential geometry it is known that

$$\left(\frac{\partial V^\beta}{\partial A}\right) = \left(\frac{1}{R_1} + \frac{1}{R_2}\right)^{-1}$$ 8.59

Here R_1 and R_2 are the two orthogonal radii of curvature of the surface as shown in Fig. 8.4 (a)

Now, $$\gamma = \left(\frac{dF}{dA}\right) = \left(\frac{\partial F}{\partial V^\beta}\right)\left(\frac{\partial V^\beta}{\partial A}\right) = -\left(P^\beta - P^\alpha\right)\left(\frac{1}{R_1} + \frac{1}{R_2}\right)^{-1}$$

Or $$(P^\alpha - P^\beta) = \Delta P = \gamma\left(\frac{1}{R_1} + \frac{1}{R_2}\right)$$ 8.60

Equation 8.60 is the well-known the Young–Laplace equation.

In the case of an interface of spherical shape (i.e., liquid drop) $R_1 = R_2 = R$ and $\Delta P = \dfrac{2\gamma}{R}$, i.e.,

the pressure inside the drop exceeds the external pressure by $\dfrac{2\gamma}{R}$.

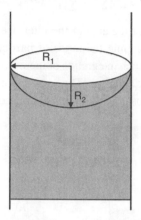

Fig. 8.3 (a) Curved interface and the two orthogonal radii

(d) *Gibbs function*

The variation in Gibbs function of the system when the area of the interface is increased by dA is given as,

$$dG = -SdT + V^\alpha dP^\alpha + V^\beta dP^\beta + \Sigma_i \mu_i dN_i + \gamma dA \qquad 8.61$$

If it is assumed that the interface is plane which means that $P^\alpha = P^\beta = P$, Eq. 8.61 reduces to

$$dG = -SdT + VdP + \Sigma_i \mu_i dN_i + \gamma dA \qquad 8.62$$

When the area of the interface is increased keeping T, P and the number of particles constant then,

$$\gamma \equiv \left(\frac{\partial G}{\partial A} \right)_{T,P,N} \qquad 8.63$$

The surface tension may also be defined as the rate of change of the Gibbs function with surface area at constant pressure, temperature and total number of particles. This is another thermodynamic definition of the surface tension.

8.3.2   Thermodynamic functions for the interface (surface film)

(a) *Internal energy for the interface (surface film)*

The change in internal energy of the interface when the area of the surface is increased by dA is given as,

$$dU^\sigma = TdS^\sigma + \Sigma_i \mu_i^\sigma dN_i^\sigma + \gamma dA \qquad 8.64$$

Equation 8.64 shows that U^σ is a linear and homogeneous function of extensive variables S^σ, N_i^σ and A. One can apply Euler's theorem to a linear homogeneous equation and can integrate it. Integration of Eq. 8.64, using Euler's theorem gives,

$$U^\sigma = TS^\sigma + \Sigma_i \mu_i^\sigma N_i^\sigma + \gamma A \qquad 8.65$$

Integration of Eq. 8.64 means that the area of the interface may be increased keeping the total number of particles constant and keeping the ratios of particles in different bulk phases and the interface unaltered. One example of such integration or increasing the surface area of the surface film is to take a liquid in a sealed test tube and tilt it. The surface area of the interface is increased but the total number of particles and the ratios of particles in bulk phases and the interface remains constant.

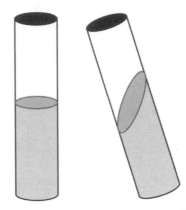

Fig. 8.4   Increase in the area of the interface by tilting the test tube

(b) Helmholtz function for the interface

The Helmholtz function is defined as $F = U - TS$ and so,

$$F^\sigma = U^\sigma - TS^\sigma = \Sigma_i \mu_i^\sigma N_i^\sigma + \gamma A \qquad 8.66$$

And

$$\frac{F^\sigma}{A} = \gamma + \Sigma_i \mu_i^\sigma \Gamma_i^\sigma \qquad 8.67$$

Also,

$$dF^\sigma = -S^\sigma dT + \Sigma_i \mu_i^\sigma dN_i^\sigma + \gamma dA \qquad 8.68$$

The temperature dependence of the surface tension may be obtained using Maxwell relation,

$$-\left(\frac{\partial \gamma}{\partial T}\right)_{A,N_i^\sigma} = \left(\frac{\partial S^\sigma}{\partial A}\right)_{T,N_i^\sigma} \qquad 8.69$$

(c) Gibbs function for the interface

The excess Gibbs function for the interface G^σ may be obtained from the relation $G = F + (PV - \gamma A)$ as,

$$G^\sigma = F^\sigma + (0 - \gamma A) = \Sigma_i \mu_i^\sigma N_i^\sigma \qquad 8.70$$

And
$$dG^\sigma = -S^\sigma dT + \sum_i \mu_i^\sigma dN_i^\sigma - A d\gamma \qquad 8.71$$

(d) Equation of state for interface

Like all other thermodynamic functions, the surface film which may be treated as an independent thermodynamic system, also has an equation of state. A surface film is characterized by its area A, surface tension σ and temperature T. In principle a relation between these three variables will constitute the equation of state of the interface. However, experiments have suggested that surface tension does not depend on the area of the film and, therefore, the equation of state for the surface film has the form,

$$\sigma = \sigma(T) \qquad 8.71(a)$$

Though in principle it should be possible to obtain the temperature dependence of the surface tension from the intermolecular forces, but more frequently it is determined through experiments which show that σ decreases almost linearly with temperature and vanishes for some value of temperature called critical temperature T_c. J. D. Van der Waals gave the following empirical relation for the surface tension of simple molecular substances,

$$\sigma = \sigma_0 \left(1 - \frac{T}{T'}\right)^{1+\alpha} \qquad 8.71 \text{ (b)}$$

Here, σ_0 and α are constants and $T' = T_c - T''$ with the free parameter T'' may be adjusted to reproduce the experimental data and is found to vary between 6 to 8 °C. Constant α has small value of the order of 0.23.

(e) Application to pure liquids

The situation becomes much simpler in case of pure liquids as there is only one value of 'i' and the interface is assumed to be placed such that Γ^σ is zero. The surface tension γ for pure liquids from Eq. 8.67 becomes

$$\gamma_{pure} = \frac{F^\sigma}{A} \qquad 8.72$$

Equation 8.69 may be used to determine the entropy of the interface from the temperature dependence of the surface tension. Assuming that the interface is homogeneous and contains negligible number of particles as compared to the bulk phases, Eq. 8.72 may be written as,

$$-\left(\frac{\partial \gamma}{\partial T}\right)_{A,P} = \left(\frac{\partial S^\sigma}{\partial A}\right)_{T,P} = s_{pure}^\sigma \text{ (specific entropy of the surface)} \qquad 8.73$$

Thus the rate of change of surface tension with temperature gives the entropy per unit area of the interface of pure liquids.

Relation given by Eq. 8.65 may be used to determine the internal energy per unit area (surface specific energy) of the interface. In case of pure liquids N^σ is negligible and, Eq. 8.65 may be written as,

$$U^\sigma = TS^\sigma + \Sigma_i \mu_i^\sigma N_i^\sigma + \gamma A \Delta U_{pure}^\sigma = TS_{pure}^\sigma + \gamma A$$

Or
$$\frac{U_{pure}^\sigma}{A} = T \frac{S_{pure}^\sigma}{A} + \gamma$$

or
$$u_{pure}^\sigma = Ts_{pure}^\sigma + \gamma = -T \left(\frac{\partial \gamma}{\partial T} \right)_{A,P} + \gamma \qquad 8.74$$

It may thus be observed that both the specific entropy and the specific internal energy of the interface may be obtained experimentally from the variation in the surface tension with temperature.

8.4   Thermodynamics of an Elastic Rod under Tension

Figure 8.5 shows a rod of some elastic material like metal or may be rubber which is subjected to a tension of infinitesimal force $d\mathfrak{J}$ applied at one end keeping the other end fixed. The original length of the rod is L and an extension of dL in the length has taken place because of the tension $d\mathfrak{J}$. Two important properties associated with material of the rod are;

Isothermal Young's modulus $\quad Y_T \equiv \dfrac{stress}{strain} = \dfrac{d\mathfrak{J}/A}{dL/L} = \dfrac{L}{A} \left(\dfrac{\partial \mathfrak{J}}{\partial L} \right)_T \qquad 8.75$

And the linear expansivity at constant tension $\alpha \mathfrak{J} \equiv \dfrac{1}{L} \left(\dfrac{\partial L}{\partial T} \right)_{\mathfrak{J}} \qquad 8.76$

Here A is the area of cross section of the rod which is assumed to remain constant.

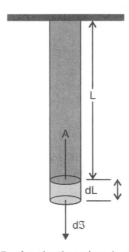

Fig. 8.5   An elastic rod under tension

It may be noted that Y_T is always positive for all materials but $\alpha_{\mathfrak{J}}$ may have positive value when the length of the rod increases on heating at constant tension and may be negative if the length decreases on heating. For example, $\alpha_\mathfrak{J}$ is positive for metallic wires while it is negative for rubber, which contracts on heating.

From the properties of the partial derivatives we know that

$$\left(\frac{\partial \mathfrak{J}}{\partial T}\right)_L \left(\frac{\partial T}{\partial L}\right)_\mathfrak{J} \left(\frac{\partial L}{\partial \mathfrak{J}}\right)_T = -1$$

And, therefore,
$$\left(\frac{\partial \mathfrak{J}}{\partial T}\right)_L = -\left(\frac{\partial L}{\partial T}\right)_\mathfrak{J} \left(\frac{\partial \mathfrak{J}}{\partial L}\right)_T = -A\,\alpha_\mathfrak{J} Y_T \qquad 8.77$$

It is obvious from Eq. 8.77 that if $\alpha_\mathfrak{J}$ is positive (metallic rod) for a fixed length of the rod, tension $\mathfrak{J}$ will decrease with the rise of temperature. This is confirmed by the fact that the metallic wires of string instruments got loose in summers.

(a)   Internal energy of the system

In the present problem work $dW = -\mathfrak{J}dL$ is performed by the force of tension $\mathfrak{J}$. The first and the second laws of thermodynamics give, $TdS = dU + dW$

Hence, $\qquad\qquad\qquad\qquad Tds = dU - \mathfrak{J}dL$

And $\qquad\qquad\qquad\qquad dU = TdS + \mathfrak{J}dL \qquad\qquad\qquad\qquad 8.78$

Therefore, $\qquad\qquad\qquad \left(\frac{\partial U}{\partial L}\right)_T = T\left(\frac{\partial S}{\partial L}\right)_T + \mathfrak{J} \qquad\qquad\qquad 8.79$

(b)   Helmholtz function of the system

Also $\qquad\qquad\qquad\qquad F = U - TS \text{ and } dF = dU - TdS - SdT$

Putting the value of dU in the above equation from Eq. 8.78, one gets:

$$dF = -SdT + \mathfrak{J}dL \qquad\qquad\qquad\qquad 8.80$$

(c)   Entropy of the system

Equation 8.80 gives,

$$S = -\left(\frac{\partial F}{\partial T}\right)_L \text{ and } \mathfrak{J} = \left(\frac{\partial F}{\partial L}\right)_T \qquad\qquad 8.81$$

Now using the Maxwell relation one gets,

$$\left(\frac{\partial S}{\partial L}\right)_T = -\left(\frac{\partial \mathfrak{J}}{\partial T}\right)_L \qquad\qquad\qquad\qquad 8.82$$

But $\left(\frac{\partial \mathfrak{J}}{\partial T}\right)_L$ from Eq. 8.77 is equal to $-A\,\alpha_\mathfrak{J} Y_T$. Substituting this value in Eq. 8.82 the rate of increase of entropy with length at constant temperature for the system is given as,

$$\left(\frac{\partial S}{\partial L}\right)_T = A\alpha_{\jmath} Y_T \qquad\qquad 8.83$$

Equation 8.83 tells that the entropy of the system will increase with the increase of the length at constant temperature provided $\alpha_{\jmath}$ is positive. Since $dS = dQ/T$, at fixed T increase of entropy means that heat dQ will be absorbed in the process. On the other hand for those substances which have negative value of $\alpha_{\jmath}$, like rubber, entropy will decrease with the increase of length at constant temperature and heat dQ will be released in the process.

Suppose at constant temperature T the length of the rod has increased by ΔL. The corresponding increase ΔS in the entropy is given by,

ΔS = *rate of change of entropy with length x change in length*

Or
$$\Delta S = \left(\frac{\partial S}{\partial L}\right)_T x\,\Delta L = A\,\alpha_{\jmath} Y_T\,\Delta L \qquad\qquad 8.84$$

And the quantity of heat ΔQ released in the isothermal extension in the length ΔL is,

$$\Delta Q = T\Delta S = TA\,\alpha_{\jmath} Y_T\,\Delta L \qquad\qquad 8.85$$

Also from Eq. 8.79 after substituting the value of $\left(\dfrac{\partial S}{\partial L}\right)_T$ from Eq. 8.83, one gets;

$$\left(\frac{\partial U}{\partial L}\right)_T = T\left(\frac{\partial S}{\partial L}\right)_T + \Im = TA\,\alpha_{\jmath} Y_T + \Im \qquad\qquad 8.86$$

The last four equations give relations of thermodynamic parameters for the system of a rod under tension, with the experimental observables of the system.

8.5   Some Engineering Applications of Thermodynamics

It may be recalled that in its initial stages the science of thermodynamics was developed with the principle objective of better understanding the process of heat conversion into work and the vice versa. Even today it is extensively used for engineering applications like designing and analyzing the operation of all sorts of engines and refrigerators. From the point of view of engineering applications the three system parameters, namely, pressure P, specific enthalpy h and specific entropy s are the most important and are frequently required for calculations. Engineers, therefore, use tables and graphs giving the values of these parameters for different materials used as working substance in engines and refrigerators. Since water/steam is very often used in heat engines, tables giving these parameters for water over wide range of pressures and temperatures, etc., are available in different systems of units and are generally called steam tables. Similar tables for other working substances are also available. Three dimensional graphs showing the interdependence of h, s and P for different working substances are also of interest for engineers. The typical P–s–h graph for water is shown in Fig. 8.7. A projection of P–s–h surface on h–s surface, called the Mollier diagram, is frequently used by engineers. All engines, devices that convert heat energy into work and refrigerators that transfer heat from a cold body to a hotter environment at the cost of work, utilize some thermodynamic cyclic operation. The reversible Carnot cycle has already been discussed. Some other cycles used in different engineering applications are discussed below.

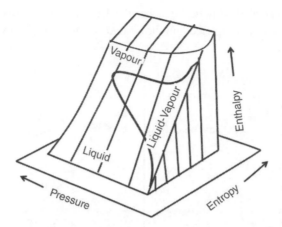

Fig. 8.6   A typical P-s-h graph for water

8.5.1   The Otto cycle (used in petrol or gasoline engines)

A schematic diagram of the Otto engine cylinder is shown in Fig. 8.7 (a). At the top end called the cylinder head, there are fuel inlet, waste fuel outlet and a spark plug that produces an electric spark at the right moment to ignite the fuel. The fuel used is a mixture of air and gasoline. A piston moves freely in the cylinder. The piston turns a crankshaft and a flywheel of large moment of inertia that stores the mechanical energy transferred to it.

The cycle begins with step-1 called the "dead center" when the cylinder is closest to the cylinder head. At this instant fuel mixture is present between the piston and the cylinder head and both valves are closed. An electric spark by the spark plug ignites the fuel. The P-V diagram of the cycle is shown in Fig. 8.7 (b). Ignition of the fuel suddenly increases the pressure to a high value. The process of sudden increase of pressure may be considered to be isochoric, i.e., at constant volume because the piston on account of large moment of inertia could not move in the initial phase of fuel explosion. This is shown by stroke AB in Fig. 8.7 (b). Eventually, however, the hot gases produced by fuel explosion expand pushing the piston down, almost adiabatically, in step-2 that is called the power stroke and is shown by BC in Fig. 8.7 (b). During power stroke a part of the heat energy is converted into mechanical energy and is transferred to the flywheel which stores it. At the end of the power stroke when the piston is at the lower end, the waste release valve at the cylinder head opens and cooler gases are released in the atmosphere. The mechanical energy stored in the flywheel during the power stroke and the crankshaft arrangement reverses the direction of motion of the piston which after reaching the bottom at the end of the power stroke starts moving up towards the cylinder head. The upwards moving piston forces the spent fuel and air mixture to exhaust through the waste release valve into the atmosphere and constitutes step-3 called the "exhaust stroke". This whole process may be represented by the isochoric process represented in the figure by CD. When the piston again reaches the dead end (the cylinder head) the exhaust release valve is closed and the fuel intake valve is opened to allow a fresh batch of fuel air mixture to enter as the piston moves down in step-4 sucking the fresh fuel mixture till it reaches the bottom end of the cylinder. The intake valve is closed when piston reaches the bottom. Both valves closed the crank-shaft arrangement reverses the direction of motion of the piston once again, that now moving upwards, in step-5, compresses the fuel mixture almost adiabatically till it reaches the dead end again. This is represented by DA in Fig. 8.7 (b). The cycle is now completed and the system is ready for the next cycle when spark plug produces another spark.

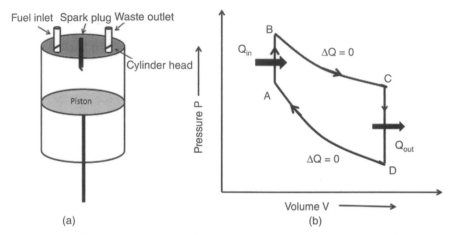

Fig. 8.7   (a) Schematic diagram of engine cylinder; (b) P-V diagram for Otto cycle

In this idealized version of the cycle, energy in the form of heat is absorbed in the first stage when the volume of the gas V_i is a minimum, and is rejected to the environment in the third stage when the volume V_f is a maximum. It may be noted that both these stages are of constant volume. The ratio $b = \dfrac{V_f}{V_i}$ is called the compression ratio and is a measure to what extent the fuel mixture is compressed before explosion by the spark.

Work is performed in the adiabatic stages BC and DA. The amount of work done is,

$$W = \int_{V_i}^{V_f} P_{BC}\,dV + \int_{V_f}^{V_D} P_{DA}\,dV \qquad\qquad 8.87$$

Here P_{BC} and P_{DA} are the pressures in the two stages BC and DA respectively. Since for an adiabatic process $PV^{\gamma} = constant$, we have $P_{BC} = \dfrac{K_1}{V^{\gamma}}$ and $P_{DA} = \dfrac{K_2}{V^{\gamma}}$, here K_1 and K_2 are two different constants. With these substitution Eq. 8.87 becomes,

$$W = K_1 \int_{V_i}^{V_f} V^{-\gamma}\,dV + K_2 \int_{V_f}^{V_D} V^{-\gamma}\,dV = \frac{1}{1-\gamma}\left[K_1 V^{1-\gamma}\right]_{V_i}^{V_f} + \frac{1}{1-\gamma}\left[K_2 V^{1-\gamma}\right]_{V_f}^{V_i}$$

$$= \frac{NR}{\gamma-1}\left[(T_B - T_C) + (T_D - T_A)\right] = \frac{NR}{\gamma-1}\left[(T_B - T_A) - (T_C - T_D)\right] \qquad 8.88$$

where T_B, etc., refer to the temperature at B, etc.

On the other hand heat absorbed in stage AB is $Q = (U_B - U_A) = C_V(T_B - T_A)$ \qquad\qquad 8.89

The efficiency of the engine $\eta = \dfrac{W}{Q} = \dfrac{NR}{c_V(\gamma-1)}\left[1 - \dfrac{(T_C - T_D)}{(T_B - T_A)}\right]$ \qquad\qquad 8.90

But for the two adiabatic stages $T_B V_i^{\gamma-1} = T_C V_f^{\gamma-1}$ and $T_D V_f^{\gamma-1} = T_A V_i^{\gamma-1}$

Hence,
$$\frac{(T_C - T_D)}{(T_B - T_A)} = \left(\frac{V_i}{V_f}\right)^{\gamma-1} = \left(\frac{1}{b}\right)^{\gamma-1} = b^{1-\gamma} \qquad 8.91$$

Also,
$$\frac{\mathbb{N}R}{c_V(\gamma-1)} = \frac{\mathbb{N}R}{c_V\left(\frac{c_P}{c_V} - 1\right)} = \frac{\mathbb{N}R}{(c_P - c_V)} = \frac{\mathbb{N}R}{\mathbb{N}R} = 1 \qquad 8.92$$

Therefore from Eq. 8.90, the efficiency of the engine is

$$\eta_{Otto} = \frac{\mathbb{N}R}{c_V(\gamma-1)}\left[1 - \left(\frac{1}{b}\right)^{\gamma-1}\right] = \left[1 - b^{1-\gamma}\right] \qquad 8.93$$

It is obvious from Eq. 8.93 that the efficiency of the engine, η, lies between 0 and 1 as $(1-\gamma) < 1$ and b > 1. Further, in this idealized modeling the efficiency depends only on the compression ratio b, larger the compression ratio, more will be efficiency. However, the compression ratio cannot be increases beyond a certain limit. The upper limit to compression ratio is put by the volatility of the fuel mixture. An excessively large compression ratio may ignite the fuel mixture before the piston has reached the dead end, resulting in the loss of power and efficiency.

8.5.2   The diesel engine

The diesel engine differs from an Otto engine that it does not have any spark plug, the combustion of the fuel occurs because of the large compression and then the fuel is allowed to burn slowly. The P–V diagram of the idealized diesel engine is shown in Fig. 8.9. The combustion of the fuel mixture starts at point A in the PV diagram. The gas burns and expands pushing the piston in power stroke, which consists of two parts, an isobaric part AB and then an adiabatic part BC when fuel is spent. Isochoric exhaust stroke CD takes place when the outlet valve is opened and spent fuel is eliminated into the atmosphere. A fresh batch of air fuel mixture is then allowed to enter the engine which is adiabatically compressed during the stroke DA to complete the cycle.

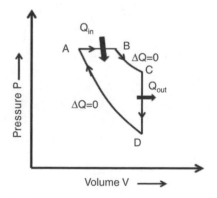

Fig. 8.8   P-V diagram for a diesel engine

It may be noted that in diesel engine work is performed in three out of the four strokes. To calculate the efficiency we calculate the heat input Q_{in} and heat rejected Q_{out}. Let T_A, T_B, T_C, T_D be the temperatures respectively at points A, B, C and D of the cycle. Then,

$$Q_{in} = C_P (T_B - T_A) \text{ and } Q_{out} = C_V (T_C - T_D) = -C_V (T_D - T_C) \qquad 8.94$$

Hence, the efficiency of the engine

$$\eta = \frac{W}{Q_{in}} = \frac{Q_{in} - Q_{out}}{Q_{in}} = 1 - \frac{C_V (T_D - T_C)}{C_P (T_B - T_A)} = 1 + \frac{1}{\gamma} \frac{(T_D - T_C)}{(T_B - T_A)} \qquad 8.95$$

If V_A, V_B, V_C, V_D respectively represent the volumes at points A, B, C and D in the PV-graph, then for the adiabatic strokes BC and DA one may write,

$$T_B V_B^{\gamma-1} = T_c V_c^{\gamma-1} \text{ and } T_D V_D^{\gamma-1} = T_A V_A^{\gamma-1} \text{ or } T_D V_C^{\gamma-1} = T_A V_A^{\gamma-1} \text{ as } V_D = V_C \qquad 8.96$$

From Eq. 8.96 one gets, $(T_D - T_C) V_C^{\gamma-1} = T_A V_A^{\gamma-1} - T_B V_B^{\gamma-1}$

Or
$$(T_D - T_C) \left(\frac{V_C}{V_A} \right)^{\gamma-1} = T_A - T_B \left(\frac{V_B}{V_A} \right)^{\gamma-1} \qquad 8.97$$

A quantity 'c' called the cut-off ratio is defined as, $c \equiv \dfrac{V_B}{V_A}$

Now for the isobaric stroke A to B,

$$\frac{V_B}{V_A} = \frac{T_B}{T_A} = c \text{ and hence } T_B = c T_A \qquad 8.98$$

Substituting the above value in Eq. 8.97, one gets;

$$(T_D - T_C) \left(\frac{V_C}{V_A} \right)^{\gamma-1} = T_A - c T_A (c)^{\gamma-1} = T_A (1 - c^{\gamma})$$

And $(T_D - T_C)(b)^{\gamma-1} = T_A (1 - c^{\gamma})$, therefore, $\dfrac{T_D - T_C}{T_B - T_A} = \dfrac{(c^{\gamma} - 1)(b)^{1-\gamma}}{(c - 1)}$ $\qquad 8.99$

Substitution of the value of the ratio $\dfrac{T_D - T_C}{T_B - T_A}$ from Eq. 8.99 into Eq. 8.95 gives the efficiency as,

$$\eta_{Diesel} = 1 + \frac{1}{\gamma} \frac{(T_D - T_C)}{(T_B - T_A)} = 1 - \frac{(c^{\gamma} - 1)(b)^{1-\gamma}}{\gamma(c - 1)} \qquad 8.100$$

Thus the efficiency of a diesel engine depends both on the compression and the cut-off ratios.

8.5.3 Rankine cycle

Most large electricity generating plants and big ship engines use water vapors (steam) as the working substance following some variations in basic Rankine cycle. (William John Macquorn Rankine, 5

July, 1820–24 December, 1872, was a Scottish Civil engineer, a physicist and a mathematician. He wrote the first book on thermodynamics in 1859).

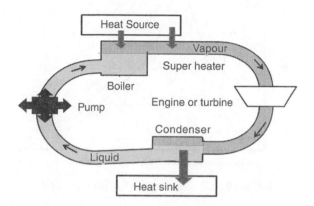

Fig. 8.9 Schematic diagram of a Rankine cycle

The schematic diagram of the modified Rankine cycle is shown in Fig. 8.10. The source of heat may be burning of coal, gas or the heat from a nuclear reactor. The boiler receives heat from the heat source and heats the working substance, saturated water, up to its boiling point when it is converted in to saturated steam (vapors). The steam is further heated in the super heater by the same source of heat. The pressure in both the boiler and the super heater is same. Though theoretically the steam in the super heater may be heated to the temperature of the heat source, which may be as high as 1500–2000 °C, but there is an upper limit to the temperature to which the steam may be heated, called the metallurgical limit ≈ 600 °C, that is decided by the material of the super heater and the pipes. Saturated super-heated steam is then led to the engine or the turbine where it produces mechanical work and in the process its temperature and pressure both drop, some of the steam is even converted in to boiling water. The mixture of steam and water then goes to a condenser where all vapors are converted in to liquid by giving out the heat of condensation which is rejected into sink. The sink may be a river loop, atmosphere or a cooling tower. The pressure in this part of the system is decided by the temperature of the sink. If T_s is the temperature of the sink then the pressure in the condenser must be as large as the vapor pressure of water at temperature T_s. The condensed water is then forced into the boiler by a pump completing the cycle. Since at the sink temperature, around 35 °C, steam condenses at a pressure lower than the atmospheric pressure, it is desirable to run the condenser at low pressure or under vacuum. Further, non-condensable gases produced in the system along with oxygen and carbon dioxide gases must be removed from the condensed liquid to avoid corrosion, before it is pumped into boiler. For this purpose a 'deaerator' is put in the system.

The work producing device was reciprocating cylinder–piston system in the initial stages in systems where Rankine cycle was employed for power generation. However, later on vapor turbines replaced cylinder–piston in most of the generators. In a reciprocating cylinder–piston arrangement steam produces work by giving motion to a piston in a cylinder which is transferred to a flywheel that stores it. In a turbine, on the other hand, steam is passed through a nozzle to give it large kinetic energy which is used to turn the buckets of the turbine.

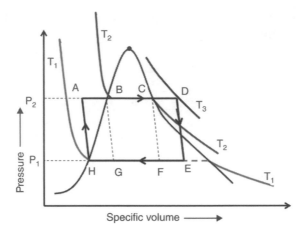

Fig. 8.10   (a) *P–v* diagram for Rankine cycle

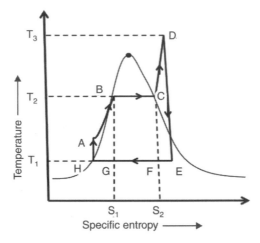

Fig. 8.10   (b) Rankine cycle in s-T plane

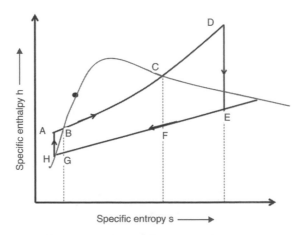

Fig. 8.10   (c) Rankine cycle in h-s plane

Projections of an ideal Rankine cycle on $(P–v)$, $(T–s)$ and $(h–s)$ planes are shown in Figs. 8.11 (a), (b) and (c), respectively. Operation of the ideal cycle starts with the isobaric and reversible heating of saturated water at temperature T_1 (point A) and its conversion into saturated steam at temperature T_3 (point D) in the boiler and the super heaters. In actual cycle the heating is isobaric but not reversible. For reversible heating infinite number of heat reservoirs from temperature T_1 to temperature T_3 in infinitesimal steps need to be employed. This is not practical and, therefore, conversion of saturated water into saturated super-heated steam is done in a single irreversible isobaric step. The average temperature of this process depends on T_1 and T_3, and has a larger value if T_3 is large. The efficiency of an actual Rankine cycle is definitely lower than that of an ideal one, but increases with the increasing value of T_3. In the second step of the ideal cycle, work is drawn from the system in the reversible adiabatic step shown by DE in Figs 8.11. In the third step of the cycle, shown by EG, heat is rejected to the sink and the mixture of super-heated steam and hot condensed water left after the second step are all converted to saturated and vapor free water at the sink temperature in a reversible isobaric step. In the final step, GA saturated water is reversibly and adiabatically compressed by a pump to the boiler. Since rise of temperature in an adiabatic process is not much, the saturated water is heated in the boiler and subsequently in the super heater to initiate the new cycle of operation. In ideal Rankine cycle both: the step one in which heat (say Q_2) is taken from the boiler and super heater and step-3 in which heat (say Q_1) is rejected at the sink, are isobaric and reversible. Therefore, Q_{in} and Q_{out} are respectively equal to the change in the enthalpy of the system in these steps. Mollier diagram, the projection of which on h–s plane is shown in Fig. 8.11 (c), is of much use and help for engineers in calculate the efficiency of the system, as follows:

$$\text{Heat absorbed (per mole) in step AD } Q_{in} = (h_D - h_A) \qquad 8.101$$

$$\text{Heat rejected (per mole) in step EH } Q_{out} = (h_E - h_H) \qquad 8.102$$

Hence,
$$\text{Work done (per mole) } w = Q_{in} - Q_{out} = (h_D - h_A - h_E - h_H) \qquad 8.103$$

$$\text{The effeciency of the Rankine cycle } \eta_{Rankine} = \frac{w}{Q_{in}} = \frac{\left(h_D - h_A - h_E + h_H\right)}{\left(h_D - h_A\right)} \qquad 8.104$$

Here h_i stands for the specific enthalpy of the working substance at point i, where i may be the point A, B, C, D, ..., H, etc. It is obvious that the efficiency of a real system based on Rankine cycle is less than the ideal one.

Reason for super heating the steam:

There are two reasons for super heating steam in Rankine cycle before extracting work from it.

1. Higher the final temperature T_3 of the steam higher will be the average temperature at which it may be assumed to have absorbed heat and hence higher the efficiency of the cycle.
2. Since most of the present day generators based on Rankine cycle employ turbines for extracting work it is required that the turbine buckets should last long. Turbine buckets are generally made of steel and their main problem is corrosion. The corrosion will be less if the steam jets hitting the buckets contain less percentage of moisture, i.e., less percentage of condensed water. The percentage of moisture in the steam when it cools in step DE is decided by point E. Lesser will

be moisture contents more the point E lies towards right hand side. Higher value of T_3 thus helps in reducing the moisture contents and the probability of corrosion of turbine buckets. Typical layout of a turbine is shown in Fig. 8.11 (d).

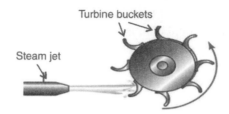

Fig. 8.10(d)   Steam jet and buckets of a turbine

Solved Examples

1. A system consists of a pure liquid in equilibrium with its vapors. What will happen to the vapor pressure of the liquid when some atoms of an inert gas that does not chemically react with the liquid vapors are introduced in the vapors at constant temperature to increase the total pressure? The vapors of the liquid and the inert gas may be assumed to be two non-interacting ideal gases.

    Solution:

    Initially the system consists of a liquid and its vapors in equilibrium at temperature T. In the initial state the vapor pressure p_{ini} of the liquid vapors is equal to the total pressure P_i. Now some atoms of another inert gas that does not chemically interacts with the vapors of the liquid are introduced in the system at same temperature T as a result of which the total pressure of the system has become P_f. It is required to investigate as to what happens to the vapor pressure p_{ini} of the liquid.

    Since the liquid and its vapors were in equilibrium initially, therefore, the specific Gibbs function of liquid state must be equal to the specific Gibbs function of the vapor state, i.e.,

    $$g_{liq} = g_{vap} \qquad\qquad\qquad \text{S-8.1.1}$$

    The system remains in equilibrium after the mixing of some atoms of the other inert gas. If dg_{liq} and dg_{vap} represents, respectively, the change in the value of the specific Gibbs functions of the liquid and the vapor phases after mixing, then the condition of equilibrium requires,

    $$g_{liq} + dg_{liq} = g_{vap} + dg_{vap}$$

    Or $$dg_{liq} = dg_{vap} \qquad\qquad\qquad \text{S-8.1.2}$$

    But $$dg_{liq} = -sdT + v_{liq}dp = v_{liq}\,dP \qquad\qquad\qquad \text{S-8.1.3}$$

    Here v_{liq} is the specific volume of the liquid and $(-sdT = 0)$, as T is constant and $dT=0$

    If it is assumed that the vapors follow the ideal gas rules, the specific Gibbs function g_{vap} will be given by $g_{vap} = g_{vap} = (\ln p + \psi\,(T))$ here p is the partial pressure of the gas

And $dg_{vap} = RT\dfrac{dp}{p}$ (as $\psi(T)$ is a function of T which remains constant)   S-8.1.4

Equating Eq. S-8.1.3 and Eq. S-8.1.4 one gets,

$$v_{liq}\, dP = RT\frac{dp}{p} \quad or \quad \frac{v_{liq}}{RT}dP = \frac{dp}{p} \qquad \text{S-8.1.5}$$

The above equation may be integrated with the limit for total pressure P going from P_i to P_f and partial pressure p going from p_{ini} to p_{fin}.

So,
$$\frac{v_{liq}}{RT}\int_{P_i}^{P_f} dP = \int_{p_{ini}}^{p_{fin}} \frac{dp}{p}$$

Or
$$\frac{v_{liq}}{RT}\left(P_f - P_i\right) = \ln\frac{p_{fin.}}{p_{ini}} \qquad \text{S-8.1.6}$$

Equation S-8.1.6 tells that the vapor pressure of the liquid will increase, which means that more liquid will be evaporated, if the total pressure of the system is increased by putting inert gas in vapors.

2. Surface tension of water decreases from 74.25×10^{-3} Nm^{-1} at 10 °C to 68.25×10^{-3} Nm^{-1} at 50 °C. Compute (i) specific Helmholtz energy of the interface at 25 °C (ii) specific entropy (per unit area) of the interface at 25 °C (iii) amount of heat required to increase the area of the interface by 0.01 m^2 at 25 °C (iv) internal energy per unit area of the interface 25 °C.

Solution:

Change in surface tension between 10°C and 50°C $\Delta\gamma = -(74.25 - 68.25) \times 10^{-3} N/m$

$$\Delta\gamma = -6.00 \times 10^{-3}\frac{N}{m} \quad \text{and} \quad \Delta T = (50 + 273) - (10 + 273) = 40$$

Hence, rate of change of surface tension with temperature

$$\frac{\partial\gamma}{\partial T} = -\frac{6.0}{40} \times 10^{-3} = 0.15 \times 10^{-3}\frac{N}{mK} \qquad \text{S-8.2.1}$$

Therefore, surface tension at 25°C $\gamma^{25} = (74.25 - 15 \times 0.15) \times 10^{-3}\frac{N}{m} = 72.0 \times 10^{-3}\frac{N}{m}$ S-8.2.2

(i) The specific Helmholtz energy of the interface at temperature 25°C $= \gamma^{25} = 72.0 \times 10^{-3}\dfrac{N}{m}$

(ii) Surface entropy at 25 °C $s_{25}^{\sigma} = -\left(\dfrac{\partial\gamma}{\partial T}\right)_{25} = -\dfrac{\partial\gamma}{\partial T} = 0.15 \times 10^{-3}\dfrac{N}{mK}$

(iii) Heat required to increase the area by A is $Q = A.q$, where q is the heat required to increase the area by 1 unit.

But $q = -TdS = -(25 + 273) \times \left(\dfrac{\partial\gamma}{\partial T}\right)_{25} = -298 \times (-0.15 \times 10^{-3}) = +44.7 \times 10^{-3}\, Jm^{-2}$

Therefore $Q = Aq = 44.7 \times 10^{-3} \times 0.01 = 44.7 \times 10^{-5}\, J$

(iv) Specific internal energy of the interface u^σ at $25^0 \, C = u^\sigma_{25} = (\gamma)_{25} - T\left(\dfrac{\partial \gamma}{\partial T}\right)_{25}$

Or $u^\sigma_{25} = 72.0 \times 10^{-3} \dfrac{N}{m} - 298 \times \left(-0.15 \times 10^{-3}\right) \dfrac{N}{m} = 116.7 \times 10^{-3} \dfrac{N}{m}$

3. N_A and N_B moles of two ideal gases A and B are filled into two parts each of volume V of a container of total volume $2V$ at same temperature T. The separating wall between the two parts is suddenly removed and the two ideal gases were allowed to diffuse into each other to establish equilibrium. Calculate the change in entropy of the system (also called the entropy of mixing).

Solution:

Figure S-8.3 (a) shows the two ideal gases A and B held in two parts each of volume V of a container of size 2V at temperature T. The gases are kept separate by a diathermic partition. Next, the partition is withdrawn and the gases are allowed to mix and finally a state of equilibrium has reached between the gases. It is required to calculate the change in entropy of the system in this process.

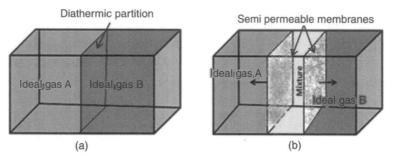

Fig. S-8.3   (a) Ideal gases before mixing (b) Reversible mixing through the motion of semi-permeable membranes

The actual process of mixing is an irreversible process. The change in entropy will, however, be same if the final equilibrium state is obtained by an irreversible process or by a reversible process. Since it is possible to calculate the entropy change in an reversible process we design a 'thought experiment' in which the final state is reached via a reversible process and make calculations for the change in entropy in this imaginary reversible process.

We assume that we have two semi permeable membranes, one with pores such that molecules of gas A only may pass through it and the other through which only the molecules of gas B can pass. The membrane through which molecules of gas B can pass is then very slowly moved from the middle towards the right hand side and the other through which the molecules of A can pass towards the left hand side. The motions of the membranes are assumed to be so slow that equilibrium is maintained at each step and the process remains reversible. The membrane which is moving towards right is under the pressure P_A due to the molecules of gas A, and similarly the other membrane is acted by a pressure P_B due to the molecules of gas B. Thus work $dW_A = (P_A dV)$ will be performed by gas A in moving the membrane by small volume dV and work $dW_B = (P_B dV)$ by the gas B. As a result of this work the temperature of the two gases will fall.

We assume that the whole system is thermally connected to a heat reservoir at temperature T, so that the gases draw heats from the reservoir as they expand and their temperatures remain constant at T. Since the gases are ideal and their temperature remains constant, there is no change in the internal energies of the gases. Ultimately the two membranes reach the end walls of the container and mixing of gases through a reversible process has taken place. The total work done in the expansion of the gases on the two sides is,

$$dW = \int_V^{2V} dW_A + \int_V^{2V} dW_B = \int_V^{2V} P_A dV + \int_V^{2V} P_B dV = N_A RT \int_V^{2V} \frac{dV}{V} + N_B RT \int_V^{2V} \frac{dV}{V}$$

Or
$$dW = \left(N_A + N_B\right) RT \ln \frac{2V}{V} = \left(N_A + N_B\right) RT \ln 2 \qquad \text{S-8.3.1}$$

But from the first law, $dQ = dU + dW$ and in the present case $dU = 0$, Hence $dQ = dW$. Therefore, the amount of heat drawn dQ is equal to work done by the gases in their expansion dW. Also from the second law $dQ = T\, dS$, or $dS = dQ/T$

Hence, the change in the entropy of the system or the entropy of mixing $= dQ/T = dW/T$

It follows from Eq. S-8.3.1, $dS = \dfrac{\left(N_A + N_B\right) RT \ln 2}{T} = \left(N_A + N_B\right) R \ln 2$

We thus see that when equal amounts of two ideal gases are mixed at constant temperature, the entropy of the system increases by $(N_A + N_B)\, R \ln 2$

4. What is Gibbs paradox? How can it be explained?

Solution:

In solved example 8.3 we have seen that the entropy of the system increases when two different ideal gases are mixed at constant temperature. Now suppose the two gases are not two different ideal gases but the same ideal gas is held in two parts of the container. Will the entropy of the system still increase when the two parts of the same ideal gas are mixed? From our derivation it appears that it will still increase but commonsense tells it should not. This divergence between the theoretical result and the outcome based on commonsense is termed as Gibbs paradox.

One way of explaining Gibbs paradox is to say that the theoretical derivation of the entropy of mixing is based on the possibility of having semi permeable membranes that may pass molecules of the selective gas. If the gas is same then it is not possible to have a semi permeable membrane which may pass molecules only of one side of the chamber. Since molecules from both side of the chamber will pass through the membrane, there will be no effect of the membrane and from the point of view of thermodynamics there will be no mixing. Consequently there will be no change in the situation and hence no change in the entropy. A more appropriate and adequate answer to Gibbs paradox is provided by quantum thermodynamics.

5. A container of total volume 3V is divided into three equal compartments. The first compartment is filled with 2 kilomole of an ideal gas A, the second by 3 kilomoles of ideal gas B and the third with 5 kilomoles of ideal gas C. The initial temperature of each compartment is 200 K and pressure $4.0 \times 10^5\ \text{Nm}^{-2}$. The partitions between the compartments are removed and the gases are allowed to mix till equilibrium is established. Compute the change in (i) the Gibbs function (ii) the entropy of the system in the process of mixing.

Solution:

It is given that the numbers of kilomole of the gases are: $N_A = 2$, $N_B = 3$, $N_C = 5$. The total number of kilomole in the mixture $N = 10$. The mole fractions $x_i (i = A,B,C)$ of the gases are:

$$x_A = \frac{N_A}{N} = \frac{2}{10} = 0.2; x_B = \frac{N_B}{N} = \frac{3}{10} = 0.3; x_C = \frac{N_C}{N} = \frac{5}{10} = 0.5 \qquad \text{S-8.5.1}$$

Let the pairs g_A^i, g_A^f; g_B^i, g_B^f; and g_C^i, g_C^f, respectively, represent the initial and final specific Gibbs functions for the gases, A, B and C and G^i and G^f the initial and final Gibbs functions of the system then,

$$G^i = N_A g_A^i + N_B g_B^i + N_C g_C^i \text{ and } G^f = N_A g_A^f + N_B g_B^f + N_C g_C^f \qquad \text{S-8.5.2}$$

Initially, each gas has the same temperature $T = 200$ K and pressure $P = 4 \times 10^5 \, Nm^{-2}$. Therefore, the specific Gibbs functions in the initial states of the three ideal gasses are,

$$g_A^i = RT \left(\ln P + \psi_A \right); \; g_B^i = RT \left(\ln P + \psi_B \right); g_C^i = RT \left(\ln P + \psi_C \right) \qquad \text{S-8.5.3}$$

Here ψ_A, ψ_B, ψ_C are functions of temperature T and since initial and final temperatures are same these functions remain constant during the mixing process.

After mixing of the gases when equilibrium has established, the specific Gibbs functions for each gas will have the same form as given in Eq. S-8.5.3 except that the pressure P that was same for all gases in the initial state will be replaced by the respective partial pressure of the gas. If p_A, p_B and p_C respectively represents the partial pressures of the gases in the final mixture then:

$$p_A = x_A P, \, p_B = x_B P, \text{ and } p_C = x_C P \qquad \text{S-8.5.4}$$

It is because the three gases are ideal and their final volume is *3V* for all gases and the temperature is *T*

Therefore,

$$g_A^f = RT \left(\ln x_A P + \psi_A \right); \; g_B^i = RT \left(\ln x_B P + \psi_B \right); g_C^i = RT \left(\ln x_C P + \psi_C \right)$$

Or $\quad g_A^f = g_A^i + RT \left(\ln x_A \right), \, g_B^f = g_B^i + RT \left(\ln x_B \right), \text{ and } g_C^f = g_C^i + RT \left(\ln x_C \right)$ (S-8.5.5)

Now from Eq. S-8.5.2 one gets,

$$\Delta G = G^f - G^i = N_A \left[g_A^f - g_A^i \right] + N_B \left[g_B^f - g_B^i \right] + N_A \left[g_C^f - g_C^i \right]$$

$$\Delta G = N_A RT \left(\ln x_A \right) + N_B RT \left(\ln x_B \right) + N_C RT \left(\ln x_C \right)$$

Or $\qquad\qquad \Delta G = RT \left[N_A \ln x_A + N_B \ln x_B + N_C \ln x_C \right] \qquad \text{S-8.5.6}$

Substituting the values of $R = 8.314 \times 10^3$ J kilomole^{-1} K^{-1}; $T = 200 \, K$ and of x' from Eq. S-8.5.1 one gets,

$$\Delta G = 8.3143 \times 103 \times 200[2 \ln 0.2 + 3 \ln 0.3 + 5 \ln 0.5] = -163.72 \times 10^5 \, J$$

As is expected the Gibbs function decreased in the mixing process.

Further, the change in entropy $\Delta S = -\dfrac{\Delta G}{T} = \dfrac{163.72 \times 10^5}{200}\ JK^{-1} = 81.86 \times 10^3\ JK^{-1}$

In the irreversible mixing process the entropy of the system has increased.

6. Blackbody radiation in an evacuated vessel of volume V in equilibrium with its walls at temperature T behaves like a photon gas with specific internal energy or energy density u = aT^4 and pressure P= $(1/3)aT^4$, where a is a constant. (i) Draw the P–V diagram for a Carnot cycle which uses blackbody radiations as the working substance and (ii) explicitly derive the efficiency of this Carnot engine.

Solution:

Following observation may be made as regards to the blackbody radiation: (a) the total internal energy depends on the volume, i.e., $U = aT^4.V$ (ii) the pressure P does not depend on V.

The Carnot cycle is made up of four steps: (i) Isothermal expansion (ii) adiabatic expansion (iii) Isothermal compression and (iv) adiabatic compression. The P–V diagram of a Carnot cycle that uses blackbody radiation as working substance is shown in Fig. S-8.6.1. Since pressure does not depend on volume, the isothermal expansion from A to B and isothermal compression from C to D are also isobaric.

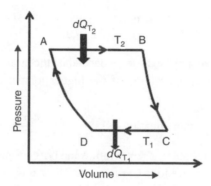

Fig. S-8.6.1 Carnot cycle in P-V plane for blackbody radiation as working substance

We denote the parameters volume V, pressure P and temperature T at point A as V_A, P_A and T_A, and at other points in the figure similarly by putting appropriate subscript to the parameter.

(A) Steps AB and CD : isothermal expansion and isothermal compression: The amount of heat dQ_{T_2} drawn from the reservoir at temperature T_2 is given by:

$$dQ_{T_2} = dU + PdV = u(V_B - V_A) + P(V_B - V_A)$$

Or $\quad dQ_{T_2} = aT_2^{\ 4}(V_B - V_A) + \dfrac{1}{3}aT_2^4(V_B - V_A) = \dfrac{4}{3}aT_2^4 V_A\left(\dfrac{V_B}{V_A} - 1\right)$ $\qquad$ S-8.6.1

Similarly for the step CD, the heat rejected at sink temperature T_1 is,

$$dQ_{T_1} = \dfrac{4}{3}aT_1^4 V_D\left(1 - \dfrac{V_C}{V_D}\right)$$ $\qquad$ S-8.6.2

(B) Adiabatic steps BC and DA: In an adiabatic process $dQ = 0 = dU + PdV$ S-8.6.3

But $U = U(V,T)$ and so $dU = \left(\dfrac{\partial U}{\partial V}\right)_T dV + \left(\dfrac{\partial U}{\partial T}\right)_V dT = \left(\dfrac{\partial aVT^4}{\partial V}\right)_T dV + \left(\dfrac{\partial aVT^4}{\partial T}\right)_V dT$

Or $dU = aT^4 dV + 4a\, VT^3\, dT$

Substituting the above value of dU in Eq. S-8.6.3 one gets:

$$aT^4 dV + 4aVT^3 dT + \frac{1}{3}aT^4 d\ = 0$$

Or $3\dfrac{dT}{T} = -\dfrac{dV}{V}$ or $\ln T^3 + \ln V = $ constant or $VT^3 = $ constant   S-8.6.4

Applying the relation $VT^3 = $ constant at the two ends of the two adiabatics one gets:

$$V_A T_2^3 = V_D T_1^3 \quad \text{and} \quad V_B T_2^3 = V_C T_1^3$$

Or $\dfrac{V_D}{V_A} = \dfrac{T_2^3}{T_1^3} = \dfrac{V_c}{V_B}$ S-8.6.5

The work done $W = dQ_{T_2} - dQ_{T_2}$

And the efficiency $\eta = \dfrac{W}{dQ_{T_2}} = 1 - \dfrac{dQ_{T_1}}{dQ_{T_2}} = 1 - \dfrac{T_1^4 V_D}{T_2^4 V_A} = 1 - \dfrac{T_1}{T_2}$

Thus the efficiency of Carnot cycle does not depend on the properties of the working substance but depends only on the temperatures of the higher temperature reservoir and the lower temperature sink.

7. A paramagnetic substance with equation of state $\mathcal{M} = \dfrac{\mathbb{N}D\mathcal{H}}{T}$ where $\mathcal{M}$ is magnetization, $\mathcal{H}$

magnetic field, $\mathbb{N}$ number of moles, T temperature and D a constant that depends on the material, is used as the working substance for a Carnot cycle. Discuss on what factors the internal energy of the system depends when magnetization is kept constant. Assuming that the specific molar thermal capacity $c_\mathcal{M}$ is constant draw the diagram of Carnot cycle in $\mathcal{M} - \mathcal{H}$ plane and calculate the amounts of heat rejected at sink, heat absorbed from the heat source, amount of work done and the efficiency of the Carnot engine.

Solution:

From the fundamental equation 8.27 we know that for a paramagnetic substance $dU = TdS + \mathcal{H}d\mathcal{M}$, when $\mathcal{M}$ is a constant, $d\mathcal{M} = 0$, and U depends only on temperature T.

Also, the internal energy $U = \mathbb{N}c_\mathcal{M}T$ and since $c_\mathcal{M}$ is constant, U is a function of temperature alone.

The Carnot cycle for paramagnetic substance is shown in Fig. S-8.7.1. It may be recalled that every molecule of a paramagnetic material behaves like a tiny magnet. The external magnetic field $\mathcal{H}$ aligns these tiny molecular magnets in its direction. Thus work is done by the external magnetic field and heat is produced when $\mathcal{H}$ is increased at constant temperature T_1, step AB in the figure. An amount of heat dQ_1 is rejected by the system at temperature T_1 of the sink. On the other hand, when $\mathcal{H}$ is decreased the molecular magnets randomize due to thermal motion and work is performed in demagnetization by the system absorbing heat from the constant temperature heat source at temperature T_2. An amount of heat dQ_2 is absorbed by the system at temperature T_2. This is indicated by the isothermal step CD in the figure. Steps BC and DA are adiabatic steps in which no heat transfer takes place.

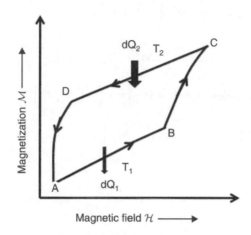

Fig. S-8.7.1   Carnot cycle for paramagnetic substance in M-H plane

We now calculate the heat rejected in step AB. From the given equation of state $\mathcal{H} = \dfrac{T\mathcal{M}}{ND}$

Isothermal steps:

The small amount of heat dQ_1 rejected when $\mathcal{M}$ increases by the amount $d\mathcal{M}$ is given by,

$$dQ_1 = dU - \mathcal{H}d\mathcal{M} = d\left(Nc_{\mathcal{M}}T\right) - \frac{T\mathcal{M}}{ND}d\mathcal{M} = Nc_{\mathcal{M}}dT - \frac{T}{ND}[\mathcal{M}d\mathcal{M}] \qquad \text{S-8.7.1}$$

In above equation $dT=0$, as the process is isothermal and $T=T_1$. The total heat liberated when $\mathcal{M}$ changes from the initial value $\mathcal{M}_A$ at point A to the final value $\mathcal{M}_B$ at point B may be obtained by integrating the above equation,

$$Q_1 = -\frac{T_1}{ND}\int_{\mathcal{M}_A}^{\mathcal{M}_B}\mathcal{M}d\mathcal{M} = -\frac{T_1}{2ND}\left[\mathcal{M}_B^2 - \mathcal{M}_A^2\right] \qquad \text{S-8.7.2}$$

Similarly, the amount of heat absorbed in the isothermal step CD when $\mathcal{M}$ changes from M_c to $\mathcal{M}_D$ at constant temperature T_2 ,is given by,

$$Q_2 = -\frac{T_2}{ND}\int_{\mathcal{M}_c}^{\mathcal{M}_D}\mathcal{M}d\mathcal{M} = -\frac{T_2}{2ND}\left[\mathcal{M}_D^2 - \mathcal{M}_C^2\right] = \frac{T_2}{2ND}\left[\mathcal{M}_C^2 - \mathcal{M}_D^2\right] \qquad \text{S-8.7.3}$$

Adiabatic steps:

Adiabatic steps are used to get a relation between $\left[\mathcal{M}_B^2 - \mathcal{M}_A^2\right]$ and $\left[\mathcal{M}_C^2 - \mathcal{M}_D^2\right]$

$dQ = 0$, in any adiabatic step. Therefore,

$$0 = dU - \mathcal{H}d\mathcal{M} \text{ or } dU = \mathcal{H}d\mathcal{M} \text{ or } \mathbb{N}c_{\mathcal{M}}dT = \frac{T}{\mathbb{N}D}[\mathcal{M}d\mathcal{M}]$$

Or

$$\mathbb{N}^2 c_{\mathcal{M}} D \frac{dT}{T} = \mathcal{M}d\mathcal{M}$$

For the step BC

$$\mathbb{N}^2 c_{\mathcal{M}} D \ln \frac{T_1}{T_2} = \frac{1}{2}\left[\mathcal{M}_C^2 - \mathcal{M}_B^2\right] \qquad \text{S-8.7.4}$$

And for step DA

$$\mathbb{N}^2 c_{\mathcal{M}} D \ln \frac{T_2}{T_1} = \frac{1}{2}\left[\mathcal{M}_A^2 - \mathcal{M}_D^2\right] \qquad \text{S-8.7.5}$$

From the above two equations $\left[\mathcal{M}_A^2 - \mathcal{M}_D^2\right] = -\left[\mathcal{M}_C^2 - \mathcal{M}_B^2\right]$ \qquad S-8.7.6

Or

$$\left[\mathcal{M}_C^2 - \mathcal{M}_D^2\right] = \left[\mathcal{M}_B^2 - \mathcal{M}_A^2\right] \qquad \text{S-8.7.7}$$

Therefore, the work done W is,

$$W = Q_2 - Q_1 = \frac{T_2}{2\mathbb{N}D}\left[\mathcal{M}_C^2 - \mathcal{M}_D^2\right] + \frac{T_1}{2\mathbb{N}D}\left[\mathcal{M}_B^2 - \mathcal{M}_A^2\right]$$

Or

$$W = \frac{\left(\mathcal{M}_C^2 - \mathcal{M}_D^2\right)}{2\mathbb{N}D}\left(\frac{T_2 - T_1}{T_2}\right)$$

And the efficiency

$$\eta = 1 - \frac{Q_1}{Q_2} = 1 - \frac{T_1}{T_2}$$

8. A rubber band of original length ℓ_0 is used to run an engine, the $(\mathfrak{I} - \ell)$ diagram (where $\mathfrak{I}$ is tension and ℓ is the length) for which is shown in Fig. (S-8.8.1). The equation of state of the rubber band is $\mathfrak{I} = k\ell T$, where k is a constant and T is the Kelvin temperature. Calculate the efficiency of the engine. Further, assume that the thermal capacity at constant length of the rubber band C_L is constant and independent of the temperature.

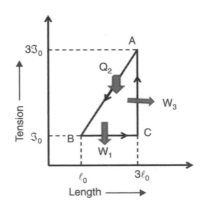

Fig. S-8.8.1

Solution:

The equation of state of the rubber band is given as $\Im = k\ell T$. Using the given equation of state we calculate the temperature at points A, B and C of the figure.

$$T_A = \frac{3\Im_0}{k3\ell_0} = \frac{\Im_0}{k\ell_0} \; ; \quad T_B = \frac{\Im_0}{k\ell_0} \; ; \text{ and } \; T_C = \frac{\Im_0}{k3\ell_0} = \frac{1}{3}T_A \; ; T_A = T_B \qquad \text{S-8.8.1}$$

Since T_A and T_B are equal, the step AB is an isothermal step.

Heat absorbed in the isothermal step AB:

$$dQ_2 = dU - \Im d\ell = C_L dT - (kT\ell) d\ell = 0 - kT\ell d\ell \; \textit{(Since temperature is constant)}$$

And $\quad Q_2 = \frac{1}{2}kT_A \left(\ell_A^2 - \ell_B^2\right) = \frac{1}{2}k\frac{\Im_0}{k\ell_0}\left[9\ell_0^2 - \ell_0^2\right] = 4\Im_0\ell_0 \qquad$ S-8.8.2

Work done in steps BC and CD :

Here $\qquad\qquad\qquad\qquad dW = dU + \Im d\ell \text{ and } dU = C_L \Delta T$

$$W_1 = C_L \left(T_B - T_C\right) - \Im_0 \left(3\ell_0 - \ell_0\right) = -C_L\left(\frac{\Im_0}{k3\ell_0} - \frac{\Im_0}{k\ell_0}\right) - 2\Im_0\ell_0 = C_L\frac{2\Im_0}{3k\ell_0} - 2\Im_0\ell_0 \text{ S-8.8.3}$$

And $W_3 = -C_L \left(T_A - T_C\right) = C_L\left(\frac{\Im_0}{k\ell_0} - \frac{\Im_0}{k3\ell_0}\right) = -C_L\frac{2\Im_0}{3k\ell_0}$ *(length is constant)* $\qquad$ S-8.8.4

Total work done $W = Q_2 + W_1 + W_3 = 4\Im_0\ell_0 + C_L\frac{2\Im_0}{3k\ell_0} - 2\Im_0\ell_0 - C_L\frac{2\Im_0}{3k\ell_0} = 2\Im_0\ell_0$

And the efficiency $\quad \eta = \frac{W}{Q_2} = \frac{2\Im_0\ell_0}{4\Im_0\ell_0} = \frac{1}{2}$

9. A gasoline engine based on Otto cycle has compression ratio of 5 and 10^6 Jkolomole^{-1} of heat is supplied to the engine in the constant volume heating step through the spark plug. The adiabatic compression process started at 300 K and one atmospheric pressure. Assuming that the fuel behaves like a diatomic ideal gas, (a) draw the P–v diagram for the operation cycle (b) compute heat and work involved at each step of the cycle (c) calculate the thermal efficiency of the engine both by the step wise heat and work computation and by the application of efficiency formula.

Solution:

Since the mass of the fuel used in the engine is not given, the pressure vs specific molar volume diagram of the operation cycle is shown in Fig. S-8.9.1

Before we carry out detailed calculations it may be pointed out that since the fuel is to be taken as a diatomic ideal gas the molar specific thermal capacity at constant volume

$c_V = \frac{5}{2}R = 2.5 \times 8.314 \times 10^3 \, J \, kilomole^{-1}K^{-1}$ and the ratio $\dfrac{c_P}{c_V} = \gamma = 1.4$.

Also, $Pv = RT$ and $Tv^{\gamma-1} = $ constant. Where the gas constant $R = 8.314 \times 10^3 \, J \, kilomole^{-1} \, K^{-1}$ and v is the molar volume. Also the compression ratio $b = 5$

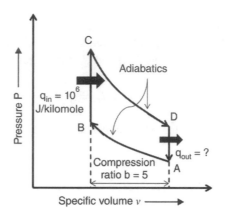

Fig. S-8.9.1

Step AB: Adiabatic Compression

Using $\quad Tv^{\gamma-1} = \text{constant}, T_B = T_A\left(\dfrac{v_A}{v_B}\right)^{\gamma-1} = T_A(b)^{0.4} = 300(5)^{0.4} = 571.09 K$

Also, $\quad \dfrac{P_B v_B}{T_B} = \dfrac{P_A v_A}{T_A}, \; P_B = P_A\left(\dfrac{v_A}{v_B}\right)\left(\dfrac{T_B}{T_A}\right) = 1.013 \times 10^5 \times 5 \times \dfrac{571.09}{300} = 9.64 \times 10^5 \, Nm^{-2}$

Since the process AB is adiabatic: No heat is drawn in the process and $dq_{AB} = 0$ and,

$$dq_{AB} - w_{AB} = du = c_V(T_A - T_B)$$

Or $\qquad w_{AB} = \dfrac{5}{2}R(300 - 571.09) = -5634.60 \times 10^3 \, J\,\text{kilomole}^{-1}$

Step BC: Constant volume combustion

Heat supplied by sparking $q_{BC} = 1 \times 10^6 \, J\,kilomole^{-1}$

Since volume remains constant hence no *Pdv* work is done and heat q_{BC} goes in raising the internal energy or temperature of the gas,

$$q_{BC} = c_V(T_C - T_B) \text{ and } T_C = \dfrac{q_{BC}}{c_V} + T_B = 616.19 \, K$$

Since it is a constant volume process $v_c = v_B$ and $P_C = P_B\left(\dfrac{T_C}{T_B}\right) = 10.45 \times 10^5 \, N\,m^{-2}$

Step CD: Adiabatic expansion:

Since $\qquad Tv^{\gamma-1} = \text{constant}, \; T_D = T_C\left(\dfrac{1}{b}\right)^{\gamma-1} = 325.26 K$

And $\quad P_D = P_C\left(\dfrac{v_C}{v_D}\right)\left(\dfrac{T_D}{T_C}\right) = 10.45 \times 10^5 \times \left(\dfrac{1}{5}\right)\left(\dfrac{325.26}{616.19}\right) = 1.097 \times 10^5 \, N\,m^{-2}$

Being an adiabatic process, $q_{CD} = 0$, and $w_{CD} = du = c_V(T_C - T_D) = 6109.34 \times 10^3 \, J\,\text{kilomole}^{-1}$

Step DA: Constant volume exhaust

Being a constant volume process $w_{DA} = 0$, and $q_{DA} = du = c_V(T_A - T_D) = -525.26 \times 103\ J\ kilomole^{-1}$

Calculating the efficiency: The net work done w_{net} may be calculated from the sum $(w_{AB} + w_{CD})$ or from the sum $(q_{BC} + q_{DA})$.

$$w_{net} = (w_{AB} + w_{CD}) = (-5634.60 \times 10^3 + 6109.34 \times 10^3) = 474.74 \times 10^3\ J\ kilomole^{-1}$$

Also, $w_{net} = (q_{BC} + q_{DA}) = (1 \times 10^6 - 525.26) = 474.74 \times 10^3\ J\ kilomole^{-1}$

The thermal efficiency $\quad \eta = \dfrac{n_{net}}{q_{in}} \times 100 = \dfrac{474.74 \times 10^3}{1 \times 10^6} \times 100 = 47.474\%$

Thermal efficiency may also be calculated from the formula

$$\eta = \left[1 - \left(\frac{1}{b}\right)^{1-\gamma}\right] \times 100 = \left[1 - \left(\frac{1}{5}\right)^{0.4}\right] \times 100 = [1 - 0.5253] \times 100 = 47.47\%$$

It may be observed that the efficiency obtained from detailed calculations and using the formula agree with each other, as expected.

10. A certain steam engine follows superheated steam Rankine cycle as shown in Fig. S-8.10.1. Superheated steam enters the turbine at 600 K and 40 bar pressure and exit from the turbine as 100% saturated vapor. After condensation the saturated liquid enters the pump at 1.0 bar pressure.

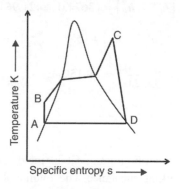

Fig. S-8.10.1

Using steam tables and superheated steam tables determine (a) rate of heat transfer into the boiler per unit mass (b) net power generated per unit mass (c) thermal efficiency of the engine.

Solution:

The following data has been taken from the superheated steam tables:

At 40 bar and 600 K $h_{in}^{sup} = 3674\ kJ\ kg^{-1}\ K^{-1}$; and $s_{in}^{sup} = 7.369\ kJ\ kg^{-1}\ K^{-1}$

Following data is taken from saturated steam tables

At 1 bar   Specific volume of saturated liquid steam $v_{exit}^{liq} = 0.001041 \ m^3 \ kg^{-1}$

Specific enthalpy of saturated liquid steam $h_{exit}^{liq} = 417.9 \ kJ \ kg^{-1} \ K^{-1}$

Specific entropy of liquid steam $s_{exit}^{liq} = 7.395 \ kJ \ kg^{-1} \ K^{-1}$

Specific enthalpy of vapor steam $h_{exit}^{vap} = 2671 \ kJ \ kg^{-1} \ K^{-1}$

Analysis: We start from point D in the figure where superheated steam after performing work on the turbine comes out as 100% saturated vapor and has specific enthalpy h_{exit}^{vap}. The condenser converts the saturated vapor into saturated liquid at point A with enthalpy h_{exit}^{liq} and specific volume v_{exit}^{liq}.

Step AB: Adiabatic compression by pump: We know that for any process $dq = dh - vdP$ but in this step dq is zero as it is adiabatic and, therefore, $dh = vdP$. This equation may be integrated with the understanding that v does not change with pressure if it is assumed that liquids are uncompressible. So,

$$(h_B - h_A) = v(P_B - P_A) = v_{exit}^{liq}(40.0 - 1.0) = 4.06 \ kJ \ kg^{-1} \ K^{-1}$$

And $h_B = (4.06 + h_A) kJ \ kg^{-1} \ K^{-1} = (4.06 + 417.9) kJ \ kg^{-1} \ K^{-1} = 421.96 \ kJ \ kg^{-1} \ K^{-1}$

Step CD: Adiabatic work on turbine:

At C: $s_{in}^{sup} = 7.369 \ kJ \ kg^{-1} \ K^{-1}$; $P_C = 40 bar$; $T_C = 600 \ K$; $h_{in}^{sup} = 3674.0 kJ \ kg^{-1} \ K^{-1}$

Rate of heat transfer $= q = \left(h_{in}^{sup} - h_B\right) = (3674.0 - 421.96) kW \ kg^{-1} = 3252.04 \ kW \ kg^{-1}$

Net power generated $= w = \left[\left(h_{in}^{sup} - h_B\right) + \left(h_{in}^{sup} - h_{exit}^{vap}\right)\right] = \left(h_{exit}^{vap} - h_B\right) = 2249.04 \ kW \ kg^{-1}$

Thermal efficiency $\eta = \dfrac{w}{q} = \dfrac{2249.04}{3252.06} = 69.15\%$

Problems

1. Figure P-8.1 shows the T–V diagram for an heat engine that uses 9.8 moles of an ideal gas for which c_V is (5/2) R as the working substance. The volume of the gas (in liter) and temperature in Kelvin are shown in the figure.

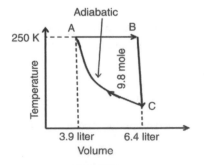

Fig. P-8.1

(a) Draw the P–V diagram for the cycle of the engine (b) determine the efficiency of the engine (c) compute the maximum efficiency for the engine working between the same temperatures (d) determine if the engine cycle is reversible or not?

2. An engine based on Otto cycle has a compression ratio of 8. 0.1 kg of fuel mixture that has the specific heat capacity at constant volume $c_V = 0.834$ kJ per kg per K and the value of $\gamma = \dfrac{c_P}{c_V} = 1.344$ is used in the engine. 80 kJ of heat is provided to the engine by the electric spark. Calculate the net work done by the engine per cycle and the efficiency of the engine.

3. A box of diathermic walls is divided into three compartments of equal volumes. One mole each of an ideal gas is kept in the two compartments at the ends and 2 mole of a different ideal gas in the compartment at the middle. The box is connected to a heat reservoir at temperature 300 K. Initially the pressure in all the three compartments is same 2 atm. The partitions between the compartments are then removed and the gases were allowed to mix and attain equilibrium. Compute the change in the entropy of the system in the mixing process.

4. A diesel engine operates with 0.5 kg of a fuel mixture at 1×10^5 N m^{-2} pressure and 300 K temperature with compression and critical ratios respectively 18 and 2. The average properties of the fuel are: $c_p = 1.121$ kJ kg^{-1} K^{-1} and $c_V = 0.834$ kJ kg^{-1} K^{-1}. Draw the P–V diagram of the Diesel engine and making a detailed analysis of the operations compute the heat input, heat rejected and the efficiency of the engine. Also show that the efficiency calculated with the standard formula agrees with the one calculated by you.

5. Two metallic tanks, each 4 m deep, and 1 m × 2 m and 1 m × 3 m size are connected to a heat reservoir at constant temperature 35°C. A certain quantity of acetic acid in thermal equilibrium in the smaller tank is transferred to the larger one and allowed to come to thermal equilibrium. Compute the heat drawn from the reservoir by the acetic acid given that the surface tension of acetic acid at 20 °C and 50°C is respectively, 27.6 mN m^{-2} and 25.3 mN m^{-2}.

6. It is known that the vapor pressure of a fluid increases if the pressure on the liquid is increased. The molar volume of a fluid at 300 K is 20×10^{-3} m^3 kilomole^{-1} and its vapor pressure at one atm. is 23.32×10^{-2} N m^{-2}. Compute the pressure at which the vapor pressure of the liquid will become 24.66×10^2 N m^{-2} at 300 K.

[Hint: use Eq.(S-8.1.6) of solved example-1 and make the approximation $\ln \dfrac{p_{fin}}{p_{ini}} =$

$\ln \dfrac{p_{ini} + \Delta p}{p_{ini}} \cong \dfrac{\Delta p}{p_{ini}}$, where p_{ini} and p_{fin} are the initial and final vapor pressures, respectively]

7. The enthalpy and the energy density for blackbody radiation contained in a cavity are respectively, 816.48×10^{-8} J and 612.36×10^{-8} Jm^{-3}. What is the temperature and the size of the cavity?

8. At 300 K temperature the inside pressure in a liquid bubble of 1 μm diameter is twice the atmospheric pressure, compute the surface tension of liquid. If the rate of decrease of surface tension with temperature at 300 K is one millionth of the surface tension, what is the internal energy density (per unit area) of the bubble surface?

Short Answer Questions

1. Compare the free expansion of an ideal gas with that of the blackbody radiation.

2. Discuss the physical reasoning for the non-existence of chemical potential for blackbody radiation.

3. Out of the four important quantities c_V, c_P, κ_T and β which are not defined for blackbody radiation and why? Symbols have their usual meanings.

4. Show that the first law of thermodynamics holds for blackbody radiation.

5. Mention difference between a real gas, an ideal gas and blackbody radiation as regards to throttling experiment.

6. Without derivation write an expression that shows that temperature will always fall in adiabatic demagnetization of a paramagnetic substance.

7. What is the reason for the rise of fluids in capillaries?

8. Define surface tension of a fluid in terms of the Helmholtz function of the system.

9. Define linear expansivity at constant tension for elastic materials and give one example each of materials for which it has positive and negative values.

10. Write an expression, without deriving, for the amount of heat transferred ΔQ in isothermal extension at temperature T of an elastic rod by ΔL, keeping the area of cross section constant, in terms of the linear expansivity at constant tension and the Yong's modulus of the material.

11. What is the advantage of using superheated steam in Rankine cycle?

12. What factors may reduce the theoretical efficiency of engines?

Long Answer Questions

1. Starting with the experimental findings that (i) the energy density of blackbody radiations u is proportional to T^4, where T is the temperature in Kelvin and (ii) that the pressure exerted by blackbody radiations $P = (1/3)$ u, obtain the equation of state and the thermodynamic functions for blackbody radiation.

2. Give with necessary details thermodynamic treatment of isothermal, adiabatic and free expansions of blackbody radiation and compare them with the corresponding process for an ideal gas.

3. Start with the equation of state for a paramagnetic salt and obtain expressions for the Helmholtz function of the system both with and without including the potential energy term. Also show how the variation of the Helmholtz function is related to the entropy of the system.

4. Use thermodynamic arguments to show that adiabatic demagnetization of a paramagnetic substance results in the production of low temperature.

5. Develop a thermodynamic treatment for the interface between bulk liquid and bulk vapor phases of a pure liquid and obtain expressions defining surface tension in terms of the Helmholtz function for the interface.

6. Define the isothermal Young's modulus and the expansivity at constant tension for an elastic string under tension and hence discuss how tension on the string changes with the increase of

temperature if the length of the string is kept constant. Also obtain expressions for the entropy and the change in entropy with length at constant temperature.

7. Draw diagrams representing Otto cycle in P–V and S–T planes and use these diagrams to analyze the working of an engine based on Otto cycle. Also derive expression for the thermal efficiency of the engine.

8. Draw Carnot cycle in (h – s) plane and drive expressions for heat transferred and work performed in each step in terms of the enthalpy and the internal energy at the corresponding vertices of the diagram.

9 With the help of a (h – s) diagram give a detailed analysis of a Rankine cycle that uses superheated steam and obtain its thermal efficiency. What are the advantages of using super heated steam?

10. In what respect a Diesel engine differs from an Otto engine? Give a detail analysis of a Diesel engine and give reason why they are not used in small vehicles. The efficiency of engines in general is less than the value derived theoretically. In your opinion what are the likely causes for this?

Multiple Choice Questions

Note: Some questions may have more than one correct alternative. All correct alternatives must be included in the answer.

1. The enthalpy of blackbody radiation at temperature T is

   (a) $\frac{4}{3}U(T)$

   (b) $\frac{3}{4}\sigma T^4$

   (c) $S\left(\frac{3P}{\sigma}\right)^{\frac{1}{4}}$

   (d) TP

2. The entropy of blackbody radiation contained in a cavity of volume V at temperature T having total internal energy U and pressure P, is given by

   (a) $\frac{4}{3}(\sigma V)^{\frac{1}{4}} U^{\frac{3}{4}}$

   (b) $\frac{4U}{3T}$

   (c) $\frac{4U}{3}$

   (d) 3PV

3. The equation of state for a paramagnetic substance, where symbols have their usual meaning is given by

   (a) $\mathcal{H} = \alpha \mathbb{N}R'\frac{T}{\mathcal{M}}$

   (b) $T = \alpha \mathbb{N}R'\frac{\mathcal{M}}{\mathcal{H}}$

   (c) $\mathcal{H} = \alpha \mathbb{N}R'\frac{\mathcal{M}}{T}$

   (d) $\mathcal{M} = \alpha \mathbb{N}R'\frac{\mathcal{H}}{T}$

4. Which thermodynamic relation holds for paramagnetic substances?
   (a) $dF = -SdT + \mathcal{H}d\mathcal{M}$
   (b) $dF = TdS - \mathcal{M}d\mathcal{H}$
   (c) $dF = -SdT - \mathcal{H}d\mathcal{M}$
   (d) $dF = Td\mathcal{H} - \mathcal{M}dS$

5. Which specific interface parameter of pure liquids is given by the rate of change of surface tension with temperature at constant pressure?
   (a) Gibbs function
   (b) Helmholtz function
   (c) entropy
   (d) enthalpy

6. If A, α_J and Y_T respectively represent the area of cross section, expansivity at constant tension and the isothermal Young's modulus of an elastic rod, then the rate of change of entropy with length at constant temperature is given by,
   (a) $A\,\alpha_J Y_T$
   (b) $\dfrac{A}{\alpha_J Y_T}$
   (c) $A\dfrac{Y_T}{\alpha_J}$
   (d) $Y_T\dfrac{\alpha_J}{A}$

7. The cycle of which engine is represented by the following P–V diagram?

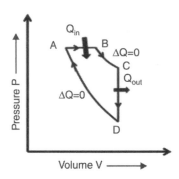

Fig.   Mc-8.7

   (a) Carnot
   (b) Rankine
   (c) Otto
   (d) Diesel

8. The thermal efficiency of Diesel engine is given by
   (a) $(b)^\gamma$
   (b) $\left[1-(b)^{\gamma-1}\right]$
   (c) $\left[1-(b)^{1-\gamma}\right]$
   (d) $\left[(b)^{\gamma-1}-1\right]$; here b is the volume compression ratio.

9. Use of superheated steam in Rankine cycle
   (a) increases the thermal efficiency
   (b) reduces the percentage of CO_2 and other uncondensed gases
   (c) reduces the power required for pumping the condensed vapors
   (d) decreases the moisture contents of the steam thereby decreasing the chance of corrosion of turbine buckets.

10. In Fig. Mc-8.7 which part of the cycle represents power stroke?

Answers to Problems and Multiple Choice Questions

Answers to problems

1. (b) 10.2% (c) 18% (d) No, change in entropy during the cycle is +0.49 J K^{-1}
2. 40.9 kJ; 51.1%
3. 17.29 J/K
4. Heat input = 454.5 kJ, Heat rejected= 192.5 kJ, efficiency= 57.64
5. 708.4 mJ
6. 7.7×10^6 N m^{-2}
7. 300 K, 1.0 m^2
8. 25.3 mJ m^{-1}; 17.71 mJ m^{-2}

Answers to multiple choice questions

1. (a), (c)	2. (a), (b)	3. (d)	4. (a)
5. (c)	6. (a)	7. (d)	8. (c)
9. (a), (d)	10. (AB and BC)		

Revision

Blackbody radiation

1. The equation of state for blackbody radiation is $P = \frac{\sigma}{3}T^4$.

2. Other thermodynamic functions for blackbody radiation are given in the table.

Table 8.1 Thermodynamic functions for blackbody radiation

P	U	S	H	F	G	μ
$\frac{1}{3}\sigma T^4$	$\sigma T^4 V$	$\frac{4}{3}\sigma V T^3$	$\frac{4}{3}\sigma V T^4$	$-\frac{1}{3}\sigma V T^4$	0	0
$\frac{1}{3}\frac{U}{V}$	$3PV$	$\frac{4}{3}\frac{U}{T}$	$\frac{4}{3}U$	$-PV$		
	$\left(\frac{3S}{4}\right)^{\frac{4}{3}}(\sigma V)^{-\frac{1}{3}}$	$4\left(\frac{\sigma}{3}\right)^{\frac{1}{4}}V P^{\frac{3}{4}}$	TS	$-\frac{1}{4}TS$		
		$\frac{4}{3}(\sigma V)^{\frac{1}{4}}U^{\frac{3}{4}}$	$4PV$	$-\frac{1}{3}U$		
			$S\left(\frac{3P}{\sigma}\right)^{\frac{1}{4}}$			

3. Relation $PV^{4/3}$ holds for an adiabatic process of blackbody radiation.

4. In contrast to the free expansion of ideal gas, temperature falls in the free expansion of blackbody radiation.
5. In contrast to both the real and ideal gas, cooling is always produced in throttling of blackbody radiation.

Paramagnetic materials

6. The equation of state for a paramagnetic substance is $M = \alpha NR' \dfrac{\mathcal{H}}{T}$

7. Thermodynamic functions for paramagnetic substances are tabulated below
   State functions for paramagnetic system

Function	Derivative of the function	Characteristic parameters
U	$dU = TdS - \mathcal{H}dM$	$U(S, M)$
$H^* = E = U + E_p$	$dE = TdS - Md\mathcal{H}$	$E(S, \mathcal{H})$
$H^* = U - \mathcal{H}M$	$dH^* = TdS - Md\mathcal{H}$	$H^*(S, \mathcal{H})$
$F = U - TS$	$dF = -SdT + \mathcal{H}dM$	$F(T, M)$
$F^* = E - TS$	$dF^* = -SdT - Md\mathcal{H}$	$F^*(T, \mathcal{H})$
$G = U - TS + \mathcal{H}M$	$dG = -SdT - Md\mathcal{H}$	$G(T, \mathcal{H})$
$G^* = E - TS + \mathcal{H}M$	$dG^* = -SdT - \mathcal{H}dM$	$G^*(T, M)$

$$\left(\frac{\partial H^*}{\partial S}\right)_{\mathcal{H}} = T\;;\quad \left(\frac{\partial H^*}{\partial \mathcal{H}}\right)_S = -M$$

$$\left(\frac{\partial F^*}{\partial T}\right)_{\mathcal{H}} = -S\;;\quad \left(\frac{\partial H^*}{\partial \mathcal{H}}\right)_T = -M$$

$$\left(\frac{\partial G^*}{\partial T}\right)_M = S\;;\quad \left(\frac{\partial G^*}{\partial M}\right)_T = \mathcal{H}$$

8.
$$C_M = \left(\frac{\partial Q}{\partial T}\right)_M = \left(\frac{\partial U}{\partial T}\right)_M$$

$$C_{\mathcal{H}} = \left(\frac{\partial Q}{\partial T}\right)_{\mathcal{H}} = \left(\frac{\partial E}{\partial T}\right)_{\mathcal{H}}$$

9. Empirical value for $C_M = \dfrac{\beta \mathbb{N} R}{T^2}$, where $\mathbb{N}$ is the number of moles, R gas constant, T temperature in Kelvin and β is constant in unit of K^2 that has different value for different materials.

10. The TdS equations are

$$TdS = C_M dT - T\left(\frac{\partial \mathcal{H}}{\partial T}\right)_M dM$$

$$TdS = C_{\mathcal{H}}dT + T\left(\frac{\partial M}{\partial T}\right)_{\mathcal{H}} d\mathcal{H}$$

11. $\left(C_{\mathcal{H}} - C_M\right) = \dfrac{\mathcal{H}M}{T}$

12. Adiabatic demagnetization of paramagnetic substance always produce cooling

$$\left(\frac{\partial T}{\partial M}\right)_S = \frac{T}{C_M}\left(\frac{\partial \mathcal{H}}{\partial T}\right)_M = \frac{TM}{C_M \alpha NR'}$$

Thermodynamics of interface or surface films: surface tension

13. Thermodynamic functions for the interface (surface film)

$$U^\sigma = TS^\sigma + \Sigma_i \mu_i^\sigma N_i^\sigma + \gamma A$$

$$dU^\sigma = TdS^\sigma + \Sigma_i \mu_i^\sigma dN_i^\sigma + \gamma dA$$

$$F^\sigma = U^\sigma - TS^\sigma = \Sigma_i \mu_i^\sigma N_i^\sigma + \gamma A$$

$$\frac{F^\sigma}{A} = \gamma + \Sigma_i \mu_i^\sigma \Gamma_i^\sigma$$

$$dF^\sigma = -S^\sigma dT + \Sigma_i \mu_i^\sigma dN_i^\sigma + \gamma dA$$

$$-\left(\frac{\partial \gamma}{\partial T}\right)_{A, N_i^\sigma} = \left(\frac{\partial S^\sigma}{\partial A}\right)_{T, N_i^\sigma}$$

$$G^\sigma = F^\sigma + \left(0 - \gamma A\right) = \Sigma_i \mu_i^\sigma N_i^\sigma$$

$$dG^\sigma = -S^\sigma dT + \Sigma_i \mu_i^\sigma dN_i^\sigma - Ad\gamma$$

14. For pure liquids

$$\gamma_{pure} = \frac{F^\sigma}{A}$$

$$-\left(\frac{\partial \gamma}{\partial T}\right)_{A,P} = \left(\frac{\partial S^\sigma}{\partial A}\right)_{T,P} = s_{pure}^\sigma \ (specific\ entropy\ of\ the\ surface)$$

$$U^\sigma = TS^\sigma + \Sigma_i \mu_i^\sigma N_i^\sigma + \gamma A \qquad \Rightarrow U_{pure}^\sigma = TS_{pure}^\sigma + \gamma A$$

$$\frac{U_{pure}^\sigma}{A} = T\frac{S_{pure}^\sigma}{A} + \gamma \ or \ u_{pure}^\sigma = Ts_{pure}^\sigma + \gamma = -T\left(\frac{\partial \gamma}{\partial T}\right)_{A,P} + \gamma$$

The specific entropy and the specific internal energy of the interface may be obtained experimentally from the variation in the surface tension with temperature.

Thermodynamics of an elastic rod under tension

15. Isothermal Young's modulus $Y_T \equiv \dfrac{stress}{strain} = \dfrac{d\mathfrak{J}/A}{dL/L} = \dfrac{L}{A}\left(\dfrac{\partial \mathfrak{J}}{\partial L}\right)_T$

16. Linear expansivity at constant tension $\alpha_\mathfrak{J} \equiv \dfrac{1}{L}\left(\dfrac{\partial L}{\partial T}\right)_\mathfrak{J}$ positive for metals and negative for rubber.

17.
$$\left(\frac{\partial \mathfrak{J}}{\partial T}\right)_L = -\left(\frac{\partial L}{\partial T}\right)_\mathfrak{J}\left(\frac{\partial \mathfrak{J}}{\partial L}\right)_T = -A\ \alpha_\mathfrak{J} Y_T$$

$$\left(\frac{\partial U}{\partial L}\right)_T = T\left(\frac{\partial S}{\partial L}\right)_T + \mathfrak{J}$$

$$dF = -SdT + \mathfrak{J}dL$$

$$S = -\left(\frac{\partial F}{\partial T}\right)_L \text{ and } \mathfrak{J} = \left(\frac{\partial F}{\partial L}\right)_T$$

$$\left(\frac{\partial S}{\partial L}\right)_T = -\left(\frac{\partial \mathfrak{J}}{\partial T}\right)_L \text{ and } \left(\frac{\partial S}{\partial L}\right)_T = A\ \alpha_\mathfrak{J} Y_T$$

$$\left(\frac{\partial U}{\partial L}\right)_T = T\left(\frac{\partial S}{\partial L}\right)_T + \mathfrak{J} = TA\ \alpha_\mathfrak{J} Y_T + \mathfrak{J}$$

Engineering applications of thermodynamics

18. Otto cycle: $\eta_{Otto} = \dfrac{\mathbb{N}R}{c_V\left(\gamma - 1\right)}\left[1 - \left(\dfrac{1}{b}\right)^{\gamma-1}\right] = \left[1 - b^{1-\gamma}\right]$ where b is the compression ratio and

$\gamma = {c_P}\Big/{c_V}$ for the fuel mixture.

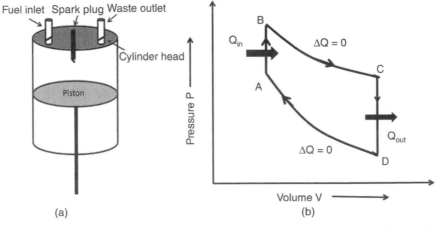

Fig.8.7(a)   Schematic diagram of engine cylinder; (b) P-V diagram for Otto cycle

19. Diesel engine

$$\eta_{Diesel} = 1 - \frac{\left(c^\gamma - 1\right)(b)^{1-\gamma}}{\gamma\left(c - 1\right)}$$ where b and c are respectively cut off and compression ratios.

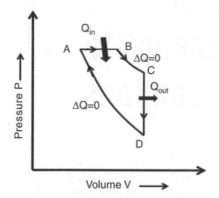

Fig. 8.8   P-V diagram for a diesel engine

20. Rankine cycle

$$\eta_{Rankine} = \frac{w}{Q_{in}} = \frac{\left(h_D - h_A - h_E + h_{II}\right)}{\left(h_D - h_A\right)}$$

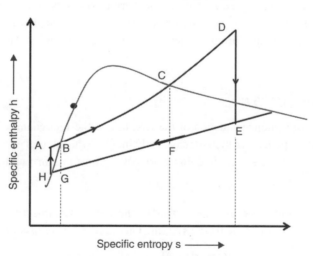

Fig. 8.10 (c)   Rankine cycle in h-s plane

21. Superheated steam increases the efficiency of the engine and reduces the moisture contents of the steam thereby reducing the chances of corrosion of turbine buckets.

Application of Thermodynamics to Chemical Reactions

9.0 Introduction

In this chapter we will learn how the laws of thermodynamics may be applied to chemical reactions. The reason for dealing with chemical reactions separately is the fact that chemists use the sign convention for thermodynamic quantities which is different than the one used by physicists.

Normally one should start by the zeroth law of thermodynamics and see how it is applied in chemical reactions. However it is not essential because in classical thermodynamics we study systems that are in thermal equilibrium, which means that the zeroth law has already been taken into account, though tacitly.

Application of the first law

We start with the energy balance equation of the first law of thermodynamics according to which the amount of heat ΔQ supplied to a system may partly increase the internal energy ΔU of the system and may partly be used by the system in doing work ΔW (against the surroundings),

$$\Delta Q = \Delta U + \Delta W \qquad\qquad 9.1$$

Physicists take ΔW positive if work is performed by the system. Chemists on the other hand define the work ΔW as negative if it is done by the system. Therefore, for chemists Eq. 9.1 becomes,

$$\Delta Q = \Delta U - \Delta W$$

Or $\qquad\qquad\qquad\qquad\qquad \Delta U = \Delta Q + \Delta W \qquad\qquad 9.2$

Equation 9.2 used by chemists, says that an amount of heat, ΔQ, supplied to the system plus a work, ΔW done on the system goes in increasing the internal energy of the system by the amount ΔU. The pictorial representation of Eq. 9.2 is given in Fig. 9.1

In chemistry, the sign convention is to take heat positive when it is absorbed by the system, which means that in a chemical reaction if cooling is produced (i.e., the system takes heat from the surroundings) the heat of the reaction is taken as positive. The reactions that produce cooling or absorb heat are called 'endothermic' reactions and for them the heat of reaction is positive. On the other hand, 'exothermic' reactions in which heat is emitted or heating is produced have a negative value of the heat of reaction.

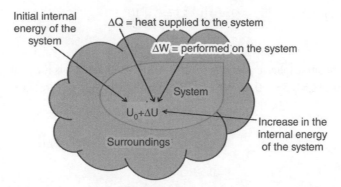

Fig. 9.1 Increase in the internal energy of a system when heat is supplied and work is done on it

Similarly, work is negative if it is performed by the system and positive if some other agency performs work on the system. For example, if the volume of a system increases by dV at constant pressure P then the work done by the system is PdV and it is negative, i.e.,

$$- \Delta W = P\Delta V \text{ (volume increases)}$$

And if some external energy (pressure of the surrounding) reduces the volume of the system by dV, then dW is positive,

$$\Delta W = P\Delta V \text{ (volume decreases)}$$

In Eq. 9.2 there are three quantities, ΔU, ΔQ and ΔW; the signs of these quantities are defined as,

$\Delta U \quad \Rightarrow \quad$ +ve when internal energy of the system increases

$\Delta Q \quad \Rightarrow \quad$ +ve when heat is absorbed by the system,

i.e., cooling is produced in the reaction

$\Delta W \quad \Rightarrow \quad$ +ve when work is done on the system

In solving numerical problems using Eq. 9.2, the two known quantities with their appropriate signs are put in the equation and the magnitude with sign of the third unknown quantity will be obtained on solving the equation. The following example will further clarify the application of Eq. 9.2.

Solved Example

1. The volume of a balloon was increased from 0.5 ℓ to 1.0 ℓ by heating the air inside the balloon. If 0.2 kJ heat is supplied to the balloon, calculate the change in the internal energy of the balloon.

Solution:

In this example balloon is our system and work is performed by the system against the atmospheric pressure in expanding and, therefore, ΔW has negative sign. But heat supplied ΔQ has positive sign as it is supplied to the system. Hence,

$$\Delta U = \Delta Q - |\Delta W| \qquad \qquad \text{S-9.1}$$

Now it is given that $\quad\quad\quad \Delta Q = +0.2\ \text{kJ} = 2 \times 10^2\ \text{J}$

The work $\quad\quad\quad\quad\quad |\Delta W| = P_{atm}\Delta V$

Here P_{atm} is the atmospheric pressure $= 1.013 \times 10^5\ \text{N m}^{-2}$ and ΔV is the change in volume of the balloon $= (1-0.5)\ \ell = 0.5\ \ell = 0.5 \times 10^{-3}\ \text{m}^3$. We convert all quantities into their basic metric units.

So, $\quad\quad\quad\quad\quad\quad |\Delta W| = P_{atm}\Delta V$

$$= 1.013 \times 10^5\ \text{Nm}^{-2} \times 0.5 \times 10^{-3}\ \text{m}^3$$

$$= 0.5065 \times 10^2\ \text{J}$$

Substituting the values of ΔQ and $|\Delta W|$ in Eq. S-9.1, we get:

$$\Delta U = \Delta Q - |\Delta W| = 2 \times 10^2\ \text{J} - 0.5065 \times 10^2\ \text{J} = 1.4935\ \text{J}$$

It is clear from calculations that ΔU is positive, which means that the internal energy of the balloon has increased by 1.4935 J

9.1   Work at Constant Pressure

Further, in case when *work is done by the system at constant pressure* Eq. 9.2 may be written as,

$$\Delta Q_P = \Delta U + P\Delta V \quad\quad\quad\quad 9.3$$

where ΔQ_P is the heat supplied to the system at constant pressure. We add $V\Delta P$ on the right hand side of Eq. 9.3. Since $V\Delta P$ is zero if pressure P is constant, and therefore, adding $V\Delta P$ does not matter.

$$\Delta Q_P = \Delta U + P\Delta V + V\Delta P$$

Or $\quad\quad\quad\quad\quad \Delta Q_P = \Delta(U + PV) = \Delta H \quad\quad\quad\quad 9.4$

where enthalpy H is defined as,

$$H \equiv U + PV \quad\quad\quad\quad 9.5$$

Therefore, for reactions at constant pressure,

$$\Delta Q_P = \Delta H \quad\quad\quad\quad 9.6$$

And dQ_P (or ΔH) is positive if heat is absorbed in the process and negative if heat is emitted in the reaction.

$\Delta Q_P = \Delta H = (H_f - H_i)$ is negative for exothermic reactions at constant pressure

$\Delta Q_P = \Delta H = (H_f - H_i)$ is positive for endothermic reactions at constant pressure

Here H_f and H_i are, respectively, the final and the initial enthalpy of the system.

9.2 Work at Constant Volume

When the volume of the system remains constant ΔV is zero and, therefore, configurational or mechanical work done by the system or on the system ($P\Delta V$) is zero and,

$$\Delta Q_V = \Delta U = (U_f - U_i) \qquad\qquad 9.7$$

Here U_f and U_i are, respectively, the final and the initial internal energies of the system. If ΔQ_V is positive, i.e., heat is absorbed in the process (endothermic reactions) final internal energy of the system will be larger than the initial internal energy.

9.3 Relation between ΔQ_P and ΔQ_V for Ideal Gas

In case of the ideal gases

$$PV = \mathbb{N}RT \quad \text{and} \quad P\Delta V = R\Delta(\mathbb{N}T) \qquad\qquad 9.8$$

where $\mathbb{N}$ is the number of moles of the gas. Now from Eq. 9.6 and Eq. 9.7

$$\Delta Q_P = \Delta Q_V + P\Delta V$$

Substituting the value of $P\Delta V$ from Eq. 9.8, one gets

$$\Delta Q_P = \Delta Q_V + R\Delta(\mathbb{N}T) \qquad\qquad 9.9$$

It may be remembered that Eq. 9.9 holds only for ideal gases. Further, $\Delta(\mathbb{N}T)$ means any change in both the number of moles $\mathbb{N}$ and temperature T or only in one of them.

9.4 Standard State

The thermodynamic standard state of a substance is the most stable pure form under specified standard pressure and temperature.

- In case the substance is in the solid or in the liquid phase the standard state is the pure form of the substance at the atmospheric pressure and at 298 K (= 25°C) temperature.
- If the substance is a pure gas then the standard state is at one atmospheric pressure and at temperature of 298 K, unless specified otherwise.
- If there is a mixture of gases than the standard state of each gas is at a partial pressure of 1 atmosphere and the specified temperature which is generally 298 K.
- For a substance in solution the standard state refers to 1-molar concentration.

9.5 The Standard Enthalpy Change, $\Delta H^{\varnothing}_{rea}$, in a Chemical Reaction

In a chemical reaction, on the left hand side of the reaction equation, are reactants and on the right hand side the reaction products.

$$\text{Reactants} \rightarrow \text{Reaction products}$$

The standard enthalpy change in such reactions refer to the change in enthalpy when specified number of moles of the reactants, all under standard states, are converted completely to the specified number of moles of products, all at standard states. The standard enthalpy change for a reaction is generally denoted by $\Delta H_{rea}^{\varnothing}$, here superscript ϕ denots that it refers to standard conditions of the reactants and the products and the subscript 'rea' denotes that it is for the specified reaction.

9.6 Standard Molar Enthalpy of Formation (Heat of Formation)

The standard molar enthalpy ΔH_{for}^{ϕ} of a substance is the change in enthalpy in the reaction in which *one mole* of the substance in specified state is formed from its elements in standard states. By convention the standard *molar enthalpy of elements in their standard states is taken to be zero.* For example, it is known that $\frac{1}{2}$ mole of H_2 in gaseous state plus $\frac{1}{2}$ mole of Br_2 in liquid state on reaction produce 1 mole of HBr_2 in gaseous state.

The above reaction in short is written as,

$$\frac{1}{2} H_2(g) + \frac{1}{2} Br_2(\ell) \rightarrow HBr_2(g)$$

The standard enthalpy change ΔH_{rea}^{ϕ} for the above reaction is -36.4 kJ mole $^{-1}$. Therefore, the heat of formation or standard molar enthalpy of formation $\Delta H_{for}^{\phi}\left(HBr\right)_{(gas)}$ is also -36.4 kJ mole $^{-1}$.

Solved Example

2. Combustion of ethanol (in liquid phase) results in the formation of carbon dioxide gas and water (in liquid phase) and the standard heat of the process at atmospheric pressure and 25°C is -13.68×10^5 J mole $^{-1}$. Calculate the change in the internal energy in the combustion process.

Solution:

The equation for the chemical process of combustion of ethanol may be written as

$$C_2H_5OH(liq) + 3O_2(gas) \rightarrow 2CO_2(gas) + 3H_2O(liq)$$

What we observe in this reaction is:

(i) Though the reaction takes place at constant atmospheric pressure, but there is a change in the molar volume during the reaction. As a convention the changes in volumes of liquid and solid phases are not counted (as they are generally negligible) but the changes in the volume of the gaseous phases are taken into account. In the above equation initially there are 3 moles of O_2 (gas) and after the reaction only 2 mole of CO_2 (gas) are formed. Thus change in the number of moles in the reaction $\Delta N = -1$

(ii) Conventionally all gases are treated as ideal gases, unless specified otherwise. The reaction has taken place at constant atmospheric pressure, denoted by P_{atm} and at temperature $T = 25°C$ which is equal to 298 K. For ideal gases,

$$PV = \mathbb{N}RT \text{ or } \Delta(P_{atm}V) = P_{atm}\,\Delta V = R\Delta(\mathbb{N}T) = RT\,(\Delta\mathbb{N}) = RT\,(-1) \qquad \text{S-9.2.1}$$

While deriving the above equation we have made use of the fact that atmospheric pressure and temperature T remains constant in the reaction.

We now apply Eq. 9.9, according to which,

$$\Delta Q_P = \Delta Q_V + R\Delta(\mathbb{N}T) \qquad \text{S-9.2.2}$$

In the above equation $\Delta Q_P = \Delta H$, and $\Delta Q_V = \Delta U$. What about the sign of ΔQ_V? It should be negative as heat is emitted in a combustion reaction. Hence, from Eq. S-9.2.2 we get,

$$\Delta H = -\,\Delta U + (-1)RT$$

$$= -13.68 \times 105 \text{ J}$$

$$= -\Delta U - (8.314 \text{ JK}^{-1}) \times 298 \text{ K}$$

Or $\qquad \Delta U = 13.68 \times 10^5 \text{ J} - 2.48 \times 10^3 \text{ J} = 13.655 \times 10^5 \text{ J}$

9.7 Relation between the Enthalpy Change in a Reaction and the Enthalpies of Formations of Reactants and Reaction Products

It is easy to show that the enthalpy change in a reaction is equal to the difference between the enthalpies of formation of products and the reactants, i.e.,

$$\Delta Q_P = \Delta H^{\phi}_{rea} = \Sigma\Delta H^{\phi}_{for}\,(\text{products}) - \Sigma\Delta H^{\phi}_{for}\,(\text{reactants}) \qquad 9.10\,(a)$$

Similarly,

$$\Delta Q_V = \Delta U = \Sigma\Delta U(\text{products}) - \Sigma\Delta U(\text{reactants}) \qquad 9.10\,(b)$$

It may further be remembered that measurements done using the bomb calorimeter are at constant volume, and, therefore, refers to the measurement of $\Delta Q_V = \Delta U$

Measurements done by a coffee calorimeter are done at constant pressure. It is, however, a common practice to specify the type of the measurement. Further, if the change in the standard enthalpy of a reaction is given it means it refers to constant pressure measurement.

Another important point to remember is that enthalpy change in a reaction and in its inverse reaction is equal in magnitude but opposite in sign.

Solved Examples

3. The ignition of CS_2 is represented by the reaction,

$$CS_2(\text{liq}) + 3O_2\,(\text{gas}) \rightarrow CO_2\,(\text{gas}) + 2SO_2\,(\text{gas})$$

Compute enthalpy change of the ignition reaction from the following data;

Standard enthalpy change for the formation of CO_2 (gas) ΔH^{ϕ}_{for} $(CO_2 (g)) = -393.5$ kJ mole^{-1}

Standard enthalpy change for the formation of SO_2 (gas) ΔH^{ϕ}_{for} $(SO_2 (g)) = -296.8$ kJ mole^{-1}

Standard enthalpy change for the formation of CS_2 (liq) ΔH^{ϕ}_{for} $(CS_2(\ell) = +87.9$ kJ mole^{-1}

Solution:

We use Eq. 9.10 (a) to calculate $\Delta H^{\phi}_{ignition}$. It may be remembered that for elements like oxygen

$$\Delta H^{\phi}_{for} O_2 = 0$$

$$\Delta H^{\phi}_{ignition} = \Sigma \Delta H^{\phi}_{for} \left(products \right) - \Sigma \Delta H^{\phi}_{for} \left(reactants \right)$$

$$= \left[\Delta H^{\phi}_{for} \left(CO_2 (g) \right) + 2\Delta H^{\phi}_{for} \left(SO_2 (g) \right) \right]$$

$$- \left[\Delta H^{\phi}_{for} \left(CS_2 (\ell) \right) + 3\Delta H^{\phi}_{for} \left(O_2 (g) \right) \right]$$

$$= \left\{ [-393.5 - 2 \times 296.8] - [87.9 + 0] \right\} \text{kJ}$$

$$= 1075.0 \text{ kJ}$$

4. Heat for the reaction $CO_2 (g) + \frac{1}{2} O_2 (g) \rightarrow CO_3$ (g) measured using bomb calorimeter at 25°C is –392.27 kJ. What is the change of enthalpy in the reaction?

Solution:

Since measurements done using a bomb calorimeter are done at constant volume, the measured heat of reaction is actually $\Delta Q_V = \Delta U = -392.27$ kJ. It is required to calculate the change in enthalpy of the reaction.

Change in enthalpy $\quad \Delta H^{\varnothing}_{rea} = \Delta Q_P = \Delta Q_V + RT\Delta(\mathbb{N}).$ $\hspace{2cm}$ S-9.4.1

Let us, therefore, calculate $\Delta(\mathbb{N}) = \left\{ 1 - \left(1 + \frac{1}{2} \right) \right\} = -\frac{1}{2}$. Substituting this value in Eq. S-9.4.1 one gets,

$$\Delta H^{\varnothing}_{rea} = \Delta Q_P = \Delta Q_V + RT\Delta(\mathbb{N}) = -392.27 \text{ kJ} + 8.314 \times (273 + 25)\left(-\frac{1}{2}\right) J$$

Or $\quad \Delta H^{\varnothing}_{rea} = -392.27 \text{ kJ} + 8.314 \times (273 + 25)\left(-\frac{1}{2}\right) J = -392.27 \text{ kJ} - 1.24 \text{ kJ}$

Or   The change in the standard enthalpy of the reaction = –393.51 kJ

5. Heat of the following reaction, measured in an experiment at constant volume and 25°C is –569.18 kJ.

$$2H_2 (g) + O_2 (g) \rightarrow 2H_2O(g)$$

Compute the heat for the formation of H_2O(g)

Solution:

Since the experiment is done at constant volume and at 25°C (= 298 K), the measured heat is ΔQ_V at 298 K. We first calculate ΔQ_P at 298 K for the raction using the relation

$$\Delta Q_P = \Delta Q_V + RT\Delta(\mathbb{N})$$

$$= -569.18 \text{ kJ} + 8.314 \times 298 \times [2 - (2 + 1)]\text{J} - 569.18 \text{ kJ} - 2.48 \text{ kJ}$$

$$= -571.18 \text{ kJ}$$

We are required to find the heat of formation of $H_2O(g)$. In the given reaction two moles of H_2O (g) are formed. Therefore, the heat of formation of one mole of H_2O (g) is $\frac{1}{2}\Delta Q_P$.

Heat of formation of $H_2O(g) = \dfrac{-571.18}{2}$ kJ = -285.59 kJ

9.8 Hess's Law

Germain Henri Hess, August 8, 1802–December 13, 1850, was Swiss born Russian chemist

Since enthalpy is a state function, the change in enthalpy depends only on the initial and the final states of the system and does not depend on the path of reaching the final state from the initial state. Hess used this principle to give his law for chemical reactions. According to Hess's law the change in standard enthalpy in a given chemical reaction is equal to the sum of the changes in standard enthalpies of all other chemical reactions that add up to the given reaction. Let $\Delta H^{\phi}_{(A+B \rightarrow C+D)}$ represent the change in standard enthalpy of the reaction $(A + B \rightarrow C+D)$ and $\Delta H^{\phi}_{(A_1+B_1 \rightarrow C_1+D_1)}$,

$\Delta H^{\phi}_{(A_2+B_2 \rightarrow C_2+D_2)}$, $\Delta H^{\phi}_{(A_3+B_3 \rightarrow C_3+D_3)}$, change in standard enthalpies, respectively, for reactions,

$(A_1 + B_1 \rightarrow C_1 + D_1)$; $(A_2 + B_2 \rightarrow C_2 + D_2)$ and $(A_3 + B_3 \rightarrow C_3 + D_3$), then $\Delta H^{\phi}_{(A+B \rightarrow C+D)} =$

$\Delta H^{\phi}_{(A_1+B_1 \rightarrow C_1+D_1)} + \Delta H^{\phi}_{(A_2+B_2 \rightarrow C_2+D_2)} + \Delta H^{\phi}_{(A_3+B_3 \rightarrow C_3+D_3)}$, provided the reactions,

$(A_1 + B_1 \rightarrow C_1 + D_1) + (A_2 + B_2 \rightarrow C_2 + D_2) + (A_3 + B_3 \rightarrow C_3 + D_3) = A + B \rightarrow C + D$

This law is frequently used to compute the change in standard enthalpy of reactions for which the changes in standard entropies of the component reactions are known.

Solved Examples

6. Using the data provided, calculate the change in the standard enthalpy for the reaction
   $H_2O(liq) \rightarrow H_2(gas) + \frac{1}{2}O_2(gas)$.

   Data: $H_2O(gas) \rightarrow H_2(gas) + \frac{1}{2}O_2(gas)$ $\Delta H^{\phi}_{rea} = 41$ kJ   (1)

   $H_2O(liq) \rightarrow H_2O(gas)$ $\Delta H^{\phi}_{rea} = 240$ kJ   (2)

Solution:

It may be observed that adding the two reactions given in the data one gets the reaction for which the change in standard enthalpy is required.

$$\cancel{H_2O}(gas) \rightarrow H_2(gas) + \frac{1}{2}O_2(gas)$$

$$H_2O(liq) \rightarrow \cancel{H_2O}(gas)$$

$$\overline{H_2O(liq) \rightarrow H_2O(gas) + \frac{1}{2}O_2(gas)}$$

Now from Hess's law

$$\Delta H^{\varnothing}_{desired\ reaction} = \Delta H^{\varnothing}_{given\ reaction-1} + \Delta H^{\varnothing}_{given\ reaction-2}$$

$$= 41\ kJ + 240\ kJ = 281\ kJ$$

The change in the standard enthalpy of the desired reaction = 281 kJ

7. Change in standard enthalpies for reactions marked (a) and (b) are respectively, $\Delta H^{\varnothing}_a = -162.0$ kJ and $\Delta H^{\phi}_b = -1012.0$ kJ. Calculate the change in the standard enthalpy of reaction (C).

   Reaction (a) $2N_2O(g) \rightarrow O_2(g) + 2N_2(g)$

   Reaction (b) $2NH_3(g) + 3N_2O(g) \rightarrow 4N_2(g) + 3H_2O(\ell)$

   Reaction (c) $4NH_3(g) + 3N_2(g) \rightarrow 2N_2(g) + 6H_2O(\ell)$

Solution:

When equation for reaction (a) is multiplied by 3 and is subtracted from the equation obtained by multiplying reaction (b) by 2, the reaction (c) is obtained as is shown below,

$2x\ (b) \Rightarrow \qquad 4NH_3(g) + 6N_2O(g) \rightarrow 8N_2(g) + 6H_2O(\ell)$

$-3x\ (a) \Rightarrow \qquad -6N_2O(g) \rightarrow -3O_2(g) - 6N_2(g)$

$\overline{\qquad\qquad\qquad 4NH_3(g) \rightarrow -3O_2(g) + 2N_2(g) + 6H_2O(\ell)}$

Or $\qquad\qquad 4NH_3(g) + 3O_2(g) \rightarrow 2N_2(g) + 6H_2O(\ell)$

Therefore, from Hess's law

$$\Delta H^{\phi}_C = 2\Delta H^{\phi}_b - 3\Delta H^{\phi}_a = 2x\ (-1012.0\ kJ) - 3(-162.0\ kJ)$$

$$\Delta H^{\phi}_C = -1532.0\ kJ$$

9.9 Heat Capacities

Specific molar heat capacities c_P and c_V are defined as:

$$c_P = \frac{\partial(Q_P)}{\partial T} = \left(\frac{\partial H}{\partial T}\right)_P \qquad\qquad 9.11\ (a)$$

$$c_V = \frac{\partial(Q_V)}{\partial T} = \left(\frac{\partial U}{\partial T}\right)_V \qquad\qquad 9.11\ (b)$$

And the molar heat capacity at constant pressure $C_P^{molar} = \mathbb{N}c_P$;

And the molar heat capacity at constant volume $C_V^{molar} = \mathbb{N}c_V$

Here $\mathbb{N}$ is the number of moles of the substance.

9.10 Temperature Dependence of Heat of Reaction

In general, the magnitude of the heat of reaction increases with temperature as is shown in Fig. 9.2.

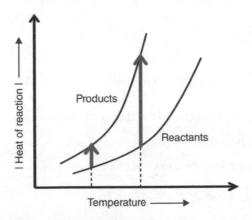

Fig. 9.2 Temperature dependence of heat of reaction

Consider the following balanced reaction equation,

$$\mathbb{N}^A A + \mathbb{N}^B B + \mathbb{N}^C C + \dots \rightarrow \mathbb{N}^x X + \mathbb{N}^Y Y + \mathbb{N}^Z Z + \dots$$

Here $\mathbb{N}^A$, $\mathbb{N}^B$, $\mathbb{N}^C$ etc are the number of moles of A, B, C … etc.

Now,

$$\Delta H_{rea} = \sum \mathbb{N}^X \Delta H_X + \mathbb{N}^Y \Delta H_Y + \mathbb{N}^Z \Delta H_z + \dots] - \sum \left[\mathbb{N}^A \Delta H_A + \mathbb{N}^B \Delta H_B + \mathbb{N}^C \Delta H_C + \dots\right] \qquad 9.12$$

Here ΔH_{rea} is the change in the enthalpy of the reaction and ΔH_A, ΔH_B, ΔH_C, … ΔH_Z etc the changes in the enthalpies of A, B, C…Z etc. We now partially differentiate Eq. 9.12 with respect to temperature T keeping the pressure constant to get,

$$\left(\frac{\partial \Delta H_{rea}}{\partial T}\right)_P = \sum \left[\mathbb{N}^X \left(\frac{\partial \Delta H_X}{\partial T}\right)_P + \mathbb{N}^Y \left(\frac{\partial \Delta H_Y}{\partial T}\right)_P + \mathbb{N}^Z \left(\frac{\partial \Delta H_Z}{\partial T}\right)_P + \dots\right]$$

$$- \sum \left[\mathbb{N}^A \left(\frac{\partial \Delta H_A}{\partial T}\right)_P + \mathbb{N}^B \left(\frac{\partial \Delta H_B}{\partial T}\right)_P + \mathbb{N}^C \left(\frac{\partial \Delta H_C}{\partial T}\right)_P + \dots\right]$$

However, from Eq. 9.11 (a), $\left(\dfrac{\partial \Delta H_A}{\partial T}\right)_P = c_P^A$, etc., here c_P^A is specific molar heat capacity for A

at constant pressure and so on,

$$\left(\frac{\partial \Delta H_{rea}}{\partial T}\right)_P = \sum \mathbb{N}^X c_P^X + \mathbb{N}^Y c_P^Y + \mathbb{N}^Z c_P^Z + ...] - \sum \left[\mathbb{N}^A c_P^A + \mathbb{N}^B c_P^B + \mathbb{N}^C c_P^C + ..\right]$$

$$= \sum \left[C_P^X + C_P^Y + C_P^Z ...\right] - \sum \left[C_P^A + C_P^B + C_P^C ...\right] \qquad 9.12$$

$C_P^X, C_P^Y, C_P^Z, ...$ etc. in Eq. 9.12 are the total heat capacities at constant pressure for X, Y, Z, ...etc.
Eq. 9.12 may be written as,

$$\left(\frac{\partial \Delta H_{rea}}{\partial T}\right)_P = \sum \left(C_P\right)_{(Products)} - \sum \left(C_P\right)_{(reactants)} = \Delta C_P \qquad 9.13$$

Integrating Eq. 9.13, one may get

$$\int_{T_1}^{T_2} \left(\frac{\partial \Delta H_{rea}}{\partial T}\right)_P dT = \int_{T_1}^{T_2} \left(\Delta C_P\right) dT$$

Or
$$\Delta H_{rea}^{T_2} - \Delta H_{rea}^{T_1} = \int_{T_1}^{T_2} \left(\Delta C_P\right) dT \qquad 9.14$$

Here, $\Delta H_{rea}^{T_2}$ and $\Delta H_{rea}^{T_1}$ are respectively the changes in reaction heats at temperature T_2 and T_1.
Now there are two distinct possibilities:

(a) (ΔC_P) does not depend on temperature and in that case,

$$\Delta H_{rea}^{T_2} - \Delta H_{rea}^{T_1} = \Delta C_P \left(T_2 - T_1\right) \qquad 9.15 \text{ (a)}$$

(b) (ΔC_P) varies slowly with temperature but an average value $(\Delta C)_P^{average}$ may be used in the temperature range $T_2 - T_1$, to give

$$\Delta H_{rea}^{T_2} - \Delta H_{rea}^{T_1} = \left(\Delta C\right)_P^{average} \left(T_2 - T_1\right) \quad 9.15 \text{ (b)}$$

Equations 9.15 (a) and 9.15 (b) are called *Kirchhoff equations*.

Similarly, $\left(\dfrac{\partial \Delta U}{\partial T}\right)_V = \Delta C_V$ and $\Delta U_{T_2} - \Delta U_{T_1} = \Delta C_V \left(T_2 - T_1\right)$ \qquad 9.16

9.11 Bond Energies

Chemical bonds of reactants break and new bonds in reaction products get formed in a chemical reaction. It is possible to calculate bond energies by proper bookkeeping of all energies in a balanced chemical reaction. The change in standard enthalpy of a reaction is related to the bond energies by the following relation.

$$\Delta H_{rea}^{\varnothing} = \sum \left(energies\, of\, bonds\, broken\right) - \sum \left(energies\, of\, bonds\, formed\right) \qquad 9.17$$

Solved Examples

8. Compute the heat of formation of NH_3 (g) at 1600 K given that the standard heat of formation for NH_3 (g) at 25°C is −46.10 kJ mole⁻¹ and $C_P[N_2(g)]$, $C_P[H_2(g)]$ and $C_P[NH_2(g)]$ are respectively, 27.30, 29.0, and 35.3 J k⁻¹. Assume that C_P remains constant during the temperature interval.

Solution:

The reaction for the formation of NH_3 may be written as,

$$\frac{1}{2}N_2(g) + \frac{3}{2}H_2(g) \rightarrow NH_3(g)$$

$$\Delta C_P = \left[35.3 - \left\{\frac{1}{2}(27.30) + \frac{3}{2}(29.0)\right\}\right] = -21.85 \text{ kJ mole}^{-1}$$

$$\Delta U_{1600\,K} - \Delta U_{298\,K} = -21.85\left[1600 - (25 + 273)\right] = -28.45 \text{ kJ}$$

Therefore, $\Delta U_{1600\,K} = (-46.10 - 28.45) = 74.55 \text{ kJ mole}^{-1}$

9. From the following data calculate the heat of formation of NH_3 (g) at 500 K.

Data:

(a) Heat of formation of NH_3 (g) at 298 K = − 46.10 kJ mole⁻¹.

(b) $C_P(NH_3, g) = 35.20 + 8.60 \times 10^{-3}T - 9.0 \times 10^{-7}T^2$

(c) $C_P(H_2, g) = 29.00 - 0.85 \times 10^{-3}T + 20.00 \times 10^{-7}T^2$

(d) $C_P(N_2, g) = 27.00 + 5.20 \times 10^{-3}T - 0.05 \times 10^{-7}T^2$

Solution:

The equation for the formation of NH_3 is

$$\frac{1}{2}N_2(g) + \frac{3}{2}H_2(g) \rightarrow NH_3(g)$$

$$\Delta C_P = C_P(NH_3, g) - \left[\frac{1}{2}C_P(N_2, g) + \frac{3}{2}C_P(H_2, g)\right]$$

$$= \left(35.20 + 8.60 \times 10^{-3}T - 9.0 \times 10^{-7}T^2\right) - \left\{\frac{1}{2}\left(27.00 + 5.20 \times 10^{-3}T - 0.05 \times 10^{-7}T^2\right)\right.$$

$$\left. + \frac{3}{2}\left(29.00 - .85 \times 10^{-3}T + 20.00 \times 10^{-7}T^2\right)\right\}$$

$$= \left(35.20 + 8.60 \times 10^{-3}T - 9.0 \times 10^{-7}T^2\right) - \left\{57.0 + 1.325 \times 10^{-3}T + 29.975 \times 10^{-7}T^2\right\}$$

$$= \left(-21.80 + 7.275 \times 10^{-3}T - 38.975 \times 10^{-7}T^2\right)$$

Now from Eq. 9.14,

$$\Delta H_{form}^{T_2} - \Delta H_{form}^{T_1} = \int_{T_1}^{T_2} \Delta C_p dT = \int_{T_1}^{T_2} \left(-21.80 + 7.275 \times 10^{-3}T - 38.975 \times 10^{-7}T^2\right)dT$$

$$= -21.8(T_2 - T_1) + 7.275 \times 10^{-3}\left\{\frac{1}{2}\left(T_2^2 - T_1^2\right)\right\} - 38.975\left\{\frac{1}{3}\left(T_2^3 - T_1^3\right)\right\}$$

Substituting the value of $T_2 = 500$ K and $T_1 = 298$ K in the above equation one gets,

$$\Delta H_{form}^{T_2} - \Delta H_{form}^{T_1} = -21.80(202) + 7.275 \times 10^{-3}\left(8.06 \times 10^4\right) - 38.975 \times 10^{-7}\left(32.85 \times 10^6\right)$$

$$= -4403.60 + 586.37 - 128.03 = -3.95 \text{ kJ}$$

Or $\Delta H_{form}^{500\,K}\left(NH_3, g\right) = \Delta H_{form}^{298\,K}\left(NH_3, g\right) - 3.95 \text{ kJ mole}^{-1} = (-46.10 - 3.95) \text{ kJ mole}^{-1}$

The heat of formation of $NH_3(g)$ at 500 K $= -50.50$ kJ mole^{-1}

9.12   The Explosion and the Flame Temperatures

The concept of explosion and flame temperatures are applicable only to those chemical reactions in which heat is liberated, i.e., exothermic reactions. Exothermal chemical reactions may take place under two different conditions: (i) at constant temperature, in this case the heat produced is given to the surroundings that have large thermal capacity and the temperature of the reaction products remains constant. This is the isothermal reaction. (ii) If, however, the reaction products are isolated from the surroundings and heat is not allowed to leave to the surroundings, the temperature of the reaction products will increase. This is called the adiabatic exothermal reaction. An adiabatic exothermal reaction may take place at constant volume in that case the heat liberated in the reaction ΔQ_V comes from the difference between the internal energies of the reaction products and reactants. The heat liberated is then taken up by the reaction products and their temperatures increase. The maximum temperature of reaction products reached in an adiabatic exothermal reaction at constant volume is called the *explosion temperature*. When an adiabatic exothermal reaction takes place at constant pressure, the heat liberated in the process is $\Delta Q_P = \Delta H_{rea}^{\varnothing}$, and is shared by the reaction products. The maximum temperature attained by reaction products in an adiabatic exothermal reaction at constant pressure is termed as the *flame temperature*. The flame and explosion temperatures may be computed from the heat of reaction if C_V and C_P for the reaction products are known. The following example will further clarify the point.

Solved Example

10. In an adiabatic constant volume experiment a mixture of hydrogen gas with a theoretical amount of air at 298 K temperature and 1 atm. pressure is exploded in a constant volume vessel. Use the data provided below to calculate the maximum explosion temperature of the reaction.

   Data: $H_2(g) + \frac{1}{2}O_2(g) \rightarrow H_2O(g)$ $\Delta H_{rea}^{\varnothing} = -240.6$ kJ mol^{-1}

The average molar heat capacities at constant volume for nitrogen gas and water vapors are, respectively, 26.36 J mol^{-1} K^{-1} and 38.91 J mol^{-1} K^{-1}.

Solution:

It is given that the heat liberated in the explosion reaction is –240.6 kJ mol^{-1}. This will now go in heating of the reaction products and will raise their temperature from 298 K to some higher value T. Now let us see what are the reaction products? It is given in the question that a theoretical amount of air was used for the explosion. The theoretical amount of air means that it has $\dfrac{1}{2}$ mole of O_2 gas. It is known that in air the ratio of oxygen to nitrogen is 1:4. Therefore, after the burning of $\dfrac{1}{2}$ mole of oxygen $4 \times \dfrac{1}{2}$ = 2 mole of nitrogen will be left in the vessel. Thus the reaction products are: 1 mole of water vapours at temperature 298 K and 2 moles of nitrogen at 298 K. Therefore, 240.6 kJ mol^{-1} of heat will go in increasing the temperature of 1 mole of water vapors and 2 moles of residual nitrogen from 298 K to some value T. The values $C_V^{average}$ (nitrogen) and $C_V^{average}$ (water vapour) are provided and it is assumed that they do no change with temperature.

Then $240600 = 2 \times C_V^{average}(\text{nitrogen})(T - 298) + 1 \times C_V^{average}(\text{water vapour})(T - 298)$

Or $240600 = (T - 298)[2 \times 26.36 + 1 \times 38.91] = (T - 298)(91.63)$

Or $T = 2923.78 \text{ K}$

The maximum explosion temperature is 2923.78 K

Application of First and Second Laws Together

9.13   Change of Entropy in Chemical Reactions

Second law of thermodynamics deals with the entropy and the change in entropy in a chemical reaction. It is important to note that in thermodynamic analysis of chemical reactions all gases and vapors are treated as ideal gas, unless specified otherwise.

Let us consider an ideal gas that undergoes a reversible compression so that its volume and temperature change from initial values T_i, V_i to final values T_f, V_f. According to the first law,

$$dQ_r = dU + PdV \qquad\qquad 9.18$$

Here dQ_r is the heat evolved in the reversible process.

Also, $dU = C_V dT$ and for an ideal gas $P = \dfrac{NRT}{V}$. With these substitutions Eq. 9.18 becomes

$$dQ_r = C_V dT + NRT \dfrac{dV}{V}$$

Dividing the above equation by T throughout, one gets

$$\frac{dQ_r}{T} = C_V \frac{dT}{T} + \mathbb{N}R \frac{dV}{V} \qquad \text{But} \quad \frac{dQ_r}{T} \equiv dS$$

Hence, $\displaystyle \int_{S_i}^{S_f} dS = C_V \int_{T_i}^{T_f} \frac{dT}{T} + \mathbb{N}R \int_{V_i}^{V_f} \frac{dV}{V}$

Or
$$S_f - S_i = \Delta S = C_V \ln \frac{T_f}{T_i} + \mathbb{N}R \ln \frac{V_f}{V_i} \qquad\qquad 9.19$$

Though Eq. 9.19 gives the change in the entropy when a system goes from initial equilibrium state (T_i, V_i) to the final equilibrium state (T_f, V_f) by a reversible process, but the same expression may be used for calculating change in entropy if the system reaches the same final state by an irreversible process as entropy is a state function. Further, it may be shown that change in entropy may also be written as,

$$S_f - S_i = \Delta S = C_P \ln \frac{T_f}{T_i} - R \ln \frac{P_f}{P_i} \qquad\qquad 9.20$$

Equations 9.19 and 9.20 may be used to calculate the change in entropy of a given system.

9.14 Spontaneity of a Chemical Reaction

According the second law, the sum of the changes in entropies of a system and its surrounding must increase in an irreversible process. Since all spontaneous processes in nature are irreversible, a chemical reaction will be spontaneous if the sum of the changes in the entropy of the system and its surroundings is positive. The change in the entropy of the system (the reactants and the reaction products) may be determined using Eqs. 9.19 and 9.20. To determine the change in the entropy of the surroundings one may use the following relation,

$$(\Delta S)_{surroundings} = \frac{(\Delta H_{rea})_T}{T} \qquad\qquad 9.21$$

In Eq. 9.21 $(\Delta S)_{surroundings}$ is the change in the entropy of the surroundings and $(\Delta H_{rea})_T$ is the heat of reaction or the change in the enthalpy of the reaction at temperature T. As such for a chemical reaction to be spontaneous,

$$(\Delta S)_{surroundings} + (\Delta S)_{system} > 0 \qquad\qquad 9.22$$

9.15 Other State Functions and Changes in their Values

The first and the second laws put together gives,

$$TdS = dU + PdV \qquad\qquad 9.23$$

And the three important state functions are defined as follows:

(a) Enthalpy: $H \equiv U + PV$

$$dH = dU + d(PV) = dU + RTd(N_m) \text{ for an ideal gas}$$

Or $$\Delta H = \Delta U + RT\Delta(\mathbb{N})$$ 9.24

Here $\Delta(\mathbb{N})$ is the change in the number of moles in the chemical reaction.

(b) Helmholtz function: $F \equiv U - TS$

$$dF = -PdV - SdT$$

For a reaction at constant temperature $dT = 0$, hence

$$(dF)_T = -PdV = -\mathbb{N}RT\frac{dV}{V}, \text{ for an ideal gas}$$

Or $$(\Delta F)_T = \left(F_f - F_i\right)_T = -\mathbb{N}RT \ln\frac{V_f}{V_i}$$ 9.25

(c) Gibbs function: $G \equiv H - TS = U + PV - TS$

Or $$G = F + PV$$

And $$dG = VdP - SdT$$

At constant temperature $dT = 0$, and assuming that all gases in chemical reactions behave like ideal gas, the above equation reduces to

$$(dG)_T = VdP = \mathbb{N}RT\frac{dP}{P}$$

And $$(\Delta G)_T = \left(G_f - G_i\right)_T = \mathbb{N}RT \ln\frac{P_f}{P_i}$$ 9.26

Equations 9.24, 9.25 and 9.26 are often used to calculate the change in the value of state functions in chemical reactions.

The magnitude of the change in the value of Helmholtz function $|\Delta F|$ in a chemical reaction is equal to the maximum energy that may be drawn in the form of work while $|\Delta G|$, the magnitude of the decrease in the value of Gibbs function in a reaction gives the maximum energy that may be drawn as non-mechanical (non-PdV) work.

Solved Examples

11. The change in standard enthalpy $\Delta H_{rea}^{\varnothing}$ at 25°C and constant pressure for a chemical reaction is -110.0 kJ mol^{-1} and the standard change in the entropy of the reaction is -209 J K^{-1}. Check if the reaction is spontaneous or not.

Solution:

The change in the entropy of the surroundings $(\Delta S)_{surrounding} = -\dfrac{\Delta H_{rea}^{\varnothing}}{25 + 273}$

Or $$(\Delta S)_{surroundings} = -\frac{-110 \times 10^3 \, \text{J} \, \text{mol}^{-1}}{298} = 369.13 \, \text{JK}^{-1}$$

The change in standard entropy of the reaction $(\Delta S)_{system} = -209 \, \text{J} \, \text{K}^{-1}$

Total change of entropy per mole $= (\Delta S)_{surrounding} + (\Delta S)_{system} = (369.13 - 209) \, \text{J} \, \text{K}^{-1}$

$$= +160.13 \, \text{J} \, \text{K}^{-1}$$

Since the total change of entropy is positive, the reaction is spontaneous.

12. The change in the standard enthalpy and standard entropy for the following reaction of methane combustion are respectively, $-890.0 \, \text{kJ} \, \text{mol}^{-1}$ and $-243.0 \, \text{JK}^{-1}$. Compute the change in Helmholtz function, Gibbs function and the energy that can be drawn as work.

$$CH_4(g) + 2O_2(g) \rightarrow CO_2(g) + 2H_2O(l)$$

Solution:

The change in the number of moles in the reaction $(\Delta \mathbb{N}) = [1 - (1 + 2)] = -2$.

$$\Delta U = \Delta H^{\emptyset}_{reac} - RT(\Delta \mathbb{N}) = -890.0 \times 10^3 + 2 \times 8.314 \times (25 + 273) = -885.04 \, \text{kJ} \, \text{mol}^{-1}$$

Also $(\Delta F)_T = \Delta U - T(\Delta S) = -885.04 - (25 + 273)(-243.0) = -812.63 \, \text{kJ} \, \text{mol}^{-1}$

And $(\Delta G)_T = \Delta H^{\emptyset}_{reac} - T(\Delta S) = -890 \times 10^3 - (25 + 273)(-243.0) = -817.60 \, \text{kJ} \, \text{mol}^{-1}$

The energy that can be drawn as work is equal to $\left|(\Delta F)_T\right| = 812.63 \, \text{kJ} \, \text{mol}^{-1}$

9.16   Standard Gibbs Energy of Formation

Like the standard heat of formation the standard Gibbs free energy for formation of compounds $\Delta G^{\emptyset}_{form}$ from their basic elements may be obtained from the difference of the standard Gibbs energy of the reaction products and the reactants.

$$\Delta G^{\emptyset}_{form} = \sum \mathbb{N} \Delta G_{form}(products) - \sum \mathbb{N} \Delta G_{form}(reactants) \qquad 9.27$$

$\mathbb{N}$ in Eq. 9.27 stands for the number of moles of different substances.

9.17   Phases in Equilibrium

When two phases of a substance are in equilibrium, not only the Gibbs functions for the two phases are equal but the derivatives of Gibbs functions at the phase boundary are also equal. If G_1 and G_2 are respectively, Gibbs functions for the phase 1 and phase 2, then for equilibrium,

$$dG_1 = dG_2.$$

But $dG = VdP - SdT$, hence

$$V_1 dP - S_1 dT = V_2 dP - S_2 dT$$

Or

$$\frac{dP}{dT} = \frac{(S_1 - S_{2)}}{(V_1 - V_2)} = \frac{(S_2 - S_{1)}}{(V_2 - V_1)}$$

If one considers molar volumes and molar entropies then the above equation may be written as,

$$\frac{dP}{dT} = \frac{(s_1 - s_{2)}}{(v_1 - v_2)} = \frac{T(s_2 - s_1)}{T(v_2 - v_1)} = \frac{\Delta H_{phase\ change}}{T(v_2 - v_1)} \qquad 9.28$$

$\Delta H_{Phase\ change}$ in Eq. 9.28 is the change of enthalpy per mole for the change of phase-1 to phase-2.

(a) **Melting or fusion:** When a substance melts and changes from solid to liquid phase, Eq. 9.28 becomes,

$$\frac{dP}{dT} = \frac{\Delta H_{Melting}}{T\left(v_{liquid} - v_{solid}\right)} = \frac{\Delta H_{Melting}}{T(\Delta V)} \qquad 9.29$$

Here (ΔV) is the change in the specific volume in going from solid to liquid phase, which is generally small and may be treated as independent of temperature. Equation 9.29 may then be integrated as follows

$$\int_{P_1}^{P_2} dP = \frac{\Delta H_{Melting}}{(\Delta V)} \int_{T_1}^{T_2} \frac{dT}{T}$$

Or

$$(P_2 - P_1) = \frac{\Delta H_{Melting}}{(\Delta V)} \ln \frac{T_2}{T_1} \qquad 9.30$$

Equation 9.30 gives a relation between the pressures and temperatures at which the solid-liquid phases will be in equilibrium.

(b) **Evaporation:** For thermodynamic equilibrium in liquid-vapor phases of a pure substance, Eq. 9.28 may be written as

$$\frac{dP}{dT} = \frac{\Delta H_{evap}}{T\left(v_{vapour} - v_{liquid}\right)} = \frac{\Delta H_{evap}}{T\left(v_{vapour}\right)} \qquad 9.31$$

The last term in Eq. 9.31 is obtained by assuming that $v_{vapour} \gg v_{liquid}$. Further, if the vapors are treated as an ideal gas then, $Pv = RT$ or $v_{vapour} = \frac{RT}{P}$. Substituting this in Eq. 9.31 one gets,

$$\frac{dP}{dT} = \frac{\Delta H_{evap}}{T\left(v_{vapour}\right)} = \frac{\Delta H_{evap}}{RT^2} P$$

Or

$$\int_{P_1}^{P_2} \frac{dP}{P} = \frac{\Delta H_{evap}}{R} \int_{T_1}^{T_2} \frac{dT}{T^2}$$

Or

$$\ln \frac{P_2}{P_1} = \frac{-\Delta H_{evap}}{R} \left(\frac{1}{T_2} - \frac{1}{T_1}\right) \qquad 9.32$$

Equations 9.30 and 9.32 are the (approximate) integral forms of Clapeyron equations for the equilibrium between the solid- liquid and vapor–liquid phases, respectively.

Solved Example

13. The heat of evaporation of pure water is 40 .0 kJ mol^{-1}. Calculate the boiling temperature of water when the pressure is half of the atmospheric pressure.

Solution:

From Eq. 9.32, $\ln \dfrac{P_2}{P_1} = \dfrac{-\Delta H_{evap}}{R}\left(\dfrac{1}{T_2} - \dfrac{1}{T_1}\right)$ we have,

$P_2 = 0.5$ atm, $P_1 = 1.0$ atm, $\Delta H_{evap} = 40 \times 10^3$ J mol^{-1}, $T_1 = (100 + 273)K$, $T_2 = ?$

$$\ln \frac{0.5}{1.0} = \frac{-40.0 \times 10^3}{8.314}\left(\frac{1}{T_2} - \frac{1}{373}\right)$$

That gives $T_2 = 353.98$ K $= 80.98°C$

9.18   Thermodynamics of Electrochemical Cell

Cells are devices that generate electromotive force (in short called e.m.f.). A source of e.m.f. when connected in a conducting circuit causes electric current to flow through the circuit. Devices in which e.m.f. is produced by some chemical process are called Electrochemical Cells. There may be two types of chemical cells: (i) Cells in which e.m.f. is produced by some chemical reaction, like Daniel or Galvanic cells. (ii) Cells in which e.m.f. is generated because of the concentration difference of some chemical specie called electrolyte. Whatever may be type of the cell, some transport process taking place inside the cell produces the electromotive force.

Let ε be the magnitude of the e.m.f. of the cell. When this cell is connected to some conductor a current is established in the external circuit which is constituted by the flow of electrons. The work done by the e.m.f. dW in making a charge dQ to flow is given by,

$$dW = \varepsilon dQ$$

When 1- mole of electron (= 6.02×10^{23} electrons) passes through a circuit, charge Q = $(1.60 \times 10^{-19} \times 6.02 \times 10^{23}$ C $=9.648 \times 10^4$ C) 1F (Faraday) flows through the circuit. If $\mathbb{N}$ mole of electrons flow the charge Q = $(\mathbb{N}F)$ will pass and the work done by the e.m.f. will be

$$W = \varepsilon(\mathbb{N}F) \tag{9.33}$$

It may be noted that W is a non-PdV work and the maximum non-PdV work that may be obtained from a system is equal to the decrease in Gibbs function of the system $\Delta\Gamma$, *provided the process is reversible*. In case the process is irreversible, the non-PdV work that may be obtained is less than the decrease of the Gibbs function.

Current flow from a cell is essentially an irreversible process because of the heat loss RI^2. However, in principle it may be made negligible if current is very small. Another way of making current flow from a cell a reversible process is to apply an external electric potential across the cell E $(E > \varepsilon)$ through a potential divider such that the externally applied potential opposes the cell e.m.f., as shown in Fig. 9.3. For a particular setting of the potential divider, when the external potential is just equal to the cell e.m.f., there will be no flow of current, but on slight change of the potential divider setting on either side, the direction of current flow may be reversed, making current flow a reversible process.

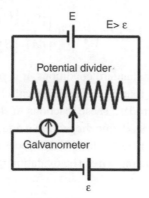

Fig. 9.3 Making current flow: a reversible process

Therefore, according to thermodynamics,

$$W = -\Delta G \qquad 9.34$$

The magnitude of both W and G depends on the conditions of the cell, i.e., the concentrations of electrolyte and other cell constituents and the temperature of the cell etc. Under standard conditions,

$$\varepsilon^0(\mathbb{N}F) = -\Delta G^0 \qquad 9.35$$

Here ε^0 is the cell potential at standard conditions

The value of ΔG is related to its value at standard conditions ΔG^0 by the relation

$$\Delta G = \Delta G^0 + RT \ln X \qquad 9.36$$

Here X is a quotient that depends on the properties of the mixture of different types. For example in case of the mixture of ideal gases X $= \dfrac{p_i}{P}$, where p_i, and P are, respectively, the partial pressure of the ith ideal gas and the total pressure.

If the chemical reaction inside the cell may be written as,

$$\alpha A + \beta B \rightleftarrows \gamma C + \delta D$$

where α, β, γ and δ are the molar concentrations of the reactant and reaction products, then the quotient X is given by,

$$X = \frac{\left[[C]^\gamma [D]^\delta \right]}{\left[[A]^\alpha [B]^\beta \right]}$$

Therefore,
$$\Delta G = \Delta G^0 + RT \ln \frac{\left[[C]^\gamma [D]^\delta \right]}{\left[[A]^\alpha [B]^\beta \right]}$$

Or
$$-\varepsilon(\mathbb{N}F) = -\varepsilon^0 (\mathbb{N}F) + RT \ln \frac{\left[[C]^\gamma [D]^\delta \right]}{\left[[A]^\alpha [B]^\beta \right]}$$

Or
$$\varepsilon = \varepsilon^0 - \frac{RT}{(\mathbb{N}\,F)} \ln \frac{\left[[C]^\gamma [D]^\delta \right]}{\left[[A]^\alpha [B]^\beta \right]} \qquad 9.37$$

To determine ε^0, the cell e.m.f. at standard conditions, it is required to understand the working of the cell. In electrochemical cells like Daniel cell, typical chemical reactions of oxidation and reduction take place at the two electrodes. These reactions in short are called 'redox' reactions. A Daniel cell consists of an anode of metal zinc dipped in an aqueous solution of Zn^{++} and a copper cathode dipped in aqueous solution of Cu^{++}. The two electrodes may be connected externally when a switch is made on through a conducting wire. The oxidation reaction (losing of electrons) occurs at the anode

$$Zn(s) \rightarrow Zn^{++}(aq) + 2e^-$$

Since anode loses electrons it acquirers positive charge. When the switch is made on these electrons travel through the wire and reach the cathode where they are captured by the cathode in the reduction reaction

$$Cu^{++}(aq) + 2e^- \rightarrow Cu(s)$$

As a result neutral copper atoms deposit on the cathode while zinc get dissolved in the solution at anode. The total reaction may be written as,

$$Zn(s) + Cu^{++}(aq) \rightarrow Zn^{++}(aq) + Cu(s)$$

The force required to push electrons from the anode to the cathode for the above reaction to take place is provided by the e.m.f. of the cell.

If the concentrations of the ions are maintained at 1 mole kg^{-1} (or 1 *m*), the cell is called the *standard Daniel cell* and the cell e.m.f. is called the *standard cell e.m.f.*, denoted as ε^0. The fact that the spontaneous cell reaction leads to the oxidation of Zn(s) to $Zn^{++}(aq)$ and the reduction of $Cu^{++}(aq)$ to Cu(s) indicates that the tendency for $Cu^{++}(aq)$ ions to accept electrons (to get reduced) is greater than that for $Zn^{++}(aq)$ ions. In the terminology of electrochemistry, we say that $Cu^{++}(aq)$ has a greater *reduction potential* than $Zn^{++}(aq)$.

There are a few conventions and notations used for short-hand representation of electrochemical cells.

An electrochemical cell is viewed as the combination of two "half-cells." Each half-cell consists of an electrode, which may participate in the cell reaction, and chemicals in contact with that electrode.

Vertical lines are used to indicate important phase boundaries across which half-cell reactions take place. Often, the electrolyte concentration is also shown.

For example, in Daniel cell anode is shown as: Zn(s)|Zn⁺⁺(aq), and the cathode as: Cu⁺⁺(aq)|Cu(s).

When two half-cells are combined to make an electrochemical cell, the half-cell in which oxidation takes place is written on the left-hand side and the half-cell in which reduction takes place is written on the right. The electrodes are written on the "outside" with the electrolytes in-between. A double vertical line is drawn between the two electrolytes to indicate a physical separation, either by a porous membrane or a salt-bridge. The electrode arrangement and short hand representation of a Daniel cell is given Fig. 9.4.

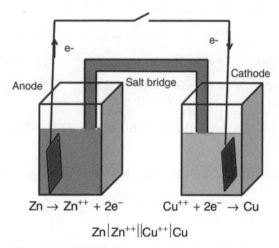

$$Zn \rightarrow Zn^{++} + 2e^- \qquad Cu^{++} + 2e^- \rightarrow Cu$$

$$Zn|Zn^{++}||Cu^{++}|Cu$$

Fig. 9.4 Half cell representation of a Daniel cell

In electrochemical cells, the anode is negatively charged because electrons are given to the anode by the species undergoing oxidation. The cathode is positively charged because electrons are removed from it by the species undergoing reduction. So, when an electrochemical cell is written as indicated above, the electron-flow in the external circuit is from the electrode on the left to the electrode on the right.

The difference in the reduction potentials of the two electrodes chosen to make up the cell determines the total cell e.m.f. A scale of reduction potentials is needed to quantitatively assign the reduction potentials of various electrodes used to make up electrochemical cells. Since there is no way to measure the reduction potential of a single electrode, the hydrogen electrode is chosen as the ultimate standard. The Hydrogen electrode operating at 25°C and 1 bar pressure is assigned a reduction potential of 0.0 V. If we were to construct a cell using a Cu(s)|Cu⁺⁺(1 m) electrode and the standard hydrogen electrode

[Pt, H₂(1 bar, 25°C)|H⁺(1 m)], the electrode reactions would be

$$H_2(g) \rightarrow 2H^+(aq) + 2e^-$$

$$Cu^{++}(aq) + 2e^- \rightarrow Cu(s).$$

The measured e.m.f. of such a cell would be 0.3419 V. Therefore, we say that the standard reduction potential of the Cu(s)|Cu⁺⁺(1 m) electrode is +0.3419 V. The fact that Cu⁺⁺ ions undergo spontaneous reduction indicates that the tendency for Cu⁺⁺(aq) ions to accept electrons (to get reduced) is greater than that for H^+ ions. In contrast, in the cell $Zn(s)|Zn^{++}(1\ m)||H^+(aq)|H_2(1$ bar, 25°C),Pt, the cell reactions will be

$$Zn(s) \rightarrow Zn^{++}(aq) + 2e^-,$$

$$2H^+(aq) + 2e^- \rightarrow H_2(g),$$

and the measured cell e.m.f. will be –0.7618 V. In this case, it is clear that the tendency for H^+ ions to accept electrons is greater than for the $Zn^{++}(aq)$ ions. Therefore, the reactions and electrode potentials in the cell $Zn(s)|Zn^{++}(1\ m)||Cu^{++}(1\ m)|Cu(s)$ will be

$$Zn(s) \rightarrow Zn^{++}(aq) + 2e^-, \qquad \varepsilon^0 = -0.7618\ V$$

$$Cu^{++}(aq) + 2e^-(aq) \rightarrow Cu(s), \qquad \varepsilon^0 = +0.3419\ V$$

Overall: $\qquad Zn(s) + Cu^{++}(aq) \rightarrow Zn^{++}(aq) + Cu(s). \qquad \varepsilon^0 = +1.1037\ V$

9.18.1 Temperature dependence of cell e.m.f. and other thermodynamic functions

By using the basic relation between the cell e.m.f. and the Gibbs free energy change, it follows,

$$dG = -SdT - VdP$$

and at constant pressure,

$$S = -\left(\frac{\partial G}{\partial T}\right)_P$$

Therefore, $\qquad \Delta S = -\left(\frac{\partial \Delta G}{\partial T}\right)_P = \left(\frac{\partial \varepsilon(NF)}{\partial T}\right)_P = (NF)\left(\frac{\partial \varepsilon}{\partial T}\right)_P \qquad$ 9.38

Also, $\qquad G = H - TS$ and $\Delta G = \Delta H - T\Delta S$

Or $\qquad \Delta H = \Delta G + T\Delta S = -\varepsilon NF + T(NF)\left(\frac{\partial \varepsilon}{\partial T}\right)_P \qquad$ 9.39

Equations 9.38 and 9.39 show that if the temperature coefficient of cell e.m.f. $\left(\frac{\partial \varepsilon}{\partial T}\right)_P$ is known the other two state functions entropy S and enthalpy H for the cell may be determined.

Also from Eq. 9.38 it follows that,

$$\left(\frac{\partial \varepsilon}{\partial T}\right)_P = \frac{\Delta S}{(NF)}$$

And if ΔS is independent of temperature (which means C_P is small)

$$\varepsilon(T) = \varepsilon(T_0) + \frac{\Delta S}{(NF)}(T - T_0)$$

For any redox reaction ΔS is small, < 50 J/K, and this leads to a change in cell e.m.f. of not more than 10^{-4} to 10^{-5} volt per kelvin.

It can be shown that the equilibrium constant K for the reaction taking place in the cell is given as

$$\ln K = -\frac{\Delta G^0}{RT}$$ 9.40

Problems

1. Following reaction takes place when potassium hydroxide pallets are added to water

$$KOH(s) \rightarrow H_2O(l) + 43\,kJ\,mole^{-1}$$

Indicate whether the reaction is exothermic or endothermic and calculate the heat of dissolution of 14 kg of KOH.

2. Calculate the change in standard enthalpy for the following reaction

$$3Al(s) + 3NH_4ClO_4(s) \rightarrow Al_2O_3(s) + AlCl_3(s) + 3NO(g) + 6H_2O(g)$$

Given the following data:

Substance	Heat of formation in kJ mol⁻¹
$NH_2ClO_4(s)$	−295.0
$Al_2O_3(s)$	−1676.0
$AlCl_3(s)$	−704.0
$H_2O(g)$	−242.0
$NO(g)$	+90.0

3. In the reaction $C_6H_{12}O_6(s) + 6O_2(g) \rightarrow 6CO_2(g) + 6H_2O(l) + 2800\,kJ$, the heats of formation for $C_6H_{12}O_6(s)$ and $CO_2(g)$ are respectively, −1276 kJ mol⁻¹ and -393.5 kJ mol⁻¹. Calculate the heat of formation for $H_2O(l)$.

4. Compute the change in standard enthalpy for the reaction,

$$H_3BO_3(aq) \rightarrow \frac{1}{2}B_2O_3(s) + \frac{3}{2}H_2O(l),$$

Given that

$H_3BO_3(aq) \rightarrow HBO_2(aq) + H_2O(l)$ $\qquad \Delta H^{\phi}_{rea} = -0.02\,kJ\,mol^{-1}$

$H_2O(l) + H_2B_4O_7(aq) \rightarrow 4HBO_2(aq)$ $\qquad \Delta H^{\phi}_{rea} = -11.30\,kJ\,mol^{-1}$

$H_2B_4O_7(aq) \rightarrow 2B_2O_3(s) + H_2O(l)$ $\qquad \Delta H^{\phi}_{rea} = -17.50\,kJ\,mol^{-1}$

5. Bond energies (in kJ mol⁻¹) for H-F, H-H and F-F bonds are respectively, 565.0, 432.0, and 154. Compute the change in the standard enthalpy for the reaction,

$$H_2(g) + F_2(g) \rightarrow 2HF(g)$$

6. From the following data calculate the change in standard enthalpy of the reaction

$$2NO(g) + O_2(g) \rightarrow 2NO_2(g)$$

Data: Standard heat of formation of $N_2O_5(g) = 11.3\,kJ\,mol^{-1}$

$$N_2(g) + O_2(g) \rightarrow 2NO(g) \qquad\qquad \Delta H^{\varnothing}_{rea} = +180.5 \text{ kJ mol}^{-1}$$

$$4NO_2(g) + O_2(g) \rightarrow 2N_2O_5(g) \qquad\qquad \Delta H^{\varnothing}_{rea} = -110.2 \text{ kJ mol}^{-1}$$

7. One mole of methane is completely burnt at constant pressure and 298 K temperature in air just sufficient for combustion. From the data given find the maximum flame temperature.

   Data: $CH_4(g) + 2O_2(g) \rightarrow CO_2(g) + 2H_2O(l)$ $\Delta H^{\varnothing}_{rea} = -881.25 \text{ kJ mol}^{-1}$

          $H_2O(l) \rightarrow H_2O(g)$ $\Delta H^{\varnothing}_{vapour} = +43.60 \text{ kJ mol}^{-1}$

          $C_P(CO_2, g) = 26.00 + 43.5 \times 10^{-3}T - 148.3 \times 10^{-7}T^2 \text{ J mol}^{-1} \text{ K}^{-1}$

          $C_P(H_2O, g) = 30.36 + 9.61 \times 10^{-3}T + 11.8 \times 10^{-7}T^2 \text{ J mol}^{-1} \text{ K}^{-1}$

          $C_P(N_2, g) = 27.30 + 5.23 \times 10^{-3}T - 0.04 \times 10^{-7}T^2 \text{ J mol}^{-1} \text{ K}^{-1}$

8. Calculate the boiling point of pure water in a pressure cooker where the pressure is 5.065×10^5 Nm^{-2}.

9. Calculate the ΔG^0 and equilibrium constant K at 25^{0}C for the following cell

   $$Pt(s)|Cl_2(g)|Cl^-(aq); \qquad \varepsilon^0 = 1.3595$$

   $$Hg(l) | Hg_2Cl_2(s), Cl^-(aq); \quad \varepsilon^0 = 0.2680$$

10. For an electrochemical cell $\varepsilon^0 = 1.01463$ at 25°C, number of mole $N_m = 2$ and the thermal coefficient of cell e.m.f. $\left(\dfrac{\partial \varepsilon}{\partial T}\right)_P = -5.0 \times 10^{-5}$ VK^{-1}. Compute ΔG, ΔS, ΔH for the cell.

Short Answer Questions

1. What causes the generation of e.m.f. in a cell?
2. How much charge is carried by one kilomole of electrons?
3. How the spontaneity of a chemical reaction may be evaluated?
4. What are maximum flame and explosion temperatures?
5. State Hess's law for the reaction energies.

Long Answer Questions

1. Drive expressions for (i) the change in entropy of a cell and (ii) temperature dependence of the cell e.m.f.
2. Discuss the working of an electrochemical cell and obtain expressions for its thermodynamic state functions.
3. Obtain expression for the change in boiling point of a pure liquid with pressure.
4. Establish a relation between the change in entropy in a chemical reaction and the initial and final temperatures and pressures.

5. Why ΔQ_P, the heat of reaction at constant pressure, is different than the heat of reaction at constant volume ΔQ_V? Establish a relation between them.

Multiple Choice Questions

Note: Some of the following questions may have more than one correct alternative. All correct alternatives must be ticked for the complete answer in such cases.

1. The correct half-cell representation of Daniel cell is
   - (a) $Zn|Zn^{++} \parallel Cu^{++}|Cu$
   - (b) $Cu|Cu^{++}\|Zn^{++}|Zn$
   - (c) $Zn^{++}|Zn \|Cu|Cu^{++}$
   - (d) $Cu^{++}|Cu\|Zn|Zn^{++}$

2. The temperature coefficient of cell e.m.f. $\left(\dfrac{\partial \mathcal{E}}{\partial T}\right)_P$ is given as
   - (a) $\dfrac{(NF)}{\Delta S}$
   - (b) $\dfrac{F\Delta S}{(N)}$
   - (c) $\dfrac{\Delta S}{(NF)}$
   - (d) $\dfrac{(N\Delta S)}{F}$

3. Tick all correct relations for an electrochemical cell
   - (a) $\Delta H = -\mathcal{E}NF + T\left(NF\right)\left(\dfrac{\partial \mathcal{E}}{\partial T}\right)_P$
   - (b) $\Delta H = \mathcal{E}NF + T\left(NF\right)\left(\dfrac{\partial \mathcal{E}}{\partial T}\right)_P$
   - (c) $\Delta S = \left(NF\right)\left(\dfrac{\partial \mathcal{E}}{\partial T}\right)_P$
   - (d) $\Delta S = -\dfrac{1}{(NF)}\left(\dfrac{\partial \mathcal{E}}{\partial T}\right)_P$

4. Change in Gibbs function at constant temperature in a chemical reaction is
   - (a) $NRT \ln \dfrac{P_f}{P_i}$
   - (b) $NRP \ln \dfrac{T_f}{T_i}$
   - (c) $\dfrac{RT}{N} \ln \dfrac{P_f}{P_i}$
   - (d) $\dfrac{N}{RT} \ln \dfrac{P_f}{P_i}$

5. Assuming that gases emitted in chemical reactions obey ideal gas laws, the change in Helmholtz function at constant temperature in chemical reactions is given by,
   - (a) $-\dfrac{1}{NRT} \ln \dfrac{V_f}{V_i}$
   - (b) $NRT \ln \dfrac{P_f}{P_i}$
   - (c) $-\dfrac{RT}{N} \ln \dfrac{P_f}{P_i}$
   - (d) $-NRT \ln \dfrac{V_f}{V_i}$

6. Tick the correct relationship (s). symbols have their usual meanings,
   - (a) $\ln K = -\dfrac{\Delta G^0}{RT}$
   - (b) $\ln K = -\dfrac{\Delta G^0}{RP}$
   - (c) $\ln K = \dfrac{\Delta G^0}{\Delta S^0}$
   - (d) $\log_{10} K = -\dfrac{\Delta G^0}{2.303xRT}$

7. If ΔH^{rea} and ΔH^{inv} respectively represents the heats of a reaction and its inverse reaction, then
   (a) $\Delta H^{rea} = \Delta H^{inv}$
   (b) $\Delta H^{rea} = -\Delta H^{inv}$
   (c) $\Delta H^{rea} = 0.5\ \Delta H^{inv}$
   (d) $\Delta H^{rea} = -0.5\ \Delta H^{inv}$

8. The magnitude of the thermal coefficient of change in cell e.m.f. (in Volts per Kelvin) is of the order of,
   (a) 10^2
   (b) 10^0
   (c) 10^{-2}
   (d) 10^{-4}

Answers to Numerical and Multiple Choice Questions

Answers to problems

1. Exothermic, -10.7×10^6 J mole^{-1}
2. -2677 kJ mol^{-1}
3. -285.8 kJ mol^{-1}
4. $+7.2$ kJ mol^{-1}
5. -544 kJ mol^{-1}
6. -114.1 kJ mol^{-1}
7. ≈ 2250 K
8. $172.24°$C
9. $\Delta G^0 = -105.33$ kJ mol^{-1}; $K = 3.26 \times 10^{42}$
10. $\Delta G = -195.81$ kJ; $\Delta S = -9.65$ JK^{-1}; $\Delta H = -198.69$ kJ

Answer to multiple choice questions

1. (a)   2. (c)   3. (a), (c)   4. (a)
5. (d)   6. (a), (d)   7. (b)   8. (d)

Revision

1. ΔW as negative if it is done by the system
2. $\Delta Q_P = \Delta H = (H_f - H_i)$ is negative for exothermic reactions and positive for endothermic reactions at constant pressure

$$\Delta Q_P = \Delta Q_V + R\Delta(\mathbb{N}T)$$

3. Standard state
   The thermodynamic standard state of a substance is the most stable pure form under specified standard pressure and temperature.
   • In case the substance is in solid or liquid phases the standard state is the pure form of the substance at the atmospheric pressure and at 298 K (=25°C) temperature.
   • If the substance is a pure gas then the standard state is at one atmospheric pressure and at temperature of 298 K, unless specified otherwise.

- If there is a mixture of gases than the standard state of each gas is at a partial pressure of 1 atmosphere and the specified temperature which is generally 298 K..
- For a substance in solution the standard state refers to 1- molar concentration.

4. The standard molar enthalpy ΔH^{ϕ}_{for} of a substance is the change in enthalpy in the reaction in which one mole of the substance in specified state is formed from its elements in standard states. By convention the standard molar enthalpy of elements in their standard states is taken to be zero.

5. Enthalpy change in a reaction and in its inverse reaction is equal in magnitude but opposite in sign.

6. Hess's Law: The change in standard enthalpy in a given chemical reaction is equal to the sum of the changes in standard enthalpies of all other chemical reactions that add up to the given reaction.

7. Heat capacities: $c_P = \dfrac{\partial(Q_P)}{\partial T} = \left(\dfrac{\partial H}{\partial T}\right)_P$; $c_V = \dfrac{\partial(Q_V)}{\partial T} = \left(\dfrac{\partial U}{\partial T}\right)_V$

8. Temperature dependence of heats of reactions: For the following balanced reaction equation,

$$\mathbb{N}^A A + \mathbb{N}^B B + \mathbb{N}^C C + \ldots \rightarrow \mathbb{N}^X X + \mathbb{N}^Y Y + \mathbb{N}^Z Z + \ldots$$

Here $\mathbb{N}^A$, $\mathbb{N}^B$, $\mathbb{N}^C$, etc., are the number of moles of A, B, C, …, etc.

$$\Delta H^{T_2}_{rea} - \Delta H^{T_1}_{rea} = \int_{T_1}^{T_2} (\Delta C_P) dT$$

Here, $\Delta H^{T_2}_{rea}$ and $\Delta H^{T_1}_{rea}$ are respectively the changes in reaction heats at temperature T_2 and T_1.

9. Bond energies: $\Delta H^{\varnothing}_{rea} = \sum (energies\ of\ bonds\ broken) - \sum (energies\ of\ bonds\ formed)$

10. The Explosion and the flame temperatures: The maximum temperature of reaction products reached in an adiabatic exothermal reaction at constant volume is called the explosion temperature. The maximum temperature attained by reaction products in an adiabatic exothermal reaction at constant pressure is termed as the flame temperature.

11. Change of entropy in chemical reactions:

$$S_f - S_i = \Delta S = C_V \ln \frac{T_f}{T_i} + \mathbb{N}R \ln \frac{V_f}{V_i} \ ; \ S_f - S_i = \Delta S = C_P \ln \frac{T_f}{T_i} - R \ln \frac{P_f}{P_i}$$

12. Spontaneity of a chemical reaction: For a chemical reaction to be spontaneous,

$$(\Delta S)_{surroundings} + (\Delta S)_{system} > 0$$

13. Change in the value of thermodynamic functions in chemical reactions:

$$\Delta H = \Delta U + RT\Delta(\mathbb{N}); \ dF = -PdV - SdT; \ dG = VdP - SdT$$

The magnitude of the change in the value of Helmholtz function $|\Delta F|$ in a chemical reaction is equal to the maximum energy that may be drawn in the form of work while $|\Delta G|$, the magnitude of the decrease in the value of Gibbs function in a reaction gives the maximum energy that may be drawn as non-mechanical (non-PdV) work.

14. Standard Gibbs energy of formation: $\Delta G^{\varnothing}_{form} = \Sigma N \Delta G_{form}(products) - \Sigma N \Delta G_{form}(reactants)$

15. Phases in equilibrium

$$\frac{dP}{dT} = \frac{\Delta H_{Melting}}{T\left(v_{liquid} - v_{solid}\right)} = \frac{\Delta H_{Melting}}{T(\Delta V)} \; ; \; (P_2 - P_1) = \frac{\Delta H_{Melting}}{(\Delta V)} \ln \frac{T_2}{T_1}$$

$$\ln \frac{P_2}{P_1} = \frac{-\Delta H_{evap}}{R}\left(\frac{1}{T_2} - \frac{1}{T_1}\right)$$

16. Thermodynamics of electrochemical cell: Some transport process taking place inside the cell produces the electromotive force. If the chemical reaction taking place in the cell is represented as: $\alpha A + \beta B \rightleftarrows \gamma C + \delta D$

$$\varepsilon^0\left(N_m F\right) = -\Delta G^0 \; ; \; \Delta G = \Delta G^0 + RT \ln X, \text{ where } X = \frac{\left[[C]^{\gamma}[D]^{\delta}\right]}{\left[[A]^{\alpha}[B]^{\beta}\right]}$$

$$\varepsilon = \varepsilon^0 - \frac{RT}{(NF)} \ln \frac{\left[[C]^{\gamma}[D]^{\delta}\right]}{\left[[A]^{\alpha}[B]^{\beta}\right]}$$

In half-cell representation Daniel cell may be represented as: Zn |Zn++||Cu++|Cu
For standard Daniel cell,

$$\text{Zn}(s) \rightarrow \text{Zn}^{++}(aq) + 2e^-, \qquad\qquad \varepsilon^0 = 0.7618 \text{ V}$$

$$\text{Cu}^{++}(aq) + 2e^-(aq) \rightarrow \text{Cu}(s), \qquad\qquad \varepsilon^0 = +0.3419 \text{ V}$$

Over all Zn(s) + Cu++(aq) → Zn++(aq) + Cu(s). $\varepsilon^0 = +1.1037$ V

17. Temperature dependence of cell e.m.f. and other thermodynamic functions

$$\Delta S = -\left(\frac{\partial \Delta G}{\partial T}\right)_P = \left(\frac{\partial \varepsilon(NF)}{\partial T}\right)_P = (NF)\left(\frac{\partial \varepsilon}{\partial T}\right)_P$$

$$\Delta H = \Delta G + T\Delta S = -\varepsilon N F + T(NF)\left(\frac{\partial \varepsilon}{\partial T}\right)_P$$

$$\left(\frac{\partial \varepsilon}{\partial T}\right)_P = \frac{\Delta S}{(NF)} \; ; \; \varepsilon(T) = \varepsilon(T_0) + \frac{\Delta S}{(NF)}(T - T_0) \; ;$$

Change in cell e.m.f. with temperature is not more than 10^{-4} to 10^{-5} volt per kelvin.

The equilibrium constant K for the reaction taking place in the cell is given as $\ln K = -\dfrac{\Delta G^0}{RT}$

Quantum Thermoynamics

10.0   Introduction

Classical thermodynamics, developed into its formal and phenomenological form essentially by Clausius, is a theory based on the observations of the behavior of macroscopic systems that has unexpectedly wide range of validity. Classical thermodynamics in general is applicable to all systems from classical gases and liquids, through quantum systems such as superconductors and nuclear matter, to black holes and elementary particles in the early universe.

Quantum thermodynamics, on the other hand gives a rational understanding of thermodynamics in terms of microscopic particles and their interactions. It deals with systems that contain large number of identical particles or entities, like a piece of crystalline solid that has large number of identical atoms or molecules arranged in a specific way, or a gas contained in a volume that also has large number of molecules that are in random motion. Systems of identical entities or particles are called *assemblies* and a system that has large number of subsystems or assemblies of identical entities is termed as an *ensemble*.

An assembly that has one microgram of Argon gas at some finite temperature and pressure contains more than 1.5×10^{16} Argon atoms moving randomly, colliding with each other and with the walls of the container. In such a situation it is neither possible nor required to know the velocity or energy, etc., of each gas molecule at each instant. What is, however, important to know is the number of gas molecules that have their velocities or energies, etc., in a given range, say, velocities between v and $(v + \Delta v)$, or energies between some value ϵ and $(\epsilon + \Delta \epsilon)$, and how this number changes with time. Conversely, one will like to know as to how a given amount of energy E will get distributed into different groups of particles in the system. This information is contained in what is called the *distribution function*, so there may be several types of distribution functions like the velocity distribution function or energy distribution function, etc., for a system. *Statistical mechanics* or *Quantum statistics* is the tool to obtain these distribution functions. Quantum statistics uses the theory of probability, like the classical statistics, but assumes discrete, i.e., non-continuous values for physical variables like velocity, energy, momentum etc. Quantum statistics further assumes that an assembly of identical particles or entities may follow different kinds of statistics, like Fermi–Dirac, Bose–Einstein or Maxwell–Boltzmann statistics. These statistics differ from each other as to how the entities of the system may be distributed into various energy levels and energy states in a level. Statistical mechanics, the science of bulk matter is an incomplete and evolving science. New ideas and concepts permit a fresh approach to old problems. With new concepts one looks for features ignored in past and expect exiting results. Important new concepts are: *deterministic chaos, fractals, self- organized criticality (SOC), turbulence* and *intermittency*. These words represent large fields of study, all using quantum statistics, which have changed how we view Nature. Disordered systems, percolation theory and fractals find applications not only in physics and engineering but also in economics and other social sciences.

Thus having obtained the required distribution function from the quantum statistics, thermodynamics analyzes the distribution function to obtain the value of a parameter called the 'Partition function'. Partition function, which depends on the type of the statistics obeyed by the constituent particles of the system, is the most important parameter from the point of view of quantum thermodynamics. Considerable efforts are put in obtaining an appropriate partition function for a given system. Partition function, which is like the heart of quantum thermodynamics, may be used to obtain physical observables of the system, i.e., the quantities like Temperature, Pressure, Volume, Specific heat capacities, Entropy, etc., that may be measured experimentally. In this chapter we will see how one can use statistical mechanics (or quantum statistics) to get distribution functions and also how these distribution functions can be further analyzed using the tools of quantum thermodynamics to yield the all-important partition functions which in turn provide values of the required system observables.

10.1   Application of Quantum Statistics (Statistical Mechanics) to an Assembly of Non-interacting Particles

Statistical mechanics may be applied to solve problems related to real systems that contain large number of identical entities or particles. The formalism of statistical physics may be developed both for the classical systems as well as for quantum systems. The resulting energy distribution and calculating the values of physical observables is simpler in the classical case. However, the formulation of the method is more transparent in the quantum mechanical formalism. In addition, the absolute value of the entropy without any undermined constant and the behavior of the entropy when absolute temperature approaches zero, may be obtained only in the quantum mechanical treatment. In the following sections we will see how quantum statistics may be applied to an assembly of non-interacting (or free) particles.

10.2   Energy Levels, Energy States, Degeneracy and Occupation Number

From quantum mechanics it is known that for an assembly of identical particles each of mass m, that do not interact with each other (free particles) and are confined to a volume V, each particle independent of the other may have several discrete energy values $\epsilon_j \left(= \dfrac{p_j^2}{2m} \right)$ given by;

$$\epsilon_j = n_j^2 \frac{\hbar^2 (2\pi)^2}{8m} V^{-\frac{2}{3}}$$

10.1

In Eq. 10.1, $\hbar$ is rationalized Planck's constant $\left(= \dfrac{h}{2\pi}\right) = 1.05457 \times 10^{-34}$ J-s.

The integer n_j is made up of three independent integers n_x, n_y and n_z, called the quantum numbers, such that

$$n_j^2 = n_x^2 + n_y^2 + n_z^2 \qquad\qquad 10.2$$

Each of these n_x, n_y and n_z can have non-zero integer values like 1, 2, 3, ..., etc.

The value of n_j^2 defines an energy level of the system. Each energy level may have one or more states. A set of the different values of quantum numbers n_x, n_y and n_z subject to the condition given by Eq. 10.2, defines the number of states of the given energy level.

Since the minimum value that n_x, n_y and n_z may have is 1, the minimum value of $n_j^2 = 1^2 + 1^2 + 1^2 = 3$ and, therefore, the lowest energy level has the energy, ϵ_1, given as,

$$\epsilon_1 = 3\frac{\hbar^2 (2\pi)^2}{8m} V^{-\frac{2}{3}} \qquad\qquad 10.3$$

It may be observed that only one set of n_x, n_y and n_z can give the value 3 to n_j^2. In the language of quantum mechanics it is said that the level ϵ_1 has only one energy state. A level that has only one energy state is called a non-degenerate level. The degeneracy of a level is denoted by g_j, and is equal to the number of energy states in the level. The degeneracy g_1 of level at energy ϵ_1 is 1, i.e., $g_1 = 1$.

The next level will be one in which one of the quantum numbers n_x, n_y or n_z has the value 2 and the other two have values 1. This gives rise to three different sets of quantum numbers, giving the same value of $n_j^2 = 2^2 + 1^2 + 1^2 = 6$. These three sets are

$$(n_x = 1, n_y = 1, n_z = 2); \; (n_x = 1, n_y = 2, n_z = 1) \text{ and } (n_x = 2, n_y = 1, n_z = 1)$$

All the three different energy states mentioned above have the same energy $\epsilon_2 = 6\dfrac{\hbar^2 (2\pi)^2}{8m} V^{-\frac{2}{3}}$.

The level with energy ϵ_2 has three states and the degeneracy of this level $g_2 = 3$.

Let us consider the level with energy $\epsilon = 14\dfrac{\hbar^2 (2\pi)^2}{8m} V^{-\frac{2}{3}}$. The six different sets of n_x, n_y and n_z shown in Table 10.1, give the same value of $n_j^2 = 14$ and hence the same energy,

Table 10.1 Different sets of n_x, n_y and n_z that give the same energy

n_x	n_y	n_z
3	2	1
3	1	2
2	3	1
2	1	3
1	3	2
1	2	3

This level, therefore, has six fold degeneracy or g = 6 for this level. In general the energy levels are non-equidistant and have different folds of degeneracy.

In particular it may be observed that the three dimensional energy expression 10.1 is an equation of sphere of radius R,

$$\epsilon_j = n_j^2 \frac{\hbar^2 (2\pi)^2}{8m} V^{-\frac{2}{3}}$$

Or
$$n_j^2 = \left(n_x^2 + n_y^2 + n_z^2 \right) = \frac{8m\epsilon_j}{\hbar^2 (2\pi)^2} V^{\frac{2}{3}} = R^2$$

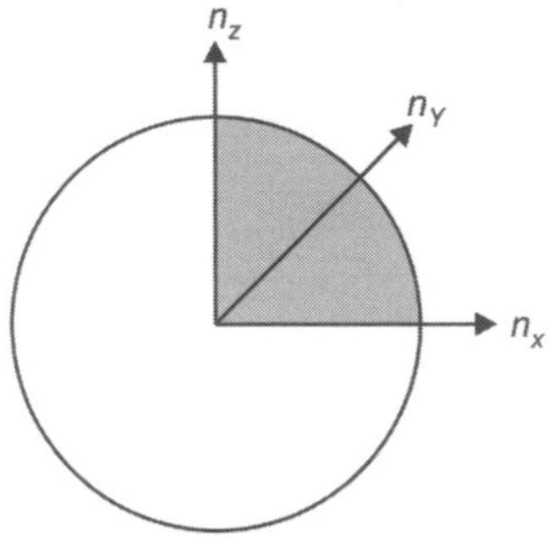

Fig. 10.1   Positive non-zero values of n_x, n_y, n_z lie in 1/8 quadrant of the sphere

The non-zero positive values of n_x, n_y, n_z lie in 1/8 quadrant of the sphere. For large values of energies the density of non- zero positive n_x, n_y, n_z points essentially fill the volume of the 1/8 quadrant. One may now treat R or energy ϵ as continuous variable and may obtain the number of lattice points consistent with energy $\leq \epsilon_j$, which is essentially the volume of the 1/8 of the sphere. The number of energy states

$$\mathcal{G}(\epsilon) = \frac{1}{8}\left(\frac{4}{3}\pi R^3 \right) = \frac{1}{6}\pi \left(R^2 \right)^{3/2} = \frac{1}{6}\pi \left[\frac{8m\epsilon_j}{\hbar^2 (2\pi)^2} V^{\frac{2}{3}} \right]^{3/2} \qquad 10.4$$

And the number of energy states in a thin shell of energy $\Delta\epsilon$ is given by

$$\mathcal{G}(\epsilon, \Delta\epsilon) = \frac{\pi}{4}\left(\frac{8mV^{\frac{2}{3}}}{\hbar^2 (2\pi)^2} \right)^{3/2} \epsilon_j^{1/2} \Delta\epsilon_j \qquad 10.5$$

If one calculates the degeneracy using Eq. 10.5 for molecules/atom moving in a room of size 10 m × 10 m × 10 m at temperature 300 K, taking $m \approx 10^{-25}$ kg, and $\Delta \epsilon_j = 0.01\epsilon_j$, it comes out to be of the order of 10^{30}.

A quantum mechanical system of N identical non-interacting particles confined in a given volume of space have many energy levels each level having a certain fold of degeneracy. Depending on the properties of the particles, each energy level accommodates a certain number of particles. If the jth level contains N_j particles, then N_j is called *the occupation number* of the level. The occupation numbers and the energy of different levels satisfy the following conditions,

$$\Sigma N_j = N \quad \text{and} \quad \Sigma \epsilon_j N_j = E \qquad\qquad 10.6$$

Here, E is the total energy of all particles and N their total number.

Figure 10.2 is a schematic representation of the energy levels, energy states, folds of degeneracy and the occupation numbers of an imaginary assembly.

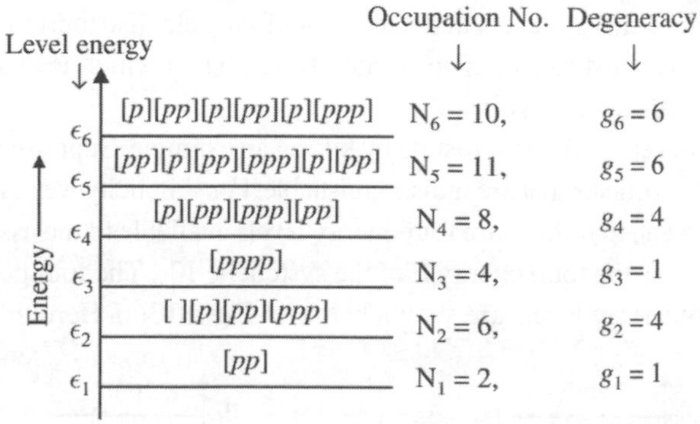

Level energy, Occupation No., Degeneracy

Level energy	States	Occupation No.	Degeneracy
ϵ_6	$[p][pp][p][pp][p][ppp]$	$N_6 = 10,$	$g_6 = 6$
ϵ_5	$[pp][p][pp][ppp][p][pp]$	$N_5 = 11,$	$g_5 = 6$
ϵ_4	$[p][pp][ppp][pp]$	$N_4 = 8,$	$g_4 = 4$
ϵ_3	$[pppp]$	$N_3 = 4,$	$g_3 = 1$
ϵ_2	$[\ ][p][pp][ppp]$	$N_2 = 6,$	$g_2 = 4$
ϵ_1	$[pp]$	$N_1 = 2,$	$g_1 = 1$

Energy →

Fig. 10.2 Schematic representation of energy levels, states, degeneracy and occupation number

Energy levels of a hypothetical system in Fig. 10.2 are shown by horizontal lines while their energies are written along the vertical energy axis. Brackets on each energy level give the different energy states in that level. The number of brackets on a level is the degeneracy of the level. The letter 'p' represents a particle and the number of particles in different energy states is equal to the number of p's in different brackets. The total number of p's in a level is the occupation number of the level. The energy of each level is given on the left while the corresponding occupation number and degeneracy are given on the right in Fig. 10.2. As may be observed in the figure it is possible that some state(s) of a level remain empty.

10.2.1   Distinguishable and indistinguishable particles

Properties of particles in a system decide their distribution in different levels and energy states. One of the important characteristic of a system of identical particles is whether particles are distinguishable or not. Identical particles may be indistinguishable from each other, like the molecules of a gas. Identical molecules of a gas are in constant motion. Therefore, it is not possible to put a mark on one particular molecule and identify it at all times. Hence, molecules of a gas are indistinguishable because they are *non-localized*. On the other hand, in a crystalline solid, atoms or molecules are identical but they may be differentiated or distinguished from one another on the basis of their location in the crystalline lattice. As such, the atoms or molecules in a crystalline solid are distinguishable as they are *localized*. In general non-localized particles are indistinguishable while localized entities are distinguishable because of their fixed location. Atoms/molecules of a paramagnetic salt if put in an external magnetic field align either parallel or anti-parallel to the applied field and therefore, may be distinguished from each other through their orientation in the external field. Similarly, nucleons (neutrons and protons) in a nucleus may be distinguished from each other on the basis of the orientation of their spins and so they are distinguishable. The fact that particles of an assembly are indistinguishable or distinguishable, as we will see, plays an important role in quantum statistics.

10.2.2   Macrostate and microstate

A given assembly of identical non-interacting particles has a specific structure of energy levels and energy states in each level. If the total number of particles N and total energy E of the system is known it is possible to distribute particles in different energy levels such that the conditions put by Eq. 10.6 are fulfilled. It is evident that in general there may be several different ways in which particles may be distributed in different energy levels and each way of particle distribution satisfy Eq. 10.6. Each such particle distribution constitutes a macrostate of the system which is characterized by a set of occupation numbers for different levels.

Let us explain the concept of the macrostate by taking an example. Suppose there is an assembly of non-interacting identical particles that are indistinguishable. For simplicity we assume that there are only 5 particles, that is N = 5 and that there are four energy levels available to the assembly, respectively, at energies 0, ϵ, 2ϵ and 3ϵ. Let the total energy E of the system be 10ϵ. The four possible ways of particle distribution in different energy levels are shown below in Fig. 10.3(a). Here 'p' denotes a particle.

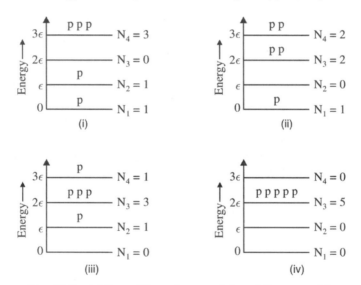

Fig. 10.3   (a) Four different macrostates of the assembly

It is easy to verify that for each macrostate shown in the figure total energy of all particles is 10ϵ and that conditions put by Eq. 10.6 are fulfilled by each macrostate. Each of the four macrostates is characterized by the occupation numbers of its energy levels. Thus it may be said that "*The macrostate of a system is defined by the number of particles in each energy level of the system*". This essentially means that if the occupation numbers of all energy levels of a system are known, the macrostate of the system is completely defined or known. For example, the macrostate (i) in Fig. 10.3 (a) is completely defined by the set of occupation numbers ($N_1 = 1$, $N_2 = 1$, $N_3 = 0$, $N_4 = 3$) and macrostate (iii) by the set of occupation numbers ($N_1 = 0$, $N_2 = 1$, $N_3 = 3$, $N_4 = 1$).

It may be observed that while defining macrostates no consideration is given to the number of states in each level (i.e., the degeneracy g) and to the number of particles that may be accommodated in each state. These characteristics of levels and states are decided by the properties of the particles of the assembly. For simplicity let us assume that all energy levels of the assembly of five particles that we have considered, are two- fold degenerate, it means that $g_1 = g_2 = g_3 = g_4 = 2$, that is, there are two energy states in each level. Also, let us further assume that any number of particles may be accommodated in each energy state of each level. Let us denote an energy state by a bracket with subscripts 'a' and 'b' and the superscript identifying the energy level 'n'. It may once again be stressed that the actual number of states in different energy levels and how many particles may be accommodated in each state is determined by the properties of the system particles. With these rather ad hoc assumptions, the macrostate (i) of Fig. 10.3 (a) may have several different ways of particle distributions in different states of the levels as shown in Fig. 10.3 (b).

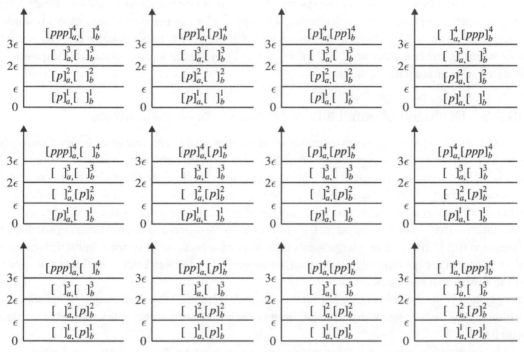

Fig. 10.3   (b) Twelve microstates of the macrostate ($N_4 = 3$, $N_3 = 0$, $N_2 = 1$, $N_1 = 1$)

Each different way in which particles in a given level may be distributed in its energy states gives rise to a new microstate. Twelve microstates corresponding to the macrostate defined by set of occupation numbers ($N_4 = 3$, $N_3 = 0$, $N_2 = 1$, $N_1 = 1$) are shown in Fig. 10.3 (b).

A microstate of the system is defined not only by the number of particles in a level but also by the number of particles in each energy state of each level. If the occupation number of a level remains constant but particles in the level shift from one energy state to another energy state, the microstate of the system gets changed. It is simple to realize that the number of macrostates and that of microstates corresponding to a given macrostate increases rapidly with the increase in the number of particles in the assembly and with the increase in the available energy.

Thus, corresponding to a given macrostate there may be a large number of microstates. In a system of indistinguishable identical particles a new macrostate of the system is formed when particles shift from one energy level to the other while a new microstate is formed when particles move from one energy state to another energy state in the same level.

Particles in a system that has a fixed energy E, may move from one energy state to another energy state or from one energy level to another level due to their collisions with each other and/or with the walls of the container, in case they are non-interacting, like the molecules of an ideal gas. If particles interact with each other, mutual interaction also contributes to the movement of particles. Generally, particles may go from one energy state to another in the same level when they gain or lose a relatively very small amount of energy, while it requires comparatively larger change in energy for a particle to move from one level to another. Since small energy changes are more frequent in collisions, system pass through large number of microstates before a new macrostate is formed. If one looks to the time evolution of the system, it dwells in a given macrostate for a relatively long time, samples a large number of microstates of this particular macrostate and then moves to a new macrostate, again stays in this new macrostate for sufficient time to pass through a large number of microstates of the new macrostate and then switches to another new macrostate. This sequence is repeated with time again and again.

10.2.3   Postulate of equal a prior probability of all microstates

A fundamental axiom of quantum thermodynamics is that *each microstate of an isolated system is equally likely or has same probability of occurrence.* Since this assumption is made before hand and is for all microstates irrespective of their macrostate, it is called the 'a prior' assumption. There may be three aspects of the assumption: (i) the time for which the system lives in a microstate is same for all microstates, irrespective of to which macrostate the microstate belongs; (ii) over a given time interval that is sufficiently large, the system passes through each microstate same number of times; and (iii) if there are very large number say N, of exactly identical systems and at a given time N_1 out of them are in some microstate, then the number of systems in each of the other microstate of the system will also be N_1.

A system is most stable in its equilibrium state and, therefore, stays in the equilibrium state for infinitely long time. It means that the equilibrium state has largest number of microstates associated with it.

The assumption of equal a prior probability can neither be derived from some other more fundamental principle nor can it be verified by any experiment. The validity of the principle is based on the correctness of the results derived from it.

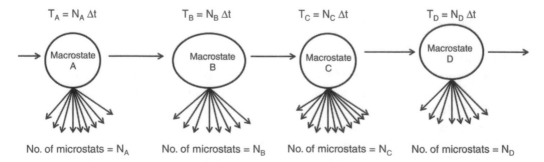

$T_A = N_A \Delta t$ $T_B = N_B \Delta t$ $T_C = N_C \Delta t$ $T_D = N_D \Delta t$

Macrostate A   Macrostate B   Macrostate C   Macrostate D

No. of microstats = N_A No. of microstats = N_B No. of microstats = N_C No. of microstats = N_D

Fig. 10.4   Schematic representation of the movement of a system from one to the other macrostate

Schematic movement of a system from one macrostate to another is shown in Fig. 10.4. Different shapes of macrostates represent the fact that the properties of each macrostate are different from the other. Arrows at the bottom of macrostate shows the number of microstates associated with it. If each microstate lives for a time Δt, the dwell time for a macrostate is equal to the number of its microstates multiplied by the time Δt. The system stays for a longer time in those macrostates which have larger number of microstates. Equilibrium state of a system, where the system stays for infinitely long time, has the largest number of microstates associated with it.

10.3 Quantum Thermodynamic Probability of a Macrostate

Each macrostate of a system is assigned a thermodynamic probability or statistical count. The number of equally likely microstates associated with a given k-th macrostate is called the thermodynamic probability or statistical count of the k-th macrostate and is denoted by W_k. *It may be noted that in quantum thermodynamics statistical count W_k or probability of a macrostate is a number, while generally probability is a ratio or fraction.*

The thermodynamic probability of the system as a whole is equal to the sum of all microstates and is denoted by Ω,

$$\Omega = \sum_k W_k$$

System properties and average occupation number

As shown in Fig. 10.4, the microscopic structure of the system changes almost continuously with time as the system moves through different macrostates. Since macrostates live for short times, it is not possible to measure the physically important system properties in a particular macrostate. Actually whenever some measurement is done the system passes through a large number of macrostates during the process of measurement. Therefore, the measured physical quantity is the average value over a large number of macrostates of the system. A change in the macrostate means change in the occupation numbers of the levels. The measured value of the physical quantity, therefore, depends on the average values of occupation numbers of different energy levels of the system. As such values of occupation numbers for different levels averaged over a large number of macrostates are of paramount importance from the point of experimental determination of system properties.

Quantum statistics provides a method to compute the average occupation number of some level, say level j, of the macrostate k, which is denoted by $\overline{\mathcal{N}_j^k}$. It is quite obvious that the value of $\overline{\mathcal{N}_j^k}$ will depend on the probability W_k of the macrostate k. The value of W_k depends on the nature of the particles of the system. Quantum mechanics classify particles according to the statistical distribution law that is followed by a group of large number of identical particles. There are three different statistical distribution laws, namely, Bose–Einstein, Fermi–Dirac and Maxwell–Boltzmann distribution law, one of which is followed by a group of large number of identical particles. Each distribution law gives a different value of W_k. We first calculate the value of W_k for systems that follow these three distribution laws.

10.3.1   The Bose–Einstein statistical distribution

Bose–Einstein distribution law is applicable when:

1. particles are indistinguishable;
2. any number of particles may be accommodated in a given energy state; and
3. states are distinguishable.

Let us consider the jth level of the system and assume that there are g_j states in this level in which N_j particles are distributed. Since there is no restriction on the number of particles in a state, there may be a large number of ways in which N_j particles may be distributed in g_j states. We will now calculate the number of these possible ways of distribution of particles. In order to make these calculations we designate particles by lower case letters a, b, c, ... As a matter of fact this representation of identical and indistinguishable particles by different distinguishable letters is wrong but it is being done to make calculations simple, further, as you will see the indistinguishable nature of particles will be maintained in calculations. The different states in level j are represented by numbers 1, 2, ..., g_j as shown in Fig. 10.5. One possible distribution of particles in level j may be the one shown in Fig. 10.5, which may be written as,

$$\{(1)ab\}\{(2)\}\{(3)cde\}...\{(g_j)lm\} \qquad 10.7$$

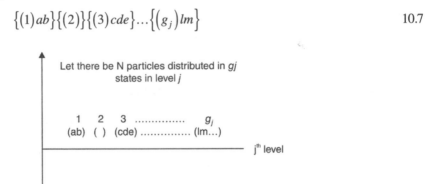

Let there be N particles distributed in *gj* states in level *j*

```
1    2    3   ..............    gⱼ
(ab) ( )  (cde) ..............  (lm...)
```
                                        j^{th} level

Fig. 10.5 j^{th} level has g_j states in which *N* particles are distributed

In Eq. 10.7 curly brackets represent states and lower case letters the particles. The sequence shown in Eq. 10.7 is made up of g_j numbers (representing states) and N_j particles that mean a total number of elements in the sequence are $(g_j + N_j)$. Any sequence of these $(g_j + N_j)$ elements of the type indicated in Eq. 10.7 give a way of distribution of particles in different states of the level. But there is one condition on the valid sequence that represents particle distribution is that the sequence must start with a number representing the state. A sequence of the type $\{ab(2)\}\{(1)fgl\}...\{(g_j)lm\}$ is not valid as it starts with a letter and not numbers. If the sequence starts with one out of the g_j numbers, the number of remaining elements becomes $[(g_j + N_j) -1]$. The number of different ways in which these remaining elements $[(g_j + N_j) -1]$ may be arranged is factorial $[(g_j + N_j) -1]$, i.e., $[(g_j + N_j) -1]!$. To compute the total number of valid sequences it is required to multiply the number of different ways in which $[(g_j + N_j) -1]$ elements can be arranged (i.e., $[(g_j + N_j) -1]!$) by g_j, any one of which may be the first element of the valid sequence. Thus the total number of different ways in which N particles may be distributed in g_j states is,

$$N^{total} = g_j \left[\left(N_j + g_j - 1 \right)! \right] \qquad 10.8$$

The number N^{total} contains sequences in which arrangements like the following $\{(1)ab\}\{(2)\}\{(3)$ $cde\}...\{(g_j)lm\}$ and $\{(1)ba\}\{(2)\}\{(3)cde\}...\{(g_j)lm\}$, etc., are treated as different sequences. However, the particles are indistinguishable hence such sequences are identical and do not give a new way of particle distribution. We thus observe that in N^{total} some otherwise identical sequences have been counted as different sequences. This has happened because we assigned distinguishable letters a, b, c., etc., to the undistinguishable particles. It is, therefore, required to correct N^{total} for this over counting. The number of particles is N, and the number of different ways in which these particles can be arranged is $N!$. Hence, correction for this over counting may be applied by dividing N^{total} by $N_j!$.

Similarly, sequences like $\{(1)ab\}\{(2)\}\{(3)cde\}...\{(g_j)lm\}$ and $\{(2)\}\{(3)cde\}\{(1)ab\}...\{(g_j)lm\}$ have also been counted as different sequences in N^{total}. But actually these are not two different sequences. It may be noted that states are distinguishable, which means that number 1, 2, 3.. ..., etc., are different but at which location in the sequence a particular number occurs is not important. As shown in the two sequences above the state $\{(1)ab\}$ appears in the first location of the sequence or it appears at any other location does not matter so long the number of particles in the state remains same. As such, the two sequences shown above refer to the same distribution. There are g_j numbers and the possible ways in which they may be arranged are $g_j!$. Correction for this over counting may be applied by dividing N^{total} by $g_j!$

Finally, the corrected number of different ways in which N_j indistinguishable particles may be distributed in g_j states or the number of different distributions for the jth level ω_j is,

$$\omega_j = N^{total}_{corrected} = \frac{N^{total}}{(g_j!)(N_j!)} = \frac{g_j(g_j + N_j - 1)!}{(g_j!)(N_j!)} = \frac{(g_j + N_j - 1)!}{\{(g_j - 1)!\}(N_j!)} \qquad 10.9$$

Before proceeding further let us check the correctness of Eq. 10.9. For simplicity we assume that there are only three particles ($N_j = 3$) and only 3 states ($g_j = 3$); then according to Eq. 10.9, the number of different ways in which these 3 indistinguishable particles may be distributed in 3 states are,

$$\omega(3, 3) = \frac{5!}{(2!)(3!)} = 10$$

These ten different ways of particle distribution are shown in Fig. 10.6, where p- represents a particle.

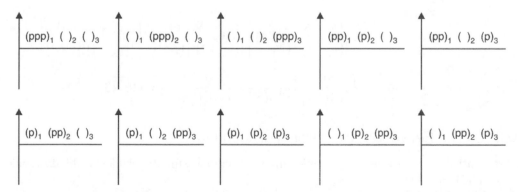

Fig. 10.6 Ten different ways of distributing three undistinguishable particles (p) in three distinguishable states

Application of formula given by Eq. 10.9 for calculating the number of ways of particle distribution to the case of a non-degenerate level needs a mention. For a non-degenerate level $g_j = 1$ and, therefore,

in the denominator of expression $\dfrac{(g_j + N_j - 1)!}{\{(g_j - 1)!\}(N_j!)}$ one gets $\dfrac{(1 + N_j - 1)!}{\{0!\}(N_j!)} = \dfrac{1}{0!}$. Now for a non-

degenerate level there is only one way of distributing indistinguishable particles that is all particles are in the same state. Hence, in order to match formula given by Eq. 10.9 with the experimental fact we should use the convention that $0! = 1$. This convention will also make the formula valid for the

state which is empty and has no particle. For an empty state $N_j = 0$ and $w_{empty} = \dfrac{(0 + g_j - 1)!}{(g_j - 1)!(0!)} = \dfrac{1}{0!}$

$= 1$.

Eq. 10.9 gives the number of possible ways in which particles may be distributed in any level. Suppose in one of the levels particles are distributed in one of the ways given by Eq. 10.9. Then in each of the remaining levels particles may be distributed according to any one of the distribution out of those specified by Eq. 10.9. Therefore, the total number of possible distributions or the thermodynamic probability of any macrostate in Bose–Einstein distribution is given by,

$$W^{Bose-Ein} = \Pi_j \frac{(g_j + N_j - 1)!}{\{(g_j - 1)!\}(N_j!)} \tag{10.10}$$

The symbol $\Pi_j f(j)$ means the forming of products of each term of function $f(j)$ for each value of j.

Suppose there is a system of three energy levels, such that in level-1 there are 2 particles and 2-states, in level-2: 3 particles and 3- states and in level-3: 1 particle and 3 states. In other words the occupation numbers and the folds of degeneracy of level-1, level-2 and level-3 are respectively, $N_1 = 2$, $g_1 = 2$; $N_2 = 3$, $g_2 = 3$ and $N_3 = 1$, $g_3 = 3$. Once the occupation numbers of the levels have been fixed it means that the macrostate is fixed. The thermodynamic probability of this macrostate for Bose–Einstein distribution can be calculated using formula given by Eq. 10.10, as follows,

$$W^{Bose-Ein} = \Pi_j \frac{(g_j + N - 1)!}{\{(g_j - 1)!\}(N!)}$$

$$= \left\{\frac{(g_1 + N_1 - 1)!}{\{(g_1 - 1)!\}(N_1!)}\right\}\left\{\frac{(g_2 + N_2 - 1)!}{\{(g_2 - 1)!\}(N_2!)}\right\}\left\{\frac{(g_3 + N_3 - 1)!}{\{(g_3 - 1)!\}(N_3!)}\right\}$$

$$= \left\{\frac{3!}{(1!)(2!)}\right\}\left\{\frac{5!}{(2!)(3!)}\right\}\left\{\frac{3!}{(2!)(1!)}\right\} = \{3\}\{10\}\{3\} = 90$$

10.3.2   The Fermi–Dirac statistical distribution

Indistinguishable particles that obey Fermi–Dirac statistics follow this distribution. In this distribution it is assumed that:

1. particles are indistinguishable;

2. not more than one particle can be accommodated in one state. This means that the number of particles N_j in j^{th} level cannot be larger than the number g_j of states in level j; and

3. states are distinguishable.

Let there be N_j particles in level j which are distributed in g_j states, where $g_j \geq N_j$. Again we assign distinguishable numbers 1, 2, 3 ..., etc., to distinguishable states in a level and lower case letters a, b, c...., etc., to indistinguishable particles. Correction for choosing distinguishable letters for indistinguishable particles will be applied later on. A possible sequence representing the arrangement of particles in the level j may be

$$\left[1(\ )2(a)3(c)4(\ )...g_j(m)\right]$$

The distribution shown above has no particle in first state, a particle in state-2, a particle in state -3, no particle in state-4.....and one particle in the last state g_j.

The problem of particle distribution may be looked in the following way:

Suppose initially all the g_j states in level-j are empty. We now take one particle (out of the total N_j particles) and put it in one of the states. This first particle may be put in any of the g_j states. That means that for the first particle there are g_j different ways of filling the states. If n_i denotes the number of ways in which the ith particle can be filled, then $n_1 = g_j$. After putting the first particle in any state, the number of empty states left is $(g_j - 1)$. Second particle may now be put in one of the $(g_j - 1)$ states. It means that the number of different ways n_2 in which second particle can be filled in remaining states is $n_2 = (g_j - 1)$. Continuing the same argument, the number of ways in which the third, the fourth and so on up to n^{th} particle filling will be respectively, $n_3 = (g_j - 2)$, $n_4 = (g_j - 3)$, ... $n_n = (g_j - n + 1)$. The number of ways in which the last N_j^{th} particle can be filled is $n_{N_j} = \left(g_j - N_j + 1\right)$.

The total number of ways of distributing particles in all states $n^{total} = n_1 \times n_2 \times n_3 \times ... n_{N_j}$

Or
$$n^{total} = g_j \times \left(g_j - 1\right) \times \left(g_j - 2\right) \times ... \left(g_j - N_j + 1\right)$$

$$n^{total} = \frac{g_j!}{\left(g_j - N_j\right)!} \qquad\qquad 10.11$$

Corrections for over-counting of ways of distribution

(i) The particles are indistinguishable and, therefore, sequences like [1(a) 2( ) 3(b)....] and [1 (b) 2 ( ) 3(a).....] and [1(d) 2( ) 3 (f) .... } etc. that have been counted as different sequences in n^{total} basically refer to only one way of distribution. Correction for this over counting may be applied by dividing n^{total} by $N_j!$ which is the number of different combinations of a, b, c, ... N_j.

It may be noted that in present counting of different ways we have not counted [1(a) 2( ) 3(b)...] and [2( ) 1(a) 3(b)......], etc., as different sequences. What has been done is to fix the locations of all energy states and fill them with particles one after another. Hence no correction for this needs to be applied

Therefore,

Number of ways in which N_j indistinguishable particles obeying Fermi–Dirac distribution maybe distributed in g_j states in level j is,

$$\omega_j = \frac{n^{total}}{\{N_j!\}} = \frac{g_j!}{\{N_j!\}(g_j - N_j)!} \qquad 10.12$$

Once again, to test the correctness of Eq. 10.12 we calculate the number of different ways in which three particles can be distributed in three energy states of level j, when particles follow Fermi–Dirac statistics. In this case, $g_j = 3\,and\,N_j = 3$. Substituting these values in Eq. 10.12 one gets,

$$\omega_j = \frac{3!}{3!0!} = 1$$

It may be noticed that if particles obey Bose–Einstein statistics then three particles may be distributed in ten different ways in three energy levels. On the other hand they may have only one way of distribution if they obey Fermi–Dirac statistics.

Now for any one of the ω_j arrangement of particles in a given level, there are ω_j ways of distribution of particles in any other level. Therefore, the number of ways in which fixed number of particles in each level may be distributed in different energy states of each level, that is the thermodynamic probability of a macrostate in Fermi–Dirac statistics is,

$$W^{Fermi-Dirac} = \Pi_j \frac{g_j!}{\{N_j!\}(g_j - N_j)!} \qquad 10.13$$

10.3.3   The Maxwell–Boltzmann statistical distribution

Particles that obey Maxwell–Boltzmann statistics are

1. distinguishable.
2. Any number of particles can be accommodated in an energy state.
3. Energy states are also distinguishable.

When particles are distinguishable, the number of different ways of distributing particles in different levels and in states of the same level becomes very large. It is because the same number of particles in a given state may be put in many different ways if particles are distinguishable while there is only one way of putting a given number of particles in a given state if particles are indistinguishable. For example if in state-1, which we denote by number 1, there are three particles that are indistinguishable then sequences 1(abc), 1(bcd), 1(mna), 1(cba), etc., are all equivalent because there is no way to make a distinction between a, b, c, ... m, n, etc. However, if particles are distinguishable, which means that a, b, c, d, m, n, ..., are all different then first three sequences are different, and each of them corresponds to a new way of particle distribution. However, sequences 1(abc), and 1(cba) are equivalent both when particles are distinguishable or not, because ordering of particles in a state does not create a new microstate.

Another reason for the large number of ways in which distinguishable particles may be distributed in different levels and states of a level is the fact that if a particle say 'm' moves from an energy level j to another level k, and a particle 'n' from level k goes to level j, a new micro state is created even if the number of particles (occupation numbers) in levels j and k remains same. On the other hand no new microstate is created if the occupation numbers of levels do not change when particles are indistinguishable.

Suppose there are N distinguishable particles in all and they are to be distributed say, in levels j_1, j_2, j_3, ... such that level j_1 has N_1 particles; level j_2, N_2 particles; level j_3, N_3 particles, and so on. The total number of particles in all levels is N, so that $N_1 + N_2 + N_3 + ... = N$. We now calculate the number of different ways in which this distribution can be done. We mark N locations and try to place N distinguishable particles in these locations one after another and count in how many different ways this can be done. Suppose we take a particular particle and place it in one of the N locations. The first particle can be placed in N different ways in these N locations. The second particle will have now $(N-1)$ locations and can be placed in one of these available locations in $(N-1)$ different ways. The third particle will have only $(N-2)$ vacant locations and can be placed in one of these locations in $(N-2)$ different ways. The 4th, the 5th and so on successive particles may be put, respectively, in $(N-3)$, $(N-4)$, $(N-5)$,..., different ways. The last particle will have only one vacant place and can be but only in 1 way. Therefore, total number of different ways in which these N particles may be placed in N locations is $N (N-1) (N-2) (N-3) ... 1 = N!$. Since each arrangement of N particles makes a sequence, the number of different sequences of N particles is $N!$.

Next suppose that out of these N particles placed in some sequence, the first N_1 particles go to fill states in level j_1, next N_2 particles in level j_2, and next N_3 particles in level j_3 and so on. We have calculated that the number of different possible sequences of N different particles is $N!$, and if each sequence is broken down into groups of $N_1, N_2, N_3, ...$ Particles, the number of different sequences for each group of particles will also be $N!$. Now in $N!$ different sequences for each group of particles, some sequences will be those in which same particles will be placed at different locations. For example, if $N_1 = 5$, then sequences of the type (i) a c d b e, (ii) c a d b e (iii) b a d e c ....., etc., though counted as different sequences in the number $N!$, are not different as ordering of distinguishable particles does not make a new sequence. Therefore, for each group of $N_1, N_2, N_3, ...$ particles over or excessive counting of different sequences have occurred. In order to apply correction for this over counting, let us calculate the number of different sequences that can be made by putting same particles at different positions. For example, in the case $N_1 = 5$, letters a b c d e may be arranged in $5! = 120$ different sequences, only three of which are shown above. If the group of particles consists of N_1 particles, the number of different sequences of same particles will be $N_1!$. Similarly group of N_2 particles will have $N_2!$ Identical sequences and so on. Correction for excess counting may, therefore, be applied by dividing $N!$ by $(N_1! \times N_2! \times N_3! \times ...)$.

Hence, the number of different ways in which $N_1, N_2, N_3....$ particles out of total N distinguishable particles, may be filled, respectively, in levels $j_1, j_2, j_3,$ ....without repeating the sequence of same particles is given by,

$$N^{level} = \frac{N!}{\left[(N_1!)(N_2!)(N_3!).....\right]} = \frac{N!}{\Pi_j N_j!} \qquad 10.14$$

Next we calculate the number of different ways in which N_j distinguishable particles in level-1 may be distributed in g_1 energy states. If we assume that each energy state is like a box, then we have g_1 distinguishable or different boxes and N_1 distinguishable particles. At random we take one particle and put it in one of the box, this may be done in g_1 different ways. Next we take any other particle, this particle can also be put in box-1, or box-2 or in any one of the g_1 boxes. Since there is no restriction on the number of particles in any state, including the state in which first particle has already been put; the second particle also has g_1 different ways of filling the energy states. As a matter of fact each of N_1 particles has g_1 different ways of filling the energy states. Therefore, the total number of ways in which N_1 particles may be filled in g_1 states in level-1 is given by,

$$N^{l-1} = (g_1) \times (g_1) \times (g_1) \times \dots N_1 \text{ terms } = g_1^{N_1}$$

Similarly, the number of different ways in which N_2 particles in level-2 may be distributed in g_2 states is $N^{l-2} = g_2^{N_2}$, and so on. Therefore, the total number of different ways in which particles may be distributed in different states of different levels, N^{state} is given by,

$$N^{state} = N^{l-1} \times N^{l-2} \times N^{l-3} \dots = g_1^{N_1} \times g_2^{N_2} \times g_3^{N_3} \dots = \Pi_j g_j^{N_j} \quad 10.15$$

The thermodynamic probability $W^{Maxwell-Boltzmann}$ (which is equal to the total number of microstates) of a macrostate in Maxwell–Boltzmann distribution may be obtained by multiplying N^{level} with N^{state}.

$$W^{Maxwell-Boltzmann} = \left(\frac{N!}{\Pi_j N_j!} \right) \left(\Pi_j g_j^{N_j} \right) = N! \Pi_j \frac{g_j^{N_j}}{N_j!} \quad 10.16$$

As an example, let us calculate the thermodynamic probability of a macrostate for a system that obeys Maxwell–Boltzmann statistics and in which five distinguishable particles are distributed in two levels with $N_1 = 3$, $g_1 = 3$ and $N_2 = 2$, $g_2 = 4$.

Using Eq. 10.16 one gets,

$$W^{Maxwell-Boltzmann} = 5! \left[\frac{3^3 \times 2^4}{3! \times 2!} \right] = 1440$$

10.4   Relation between Entropy and Thermodynamic Probability

Suppose there are two independent systems A and B with entropies S_A and S_B. Let the thermodynamic probabilities of the two systems be Ω_A and Ω_B. It is known that entropies are additive, therefore, the total entropy S_{total} of the two systems put together is

$$S_{total} = S_A + S_B \quad 10.17$$

The thermodynamic probabilities, Ω_A and Ω_B, of the two systems give, respectively, the total number of microstates of systems A and B. Since for each microstate of system A, there will be Ω_B number of microstates of system B, the total number of microstates Ω_{total} of the two system put together will be

$$\Omega_{total} = \Omega_A \times \Omega_B \quad 10.18$$

It may be noticed that while entropies are additive, thermodynamic probabilities are multiplicative. As such, there cannot be one to one correspondence between entropy S and thermodynamic probability Ω. That means,

$$S_A \neq \text{constant} \times \Omega_A \text{ and } S_B \neq \text{constant} \times \Omega_B$$

However, it is possible that entropy S of a system is some function of thermodynamic probability Ω of the system.

Let us assume that

$$S = f(\Omega)$$ 10.19

Where, f represents some function. Our task is to explore the nature of function f.

It follows from Eqs. 10.17 and 10.18, that

$$f(\Omega_A) + f(\Omega_B) = f(\Omega_A\Omega_B)$$ 10.20

We differentiate Eq. 10.20 with respect to Ω_A to get

$$\frac{df(\Omega_A)}{d\Omega_A} + 0 = \frac{df(\Omega_A\Omega_B)}{d\Omega_A}$$

Or

$$\frac{df(\Omega_A)}{d\Omega_A} = \frac{df(\Omega_A\Omega_B)}{d(\Omega_A\Omega_B)}\frac{d(\Omega_A\Omega_B)}{d\Omega_A} = \Omega_B\frac{df(\Omega_A\Omega_B)}{d(\Omega_A\Omega_B)}$$

Or

$$\frac{df(\Omega_A)}{d\Omega_A} = \Omega_B\frac{df(\Omega_A\Omega_B)}{d(\Omega_A\Omega_B)}$$ 10.21

Similarly, when Eq. 10.20 is differentiated with respect to Ω_B, one gets

$$\frac{df(\Omega_B)}{d\Omega_B} = \Omega_A\frac{df(\Omega_A\Omega_B)}{d(\Omega_A\Omega_B)}$$ 10.22

When Eqs. 10.21 and 10.22 are multiplied, respectively, by Ω_A and Ω_B one gets,

$$\Omega_A\frac{df(\Omega_A)}{d\Omega_A} = \Omega_B\frac{df(\Omega_B)}{d\Omega_B}$$ 10.23

The two sides of Eq. 10.23 contain functions of two independent variables and hence this equation will be true only when the two sides are equal to some constant. Let this constant be denoted by k_B. So that

$$\Omega_A\frac{df(\Omega_A)}{d\Omega_A} = \Omega_B\frac{df(\Omega_B)}{d\Omega_B} = \ldots = \Omega\frac{df(\Omega)}{d\Omega} = k_B$$

Or

$$df(\Omega) = k_B\frac{d\Omega}{\Omega}$$

Or

$$f(\Omega) = k_B \ln \Omega$$

Or

$$S(\Omega) = k_B \ln \Omega$$ 10.24

It may thus be observed that the only function of Ω that may satisfy the requirements that entropies are additive and thermodynamic probabilities are multiplicative is $\ln \Omega$.

Though Eq. 10.24 in its present form was put by Max Planck but Planck acknowledged that the equation had already been derived by Ludwig Boltzmann (20 Feb, 1844–5 Sept, 1906; he was an Austrian physicist who laid the foundation of statistical mechanics) almost twenty five years earlier in connection with the microscopic motion of the molecules of an ideal gas. The numerical value of constant k_B, called the Boltzmann constant, has been determined by actually matching the value of entropy with ln Ω and has been found to be,

$$k_B = \frac{R\,(Gas\,constant)}{A\,(avogadro's\,number)} = 1.38062 \times 10^{-23}\ \text{J K}^{-1}$$

10.4.1   Entropy as a measure of the order of a system

In quantum thermodynamics the order of a system is determined by the number of its microstates. A perfectly ordered system is one in which the state of every particle of the system is well defined. A perfectly ordered system will have only one microstate and zero entropy (S = ln1 = 0). Disorder in a system increases with the increase in the number of its microstates, which is accompanied with the increase of its entropy. Therefore, from the point of view of quantum thermodynamics, a larger value of entropy of a system means larger number of available microstates of the system and hence more disorder.

10.4.2   Quantum thermodynamic interpretation for the change of entropy of a system

We will consider some cases here, in which the nature of the change in entropy of the system, from the point of view of classical thermodynamics, is known. We will discuss how this entropy change can be explained in the frame work of quantum thermodynamics.

Any change ΔS in entropy is equal to $k_B\,\Delta(\ln \Omega)$. Now, $\Delta(\ln \Omega)$, change in the value of natural logarithm of the number of microstates, depends on two factors (i) the spacing between the levels given by $\epsilon = n^2 \frac{\hbar^2 (2\pi)^2}{8m} V^{-\frac{2}{3}}$, that varies inversely with the volume V of the system and (ii) the number of energy levels available to the system, which depends on level spacing ϵ and the total energy $E (= \sum_j \epsilon_j N_j)$ of the system. The maximum number of levels that may be available to a system of total energy E and level spacing ϵ is approximately $\approx E/\epsilon$.

(a) Heat flow in a system of non-interacting particles at constant volume

According to quantum thermodynamics if energy E of the system of indistinguishable and non-interacting particles is increased keeping its volume constant, level spacing does not change but larger number of energy levels become available to the system because of the increase of energy E. Hence the number of microstates Ω of the system increases, which in turn increases the value of (ln Ω) and the entropy. This happens when heat Q is supplied to an ideal gas at constant temperature T and volume V. It is known from classical thermodynamics that entropy of the system increases by an amount $\frac{Q}{T}$ and since no PdV work is done (as volume remains constant $dV = 0$), Q goes in increasing the internal energy of the system. It may thus be observed that quantum mechanical approach can correctly predict the nature of change in entropy in the above case.

(b) Free expansion of a system of non-interacting particles

Next let us consider the case when volume of a system of indistinguishable and non-interacting particles is increased keeping the total energy E of the system constant. In this case, from the point of view of quantum thermodynamics, the level spacing will decrease. Smaller level spacing will generate larger number of levels for the same energy E. When larger number of energy levels is available to the system, the number of possible ways of distributing particles in levels will increase which means increase in quantum thermodynamic probability Ω. Therefore, from the point of view of quantum thermodynamics entropy of the system should increase. Increase in volume at constant energy corresponds to free expansion of an ideal gas. It is know from classical thermodynamics that the entropy of the ideal gas increases in free expansion, as predicted by quantum thermodynamics.

(c) Adiabatic expansion of a system of non-interacting particles

In case of the adiabatic expansion of an ideal gas against constant pressure, work is done by the gas against the external pressure which reduces the energy of the system from some initial value E to E' (E > E'). On the other hand, from quantum point of view, an increase in the volume of the system reduces level spacing from ϵ to a smaller value ϵ'. However, the expected increase in the number of available energy levels due to the decrease in the level spacing is exactly counter balanced by the decrease in the number of available energy levels due to the reduction in energy from E to E'. As a result there is no net change in the thermodynamic probability Ω. The entropy of the system of non-interacting particles (ideal gas), therefore, does not change in adiabatic processes. The same result for adiabatic changes is obtained from the classical thermodynamics.

(d) Isothermal expansion of a system of non-interacting particles

In isothermal expansion the temperature of the system is kept constant by supplying heat from a reservoir which increases the energy of the system. However, the system performs work in expanding against constant external pressure and consumes a part of the energy supplied by the heat source. But in the overall process there is a net increase in the energy and in the volume of the system. With the increase of energy the number of available energy levels to the system increases which in turn increases the number of microstates. Increase in volume of the system decreases the level spacing. Therefore, both the increase in energy and the increase in volume results in larger number of available energy levels and larger number of microstates of the system. An increase in the number of microstates means increase in entropy of the system. The same result is obtained when the process is considered in the frame work of classical thermodynamics.

Many more examples may be given to show that quantum thermodynamics can satisfactorily explain the change in entropy of a system in terms of the change in the value of quantum thermodynamic probability Ω of the system.

10.5   The Distribution Function

It has already been discussed that the average occupation numbers play an important role so far as the system observables are concerned. Since even very small physical systems at room temperature contain very large number of particles and available energy levels, it is almost impossible to calculate average occupation numbers by counting levels and calculating possible ways of particle distributions. The average occupation numbers are, therefore, determined theoretically, using the distribution laws of quantum thermodynamics and expressions for entropy change are borrowed from classical thermodynamics. The expression for average occupation number per energy state $\dfrac{W}{g}$ is called the distribution function. Distribution functions for different statistical distributions will be derived in the following.

The basic technique of obtaining the distribution function for a given statistical distribution is to find out the ratio of the thermodynamic probabilities for two systems that are very nearly identical except that in one of them the number of particles in one of the levels is slightly different from the other. Natural logarithm of the ratio of thermodynamic probabilities of the two systems gives the difference in entropies of the two systems, which may be related to the chemical potential using expressions of classical thermodynamics. Assuming that the average occupation numbers in the two systems are same, an expression for the distribution function in terms of the chemical potential of the system may be obtained. The following derivations will further clarify the procedure.

10.5.1   Distribution function for Bose–Einstein distribution

Let there be two systems A and B, each of indistinguishable particles which obey Bose–Einstein statistics, having slightly different energies, E_A and E_B, respectively. The systems A and B are almost identical, having same number of levels except that in system B the number of particles in some, say r^{th} level is one less than in the rth level of system A. In all other levels, except the rth level, the distribution of particles is identical in both systems. Therefore,

$$E_B = E_A - \epsilon_r \text{ and } N_B = N_A - 1 \qquad\qquad 10.25$$

Here, E_A, E_B, N_A, N_B, and ϵ_r are, respectively, the energies, number of particles and the energy of rth level in systems A and B. The quantum thermodynamic probability of some macrostate k in the two systems may be written as,

$$W_k^A = \Pi_i \frac{\left(g_j + N_{jk}^A - 1\right)!}{\left(g_j - 1\right)! N_{jk}^A!} \qquad\qquad 10.26$$

$$W_k^B = \Pi_i \frac{\left(g_j + N_{jk}^B - 1\right)!}{\left(g_j - 1\right)! N_{jk}^B!} \qquad\qquad 10.27$$

In Eqs. 10.26 and 10.27 N_{jk}^A and N_{jk}^B are, respectively, the number of particles in in j^{th} level of kth macrostate of systems A and B.

Dividing Eq. 10.27 by Eq. 10.26 one gets,

$$\frac{W_k^B}{W_k^A} = \Pi_j \frac{N_{jk}^A! \left(g_j + N_{jk}^B - 1\right)!}{N_{jk}^B! \left(g_j + N_{jk}^A - 1\right)!}$$

$$= \frac{\left(N_1^A! N_2^A! N_3^A! \ldots N_r^{A!} \ldots\right)\left[\left(g_1 + N_{1k}^B - 1\right)!\left(g_2 + N_{2k}^B - 1\right)! \ldots\left(g_r + N_{rk}^B - 1\right)! \ldots\right]}{\left(N_1^B! N_2^B! N_3^B! \ldots N_r^B! \ldots\right)\left[\left(g_1 + N_{1k}^A - 1\right)!\left(g_2 + N_{2k}^A - 1\right)! \ldots\left(g_r + N_{rk}^A - 1\right)! \ldots\right]}$$

Since both in systems A and B number of particles in all other level except the rth level are equal and therefore, all terms except the terms corresponding to the rth level in the two systems cancel out. Thus,

$$\frac{W_k^B}{W_k^A} = \frac{\left(N_r^A\right)!\left[\left(g_r + N_{rk}^B - 1\right)\right]!}{\left(N_r^B\right)!\left[\left(g_r + N_{rk}^A - 1\right)\right]!} = \frac{\left(N_r^A\right)\left\{\left(N_r^A - 1\right)!\right\}\left[\left(g_r + N_{rk}^B - 1\right)!\right]}{\left(N_r^B\right)!\left[\left\{g_r + N_{rk}^B\right\}\left[\left(g_r + N_{rk}^B - 1\right)!\right]\right]} = \frac{\left(N_r^A\right)}{\left\{g_r + N_{rk}^B\right\}}$$

Or
$$\frac{W_k^B}{W_k^A} = \frac{\left(N_r^A\right)}{\left\{g_r + N_{rk}^B\right\}} \qquad 10.28$$

In deriving Eq. 10.28, following facts have been used, $N_r^B = \left(N_r^A - 1\right); (N_r^A! = N_r^A\left(N_r^B!\right)$

Or
$$W_k^B\left\{g_r + N_{rk}^B\right\} = W_k^A\left(N_r^A\right) \qquad 10.29$$

Summing Eq. 10.29 over all macrostates one gets,

$$\Sigma_k W_k^A\left(N_r^A\right) = \Sigma_k W_k^B N_{rk}^B + g_r \Sigma_k W_k^{B\prime} \qquad 10.30$$

Summation in Eq. 10.30 is over all the possible macrostates of systems A and B. If $\overline{N_r^A}$ and $\overline{N_r^B}$ are, respectively, the average values of the number of particles in r^{th} energy levels of different macrostates of systems A and B and Ω_A and Ω_B the thermodynamic probabilities of system A and B, then Eq. 10.30 may be written as

$$\overline{N_r^A}\,\Omega_A = \overline{N_r^B}\,\Omega_B + g_r\Omega_B \qquad 10.31$$

Since the number of particles in systems A and B differ only by one, $\overline{N_r^A} \approx \overline{N_r^B} = \overline{N_r}$. Making this substitution in Eq. 10.31 one gets,

Or
$$\frac{\Omega_B}{\Omega_A} = \frac{\overline{N_r}}{\overline{N_r} + g_r} \qquad 10.32$$

Taking log of the two sides of Eq. 10.32

$$\frac{k_B}{k_B}\left(\ln\Omega_B - \ln\Omega_A\right) = \ln\frac{\overline{N_r}}{\overline{N_r} + g_r}$$

Or
$$\frac{S_B - S_A}{k_B} = \frac{\Delta S}{k_B} = \ln\frac{\overline{N_r}}{\overline{N_r} + g_r} \qquad 10.33$$

Here ΔS is the difference between the entropies of system B and system A and k_B Boltzmann constant.

However, it is known from classical thermodynamics that the difference in entropies of two systems that are made of same type of non-interacting particles at same temperature T but have different number of particles N_1 and N_2 and energies E_1 and E_2 is given by,

$$T\Delta S = (E_1 - E_2) - \mu(N_1 - N_2)$$

Here we have $\left(E_1 - E_2\right) = \left(E_B - E_A\right) = -\epsilon_r$ and $\left(N_1 - N_2\right) = \left(N_B - N_A\right) = -1$

Therefore, $T\left(k_B \ln \dfrac{\overline{N_r}}{\overline{N_r} + g_r}\right) = -\epsilon_r - \mu(-1) = \mu - \epsilon_r$

Or $\ln \dfrac{\overline{N_r}}{\overline{N_r} + g_r} = \dfrac{\mu - \epsilon_r}{Tk_B}$

Or $\dfrac{\overline{N_r}}{\overline{N_r} + g_r} = e^{\frac{\mu - \epsilon_r}{Tk_B}}$

Or $\dfrac{\overline{N_r} + g_r}{\overline{N_r}} = e^{-\left(\frac{\mu - \epsilon_r}{Tk_B}\right)}$

Or $1 + \dfrac{g_r}{\overline{N_r}} = e^{-\left(\frac{\mu - \epsilon_r}{Tk_B}\right)}$

Or $\dfrac{g_r}{\overline{N_r}} = \left[e^{-\left(\frac{\mu - \epsilon_r}{Tk_B}\right)} - 1\right]$

Or $\dfrac{\overline{N_r}}{g_r} = \dfrac{1}{\left[e^{-\left(\frac{\mu - \epsilon_r}{Tk_B}\right)} - 1\right]}$

Or $\dfrac{\overline{N_j}}{g_j} = \dfrac{1}{\left[e^{-\left(\frac{\mu - \epsilon_j}{Tk_B}\right)} - 1\right]}$ 10.34

Equation 10.34 gives the Bose–Einstein distribution function. It may be realized that r^{th} level may be any level and, therefore, the subscript r can be j, or I or m or any other level.

10.5.2   Distribution function for Fermi–Dirac distribution

We follow exactly the same method as was used to find the distribution function for Bose–Einstein distribution for obtaining the distribution function of Fermi–Dirac distribution. We choose two systems A and B both containing non-interacting indistinguishable identical particles that obey Fermi–Dirac statistics. The two systems are almost identical except that in some level, say in m^{th} level the number of particles N_m^B in system B is one less than particles N_m^A in the mth level of system A. Except the mth level, the numbers of particles in all other levels of the two systems are same. If in both the systems the energy of level-m is denoted by ϵ_m and the energies of the two systems by U_A and U_B, then

$$U_A - \epsilon_m = U_B \text{ and } N_m^B = N_m^A - 1$$ 10.35

The quantum thermodynamic probabilities for some macrostate denoted by k for systems A and B are given by,

$$W_k^A = \Pi_j \frac{g_j^A!}{\left\{N_{jk}^A!\right\}\left\{g_j^A - N_{jk}^A\right\}!} \text{ and } W_k^B = \Pi_j \frac{g_j^B!}{\left\{N_{jk}^B!\right\}\left\{g_j^B - N_{jk}^B\right\}!}$$

Now in the two systems A and B the numbers of states in each level, including the mth level, are same, therefore, we drop the superscripts A and B and write $g_j^A = g_j^B = g_j$.

$$\frac{W_k^B}{W_k^A} = \Pi_j \frac{g_j!\left\{N_{jk}^A!\right\}\left\{g_j - N_{jk}^A\right\}!}{g_j!\left\{N_{jk}^B!\right\}\left\{g_j - N_{jk}^B\right\}!}$$

$$= \frac{\left(N_{mk}^A!\right)\left\{g_j - N_{mk}^A\right\}!}{\left(N_{mk}^B!\right)\left\{g_j - N_{mk}^B\right\}!}$$

$$= \frac{\left(N_{mk}^A\right)\left(N_{mk}^B\right)!\left\{g_j - N_{mk}^A\right\}!}{\left(N_{mk}^B!\right)\left(g_j - N_{mk}^A + 1\right)\left\{g_j - N_{mk}^A\right\}!} \qquad 10.36$$

Equation 10.36 is obtained using the relations $\left\{N_{mk}^A!\right\} = \left(N_{mk}^A\right)\left(N_{mk}^A - 1\right)! = \left(N_{mk}^A\right)\left(N_{mk}^B!\right)$ and

$\left\{g_j - N_{mk}^B\right\}! = \left(g_j - (N_{mk}^A - 1)\right)! = \left\{g_j - N_{mk}^A + 1\right\}! = \left(g_m - N_{mk}^A + 1\right)\left\{g_j - N_{mk}^A\right\}!$.

Cancelling the common terms in the enumerator and denominator of Eq. 10.36, one gets

$$\frac{W_k^B}{W_k^A} = \frac{\left(N_{mk}^A\right)}{\left(g_m - N_{mk}^A + 1\right)} = \frac{\left(N_{mk}^A\right)}{\left[g_m - \left(N_{mk}^A - 1\right)\right]} = \frac{\left(N_{mk}^A\right)}{\left(g_m - N_{mk}^B\right)}$$

Or
$$\left(g_m - N_{mk}^B\right)W_k^B = \left(N_{mk}^A\right)W_k^A \qquad 10.37$$

Summing Eq. 10.37 over all macrostates of the two systems,

$$\Sigma_k\left(g_m - N_{mk}^B\right)W_k^B = \Sigma_k\left(N_{mk}^A\right)W_k^A$$

Or
$$g_m\Sigma_k W_k^B - \Sigma_k N_{mk}^B W_k^B = \Sigma_k\left(N_{mk}^A\right)W_k^A$$

Or
$$g_m\Omega_B - \overline{N_m^B}\Omega_B = \overline{N_m^A}\Omega_A \qquad 10.38$$

Here $\overline{N_m^A}$ and $\overline{N_m^B}$ are the average value of the number of particles in m^{th} levels of different macrostates, respectively, for system A and B. Since the number of particles in any of the mth level of system B is one less than the corresponding level of system A, the average values are nearly equal and dropping the superscripts A and B, we write $\overline{N_m^A} = \overline{N_m^B} = \overline{N_m}$. Making this substitution Eq. 10.38 gives,

$$\frac{\Omega_B}{\Omega_A} = \frac{\overline{N_m}}{\left(g_m - \overline{N_m}\right)} \qquad 10.39$$

Taking log of both sides of Eq. 10.39 one gets

$$\frac{k_B \ln \Omega_B - k_B \ln \Omega_A}{k_B} = \ln\left(\frac{\overline{N_m}}{\left(g_m - \overline{N_m}\right)}\right)$$

Or

$$S_B - S_A = \Delta S = k_B \ln\left(\frac{\overline{N_m}}{\left(g_m - \overline{N_m}\right)}\right) \qquad 10.40$$

But from classical thermodynamics, $T\Delta S = \Delta U - \mu \Delta N$, where $\Delta U = -\epsilon_m$ and $\Delta N = -1$. Putting these values in Eq. 10.40, one gets

$$Tk_B \ln\left(\frac{\overline{N_m}}{\left(g_m - \overline{N_m}\right)}\right) = -\epsilon_m - \mu(-1) = \mu - \epsilon_m$$

Or

$$\left(\frac{\overline{N_m}}{\left(g_m - \overline{N_m}\right)}\right) = e^{\frac{(\mu - \epsilon_m)}{k_B T}}$$

Or

$$\frac{g_m}{\overline{N_m}} - 1 = e^{-\frac{(\mu - \epsilon_m)}{k_B T}}$$

And

$$\frac{\overline{N_m}}{g_m} = \frac{1}{1 + e^{\frac{(\epsilon_m - \mu)}{k_B T}}}$$

Since m^{th} state may be any state and therefore, the above equation may be written in more general way for any level j as,

$$\frac{\overline{N_j}}{g_j} = \frac{1}{1 + e^{\frac{(\epsilon_j - \mu)}{k_B T}}} \qquad 10.41$$

10.5.3 Distribution function for classical distribution

If in a system of indistinguishable particles the number of particles per energy state is very small, the denominator of Eq. 10.41 is obviously very large. Then 1 may be neglected in comparison to $e^{\frac{(\epsilon_j - \mu)}{k_B T}}$, the resulting distribution is called classical distribution and the distribution function for classical distribution is given by,

$$\frac{\overline{N_j}}{g_j} = \frac{1}{e^{\frac{(\epsilon_j - \mu)}{k_B T}}} = e^{\frac{(\mu - \epsilon_j)}{k_B T}} \qquad 10.42$$

10.5.3.1 *The partition function for classical distribution*

Equation 10.42 may be written as

$$\overline{N_j} = g_j e^{\frac{\mu}{k_B T}} e^{-\frac{\epsilon_j}{k_B T}} \qquad 10.42(a)$$

And if it is summed over j, the left hand side will give the total number of particles N. So,

$$N = \sum_j \overline{N_j} = e^{\frac{\mu}{k_B T}} \sum_j g_j e^{-\frac{\epsilon_j}{k_B T}} \qquad 10.43$$

The quantity $\sum_j g_j e^{-\frac{\epsilon_j}{k_B T}}$ is called the partition function and denoted by Z, i.e.;

$$Z \equiv \sum_j g_j e^{-\frac{\epsilon_j}{k_B T}} \qquad 10.44$$

Equation 10.43 may be written as,

$N = Z e^{\frac{\mu}{k_B T}}$ and $e^{\frac{\mu}{k_B T}} = \dfrac{N}{Z}$ Substituting this value in Eq. 10.42(a) one gets,

$$\overline{N_j} = g_j e^{\frac{\mu}{k_B T}} e^{-\frac{\epsilon_j}{k_B T}} = g_j \frac{N}{Z} e^{-\left(\frac{\epsilon_j}{T k_B}\right)}$$

Or

$$\frac{\overline{N_j}}{g_j} = \frac{N}{Z} e^{-\left(\frac{\epsilon_j}{T k_B}\right)} \qquad 10.45$$

Equation 10.45 tells that the average number of particles per state in every level exponentially decreases with the energy ϵ_j of the level, for a system of indistinguishable particles obeying classical statistics. Further, the decrease of the number of particles per state is faster at low temperature. The same result will be found in case of the systems that follow Maxwell Boltzmann distribution, though particles are distinguishable in this distribution.

10.5.4   Distribution function for Maxwell–Boltzmann distribution

Following the standard method, we write down the ratio of the thermodynamic probabilities for some macrostate k in two almost identical systems A and B, except that in system B the number of particles in some level m, N_m^B is one less than N_m^A, the number of particles in the corresponding level in system A. Both systems obey Maxwell–Boltzmann statistics: particles are distinguishable and any number of particles may be accommodated in any state. Hence, the ratio of the thermodynamic probability of kth macrostate in the two systems may be given as,

$$\frac{W_k^B}{W_k^A} = \frac{(N-1)! \Pi_j \dfrac{g_j^{N_{jk}^B}}{N_{jk}^B!}}{N! \Pi_j \dfrac{g_j^{N_{jk}^A}}{N_{jk}^A!}} = \frac{1}{N} \Pi_j \frac{g_j^{N_{jk}^B}}{N_{jk}^B!} \frac{N_{jk}^A!}{g_j^{N_{jk}^A}}$$

$$= \frac{1}{N} \frac{g_m^{N_{mk}^B}}{g_m^{N_{mk}^A}} \frac{N_{mk}^A \left(N_{mk}^A - 1\right)!}{N_{mk}^B!}$$

$$= \frac{1}{N} \frac{1}{g_m^{\left(N_{mk}^A - N_{mk}^B\right)}} \frac{N_{mk}^A}{1} = \frac{1}{N} \frac{N_{mk}^A}{g_m} \qquad 10.46$$

Here we have used the facts: total number of particles in B is one less than in A, and $(N_{mk}^A - 1) = N_{mk}^B$, where subscript mk indicates the mth level in k^{th} macrostate.

Hence from Eq. 10.46 one gets,

$$Ng_m W_k^B = N_{mk}^A W_k^A$$

Summing over all macrostates, above equation gives,

$$Ng_m \Omega_B = \overline{N_m} \Omega_A$$

Or

$$\frac{\Omega_B}{\Omega_A} = \frac{\overline{N_m}}{Ng_m}$$

Taking log of both sides of the above equation gives,

$$k_B \left[\ln \Omega_B - \ln \Omega_A \right] = k_B \ln \frac{\overline{N_m}}{Ng_m}$$

Or

$$S_B - S_A = \Delta S = k_B \ln \frac{\overline{N_m}}{Ng_m} \qquad\qquad 10.47$$

Using the relation $T \Delta S = \Delta U - \mu \Delta N$ and $\Delta U = -\varepsilon_m$ and $\Delta N = -1$, one gets

$$Tk_B \ln \frac{\overline{N_m}}{Ng_m} = -\varepsilon_m + \mu$$

Or

$$\frac{\overline{N_m}}{g_m} = Ne^{\left(\frac{\mu - \varepsilon_m}{Tk_B} \right)}$$

Above equation holds for any level of the system and generalizing it one may write

$$\frac{\overline{N_j}}{g_j} = Ne^{\left(\frac{\mu - \varepsilon_j}{Tk_B} \right)} \qquad\qquad 10.48$$

Equation 10.48 gives the desired distribution function.

10.5.4.1 *Partition function for Maxwell–Boltzmann distribution*

Equation 10.48 may be written as,

$$\overline{N_j} = Ng_j e^{\frac{\mu}{Tk_B}} e^{-\frac{e_j}{Tk_B}} \qquad\qquad 10.48(a)$$

Summing the above equation over all j values one gets,

$$N = \sum_j \overline{N_j} = Ne^{\frac{\mu}{Tk_B}} \sum_j g_j e^{-\frac{e_j}{Tk_B}} \qquad\qquad 10.49$$

We define the partition function denoted by Z as,

$$Z \equiv \sum_j g_j e^{-\frac{e_j}{Tk_B}} \qquad\qquad 10.50$$

The partition function depends on temperature T and on the parameters that determines the value

of the energy $e_j \left(= n_j^2 \dfrac{\hbar^2 (2\pi)^2}{8m} V^{-\frac{2}{3}} \right)$. Z is, therefore, a function of thermodynamic parameters of

the system (*T, V,* and *P*) and non-thermodynamic parameter like the mass of the particles. Replacing

$\sum_j g_j e^{-\frac{e_j}{Tk_B}}$ by Z in Eq. 10.50 one gets,

$$e^{\frac{\mu}{Tk_B}} = \frac{1}{Z}$$ 10.51

Using Eq. 10.51, Eq. 10.48(a) becomes,

$$\overline{N_j} = \frac{N}{Z} g_j e^{-\frac{e_j}{Tk_B}}$$

Or

$$\frac{\overline{N_j}}{g_j} = \frac{N}{Z} e^{\frac{-e_j}{Tk_B}}$$ 10.52

It is clear from Eqs. 10.45 and 10.52 that distribution functions for Classical distribution and Maxwell–Boltzmann distribution are identical if written in terms of the partition function.

10.6 Significance of Partition Function: A Bridge from Quantum to Classical Thermodynamics

Partition function describes the statistical properties of a system in thermodynamic equilibrium. It is a function of thermodynamic parameters such as the volume (*V*), the temperature (*T*), number of particles (*N*) of the system and the mass of the particles. Most of the aggregate variables of the system important from the point of view of classical thermodynamics, such as the total energy, entropy, Helmholtz function, Gibbs function, heat capacity at constant volume and pressure etc., can be expressed in terms of the partition function or its derivatives. Partition function, therefore, provides a bridge between the quantum and classical thermodynamics. Generally, the partition function is represented by the letter Z which is taken from the German language word 'zustands-summe', however, in some texts it is also represented by letter Q. In principle the partition function for a given system may be derived from the properties of the system i.e., from the distribution law followed by the system entities. In practice, however, it is easy to obtain partition function for some distributions like the classical distribution and Maxwell–Boltzmann distributions but in most other distributions like the Bose–Einstein and Fermi–Dirac distributions it is quite involved. Researchers working in the field of statistical mechanics devote a considerable time in obtaining partition functions for complicated systems.

10.6.1 Obtaining classical thermodynamic parameters for a system obeying Maxwell–Boltzmann distribution using partition function

The distribution function for both the Classical and the Maxwell–Boltzmann distributions is given as,

$$\frac{\overline{N_j}}{g_j} = \frac{N}{Z} e^{\frac{-e_j}{Tk_B}}$$

Or
$$\frac{\overline{N_j}}{N} = p_i = g_j \frac{e^{-\beta\epsilon_j}}{Z} \qquad 10.53$$

The left hand side of Eq. 10.53 gives the probability p_i of finding a particle with energy ϵ_i at temperature T. $k_B = \frac{R}{A_v}$ is the Boltzmann constant, R gas constant and A_v the Avogadro's number.

Further, we have substituted

$$\beta = \frac{1}{Tk_B} \qquad 10.54$$

The total energy of the system $U = N \times$ average energy of the particles given by,

$$U = N\Sigma_i p_i\epsilon_i = N\Sigma_i\epsilon_i g_j \frac{e^{-\beta\epsilon_j}}{Z}$$

$$= -\frac{N}{Z}\frac{\partial\Sigma_i g_j e^{-\beta\epsilon_i}}{\partial\beta} = -\frac{N}{Z}\frac{\partial Z}{\partial\beta} = -N\frac{\partial(\ln Z)}{\partial\beta}$$

Or
$$U = -N\frac{\partial(\ln Z)}{\partial\beta} \qquad 10.55$$

Also
$$U = -N\frac{\partial(\ln Z)}{\partial T}\frac{\partial T}{\partial\beta} \qquad 10.56$$

But $\beta = \frac{1}{Tk_B}$, therefore, $\frac{\partial\beta}{\partial T} = -\frac{1}{T^2 k_B}$

And $\frac{\partial T}{\partial\beta} = -T^2 k_B$, substituting back this value of $\frac{\partial T}{\partial\beta}$ in Eq. 10.56 gives,

$$U = -N\frac{\partial(\ln Z)}{\partial\beta} = NT^2 k_B \frac{\partial(\ln Z)}{\partial T} \qquad 10.57$$

Further, the heat capacity at constant volume C_V is given by

$$C_V = \left(\frac{\partial U}{\partial T}\right)_V = N\left(\frac{\partial\left\{T^2 k_B \frac{\partial(\ln Z)}{\partial T}\right\}}{\partial T}\right)_V$$

Or
$$C_V = 2NTk_B\left(\frac{\partial\ln Z}{\partial T}\right)_v + T^2 k_B N\left(\frac{\partial^2\ln Z}{\partial T^2}\right)_V \qquad 10.58$$

If the entropy of a system is taken to be zero at absolute zero of temperature, as per the third law of thermodynamics, the entropy at any other temperature T may be given as,

$$S(T) = 0 + \int_0^T \frac{C_V}{T} dT = \int_0^T \left[2k_B N \left(\frac{\partial \ln Z}{\partial T} \right)_v + NTk_B \left(\frac{\partial^2 \ln Z}{\partial T^2} \right)_v \right] dT$$

Or

$$S(T) = \left[2k_B N \ln Z \right]_0^T + \left[Tk_B \frac{\partial \ln Z}{\partial T} - k_B \int_0^T \frac{\partial \ln Z}{\partial T} \right]_0^T$$

Or

$$S(T) = \left[2k_B \ln Z \right]_0^T + N \left[Tk_B \frac{\partial \ln Z}{\partial T} - k_B \ln Z \right]_0^T$$

Or

$$S(T) = Nk_B \ln Z + NTk_B \frac{\partial \ln Z}{\partial T} \qquad 10.59$$

However, from Eq. 10.57 N $Tk_B \dfrac{\partial \ln Z}{\partial T} = \dfrac{U}{T}$, putting this value in Eq. 10.59 we get,

$$S(T) = Nk_B \ln Z + \frac{U}{T}$$

Or

$$U - TS = - Nk_B T \ln Z \qquad 10.60$$

The Helmholtz function $F = U - TS$

Hence,

$$F = - Nk_B T \ln Z \qquad 10.61$$

Also, Pressure

$$P = - \left(\frac{\partial F}{\partial V} \right)_{T,N} = Nk_B T \left(\frac{\partial \ln Z}{\partial V} \right)_{T,N} \qquad 10.62$$

And Chemical potential

$$\mu = \left(\frac{\partial F}{\partial N} \right)_{V,T} = -k_B T \left(\frac{\partial N \ln Z}{\partial N} \right)_{V,T} = -k_B T \ln Z \qquad 10.63$$

We thus see that if the partition function for a system is known all other state functions of classical thermodynamics for the system may be obtained.

10.6.2 Obtaining classical thermodynamic parameters for a system obeying Classical distribution using partition function

The distribution function for classical distribution is given by Eq. 10.43 as,

$$\frac{\overline{N_j}}{g_j} = e^{\left(\frac{\mu - \epsilon_j}{Tk_B} \right)}$$

And

$$\overline{N_j} = e^{\frac{\mu}{Tk_B}} g_j \, e^{-\frac{\epsilon_j}{Tk_B}} \qquad 10.64$$

Summing both sides of the above equation over j gives,

$$\sum_j \overline{N_j} = e^{\frac{\mu}{Tk_B}} \sum_j g_j e^{-\frac{\epsilon_j}{Tk_B}} \qquad 10.65$$

But $\sum_j \overline{N_j} = N$, the total number of particles in the system, and $\sum_j g_j \, e^{-\frac{\epsilon_j}{Tk_B}} = Z$ for the system obeying classical statistics. Eq. 10.65 may then be written as,

$$e^{\frac{\mu}{Tk_B}} = \frac{N}{Z}$$ 10.66

And the chemical potential for classical distribution may be written as,

$$\mu = -Tk_B[\ln Z - \ln N]$$ 10.67

Now from Eq.7.63 of chapter-7, the chemical potential of a system is related to its Helmholtz function through the relation,

$$\mu = \left(\frac{\partial F}{\partial N}\right)_{V,T}$$ 10.68

Hence, from Eqs. 10.67 and 10.68 one gets for classical distribution,

$$\left(\frac{\partial F}{\partial N}\right)_{V,T} = -Tk_B[\ln Z - \ln N]$$

And $$F = -Tk_B\{[\ln Z - \ln N]dN\} = -Tk_B\{N \ln Z - N \ln N + N\}$$

Or $$F = -NTk_B\{\ln Z - \ln N + 1\}$$ 10.69

But, from Eq.7.29 of chapter-7, that holds for a closed system for which the total number of particles N is constant,

$$P = -\left(\frac{\partial F}{\partial V}\right)_{T,N}$$

Hence for a system that obeys classical distribution

$$P = -\left[\frac{\partial}{\partial V}\{-NTk_B\{\ln Z - \ln N + 1\}\}\right]_{T,N}$$

Or $$P = NTk_B\left(\frac{\partial Z}{\partial V}\right)_{T,N}$$ 10.70

Also, from Eq.7.29 (Chapter-7), the entropy S of a closed system of N particles is given as,

$$-S = \left(\frac{\partial F}{\partial T}\right)_{V,N}$$

For the case of classical distribution, the above equation becomes,

$$S = -\left[\left(\frac{\partial}{\partial T}\{-NTk_B(\ln Z - \ln N + 1)\}\right)_{V,N}\right]$$

$$= Nk_B T\left(\frac{\partial \ln Z}{\partial T}\right)_{V,N} + N k_B(\ln Z - \ln N + 1)$$ 10.71

And,

$$ST = Nk_BT^2\left(\frac{\partial \ln Z}{\partial T}\right)_{V,N} + NTk_B(\ln Z - \ln N + 1)$$

$$= Nk_BT^2\left(\frac{\partial \ln Z}{\partial T}\right)_{V,N} - F$$

Or

$$ST + F = Nk_BT^2\left(\frac{\partial \ln Z}{\partial T}\right)_{V,N} \qquad 10.72$$

But $ST + F = U$, hence for a system obeying classical distribution,

$$U = Nk_BT^2\left(\frac{\partial \ln Z}{\partial T}\right)_{V,N} \qquad 10.73$$

Also,

$$S = \frac{U}{T} + Nk_B(\ln Z - \ln N + 1) \qquad 10.74$$

Thus all important parameters of classical thermodynamics may be obtained if the partition function of the system is known.

Solved Examples

1. In a system of total energy 12ϵ five indistinguishable particles obeying Bose–Einstein statistics are distributed in five equidistant levels of energies 0, ϵ, 2ϵ, 3ϵ, 4ϵ, all levels have four energy states. Calculate the number of macrostates, microstates corresponding to each macrostate, and average occupation number of each level of the system.

   Solution:

   The possible distributions of five particles in five levels subject to the conditions that the sum of their energies is *12ε* are shown in the table given below. Since particles follow Bose–Einstein statistics there is no restriction on the number of particles in a given energy state. Rows in the table show the number of particle in different energy levels. Columns numbered 1 to 9 show nine macrostates of the system. In the table, lower case letter *p* indicates one particle and the number of *p's* gives the number of particles in that level. As shown in the table,

   1-macrostate has 3-particles in level of energy 4ϵ, and 2-particle in level of energy 0. The number of microstates for this macrostate may be calculated using the formula,

   $$W^{Bose-Ein} = \Pi_j \frac{(g_j + N_j - 1)!}{\{(g_j - 1)!\}(N_j!)}$$

   It is also given that there are 4-states in each level, i.e., $g_j = 4$ for all j

   Number of microstates of macrostatye-1 =

   $$\frac{(4 + 3 - 1)!}{(4 - 1)!(3)!} \times \frac{(4 + 2 - 1)!}{(4 - 1)!(2)!} = 20 \times 10 = 200$$

Number of microstates for macrostate-2 =

$$\frac{(4+2-1)!}{(4-1)!(2)!} \times \frac{(4+2-1)!}{(4-1)!(2)!} \times \frac{(4+1-1)!}{(4-1)!(1)!} = 10 \times 10 \times 4 = 400$$

Number of microstates for macrostate-3 =

$$\frac{(4+1-1)!}{(4-1)!(1)!} \times \frac{(4+2-1)!}{(4-1)!(2)!} \times \frac{(4+1-1)!}{(4-1)!(1)!} \times \frac{(4+1-1)!}{(4-1)!(1)!} = 4 \times 10 \times 4 \times 4 = 640$$

Number of microstates for macrostate-4 =

$$\frac{(4+1-1)!}{(4-1)!(1)!} \times \frac{(4+2-1)!}{(4-1)!(2)!} \times \frac{(4+1-1)!}{(4-1)!(1)!} \times \frac{(4+1-1)!}{(4-1)!(1)!} = 4 \times 10 \times 4 \times 4 = 640$$

In this way microstates for all other macrostates have been calculated and are given in the table.

The total number of all microstates = 3080

Next we calculate the mean occupation number for different levels

Level with energy 4ϵ: It has occupation no.3 in 200 microstates of macrostate-1, occupation no.2 in 400 microstates of macrostate-2 , occupation no.2 in 640 microstates of macrostate 3, occupation no. 1 in 640 microstates of macrostate-4, Occupation no.1 in 400 microstates of macrostate-5, and occupation no.1 in 140 microstates of macrostate -6. No particle in this level is held in macrostates 7, 8 and 9.

We multiply the occupation no. with number of microstates in macrostate, sum over all macrostates that have some particle in level of energy 4ϵ and divide the sum by total number of microstates (3080)

Mean occupation no. for level-4

$$\epsilon = \frac{3 \times 200 + 2 \times 400 + 2 \times 640 + 1 \times 640 + 1 \times 400 + 1 \times 140}{3080} = \frac{3860}{3080} = 1.253$$

Mean occupation no. for level-3

$$\epsilon = \frac{1 \times 640 + 2 \times 640 + 2 \times 400 + 4 \times 140 + 3 \times 320 + 2 \times 200}{3080} = \frac{4440}{3080} = 1.441$$

Mean occupation no. for level-2

$$\epsilon = \frac{2 \times 400 + 1 \times 640 + 4 \times 140 + 1 \times 320 + 3 \times 200}{3080} = \frac{2900}{3080} = 0.941$$

Mean occupation no. for level

$$\epsilon = \frac{1 \times 640 + 2 \times 400 + 1 \times 320}{3080} = \frac{1760}{3080} = 0.571$$

Mean occupation no. for level of energy-0

$$= \frac{2 \times 200 + 1 \times 400 + 1 \times 640 + 1 \times 640 + 1 \times 140}{3080} = \frac{2220}{3080} = 0.720$$

Table S-10.1   Particle distribution in different energy levels of each macrostate

Energy of the level ↓	1	2	3	4	5	6	7	8	9	Mean occupation number
4ϵ	ppp	pp	pp	p	p	P				1.253
3ϵ			p	Pp	pp		Pppp	ppp	pp	1.441
2ϵ		pp		P		Pppp		p	ppp	0.941
ϵ			p		pp			p		0.571
0	pp	P	p	P			P			0.720
Number of microstates →	200	400	640	640	400	140	140	320	200	Total number of microstates = 3080

Note that the sum of mean occupation numbers: 1.253+1.441+0.941+0.571+0.720 = 5.0 which is the total number of particles in the system.

2. Compute the number of macrostates, number of microstates corresponding to each macrostate, and the average value of occupation numbers for each level, if the particles in problem 10.1 obey Fermi–Dirac statistics instead of the Bose–Einstein statistics, all other values, like the energies of levels, number of particles, total energy of the system and degeneracy of each level remaining same.

Solution:

In this problem, like the previous one, five indistinguishable particles are to be distributed in five equidistant levels of energies, 4ϵ, 3ϵ, 2ϵ, ϵ, 0 each of degeneracy $g_i = 4$, subject to the conditions that the total energy of all particles must add up to 12ϵ and the particles are distributed according to Fermi–Dirac statistics.

Each possible way of particle distribution in five levels such that their energies add up to 12ϵ, constitutes a macrostate of the system. The number of macrostates is decided by the number and energies of the levels, number of particles and the total energy of the system. Therefore, the number and the structure of particle distribution in levels for each macrostate remain same as it was for problem 10.1. This is shown in table-(S-10.2) given below.

As shown in the table, the macrostate-1 has three particles in the level of energy 4ϵ, and the remaining two particles in level of zero energy. The total energy of these particles is $3 \times 4\epsilon + 2 \times 0 = 12\epsilon$. Similarly, total energy of particles for each of the nine macrostates adds up to 12ϵ as required.

Macrostate-1 is made up of three particles in level-4 ϵ and two particles in level-0 distributed according to Fermi–Dirac distribution. The number of microstates for a macrostate can be calculated using the Fermi–Dirac distribution formula $\mathcal{W}^{Fermi-Dirac} = \prod_j \dfrac{g_j!}{\{N_j!\}(g_j - N_j)!}$, with $g_j = 4$, for all j's, with j = 1,2,3,4,5 for the five levels. We designate j = 1 for level of energy 0, j = 2, for level with energy ϵ, and so on.

Table S-10.2   Particle distribution in different levels of macrostates

Energy of the level ↓		1	2	3	4	5	6	7	8	9	Mean occupation number
4ϵ	j = 5	ppp	pp	pp	P	p	p				1.411
3ϵ	j = 4			p	pp	pp		pppp	ppp	pp	1.443
2ϵ	j = 3		pp		P		pppp		p	ppp	0.700
ϵ	j = 2			p		pp			p		0.625
0	j = 1	pp	P	p	P			p			0.819
Number of microstates →		24	144	384	384	144	4	4	64	24	Total number of microstates = 1176

Number of microstates for any macrostate may be calculated using the Fermi–Dirac formula as follows,

No. microstates of Macrostate-1 =

$$\left[\frac{4!}{(3!)(4-3)!}\right]_{j=5}\left[\frac{4!}{2!(4-2)!}\right]_{j=1} = 4 \times 6 = 24$$

No. microstates of Macrostate-2 =

$$\left[\frac{4!}{(2!)(4-2)!}\right]_{j=5}\left[\frac{4!}{(2!)(4-2)!}\right]_{j=3}\left[\frac{4!}{(1!)(4-1)!}\right]_{j=1} = 6 \times 6 \times 4 = 144$$

No. microstates of Macrostate-3 =

$$\left[\frac{4!}{(2!)(4-2)!}\right]_{j=5}\left[\frac{4!}{(1!)(4-1)!}\right]_{j=4}\left[\frac{4!}{(1!)(4-1)!}\right]_{j=2}\left[\frac{4!}{(1!)(4-1)!}\right]_{j=1} = 6 \times 4 \times 4 \times 4$$

$$= 384$$

In this way number of microstate for each macrostate have been calculated and shown in Table S-10.2

The total number of microstates = 1176

Average occupation numbers for different levels are calculated as follows

It may be observed from the table that each energy level has a different occupation number in each macrostate. For example, level j = 5 with energy 4ϵ has occupation numbers 3, 2, 2,

1, 1, 1, 0, 0, and 0, respectively, for macrostates,1, 2, 3, 4,........9. The occupation number of level j = 5 will remain 3 for all microstates of macrostate-1 and it will be 2 for all microstates of macrostate-2 and so on. Therefore, the average occupation number of level j = 5 may be obtained by the formula,

$$\left(N_{occu}^{average}\right)_j = \frac{\left(\Sigma_i N_i^{occu} N_i^{micro}\right)_j}{\text{Total number of microstates}}$$

Here, $\left(N_{occu}^{average}\right)_j$ is the average occupation number of level j, N_i^{occu}, the occupation number of jth level in i^{th} macrostate and N_i^{micro} the number of microstates of macrostate–i. Following calculations will further clarify the procedure.

Energy level   *Average occupation number*

$J = 5$ $\quad \dfrac{3 \times 24 + 2 \times 144 + 2 \times 384 + 1 \times 384 + 1 \times 144 + 4 \times 1}{1176} = \dfrac{1660}{1176} \quad = 1.411$

$J = 4$ $\quad \dfrac{1 \times 384 + 2 \times 384 + 2 \times 144 + 4 \times 4 + 3 \times 64 + 2 \times 24}{1176} = \dfrac{1692}{1176} \quad = 1.443$

$J = 3$ $\quad \dfrac{2 \times 144 + 1 \times 384 + 4 \times 4 + 1 \times 64 + 3 \times 24}{1176} = \dfrac{824}{1176} \quad = 0.700$

$J = 2$ $\quad \dfrac{1 \times 384 + 2 \times 144 + 1 \times 64}{1176} = \dfrac{736}{1176} \quad = 0.625$

$J = 1$ $\quad \dfrac{2 \times 24 + 1 \times 144 + 1 \times 384 + 1 \times 384 + 1 \times 4}{1176} = \dfrac{964}{1176} \quad = 0.819$

Computed values of all desired quantities are tabulated in Table S-10.2. It may be observed how the number of microstates associated with a given macrostate changes with the type of distribution law followed by the particles.

3. Compute the number of macrostates, number of microstates for each macrostate and the average occupation numbers for a system of total energy 12ε in which five distinguishable particles are distributed in five equidistant levels of energies 0, ε, 2ε, 3ε and 4ε. Particles of the system obey Maxwell–Boltzmann statistics.

Solution:

Since the number of particles, total energy and level energies are same as in examples 10.1 and 10.2, the number of macrostates and distribution of particles in these macrostates are same as in the previous two examples. However, the microstates per macrostate and the average occupation numbers of different levels will become different as particles are distinguishable and obey Maxwell Boltzmann statistics. A typical particle distribution of the system in nine macrostates is tabulated in Table S-10.3.

In Table S-10.3 different upper case alphabets A, B, C, D, E,(instead of p) have been used to represent different distinguishable particles. The number of microstates corresponding to different macrostates may be calculated using the Maxwell–Boltzmann formula,

$$\mathcal{W}^{Maxwell-Boltzmann} = N! \prod_j \frac{g_j^{N_j}}{N_j!},$$

Here, N is the total number of particles in the system which in the present case is N = 5. Also, in the present case, $N_j = 4$ for all j.

Table S-10.3 Distribution of particles in different levels of macrostates

Macrostate no. → Energy of the level ↓	1	2	3	4	5	6	7	8	9	Mean occupation number
4ϵ, j =5	ABC	DE	EC	C	E	A				1.289
3ϵ, j = 4			D	ED	CD		BCDE	ABC	DB	1.511
2ϵ, j = 3		AB		A		CDBE		E	ACE	0.844
ϵ, j = 2			B		B			D		0.622
0, j = 1	DE	C	A	B	A		A			0.733

Number of the macrostate   *Number of microstates*

1                      $5!\left[\dfrac{4^3}{3!}\right]_{j=5}\left[\dfrac{4^1}{1!}\right]_{j=1} = 5120$

2                      $5!\left[\dfrac{4^2}{2!}\right]_{j=5}\left[\dfrac{4^2}{2!}\right]_{j=3}\left[\dfrac{4^1}{1!}\right]_{j=1} = 30720$

3                      $5!\left[\dfrac{4^2}{2!}\right]_{j=5}\left[\dfrac{4^1}{1!}\right]_{j=4}\left[\dfrac{4^1}{1!}\right]_{j=2}\left[\dfrac{4^1}{1!}\right]_{j=1} = 61440$

4                      $5!\left[\dfrac{4^1}{1!}\right]_{j=5}\left[\dfrac{4^2}{2!}\right]_{j=4}\left[\dfrac{4^1}{1!}\right]_{j=3}\left[\dfrac{4^1}{1!}\right]_{j=1} = 61440$

5                      $5!\left[\dfrac{4^1}{1!}\right]_{j=5}\left[\dfrac{4^2}{2!}\right]_{j=4}\left[\dfrac{4^2}{2!}\right]_{j=2} = 30720$

Carrying out these calculations further gives the number of microstates for 6, 7, 8 and 9 macrostates, respectively, as, 5120, 5120, 20480 and 10240

Total number of microstates = 230400

To calculate the average occupation numbers for different levels we use the formula

$$\left(N_{occu}^{average}\right)_j = \frac{\left(\sum_i N_i^{occu} N_i^{micro}\right)_j}{\text{Total number of microstates}}$$

Level	Average occupation number
J = 5	$\dfrac{3 \times 5120 + 2 \times 30720 + 2 \times 61440 + 1 \times 61440 + 1 \times 30720 + 1 \times 5120}{230400}$

$$= \frac{296960}{230400} = 1.289 \,.$$

Level	
J = 4	$\dfrac{1 \times 61440 + 2 \times 61440 + 2 \times 30720 + 4 \times 5120 + 3 \times 20480 + 2 \times 10240}{230400}$

$$= \frac{348160}{230400} = 1.511$$

Further calculations give the average occupation numbers for j = 3, j = 2 and j = 1 levels as 0.844, 0.622 and 0.733 respectively.

4. A system of total energy 7ϵ contains six indistinguishable particles that are distributed amongst (a) eight (b) five equidistant energy levels starting from 0, ϵ, 2ϵ, 3ϵ, …. . Draw diagram's to show the macrostates of the system for the two cases.

Solution:

The details of 14 macrostates in case (a) are tabulated in Table S-10.4a and are given below:

Macrostate-1 has: 1 particle in level J = 8 (7ϵ); 5 particles in level j = 1 (energy 0)

Macrostate-2 has: 1 particle in level j = 7(6ϵ), 1 particle in level j = 2 (ϵ) , 4 particles in level j = 1(0)

Macrostate-3 has: 1 particle in level j = 6(5ϵ), 1 particle in level j = 3 (2ϵ), 4 particles in level j = 1(0)

…

Macrostate-13 has: 1 particle in level j = 3 (2ϵ), 5 particles in level j = 2 (ϵ)

Macrostate-14 has: 1 particle in level j = 6(5ϵ), 1 particle in level j = 3(2ϵ), 4 particles in level j = 1(0)

Table S-10.4 (a) Distribution of particles in different levels of macrostates

Number of macrostates →	1	2	3	4	5	6	7	8	9	10	11	12	13	14
Level j = 8, energy 7ϵ; No. particles →	1	0	0	0	0	0	0	0	0	0	0	0	0	0
Level j = 7, energy 6ϵ	0	1	0	0	0	0	0	0	0	0	0	0	0	0
Level j = 6, energy 5ϵ	0	0	1	0	0	0	0	0	0	0	0	0	0	1
Level j = 5, energy 4ϵ	0	0	0	1	1	1	0	0	0	0	0	0	0	0
Level j = 4, energy 3ϵ	0	0	0	1	0	0	2	1	1	1	0	0	0	0
Level j = 3, energy 2ϵ	0	0	0	0	1	0	0	2	1	0	3	2	1	1
Level j = 2, energy ϵ	0	1	2	0	1	3	1	0	2	4	1	3	5	0
Level j = 1, energy 0	5	4	3	4	3	2	3	3	2	1	2	1	0	4

For case (b) the details of macrostate are tabulated in Table S-10.4 b

Table S-10.4 (b) Details of macrostates for part (b)

Number of macrostates →	1	2	3	4	5	6	7	8	9	10
Level j = 5, energy 4ε; No. particles →	1	1	1	0	0	0	0	0	0	0
Level j = 4, energy 3ε; No. particles →	1	0	0	2	1	1	0	0	1	0
Level j = 3, energy 2ε; No. particles →	0	0	1	0	2	0	3	2	1	1
Level j = 2, energy ε; No. particles →	0	3	1	1	0	4	1	3	2	5
Level j = 1, energy 0ε; No. particles →	4	2	3	3	3	1	2	1	2	0

5. For the macrostates listed in table-(S-10.4b) of 10.4, calculate the quantum thermodynamic probability of each macrostate and the quantum thermodynamic probability of the complete system assuming that the particles are indistinguishable and follow Bose–Einstein statistics and each level has two-fold degeneracy. Total energy of the system is 7ϵ, and total number of particles is 6. Also compute the average occupation number for each level. Which macrostate is the most probable state?

Solution:

The thermodynamic probability of a macrostate is equal to the number of microstates of that macrostate and the thermodynamic probability of the system is the total number of microstates of the system. Since the system particles obey Bose–Einstein statistics the number of microstates of different macrostates may be calculated using the formula,

$$W^{Bose-Ein} = \Pi_j \frac{\left(g_j + N_j - 1\right)!}{\left\{\left(g_j - 1\right)!\right\}\left(N_j!\right)}$$

Given that $g_j = 2$(two fold degeneracy) for all j's

Calculations of the number of microstates for each macrostate are given below:

Number of   *Number of microstates or quantum thermodynamic probability*
the macrostate

1    $\left[\dfrac{(2+1-1)!}{(2-1)!(1!)}\right]_{j=5} \left[\dfrac{(2+1-1)!}{(2-1)!(1!)}\right]_{j=4} \left[\dfrac{(2+4-1)!}{(2-1)!(4!)}\right]_{j=1} = 2 \times 2 \times 5 = 20$

2    $\left[\dfrac{(2+1-1)!}{(2-1)!(1!)}\right]_{j=5} \left[\dfrac{(2+3-1)!}{(2-1)!(3!)}\right]_{j=2} \left[\dfrac{(2+2-1)!}{(2-1)!(2!)}\right]_{j=1} = 2 \times 4 \times 3 = 24$

3    $\left[\dfrac{(2+1-1)!}{(2-1)!(1!)}\right]_{j=5} \left[\dfrac{(2+1-1)!}{(2-1)!(1!)}\right]_{j=3} \left[\dfrac{(2+1-1)!}{(2-1)!(1!)}\right]_{j=2} \left[\dfrac{(2+3-1)!}{(2-1)!(3!)}\right]_{j=1} = 32$

4    $\left[\dfrac{(2+2-1)!}{(2-1)!(2!)}\right]_{j=4} \left[\dfrac{(2+1-1)!}{(2-1)!(1!)}\right]_{j=2} \left[\dfrac{(2+3-1)!}{(2-1)!(3!)}\right]_{j=1} = 3 \times 2 \times 4 = 24$

5 $\qquad \left[\dfrac{(2+1-1)!}{(2-1)!(1!)}\right]_{j=4} \left[\dfrac{(2+2-1)!}{(2-1)!(2!)}\right]_{j=3} \left[\dfrac{(2+3-1)!}{(2-1)!(3!)}\right]_{j=1} = 2 \times 3 \times 4 = 24$

6 $\qquad \left[\dfrac{(2+1-1)!}{(2-1)!(1!)}\right]_{j=4} \left[\dfrac{(2+4-1)!}{(2-1)!(4!)}\right]_{j=2} \left[\dfrac{(2+1-1)!}{(2-1)!(1!)}\right]_{j=1} = 2 \times 5 \times 2 = 20$

7 $\qquad \left[\dfrac{(2+3-1)!}{(2-1)!(3!)}\right]_{j=3} \left[\dfrac{(2+1-1)!}{(2-1)!(1!)}\right]_{j=2} \left[\dfrac{(2+2-1)!}{(2-1)!(2!)}\right]_{j=1} = 4 \times 2 \times 3 = 24$

8 $\qquad \left[\dfrac{(2+2-1)!}{(2-1)!(2!)}\right]_{j=3} \left[\dfrac{(2+3-1)!}{(2-1)!(3!)}\right]_{j=2} \left[\dfrac{(2+1-1)!}{(2-1)!(1!)}\right]_{j=1} = 3 \times 4 \times 2 = 24$

9 $\qquad \left[\dfrac{(2+1-1)!}{(2-1)!(1!)}\right]_{j=4} \left[\dfrac{(2+1-1)!}{(2-1)!(1!)}\right]_{j=3} \left[\dfrac{(2+2-1)!}{(2-1)!(2!)}\right]_{j=2} \left[\dfrac{(2+2-1)!}{(2-1)!(2!)}\right]_{j=1} = 36$

10 $\qquad \left[\dfrac{(2+1-1)!}{(2-1)!(1!)}\right]_{j=3} \left[\dfrac{(2+5-1)!}{(2-1)!(5!)}\right]_{j=2} = 2 \times 6 = 12$

Total number of microstates = 240

The quantum thermodynamic probability of the system is 240.

To calculate the average occupation number for each level we use the formula

$$\left(N_{occu}^{average}\right)_j = \frac{\left(\Sigma_i N_i^{occu} N_i^{micro}\right)_j}{\text{Total number of microstates}}$$

Level   *Average occupation number*

$J = 5 \qquad \dfrac{1 \times 20 + 1 \times 24 + 1 \times 32}{240} = \dfrac{76}{240} = 0.316$

$J = 4 \qquad \dfrac{1 \times 20 + 2 \times 24 + 1 \times 20 + 1 \times 36}{240} = \dfrac{124}{240} = 0.516$

$J = 3 \qquad \dfrac{1 \times 32 + 2 \times 24 + 3 \times 24 + 2 \times 24 + 1 \times 36 + 1 \times 12}{240} = \dfrac{248}{240} = 1.033$

$J = 2 \qquad \dfrac{3 \times 24 + 1 \times 32 + 1 \times 24 + 4 \times 20 + 1 \times 24 + 3 \times 24 + 2 \times 36 + 5 \times 12}{240} = \dfrac{436}{240} = 1.816$

$J = 1 \qquad \dfrac{4 \times 20 + 2 \times 24 + 3 \times 32 + 3 \times 24 + 3 \times 24 + 1 \times 20 + 2 \times 24 + 1 \times 24 + 2 \times 36}{240} = \dfrac{532}{240} = 2.216$

Macrostate number -9 is the most probable macrostate as it has the largest number of microstates. Further, note that the sum of mean occupation numbers for different levels is equal to 6, the total number of particles.

6. For a system of N distinguishable particles that obey either classical or Maxwell–Boltzmann distribution, show that the average number of particles $\overline{N_j}$ in level j is given by

$$\overline{N_j} = -Nk_BT\left[\frac{\partial \ln Z}{\partial \epsilon_j}\right]_T$$

Solution:

Both for classical and Maxwell–Boltzmann distributions the distribution function is given by the relation,

$$\frac{\overline{N_j}}{g_j} = \frac{N}{Z}e^{\frac{-e_j}{Tk_B}}$$

Hence,

$$\frac{\overline{N_j}}{N} = \frac{1}{Z}g_j\, e^{\frac{-e_j}{Tk_B}}$$

Or

$$\sum_j \frac{\overline{N_j}}{N} = \frac{1}{Z}\sum_j g_j\, e^{\frac{-e_j}{Tk_B}} \qquad\text{S-10.6.1}$$

Now the partition function for both the distributions is given by

$$Z = \sum_j g_j e^{\frac{-e_j}{Tk_B}}$$

We partially differentiate Z with respect to ϵ_j keeping T constant to get,

$$\left(\frac{\partial Z}{\partial \epsilon_j}\right)_T = -\frac{1}{Tk_B}\sum_j g_j\, e^{\frac{-e_j}{Tk_B}}$$

Or

$$\sum_j g_j\, e^{\frac{-e_j}{Tk_B}} = -Tk_B\left(\frac{\partial Z}{\partial \epsilon_j}\right)_T \qquad\text{S-10.6.2}$$

Substituting the value of $\sum_j g_j\, e^{\frac{-e_j}{Tk_B}}$ from Eq. S-10.6.2 in Eq. S-10.6.1 one gets,

$$\sum_j \frac{\overline{N_j}}{N} = \frac{1}{Z}\left[-Tk_B\left(\frac{\partial Z}{\partial \epsilon_j}\right)_T\right] = -Tk_B\left(\frac{\partial z/z}{\partial \epsilon_j}\right)_T = -Tk_B\left(\frac{\partial \ln Z}{\partial \epsilon_j}\right)_T \qquad\text{S-10.6.3}$$

Right hand side of Eq. S-10.6.3 does not depend on summation, hence

$$\overline{N_j} = -NTk_B\left(\frac{\partial \ln Z}{\partial \epsilon_j}\right)_T$$

7. A system has 9 distinguishable particles that obey Maxwell Boltzmann statistics. Calculate the entropy of the system when two particles of the system are in a non-degenerate excited state at E energy above the ground state. The ground state has three folds degeneracy. Also write the partition function for the system.

Solution:

Entropy of the system can be calculated if the quantum thermodynamic probability that is the number of microstates of the system is known. We, therefore, calculate the number of microstates when two particles are in excited state of $g_{ex} = 1$ and 7 particles in ground state of $g_g = 3$, using the formula

$$W^{Maxwell-Boltzmann} = N! \prod_j \frac{g_j^{N_j}}{N_j!}$$

$$\Omega = W^{Maxwell-Boltzmann} = N! \left[\left\{ \frac{g_{ex}^{N_{ex}}}{(N_{ex})!} \right\} \left\{ \frac{g_g^{N_g}}{(N_g)!} \right\} \right] \qquad \text{S-10.7.1}$$

In Eq. S-10.7.1, $N = 9$, is the total number of particles, and $N_{ex} = 2$, $N_g = 7$ are, respectively, the number of particles in the excited and ground states.

$$\Omega = 9! \left[\left\{ \frac{1^2}{2!} \right\} \left\{ \frac{3^7}{7!} \right\} \right] = 78732$$

Now entropy S is given by,

$$S = k_B \ln \Omega = k_B \ln (78732) = 11.27 k_B$$

The entropy of the system is $11.27 \, k_B$

The partition function for the system is,

$$Z = \sum_j g_j e^{\frac{-\epsilon_j}{k_B T}} = \left[g_g e^{\frac{-\epsilon_g}{k_B T}} \right] + \left[g_{ex} e^{\frac{-\epsilon_{ex}}{k_B T}} \right] = \left[3e^{-0} \right] + \left[1 e^{\frac{-E}{k_B T}} \right]$$

The partition function $Z = 3 + e^{\frac{-E}{k_B T}}$

8. A system of N particles that obey Maxwell Boltzmann distribution law, has a ground level and an excited level separated by energy E. Both these levels are non-degenerate. The system is in equilibrium at some temperature T such that N_g particles are in the ground state while N_{ex} particles are in the excited state. Compute (a) the ratios N_g/N, and N_{ex}/N (b) internal energy of the system (c) the partition function for the system and (d) the entropy of the system.

Solution:

Let us first find out the partition function Z for the system that is given by,

$$Z = \sum_j g_j e^{\frac{-\epsilon_j}{k_B T}}$$

In the present problem only two levels, the ground and the excited levels are involved and, therefore, j =2 and both the levels are non-degenerate so $g_j = 1$ for both levels. Also $\epsilon_j = E$. With these values

$$Z = \sum_j e^{\frac{-\epsilon_j}{k_BT}} = \left[1e^0\right] + \left[1e^{\frac{-E}{k_BT}}\right] = 1 + e^{\frac{-E}{k_BT}} \qquad \text{S-10.8.1}$$

Also, from Eq. 10.52 we have,

$$\frac{\overline{N_j}}{g_j} = \frac{N}{Z}e^{\frac{-e_j}{Tk_B}} \qquad \text{S-10.8.2}$$

Writing Eq. S-10.8.2 for the excited level

$$\frac{N_{excit}}{N} = \frac{1}{Z}1.e^{\frac{-E}{Tk_B}}$$

Substituting the value of Z from Eq. S-10.8.1,

$$\frac{N_{excit}}{N} = \frac{e^{\frac{-E}{Tk_B}}}{1 + e^{\frac{-E}{k_BT}}} = \frac{1}{1 + e^{\frac{E}{k_BT}}} \qquad \text{S-10.8.3}$$

Similarly,

$$\frac{N_{ground}}{N} = \frac{e^{\frac{-0}{Tk_B}}}{Z} = \frac{1}{1 + e^{\frac{-E}{k_BT}}} \qquad \text{S-10.8.4}$$

The total internal energy $U = N\sum_j \epsilon_j p_j$ where $p_j = \dfrac{N_j}{N}$

So, $$U = N\sum_j \epsilon_j p_j = N\left[p_{gr}\epsilon_{gr} + p_{exc}\epsilon_{exc}\right]$$

Or $$U = N\left[\frac{N_{exc}}{N}E + \frac{N_{gr}}{N}x0\right]$$

$$= N\left[\frac{1}{1 + e^{\frac{E}{k_BT}}}E\right] = NE\left[\frac{1}{1 + e^{\frac{E}{k_BT}}}\right] \qquad \text{S-10.8.5}$$

From Eq. 10.52 the entropy of the system may be given as,

$$S(T) = Nk_B \ln Z + NTk_B \frac{\partial \ln Z}{\partial T}$$

Substituting the expression for Z from Eq. S-10.8.1,

$$S(T) = Nk_B \ln\left(1 + e^{\frac{-E}{k_BT}}\right) + NTk_B \frac{\partial \ln\left(1 + e^{\frac{-E}{k_BT}}\right)}{\partial T}$$

$$= Nk_B \ln\left(1 + e^{\frac{-E}{k_B T}}\right) + NTk_B\left[\left(\left(\frac{1}{\left(1 + e^{\frac{-E}{k_B T}}\right)}\right)\left(\frac{E}{k_B T^2}\right)\right)\right]$$

$$= Nk_B \ln\left(1 + e^{\frac{-E}{k_B T}}\right) + \frac{NE}{T}\left(\frac{1}{1 + e^{\frac{-E}{k_B T}}}\right) \qquad \text{S-10.8.6}$$

Equations S-10.8.1, S-10.8.4, S-10.8.5, and S-10.8.6 give the values of the desired functions.

9. A system of five distinguishable particles that follow Maxwell Boltzmann distribution law has particles distributed in five equidistant levels of energies 0, ϵ, 2ϵ, 3ϵ and 4ϵ. All energy levels are 5-fold degenerate and the total energy of the system is 6ϵ. A new system is formed by removing one particle from the level of energy 2ϵ. With the help of suitable diagram/table show particle distributions in different levels of macrostates for the original and the new systems, calculate the average occupation numbers for the two systems and the change in the entropy of the system by removing one particle from level of energy 2ϵ.

Solution:

In the original system there are $N = 5$ total particles which are distributed such that energies of all particles add up to 6ϵ. Possible distributions of particles in energy levels that are consistent with the total energy of the system are shown in Table S-10.9a. As may be seen from this table, there are eight different ways, shown by eight columns, in which particles may be distributed. Each column represents a macrostate of the system. For example, macrostate-1 has one particle in level $j = 5$ of energy 4ϵ, one particle in level $j = 3$ of energy 2ϵ and the remaining three particles in level $j = 1$ of energy zero. The total energy of all the five particles add up to 6ϵ as required. The same is true for all other

Table S-10.9 (a) Distribution of particles in different levels of macrostates

Macrostate no. → Energy and number of the level ↓	1	2	3	4	5	6	7	8	Mean occupation number
4ϵ $j = 5$	1	1							0.355
3ϵ $j = 4$			2	1	1				0.533
2ϵ $j = 3$	1			1		3	2	1	0.777
ϵ $j = 2$		2		1	3		2	4	1.422
0 $j = 1$	3	2	3	2	1	2	1		1.911

macrostates also. The number of microstates in a given macrostate may be calculated using the formula for Maxwell–Boltzmann distribution $W^{Maxwell-Boltzmann} = N!\prod_j \dfrac{g_j^{N_j}}{N_j!}$. In the present case N = 5 and g = 5 for all levels. As a sample, calculations for the number of microstates for 4th macrostate that has one particle each in j = 4. J = 3 and j = 2 levels and two particles in j = 1 level are shown here.

$$\text{No. of microstates for macrostate-4} = 5!\left[\left\{\frac{5^1}{1!}\right\}_{j=4}\left\{\frac{5^1}{1!}\right\}_{j=3}\left\{\frac{5^1}{1!}\right\}_{j=2}\left\{\frac{5^2}{2!}\right\}_{j=1}\right] = 60 \times 5^5 = 187500$$

The numbers of microstates in all other macrostates have been calculated in this way and are tabulated in Table S-10.9b. The sum of all microstate which is equal to the quantum thermodynamic probability of the system is also given in the last row of the table.

Table S-10.9   (b) Number of microstates in different macrostates

Macrostate number	Number of microstates
1	62500
2	187500
3	62500
4	187500
5	62500
6	31250
7	93750
8	15625
Sum of all microstates	703125
Total number of microstates (Ω_{orig})	703125

The average occupation numbers for the for the five level may be calculated using the formula

$$\left(N_{occu}^{average}\right)_j = \frac{\left(\Sigma_i N_i^{occu}\, N_i^{micro}\right)_j}{Total\ number\ of\ microstates}$$

For example, the average occupation number for level j = 4, that has occupation number 2 from macrostate- 3, occupation number-1 from macrostate-4 and occupation number 1 from macrostate-5, is given by

$$\text{Average occupation no. of level } j = 4 = \frac{2 \times 62500 + 1 \times 187500 + 1 \times 6250}{703125} = \frac{375000}{703125} = 0.533$$

Average occupation numbers for other levels are given in the last column of Table S-2.9a.

Next we consider the new system made by removing one particle from level $j = 3$ of the original system that has energy 2ϵ. It is obvious that in the new system only those macrostates of the original system that have at least one particle in level $j = 3$ can participate. Macrostates like 2, 3 and 5, which do not contain any particle in energy level $j = 3$ cannot form a part of the new system as no particle from level $j = 3$ can be removed from these macrostates. Remaining macrostates, 1, 4, 6, 7 and 8 will form the new system but with one particle less in level $j = 3$ in each of these macrostates. The total number of particles in each macrostate of the new system will be $(5 - 1 = 4)$ four and the total energy of the new system will be $(6\epsilon - 2\epsilon = 4\epsilon)$. The distribution of particles of the new system in each level (rows) and in each macrostate (column) is shown in table-(S-10.9 c). The number of the macrostate in original system is also mentioned in this table.

In the new system the total number of particles $N = 4$, and the degeneracy of each level is still (g =) 5-fold. Sample calculations for the number of microstates in macrostate-2 of the new system that contains one particle each in levels $j = 4$, and $j = 2$ and two particles in $j = 1$ is given below,

Number of microstates for macrostate-2 (new system)

$$= 4! \left[\left\{ \frac{5^1}{1!} \right\}_{j=4} \left\{ \frac{5^1}{1!} \right\}_{j=2} \left\{ \frac{5^2}{2!} \right\}_{j=4} \right] = 12 \times 5^4 = 7500$$

Table S-10.9 (c) Details of the new system formed by removing one particle from level $j = 3$

Energy and Number of the level	Macrostate-1 (no.1 of original system)	Macrostate-2 (no.4 of original system)	Macrostate-3 (no.6 of original system)	Macrostate-4 (no.7 of original system)	Macrostate-5 (no.8 of original system)
$4\epsilon,\ j = 5$	1				
$3\epsilon,\ j = 4$		1			
$2\epsilon,\ j = 3$			2	1	
$\epsilon,\ j = 2$		1		2	4
$0,\ j = 1$	3	2	2	1	

The numbers of microstates for all other macrostates of the new system have been calculated in the similar way and are given in Table S-10.9d. The total number of microstates of the new system, which is the quantum thermodynamic probability of the new system, is also given in this table. Details of the new system formed by removing one particle from level $j = 3$

Sample calculations for the average occupation number of level $j = 1$ of the new system are as follows,

Average occupation number of level $j = 1$ of new system

$$= \frac{\begin{array}{l} 3 \times (\text{no. mic sta. of mac. state.} - 1) + 2 \times (\text{no. mic sta. of mac. sta.} - 2) \\ + 2 \times (\text{no. mic sta. of mac. state.} - 3) + 1 \times (\text{no. mic sta. of mac} - 4) \end{array}}{\text{Total number of microstates of the new system} = 21875}$$

$$= \frac{3 \times 2500 + 2 \times 7500 + 2 \times 3750 + 1 \times 7500}{21875} = \frac{37500}{21875} = 1.714$$

Number of microstates for each macrostate and the average occupation numbers for all levels of the new system are given in Table S-10.9d.

Table S-10.9 (d) Details of microstates and average occupation numbers for new system

Macrostate no.	Number of microstates	Level number	Average occupation number
1	2500	$J = 5$	0.114
2	7500	$J = 4$	0.342
3	3750	$J = 3$	0.685
4	7500	$J = 2$	1.142
5	625	$J = 1$	1.714
Total number of microstates (Ω_{new})	21875	Total of Av. occup. nos.	$3.997 \approx 4$

The entropy of the original system $S_{orig} = k_B \ln \Omega_{orig} = k_B \ln 703125 = 13.463 k_B$

The entropy of the new system $S_{new} = k_B \ln \Omega_{new} = k_B \ln 21875 = 9.93 k_B$

Change in entropy by removing one particle from level j = 3, = 3.47 k_B

10. Three particles that obey Maxwell–Boltzmann statistics are distributed in four equidistant levels of energies, 0, E, 2E, and 3E in such a way that the sum of the particle energies is 5E. Each level is threefold degenerate. Compute the increase in the entropy of the system when one particle is added in level of energy E of the assembly.

Solution:

The distribution of three particles in different possible ways so that their energies add up to 6E is shown in Table S-10.10.1.

Table S-10.10.1 Distribution of particles in different levels of macrostates A, B and C

Name of the macrostate → Energy and no. of levels ↓	A	B	C
3E j = 4	1	1	
2E j = 3	1		2
E j = 2		2	1
0 j = 1	1		

Since particles obey Maxwell–Boltzmann distribution, the number of microstates or quantum thermodynamic probability of each macrostate may be calculated using the formula $N!\prod_j \dfrac{g_j^{N_j}}{N_j!}$

Number of microstates of macrostate A $= 3!\left[\dfrac{3^1}{1!}\right]\left[\dfrac{3^1}{1!}\right]\left[\dfrac{3^1}{1!}\right] = 6 \times 3^3 = 162$

Number of microstates of macrostate B $= 3!\left[\dfrac{3^1}{1!}\right]\left[\dfrac{3^2}{2!}\right] = 3 \times 3^3 = 81$

Number of microstates of macrostate C $= 3!\left[\dfrac{3^2}{2!}\right]\left[\dfrac{3^1}{1!}\right] = 3 \times 3^3 = 81$

Total number of microstates = quantum thermodynamic probability of the system $\Omega_{orig} = 324$

Entropy of the original system $S_{orig} = \ln\Omega_{orig} = k_B \ln 324 = 5.78 k_B$ $\hspace{2cm}$ S-10.10.1

Distribution of particles in the new system obtained by adding one particle to the level of energy E is shown in Table S-10.10.2.

Table S-10.10.2   Particle distribution in different levels of macrostates

Name of the macrostate → Energy and no. of levels ↓	A'	B'	C'
3E j = 4	1	1	
2E j = 3	1		2
E j = 2	1	3	2
0 j = 1	1		

The new system has four particles and total energy of 6E. Numbers of microstates of the modified macrostates are calculated below.

Number of microstates of macrostate A' $= 4!\left[\dfrac{3^1}{1!}\right]\left[\dfrac{3^1}{1!}\right]\left[\dfrac{3^1}{1!}\right]\left[\dfrac{3^1}{1!}\right] = 24 \times 3^4 = 1944$

Number of microstates of macrostate B' $= 4!\left[\dfrac{3^1}{1!}\right]\left[\dfrac{3^3}{3!}\right] = 4 \times 3^4 = 324$

Number of microstates of macrostate c' $= 4!\left[\dfrac{3^2}{2!}\right]\left[\dfrac{3^2}{2!}\right] = 6 \times 3^4 = 486$

Total number of microstates = quantum thermodynamic probability of the system $\Omega_{new} = 2754$

Entropy of the new system $S_{new} = \ln \Omega_{new} = k_B \ln 2754 = 7.92 k_B$ S-10.10.2

Increase in entropy on adding one particle in level of energy $E = S_{new} - S_{orig} = 2.14 k_B$

Problems

1. Four particles with total energy 6E and that follow Bose–Einstein statistics are distributed amongst four equidistant and 4-fold degenerate levels of energies 0, E, 2E, and 3E. Compute (i) the number of possible macrostates, (ii) number of microstates per macrostate and (iii) average occupation numbers of energy levels.

2. Repeat the calculations for problem-1 when particles follow Maxwell–Boltzmann distribution.

3. Nine units of energy is to be sheared between six identical particles with the condition that not more than two particles are in any energy level. If levels are twofold degenerate and particles obey Fermi–Dirac statistics, compute the number of possible macrostates of the system and the average occupation numbers of different energy levels.

4. Distribution of six particles that obey Fermi–Dirac statistics, in equidistant energy levels, each having 3-fold degeneracy, is shown in Table P-10.4.1. Compute (i) the number of microstates for each macrostate (b) the average occupation number of each energy level and (iii) the entropy of the system.

Table P-10.4.1   Particle distribution in different macrostates

Macrostates → Number and energy of the level ↓	A	B	C	D	E
J = 5 4ε	1				
J = 4 3ε		1	1		
J = 3 2ε		1		3	2
J = 2 1ε	2	1	3		2
J = 1 0	3	3	2	3	2

5. Calculate (a) the change in entropy of the system (b) average occupation number, when one particle from level j = 1 (energy zero) is removed from the distribution given in table-(P-10.4.1).

6. For the macrostate shown in Fig.(P-10.6), compute (a) the number of microstates when particles obey (i) Fermi–Dirac (ii) Maxwell–Boltzmann (iii) Bose–Einstein statistics (b) number of microstates for the three cases mentioned in (a) when one particle is removed from the level

at zero energy. Also indicate for which distributions the number of microstates is smallest and largest and why?

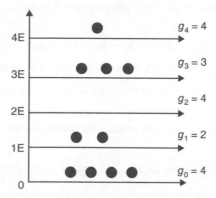

Fig. P-10.6   Distribution of particles in a macrostate

7. A new macrostate with same number of particles and total energy is created by shifting one particle from level of energy 3E to the level of energy 2E and simultaneously removing a particle from level of energy 1E and putting it at the level of energy zero, in Fig.(P-10.6) of problem-6. Compute the number of microstates for the new macrostate for Maxwell–Boltzmann distribution and the change in entropy of the system as a result of this shifting of particles.

8. Show that both Fermi–Dirac and Bose–Einstein distributions reduces to $\prod_j \dfrac{g_j^{N_j}}{N_j!}$ when $g_j \gg N_j$.

9. Four particles that follow Fermi–Dirac statistics are distributed in two energy levels each of degeneracy 4. Compute (a) the number of macrostates of the system (b) the most likely macrostate (c) the entropy of the system in units of Joule per Kelvin.

10. Six particles are to be distributed into three equidistant levels of energies 0, ϵ and 2ϵ. Each level is six-fold degenerate. Identify and calculate the thermodynamic probabilities and entropies of the most likely and the least likely distributions, when particles obey Maxwell–Boltzmann statistics.

Short Answer Questions

1. Taking suitable example explain the terms: energy level, energy state, degeneracy and occupation number for a system of identical particles.

2. State the postulate of equal a prior probability.

3. Distinguish between the macro and micro states of a system.

4. Discuss in brief the importance of average occupation numbers.

5. Write expressions for Bose–Einstein and Fermi–Dirac distributions.

6. Write the characteristics of the particles that obey Bose–Einstein, Fermi–Dirac and Maxwell–Boltzmann distributions.

7. Give the relationship between the quantum thermodynamic probability and the entropy of a system.

8. Comment on the statement, "entropy is a measure of disorder of the system".

9. What happen to the energy levels of a system when the volume of the system is increased?

10. In the frame work of quantum thermodynamics discuss the change in entropy of an ideal gas in free expansion.

11. How in quantum thermodynamics can explain the invariance of entropy in adiabatic change of volume?

12. What is the difference between the distribution formula and the distribution function for a given statistical distribution?

13. Describe the significance of distribution function.

14. Write expression for the partition function for Maxwell–Boltzmann distribution and without derivation, show that number of particles decreases with the energy of the level.

15. In what respect classical distribution differs from Maxwell–Boltzmann distribution?

16. What is the physical significance of the partition function of a distribution?

Long Answer Questions

1. Define quantum thermodynamic probability for a system of identical particles and show that only a logarithmic function of thermodynamic probability multiplied by a constant may represent entropy.

2. Name the two typical distributions in which indistinguishable particles maybe distributed in different energy levels and point out the difference(s) between the two distributions. Derive an expression for thermodynamic probability of a macrostate in one of these distributions.

3. Derive expressions for thermodynamic probability and distribution function of a system of identical particles when particles obey (a) Bose–Einstein (b) Fermi–Dirac statistics.

4. Spell out the characteristics of Maxwell–Boltzmann distribution, derive expressions for (a) quantum thermodynamic probability (b) the distribution function and (c) the partition function for this distribution.

5. Under what condition(s) both Maxwell–Boltzmann and Fermi–Dirac distribution functions approach Classical distribution? Show that all parameters of classical thermodynamics for Maxwell–Boltzmann distribution may be obtained if the partition function for the system is known.

Multiple Choice Questions

Note: Some of the multiple choice questions may have more than one correct alternative. All correct alternatives must be marked for the complete answer/full marks in such cases.

1. Sum of average occupation numbers over all energy levels of a system is equal to;
   (a) Total number of levels
   (b) total number of particles
   (c) average value of the degeneracy of levels
   (d) average value of quantum thermodynamic probability

2. Functional relationship between the entropy of an assemble and its quantum thermodynamic probability is;
   (a) Linear   (b) Binomial
   (c) exponential   (d) logarithmic

3. Experimentally determined properties of an assembly depends strongly on;
   (a) Occupation number of the level with highest energy
   (b) occupation number of the level with least energy
   (c) average occupation numbers of all levels
   (d) temperature of the assembly

4. Four identical particles that obey Maxwell–Boltzmann distribution law are distributed in two levels A and B of energies 1×10^{-3} eV and 2×10^{-3} eV respectively. Both levels are two-fold degenerate. The number of particles in level A of the most probable macrostate of the system is;
   (a) 4   (b) 3
   (c) 2   (d) 1

5. Which of the following expressions represents the quantum thermodynamic probability of an assembly of particles obeying Maxwell–Boltzmann distribution law? Symbols have their usual meaning.

   (a) $\displaystyle \prod_j \frac{(g_j + N_j - 1)!}{\{(g_j - 1)!\}(N_j!)}$

   (b) $\displaystyle \prod_j \frac{g_j!}{\{N_j!\}(g_j - N_j)!}$

   (c) $\displaystyle N! \prod_j \frac{g_j^{N_j}}{N_j!}$

   (d) $\displaystyle \frac{\overline{N_r}}{g_r} = \frac{1}{\left[e^{-\left(\frac{\frac{1}{4}-\epsilon_r}{Tk_B}\right)} - 1 \right]}$

6. A typical macrostate of a system of four identical particles has two particles each in two levels of two-fold degeneracy each. The quantum thermodynamic probability of the macrostate is 1 when particles follow statistical distribution law A and 9 when B. Statistical distribution laws A and B are respectively;
   (a) Bose–Einstein, Maxwell-Boltzmann
   (b) Fermi–Dirac, Maxwell–Boltzmann
   (c) Maxwell- Boltzmann, Bose–Einstein
   (d) Fermi–Dirac, Bose–Einstein

7. Which of the following expression represents Maxwell–Boltzmann distribution function?

(a) $\displaystyle \prod_j \frac{(g_j + N_j - 1)!}{\{(g_j - 1)!\}(N_j!)}$

(b) $\displaystyle \prod_j \frac{g_j!}{\{N_j!\}(g_j - N_j)!}$

(c) $\displaystyle N! \prod_j \frac{g_j^{N_j}}{N_j!}$

(d) $\displaystyle \frac{\overline{N_r}}{g_r} = \frac{1}{\left[e^{-\left(\frac{1/4 - \epsilon_r}{Tk_B}\right)} - 1 \right]}$

8. Four identical particles equally distributed in two two-fold degenerate levels have 96 microstates. Nature of the particles and the statistics they follow are;

(a) distinguishable, Bose–Einstein statistics

(b) indistinguishable, Fermi–Dirac statistics

(c) distinguishable, Maxwell Boltzmann statistics

(d) indistinguishable, Classical statistics

9. Pick the correct expression(s):

(a) $\displaystyle \frac{\overline{N_j}}{g_j} = \frac{N}{Z} e^{\frac{-\epsilon_j}{Tk_B}}$

(b) $\displaystyle \frac{\overline{N_j}}{g_j} = \frac{Z}{N} e^{\frac{-\epsilon_j}{Tk_B}}$

(c) $\displaystyle \frac{\overline{N_r}}{g_r} = \frac{1}{\left[e^{-\left(\frac{1/4 - \epsilon_r}{Tk_B}\right)} - 1 \right]}$

(d) $\displaystyle Z \equiv \sum_j g_j e^{-\frac{\epsilon_j}{k_B T}}$

10. Helmholtz function for a system obeying Maxwell–Boltzmann distribution is given by,

(a) $\displaystyle -\frac{Nk_B}{T} \ln Z$

(b) $\displaystyle -Nk_B T \ln Z$

(c) $\displaystyle NT^2 k_B \frac{\partial (\ln Z)}{\partial T}$

(d) $\displaystyle -NT^2 k_B \frac{\partial (\ln Z)}{\partial T}$

11. Five macrostates of a system are shown in table- (Mc-10.11) where 0 indicates a particle. Which of them have same number of microstates?

Table Mc-10.11   Particle distribution in different levels of macrostates

Name of the Macrostate → Energy levels with energy ↓	A	B	C	D	E
2ϵ	0000	000	00		000
ϵ			0	000	
0			00	0	0

12. For the particle distribution given by table-(Mc-10.11), if a, b, c, ..e, respectively, denotes the quantum thermodynamic probabilities of macrostates A, B, C....E, then the average occupation number of the level with energy ϵ is;

(a) $\dfrac{4b}{3a + b + c}$

(b) $\dfrac{4d}{a + 3b + c}$

(c) $\dfrac{4a}{a + 3b + c}$

(d) $\dfrac{4e}{a + c + 3e}$

13. What happens to a level of energy ϵ of a system when the volume of the system is doubled?

(a) Shifts to a higher energy by 0.37ϵ

(b) No shift in the energy of the level

(c) Shifts to a lower energy by 0.74ϵ

(d) Shifts to a lower energy by 0.37ϵ

14. In quantum thermodynamics the increase in entropy of an ideal gas in isothermal expansion may be explained as follows,

(a) With the increase in volume of the system energy levels come closer and more levels become available for the same internal energy which remains fixed in an isothermal process.

(b) In an isothermal process internal energy of the system increases with the increase of the volume and more energy levels become available to the particles.

(c) In an isothermal expansion collisions between particles increases increasing the entropy of the system

(d) In isothermal expansion the mean free path of particles increases which results in the increase of the entropy.

Answers to Problems and Multiple Choice Questions

Answers to problems

1. (i) 5 (ii) 100, 256, 80, 80, 100 (iii) 0.87, 1.13, 1.13, 0.87

2. (i) 5 (ii) 1536, 6144, 1024, 1024, 1536 (iii) 0.90, 1.10, 1.10, 0.90

3. Number of macrostates 5, Avg. Occupation numbers: 1.8, 1.6, 1.2, 0.8, 0.4, 0.2

4. (i) 9,27,9, 1, 27 (ii) 0.123, 0.494, 1.15, 1.73, 2.51 (iii) $4.29\,k_B$

5. (a) entropy increases by $0.70\,k_B$ (b) 1.76, 1.47, 0.98, 0.61, 0.18

6. (a) (i) 1 (ii) 348364800 (iii) 4200 (b) (i) 4 (ii) 104509440 (iii) 2400

7. 1114767360, increase in entropy $= 1.16\,k_B$

9. (a) 5 (b) macrostate with two particles in each level (c) 5.86×10^{-23}

10. Most likely distribution has two particles in each level, its entropy is $15.25\,k_B$, the least likely distribution is the one with all particles in the highest energy level and its entropy is $10.75\,k_B$

Answer to multiple choice questions

1. (b)   2. (d)   3. (c), (d)   4. (c)

5. (c)   6. (d)   7. (d)   8. (c)

9. (a), (b), (c)   10. (b)   11. B,D,E   12. (b), (d)

13. (d)   14. (a)

Revision

1. The energy values $\epsilon_j \left(= \dfrac{p_j^2}{2m} \right)$ of a free particle (not acted upon by any external force) of mass

   m confined to a volume V is given by,

   $$\epsilon_j = n_j^2 \, \frac{\hbar^2 \left(2\pi\right)^2}{8m} V^{-\frac{2}{3}}$$

   Here n_j is a non-zero integer. Also $n_j^2 = \left(n_x^2 + n_y^2 + n_z^2 \right)$ with n_x, n_y and n_z are themselves non-zero integers.

2. For a given n_j different possible combinations of n_x, n_y, and n_z gives the degeneracy of the level.

3. Identical particles may be distinguishable or indistinguishable

4. A system of large number of identical particles is called an assembly. Particles in an assembly are distributed in different discrete energy levels. The number of particles in a given energy level is called the occupation number of the level. A set of occupation numbers of all filled energy levels defines a macrostate of the system. For a given occupation number each different combination of particles in a level constitutes a microstate. The number of microstates of a given macrostate depends on the statistics followed by the particles of the assembly.

5. Indistinguishable identical particles may follow Bose–Einstein or Fermi–Dirac statistics, while distinguishable particles may follow Maxwell–Boltzmann or Classical statistics.

6. Quantum thermodynamics assumes that each microstate of a system is equally likely. This is called the equal a priori probability assumption. Total number of microstates of an assembly is called the (quantum) thermodynamic probability of the system.

7. Number of microstates $\mathcal{W}$ or quantum thermodynamic probabilities Ω for different types of statistics are:

   $$\Omega_{Bose-Einstein} = \mathcal{W}^{Bose-Ein} = \prod_j \frac{\left(g_j + N_j - 1\right)!}{\left\{\left(g_j - 1\right)!\right\}\left(N_j!\right)}$$

   $$\Omega_{Fermi-Dirac} = \mathcal{W}^{Fermi-Dirac} = \prod_j \frac{g_j!}{\left\{N_j!\right\}\left(g_j - N_j\right)!}$$

   $$\Omega_{Maxwell-Boltzmann} = \mathcal{W}^{Maxwell-Boltzmann} = N! \prod_j \frac{g_j^{N_j}}{N_j!}$$

8. The quantum thermodynamic probability Ω of a system is related to the entropy S of the system by the relation: $S = k_B \ln \Omega$ where k_B is Boltzmann constant and has values: 1.38×10^{-23} JK^{-1}; or 8.617×10^{-5} eV K^{-1} or 1.38×10^{-16} erg K^{-1}.

9. According to quantum thermodynamics entropy is a measure of the disorder of the system

10. In quantum thermodynamics change in entropy in any process may be explained in terms of the change in the number of accessible energy levels and the change in the number of possible microstates of the assembly.

11. The measured values of physical variables of an assembly depend on the average occupation numbers of the system over all its microstates. A relation that gives the average occupation number of the system is called the distribution function. Distribution functions for the three types of statistics are;

Distribution function for Bose–Einstein statistics: $\dfrac{\overline{N_j}}{g_j} = \dfrac{1}{\left[e^{-\left(\frac{\mu - \epsilon_j}{T k_B}\right)} - 1\right]}$

Distribution function for Fermi–Dirac statistics: $\dfrac{\overline{N_j}}{g_j} = \dfrac{1}{1 + e^{\frac{(\epsilon_j - \mu)}{k_B T}}}$

Distribution function for Maxwell–Boltzmann statistics: $\dfrac{\overline{N_j}}{g_j} = N e^{\left(\frac{\mu - \epsilon_j}{T k_B}\right)}$

If in the indistinguishable particle distribution like the Fermi–Dirac distribution, the number of particles per energy state is very low, the distribution reduces to the Classical distribution.

Distribution function for Classical statistics: $\dfrac{\overline{N_j}}{g_j} = e^{\frac{(\mu - \epsilon_j)}{k_B T}}$

12. Partition function Z for Maxwell–Boltzmann and Classical distributions is given by,

$$Z \equiv \sum_j g_j e^{-\frac{\epsilon_j}{k_B T}}$$

13. Partition function for a distribution provides a bridge between the quantum and classical thermodynamics. All functions of the classical thermodynamics may be obtained if the partition function of a system is known.

14. Important functions of classical thermodynamics like the internal energy U, entropy S, Helmholtz function F, etc., for Maxwell–Boltzmann distribution may be given in terms of its partition function as

$$U = -N \frac{\partial(\ln Z)}{\partial \beta} \quad NT^2 k_B \frac{\partial(\ln Z)}{\partial T}$$

$$C_V = C_V = 2NTk_B \left(\frac{\partial \ln Z}{\partial T}\right)_v + T^2 k_B N \left(\frac{\partial^2 \ln Z}{\partial T^2}\right)_V$$

$$S(T) = Nk_B \ln Z + \frac{U}{T}$$

$$F = -Nk_B T \ln Z$$

$$\text{Pressure } P = -\left(\frac{\partial F}{\partial V}\right)_{T,N} = Nk_B T \left(\frac{\partial \ln Z}{\partial V}\right)_{T,N}$$

$$\text{Chemical potential } \mu = \left(\frac{\partial F}{\partial N}\right)_{V,T} = -k_B T \left(\frac{\partial N \ln Z}{\partial N}\right)_{V,T} = -k_B T \ln Z$$

For classical distribution

$$\mu = -T k_B [\ln Z - \ln N]$$

$$F = -N T k_B \{\ln Z - \ln N + 1\}$$

$$P = -\left(\frac{\partial F}{\partial V}\right)_{T,N}$$

$$U = N k_B T^2 \left(\frac{\partial \ln Z}{\partial T}\right)_{V,N}$$

$$S = \frac{U}{T} + N k_B (\ln Z - \ln N + 1)$$

CHAPTER

Some Applications of Quantum Thermodynamics

11

11.0 Introduction

Essential tools of quantum thermodynamics were developed in chapter-10. In this chapter quantum thermodynamics will be applied to some of those systems that were earlier studied in the frame work of classical thermodynamics to illustrate the advantage of using quantum thermodynamics. We start with the quantum thermodynamic analysis of a monatomic ideal gas.

11.1 Quantum Thermodynamic Description of a Monatomic Ideal Gas

Kinetic theory defines an ideal gas as a gas that obeys Boyle's and Charles' laws at all temperatures and pressures. However, in classical thermodynamics an ideal gas is the one for which the ratio $\frac{Pv}{T}$ = R at all pressures. In quantum thermodynamics a system is characterized by the statistics followed by its entities. Following assumptions are generally made in quantum thermodynamics to describe a monatomic ideal gas.

1. All molecules of an ideal monatomic gas are identical, have one atom of gas each, have the same mass, say 'm' and are *indistinguishable*. Atoms are point particles having no volume. Since the gas is monatomic, the gas molecules have only translational motion and no energy is associated with their rotational or vibrational motions.
2. While discussing the quantum mechanical behavior of a monatomic ideal gas it is assumed that the gas is at a temperature higher than the temperature at which it liquefies, so that the system always remains in gaseous state.
3. It is further assumed that the molecules of an ideal gas do not interact with each other and with the walls of the container, except at the instant when they collide with each other or with the walls of the container. As such each molecule of an ideal gas is treated as a free, non-interacting and independent particle moving in the volume of the container. The energy level structure of the ideal gas molecule is, therefore, like that of a free particle in a box of volume V and the energy ϵ_j of the molecule in j^{th} energy level is given by $\epsilon_j = n_j^2 \frac{\hbar^2 (2\pi)^2}{8m} V^{-\frac{2}{3}}$

4. In light of the assumptions made above it may be proved that the mean occupation number or the average number of molecules in all possible energy levels is much smaller than the degeneracy of that level, i.e. $\dfrac{\overline{N_j}}{\Delta \mathcal{G}_j} \ll 1$.

The assumptions that the ideal gas molecules are assumed to be indistinguishable and that the mean number of particles in all possible free particle energy levels is small, facilitates the application of 'classical statistics' to the system consisting of ideal gas molecules (see Art. 10.6.3 of chapter-10).

11.1.1 Partition function for a monatomic ideal gas

In the previous chapter it has been shown that all thermodynamic properties of a quantum system may be obtained from its partition function. For a system that obeys 'classical statistics' the partition function Z is given by (Eq. 10.44, chapter-10)

$$Z = \sum_j g_j \, e^{-\frac{\epsilon_j}{k_B T}} \qquad\qquad 11.1$$

In Eq. 11.1 g_j is the degeneracy (number of states) of level j of energy ϵ_j, k_B the Boltzmann constant and T the Kelvin temperature of the ideal gas system. The summation on the right hand side is to be carried out over all possible energy levels of the system.

In chaper-10 we have seen that calculation of mean occupation number and the number of possible energy levels is easy if the number of particles is not very large. However, in the case of an ideal gas contained in a reasonable volume there is very large number of gas molecules and there is large number of energy levels accessible to molecules. In general, the number of energy levels of a system increases rapidly with the increase in the number of quantum numbers associated with the system. The level structures of systems with small and large quantum numbers are shown in Fig. 11.1 (a) and (b).

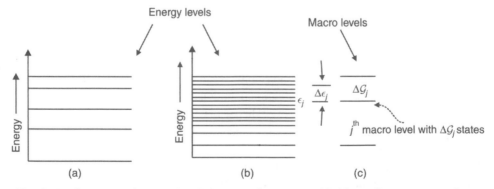

Fig. 11.1 Structure of energy levels in case of a system with (a) small quantum numbers (b) large quantum numbers and (c) grouping into macro levels

In the case of systems with large quantum numbers, like an ideal gas, often the level separations (energy difference between successive levels) become much less than the level energies. It, therefore, becomes difficult to calculate degeneracy and occupation number for each individual level. To simplify calculations in such cases, levels between certain energy ϵ_j and $(\epsilon_j + \Delta\epsilon_j)$ are lumped together and are replaced by a single level called the macro level of energy ϵ_j with the number of states (or degeneracy) $\Delta\mathcal{G}_j$, as shown in Fig. 11.1 (c). The magnitude of degeneracy $\Delta\mathcal{G}_j$ of the macro level depends on the energy interval $\Delta\epsilon_j$ and the degeneracy's $(g_j's)$ of all levels lumped together, which in turn depends on the type of the quantum statistics followed by the system. Let $\mathcal{G}_j$ be the sum of the states in all levels up to energy ϵ_j and $\Delta\mathcal{G}_j$ the number of states in the macro-level at energy ϵ_j. Applying the concept of macro- level to the case of an Ideal gas system that has large number of gas molecules and obeys 'classical distribution', the partition function given by Eq. 11.1 gets modified to,

$$Z = \sum_j \Delta\mathcal{G}_j\, e^{-\frac{\epsilon_j}{k_B T}}$$ 11.2

11.1.2   Calculation of the degeneracy $\Delta\mathcal{G}_j$ of the jth macro level

When a free particle of mass m is confined in a volume V, the energy of the jth level of the system is given by the expression,

$$\epsilon_j = n_j^2 \frac{\hbar^2 (2\pi)^2}{8m} V^{-\frac{2}{3}}$$ 11.3

Where quantum number n_j is a positive non-zero integer and is related to the other three quantum numbers, n_x, n_y and n_z by the relation,

$$n_j^2 = n_x^2 + n_y^2 + n_z^2$$ 11.4

The degeneracy of level $-j$ (or the number of states in the level- j) arises because different sets of n_x, n_y and n_z may give the same value for n_j. In order to count the number of these different sets of n_x, n_y and n_z that gives the same value of n_j, we consider a three dimensional space made up of the three mutually perpendicular axes n_x, n_y, n_z and call this space the n-space. It may be remembered that the quantum number n_x, n_y and n_z and n_z may have only positive non-zero integer values on their respective axes and that any set of these (integer values of these quantum numbers) will give a possible state of the system. If we denote a given set of positive integer values of quantum number n_x, n_y, and n_z by a black point in the n-space, then such black points lying only in the positive quadrant of the n-space will give the number of possible states of the system. We call this distribution of black points in positive quadrant of space as a grid. Since the volume of the positive quadrant is 1/8 of the total n-space, the grid of black dots denoting possible states of the system will all lie only in 1/8 of the total n-space. Though the grid of points will be a three dimensional body, however, in Fig. 11.2(a) two-dimensional section of the grid is shown.

An interesting property of this grid is that each point of the grid is separated from the adjacent point by a cubical cell of unit volume on all sides. In other words one may say that there is one point (or possible state) per unit volume of the n-space. Thus it is possible to determine the total number of states in all levels up to and including the jth level from the volume of the n-space associated with a sphere of radius n_j as shown in Fig. 11.2. We denote the total number of states of the system in all levels up to and including the jth level by $\mathcal{G}_j$, then,

$$\mathcal{G}_j = \frac{1}{8}\ \text{volume of sphere of radius } n_j$$

Or
$$= \frac{1}{8}\left(\frac{4}{3}\pi n_j^3\right) = \frac{1}{6}\pi n_j^3 \qquad 11.5$$

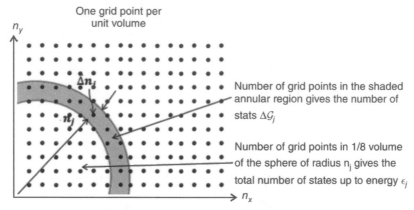

One grid point per unit volume

n_y

Δn_j

n_j

n_x

Number of grid points in the shaded annular region gives the number of stats $\Delta \mathcal{G}_j$

Number of grid points in 1/8 volume of the sphere of radius n_j gives the total number of states up to energy ϵ_j

Fig. 11.2   Each point in the figure gives one possible state of the system

Since the number of points in the grid is very large, n_j may be assumed to be a continuous variable and the number of states $\Delta \mathcal{G}_j$ in the macro level $-j$ (the grid dots lying in the strip) may be obtained by differentiating Eq. 11.5 with respect to n_j,

$$\Delta \mathcal{G}_j = \frac{1}{2}\pi n_j^2 \, dn_j \qquad 11.6$$

In geometric terms this refers to the number of grid points lying in a spherical strip of thickness dn_j around the sphere of radius n_j.

The partition function for an ideal gas may be obtained by substituting the value of $\Delta \mathcal{G}_j$ from Eq. 11.6 in Eq. 11.2 to get,

$$Z = \frac{1}{2}\pi \sum_j n_j^2 \, dn_j \, e^{-\frac{\epsilon_j}{k_B T}} \qquad 11.7$$

Since it has been assumed that n_j are continuous, the summation in Eq. 11.7 may be replaced by integration from zero to infinity and the value of energy ϵ_j in terms of n_j may be substituted from Eq. 11.3. Hence, the partition function Z for an ideal gas may be given as,

$$Z = \frac{1}{2}\pi \int_0^\infty n_j^2 \, dn_j \, e^{-\frac{n_j^2 \frac{\hbar^2 (2\pi)^2}{8m} V^{-\frac{2}{3}}}{k_B T}}$$

$$= \frac{1}{2}\pi \int_0^\infty n_j^2 \, e^{-\left(\frac{n_j^2 \hbar^2 (2\pi)^2 V^{-\frac{2}{3}}}{8 m k_B T}\right)} \, dn_j \qquad 11.8$$

Equation 11.8 may be written as,

Or
$$Z = \frac{1}{2}\pi \int_0^\infty y^2 e^{-c y^2} \, dy \qquad 11.9$$

where,
$$y = n_j \text{ and } c = \frac{\hbar^2 (2\pi)^2 V^{-\frac{2}{3}}}{8mk_B T}$$ 11.10

However, $\int_0^\infty y^2 e^{-cy^2} dy$ is a definite integral with the value $= \frac{1}{4}\left[\left(\frac{\pi}{c^3}\right)^{\frac{1}{2}}\right]$. With these substitutions

Eq. 11.9 gives,

$$Z = \left[V\left(\frac{mk_B T}{2\pi\hbar^2}\right)^{\frac{3}{2}}\right]$$ 11.11

Equation 11.11 gives the partition function of a monatomic ideal gas. It may be noted that the partition function depends both on the volume of the gas V and the temperature of the gas T. It may be recalled that V and T are the two, one extensive and other intensive, variables of classical thermodynamics.

In chapter-10 it was shown that all important functions of classical thermodynamics may be obtained from the partition function of the system. In the following we will obtain these classical thermodynamic functions using the value of partition function Z for an ideal gas.

11.2   Classical Thermodynamic Functions for the Quantum Ideal Gas

11.2.1   The pressure

Equation 10.70, given here for ready reference, may be used to obtain the pressure P of the system if the partition function Z of the system is known.

$$P = NTk_B \left(\frac{\partial Z}{\partial V}\right)_{T,N}$$ 11.12

For the quantum ideal gas,

$$Z = \left[V\left(\frac{mk_B T}{2\pi\hbar^2}\right)^{\frac{3}{2}}\right]$$

And

$$\ln Z = \ln V + \frac{3}{2}\ln\left(\frac{mk_B T}{2\pi\hbar^2}\right)$$

Therefore,
$$\left(\frac{\partial \ln Z}{\partial V}\right)_{T,N} = \frac{1}{V}$$

Hence,
$$P = Nk_B T\left(\frac{\partial \ln Z}{\partial V}\right)_{T,N} = \frac{Nk_B T}{V}$$ 11.13

However, $k_B = \dfrac{\mathbb{N}R}{N}$, where $\mathbb{N}$ is the number of moles of the gas and R the gas constant

Therefore,
$$P = N k_B T \left(\frac{\partial \ln Z}{\partial V}\right)_{T,N} = \frac{N k_B T}{V} = \frac{\mathbb{N} R \, T}{V} \qquad 11.14$$

Equation 11.14 derived using the partition function, gives the desired relation between the pressure, volume and temperature of an ideal gas (equation of state). It may be noted that the same expression for the pressure of an ideal gas was obtained in chapter-1 using the kinetic theory.

11.2.2 Internal energy U, and heat capacity at constant volume C_v, of the quantum ideal gas

The internal energy U of an ideal gas may be obtained from Eq. 10.73, as:

$$U = NT^2 k_B \frac{\partial (\ln Z)}{\partial T} = NT^2 k_B \left[\frac{\partial\left\{\ln V + \dfrac{3}{2}\ln\left(\dfrac{m k_B T}{2\pi\hbar^2}\right)\right\}}{\partial T}\right] = NT^2 k_B \left[\frac{3}{2T}\right]$$

Or
$$U = \frac{3}{2} NTk_B = \frac{3}{2} \mathbb{N}RT \qquad 11.15$$

The heat capacity at constant volume C_v may be obtained from the relation

$$C_v = \left(\frac{\partial U}{\partial T}\right)_V = \frac{3}{2} Nk_B = \frac{3}{2} \mathbb{N}R \qquad 11.16$$

And the specific molar heat capacity c_v is,

$$c_v = C_v\big/\mathbb{N} = \frac{3}{2} R \qquad 11.17$$

11.2.3 Entropy S(T) of the quantized ideal gas

The expression for the entropy of a system obeying classical distribution is given by Eq. 10.74 of chapter-10

$$S(T) = \frac{U}{T} + N k_B (\ln Z - \ln N + 1)$$

Substituting the values of U and ln Z in the above equation one gets,

$$S(T) = \frac{3}{2} Nk_B + N k_B\left(\ln V + \frac{3}{2} \ln\left(\frac{m k_B T}{2\pi\hbar^2}\right) - \ln N + 1\right)$$

$$= Nk_B\left[\frac{5}{2} + \ln\left\{\frac{V}{N}\left(\frac{m k_B T}{2\pi\hbar^2}\right)^{\frac{3}{2}}\right\}\right] \qquad 11.18$$

The above expression for the entropy may also be written in terms of the number of moles $\mathbb{N}$ and the gas constant R by replacing Nk_B by the equivalent term $\mathbb{N}R$,

Therefore,
$$S(T) = \mathbb{N}R\left[\frac{5}{2} + \ln\{\frac{V}{N}\left(\frac{mk_BT}{2\pi\hbar^2}\right)^{\frac{3}{2}}\}\right]$$

Or
$$= \frac{5}{2}\mathbb{N}R + \mathbb{N}R\ln V + \mathbb{N}R\ln\{\frac{1}{N}\left(\frac{mk_BT}{2\pi\hbar^2}\right)^{\frac{3}{2}}\} \qquad 11.19$$

An expression for the specific molar entropy s(T) for the quantum ideal gas may be obtained by dividing the Eq. 11.19 by the number of moles $\mathbb{N}$,

$$s(T) = \frac{5}{2}R + R\ln V + \frac{3}{2}R\ln T + R\left[\ln\{\frac{1}{N}\left(\frac{mk_B}{2\pi\hbar^2}\right)^{\frac{3}{2}}\}\right] \qquad 11.20$$

The coefficient $\frac{3}{2}R$ in the third term in above expression may be replaced by the molar specific heat capacity c_v (Eq. 11.17) to get the following expression often called the *Sackur–Tetrode* equation for the absolute entropy of a monatomic ideal quantum gas,

$$s(T) = \frac{5}{2}R + R\ln V + c_v\ln T + R\left[\ln\{\frac{1}{N}\left(\frac{mk_B}{2\pi\hbar^2}\right)^{\frac{3}{2}}\}\right] \qquad 11.21$$

An important property of the quantum derivation of entropy is that it gives the absolute value of entropy without any undetermined constant. On the other hand in classical thermodynamics entropy is always relative and it is only the difference in entropies of the two states of the system that can be calculated. This is a big advantage of using the quantum thermodynamic approach.

11.3   Speed Distribution of Ideal Gas Molecules

Expression for the speed distribution of the molecules of an ideal gas has been derived in chapter-1. The same may also be obtained using the formulations for the average occupation number ΔN_j for the j-th macro-level at energy ϵ_j and $(\epsilon_j + \Delta\epsilon_j)$. Equation 10.52 that holds for an isolated level at energy ϵ_j for a system that obeys either classical or Maxwell–Boltzmann distribution, may be written as,

$$\frac{\overline{N_j}}{g_j} = \frac{N}{Z}e^{-\frac{\epsilon_j}{Tk_B}}$$

$\overline{N_j}$ and g_j in the above expression are respectively the average occupation number and the degeneracy of the jth level. The corresponding expression for the jth macro-level may be written as,

$$\frac{\Delta N_j}{\Delta \mathcal{G}_j} = \frac{N}{Z}e^{-\frac{\epsilon_j}{Tk_B}} \qquad 11.22$$

In Eq. 11.22, ΔN_j is the average occupation number of particles (ideal gas molecules) in the jth macro-level and $\Delta \mathcal{G}_j$ is the degeneracy (number of states) of the jth macro-level. Our aim is to find an expression for ΔN_j in terms of the speed v of the molecule. For that we use the following procedure. Rewriting Eq. 11.22 we have,

$$\Delta N_j = \frac{N}{Z} \Delta \mathcal{G}_j \, e^{-\frac{\epsilon_j}{Tk_B}}$$

11.23

And expressing the energy ϵ_j in terms of the speed v_j of the particle in j^{th} level one has,

$$\epsilon_j = n_j^2 \frac{\hbar^2 (2\pi)^2}{8m} V^{-\frac{2}{3}} = \frac{1}{2} m v_j^2$$

Therefore,

$$n_j^2 = \frac{1}{2} m v_j^2 \times \frac{8mV^{\frac{2}{3}}}{\hbar^2 (2\pi)^2} = \frac{4m^2 v_j^2 V^{\frac{2}{3}}}{\hbar^2 (2\pi)^2}$$

11.24

And

$$n_j = \frac{2m v_j V^{\frac{1}{3}}}{2\pi \hbar}$$

11.25

Hence,

$$\Delta n_j = \frac{2m V^{\frac{1}{3}}}{2\pi \hbar} \Delta v_j$$

11.26

However, from Eq. 11.6 we have,

$$\Delta \mathcal{G}_j = \frac{1}{2} \pi n_j^2 \, \Delta n_j$$

Substituting values of n_j^2 and Δn_j in the above expression, respectively from Eq. 11.24 and Eq. 11.26 one gets,

$$\Delta \mathcal{G}_j = \frac{4\pi m^3 V}{(2\pi)^3 \hbar^3} v_j^2 \Delta v_j$$

11.27

The average number of molecules with velocity v_j (or the average occupation number) of the jth macro-level may be found by substituting the value of $\Delta \mathcal{G}_j$ from Eq. 11.27 in Eq. 11.23 as,

$$\Delta N_j^{v_j} = \frac{N}{Z} \frac{4\pi m^3 V}{(2\pi)^3 \hbar^3} e^{-\frac{\epsilon_j}{Tk_B}} v_j^2 \Delta v_j$$

11.28

It may be noted that all molecules in the jth macro-level have their speeds between v_j and $(v_j + \Delta v_j)$.

Putting the value of $Z(= \left[V \left(\frac{m k_B T}{2\pi \hbar^2} \right)^{3/2} \right])$ from Eq. 11.11 and the value of $\epsilon_j (= \frac{1}{2} m v_j^2)$ in Eq. 11.28 one gets,

$$\Delta N_j^{v_j} = 4N(\pi)^{-\frac{1}{2}} \left(\frac{m}{2k_B T} \right)^{\frac{3}{2}} v_j^2 \, e^{-\left(\frac{m v_j^2}{2k_B T} \right)} \Delta v_j$$

11.29

And

$$\frac{\Delta N_j^{v_j}}{\Delta v_j} = 4N(\pi)^{-\frac{1}{2}} \left(\frac{m}{2k_BT}\right)^{\frac{3}{2}} v_j^2 \, e^{-\left(\frac{mv_j^2}{2k_BT}\right)}$$

11.30

In Eq. 11.30 $\Delta N_j^{v_j}$ is the average number of ideal gas molecules that have their velocities between v_j and $(v_j + \Delta v_j)$ while N is the total number of molecules. Eq. 11.30 gives the desired speed distribution of ideal gas molecules. Apart from the term $\left[4N(\pi)^{-\frac{1}{2}} \left(\frac{m}{2k_BT}\right)^{\frac{3}{2}} \right]$ that is constant for a given mass of the molecule, the shape of the speed distribution curve is decided by the two terms v_j^2 and $e^{-\left(\frac{mv_j^2}{2k_BT}\right)}$, the first term increases as the speed v_j increases while the second term exponentially decreases with the increase of the speed. The resulting speed distribution curve for the ideal gas molecules is shown in Fig. 11.3.

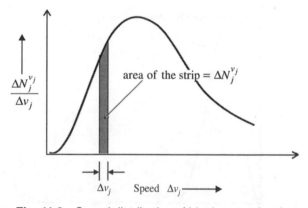

Fig. 11.3   Speed distribution of ideal gas molecules

The speed distribution curve passes through (0, 0), indicating that at a finite non-zero temperature essentially no molecule has zero speed. Further, as shown in the figure the area of a strip of width Δv_j and height $\frac{\Delta N_j^{v_j}}{\Delta v_j}$ is equal to the number of molecules $\Delta N_j^{v_j}$ in the level.

Since the jth level may be any of the levels, the subscript j may be dropped and Eq. 11.30 may also be written as

$$\frac{\Delta N^v}{\Delta v} = 4N(\pi)^{-\frac{1}{2}} \left(\frac{m}{2k_BT}\right)^{\frac{3}{2}} v^2 \, e^{-\left(\frac{mv^2}{2k_BT}\right)}$$

11.30(a)

Where N denotes total number of gas molecules and ΔN^V the number of molecules with velocities in the range v and $(v + \Delta v)$ at temperature T.

11.3.1   Temperature dependence of the speed distribution function

The temperature T appears in the multiplicative term $\left(\dfrac{m}{2k_BT}\right)^{\frac{3}{2}}$ and the exponential term $e^{-\left(\dfrac{mv_j^2}{2k_BT}\right)}$. All

other factors remaining same, the multiplicative term decreases with the rise of the temperature while the magnitude of the exponential term increases. The net result of the rise in temperature is that the maxima of the distribution curve shifts away from the origin towards the higher velocity side and the width of the curve increases making the peak of the curve broad. Speed distribution curves for three temperatures T_1, T_2, and T_3, where $T_1 < T_2 < T_3$ are shown in Fig. 11.4. Since the area of the curve gives the total number of molecules N, the area lying between the curve and the x-axis does not change with the change of the temperature. It is the height of the peak that decreases with the rise in temperature making the distribution curve broader. This is expected also as with the rise in temperature larger number of gas molecules acquires higher speed that shifts the peak of the speed distribution curve towards higher speed, however, since the total number of gas molecules remains the same, the height of the peak decreases to keep the area of the curve constant.

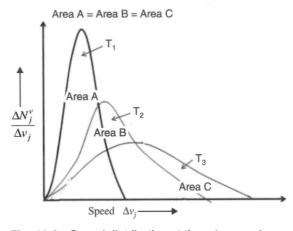

Fig. 11.4   Speed distribution at three temperatures

11.3.2   The most probable, mean and the root mean square speeds of ideal gas molecules

As has been shown in Fig. 11.1(c), the system consisting of an ideal gas may be considered to be made up of several macro-levels j, k, I, etc., at energies ϵ_j, ϵ_k, ϵ_l ..., etc. The average occupation numbers of these levels will have different values depending on the energy and the width of the macro-level. The average occupation number of one of these macro-levels will have maximum value, i.e., the macro-level will have largest number of ideal gas molecules. The speed of the molecules in this particular macro-level of largest occupation number is called the most probable speed of the molecular speed distribution and is represented by v_m. To obtain an expression for the most probable speed one may differentiate Eq. 11.30 and set it equal to zero, i.e.,

$$\frac{d}{dv_j}\left[4N(\pi)^{-\frac{1}{2}}\left(\frac{m}{2k_BT}\right)^{\frac{3}{2}}v_j^2\,e^{-\left(\frac{mv_j^2}{2k_BT}\right)}\right] = 0$$

For a non-zero finite temperature the above expression reduces to

$$\frac{d}{dv_j}\left[v_j^2\,e^{-\left(\frac{mv_j^2}{2k_BT}\right)}\right] = 0$$

Or

$$2v_j\,e^{-\left(\frac{mv_j^2}{2k_BT}\right)} - \frac{mv_j^2}{2k_BT}2v_je^{-\left(\frac{mv_j^2}{2k_BT}\right)} = 0$$

Or

$$1 - \frac{mv_j^2}{2k_BT} = 0 \ or \ v_j = \left(\frac{2k_BT}{m}\right)^{\frac{1}{2}} \qquad 11.31$$

The speed defined by Eq. 11.31 is the most probable speed denoted by v_m

$$v_m = \left(\frac{2k_BT}{m}\right)^{\frac{1}{2}} = 1.414\sqrt{\frac{k_B}{m}} \qquad 11.32$$

The speed distribution curve tells that different molecules of the gas have different speeds. Assuming the speed distribution to be continuous, Eq. 11.30 may be written as,

$$dN_j^{v_j} = 4N(\pi)^{-\frac{1}{2}}\left(\frac{m}{2k_BT}\right)^{\frac{3}{2}}v_j^2\,e^{-\left(\frac{mv_j^2}{2k_BT}\right)}dv_j \qquad 11.33$$

The average or the mean speed $\bar{v}$ may be obtained using the standard formulation

$$\bar{v} = \frac{1}{N}\int_0^\infty v_j dN_j^{v_j} = \frac{1}{N}\int_0^\infty v_j\,4N(\pi)^{-\frac{1}{2}}\left(\frac{m}{2k_BT}\right)^{\frac{3}{2}}v_j^2\,e^{-\left(\frac{mv_j^2}{2k_BT}\right)}dv_j \qquad 11.34$$

Although the fact that there is no molecule with zero speed and also with infinite speed, the lower and upper limits of the integration in Eq. 11.34 are taken as zero and infinite since the distribution curve approaches to zero very sharply both for the upper and the lower ends.

$$\bar{v} = 4(\pi)^{-\frac{1}{2}}\left(\frac{m}{2k_BT}\right)^{\frac{3}{2}}\int_0^\infty v_j^3\,e^{-\left(\frac{mv_j^2}{2k_BT}\right)}dv_j$$

The above equation may be written as,

$$\bar{v} = 4(\pi)^{-\frac{1}{2}}\left(\frac{m}{2k_BT}\right)^{\frac{3}{2}}\int_0^\infty y^3\,e^{-ay^2}\,dy \qquad 11.35$$

where $= v_j$ and $a = \dfrac{m}{2k_BT}$. The definite integral $\int_0^\infty y^3\,e^{-ay^2}\,dy$ has the value $= \dfrac{1}{2a^2} = \dfrac{1}{2\left(\dfrac{m}{2k_BT}\right)^2}$

Substituting the value of the definite integral in Eq. 11.35, one gets

$$\bar{v} = 4(\pi)^{-\frac{1}{2}}\left(\frac{m}{2k_BT}\right)^{\frac{3}{2}}\frac{1}{2\left(\frac{m}{2k_BT}\right)^2} = \frac{2}{\sqrt{\pi}}\left(\frac{m}{2k_BT}\right)^{-\frac{1}{2}} = \sqrt{\frac{8}{\pi}\frac{k_BT}{m}}$$

Therefore, the average speed of gas molecules $\bar{v} = \dfrac{2}{\sqrt{\pi}}\left(\dfrac{m}{2k_BT}\right)^{-\frac{1}{2}} = \sqrt{\dfrac{8}{\pi}\dfrac{k_BT}{m}}$

$$= \frac{2}{\sqrt{\pi}}v_m = 1.1281\,v_m$$

The root mean square speed defined as $v_{rms} \equiv \sqrt{\left(v_j^2\right)}$ may be calculated from the relation

$$(v_{rms})^2 = \frac{1}{N}\int_0^\infty v_j^2\,dN_j^{v_j} = \frac{1}{N}\int_0^\infty v_j^2\,4N(\pi)^{-\frac{1}{2}}\left(\frac{m}{2k_BT}\right)^{\frac{3}{2}}v_j^2\,e^{-\left(\frac{mv_j^2}{2k_BT}\right)}dv_j$$

$$= 4(\pi)^{-\frac{1}{2}}\left(\frac{m}{2k_BT}\right)^{\frac{3}{2}}\int_0^\infty y^4\,e^{-ay^2}\,dy\,;\text{ Where } y = v_j \text{ and } a = \frac{m}{2k_BT}$$

The definite integral $\int_0^\infty y^4\,e^{-ay^2}\,dy$ has the value $\dfrac{3}{8}\sqrt{\dfrac{\pi}{a^5}}$. Substituting this value in the above

equation gives,

$$(v_{rms})^2 = \frac{3}{2}\left(\frac{2k_BT}{m}\right) = \frac{3}{2}v_m^2$$

And
$$v_{rms}^{\;2} = \sqrt{\frac{3k_BT}{m}} = \sqrt{3/2}\,v_m = 1.225\,v_m \qquad\qquad 11.36$$

It may be noted that the expressions for the mean speed $\bar{v}$, most probable speed v_m and the root mean square speed v_{rms} presently obtained using the partition function are same as obtained earlier in chapter-1 using the kinetic theory approach. This demonstrates that the quantum thermodynamic approach is capable of reproducing results obtained earlier using kinetic theory.

11.3.3 Maxwell–Boltzmann velocity distribution function

Maxwell–Boltzmann velocity distribution function tells how many molecules of the gas with speeds between v_j and $(v_j + dv_j)$ may be found in a unit volume anywhere in the gas. In other words it tells about the density distribution of gas molecules with speeds between v_j and $(v_j + dv_j)$. To obtain an expression for this distribution we refer to Fig. 1.1 of Chapter-1 where molecular velocity vectors are transported to a common origin. Molecules with speeds between v_j and $(v_j + dv_j)$ will all lie in the velocity space in a spherical shell of inner radius v_j and outer radius $(v_j + d_{vj})$, as shown in Fig. 11.5. Further, the number $dN_j^{v_j}$ of molecules with speeds between v_j and $(v_j + d_{vj})$ is given by Eq. 11.33.

The volume of the spherical shell ΔV_j is

$$\Delta V_j = \frac{4}{3}\pi\left[\left(v_j + dv_j\right)^3 - \left(v_j\right)^3\right] = 4\pi v_j^{~2} dv_j \qquad 11.37$$

In deriving above expression terms containing higher powers of dv_j are neclected.

Hence, $\rho_{v_j} \equiv \dfrac{dN_j^{v_j}}{\Delta V_j} = \dfrac{4N(\pi)^{-\frac{1}{2}}\left(\dfrac{m}{2k_BT}\right)^{\frac{3}{2}} v_j^2 e^{-\left(\dfrac{mv_j^2}{2k_BT}\right)} dv_j}{4\pi v_j^{~2} dv_j} = \dfrac{N}{\pi^{3/2}}\left(\dfrac{m}{2k_BT}\right)^{3/2} e^{-\left(\dfrac{mv_j^2}{2k_BT}\right)}$ \qquad 11.38

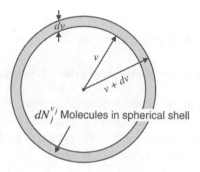

Fig. 11.5 In velocity space, $dN_j^{v_j}$ gas molecules with velocities between v and (v + dv) lie in a spherical shell of thickness dv around a sphere of radius v.

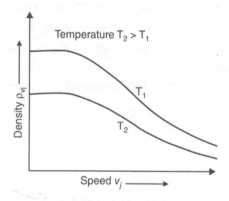

Fig. 11.6 Maxwell-Boltzmann velocity distribution function at two temperatures

Variation of the Maxwell–Boltzmann velocity function ρ_{v_j} with v_j is shown in Fig. 11.6 which may create some confusion because it appears from this figure that the density of molecules with zero speed ($v_j = 0$) is maximum while Fig. 11.4 tells that there is no molecule with speed zero. This contradiction may be understood when one considers the volume of velocity space available at $v_j = 0$. The volume is a sphere of radius dv_j, and since dv_j is negligibly small the corresponding volume is also negligible. Further, at a fixed temperature

$$dN_j^{v_j} \propto v_j^2 e^{-\left(\dfrac{mv_j^2}{2k_BT}\right)} dv_j$$

while,

$$\Delta V_j \propto v_j^2 dv_j$$

Therefore the ratio

$$\rho_{v_j} \propto e^{-\left(\dfrac{mv_j^2}{2k_BT}\right)}, \text{ as shown in Fig. 11.6.}$$

At a higher temperature both the magnitude of ρ_{vj} and its rate of exponential decay with the increase of v_j becomes smaller.

11.3.4   About Boltzmann distribution law

If one considers a system with energy E $(= \frac{1}{2}mv^2)$ that obeys either the Maxwell–Boltzmann or classical distribution and look to the expression that describes some distribution function of the system, for example the velocity distribution function (Eq. 11.30) or the density distribution function (Eq. 11.38), it will be observed that the property of the system described by the distribution function occurs with the frequency that is proportional to the factor

$$e^{-(E/k_BT)}$$

As a matter of fact all these distribution laws may be derived from a more general law called *Boltzmann distribution law which states that if the energy associated with some state of a system is E, then the frequency with which that state occurs is proportional to* $e^{-(E/k_BT)}$ *where T is the absolute temperature at which the system is in equilibrium and* k_B *is the Boltzmann constant.* In the following it will be shown how Boltzmann distribution law may be used to derive the number density, i.e., the number of molecules per unit volume of an ideal gas under gravitational field of the earth.

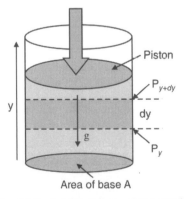

Fig. 11.6   (a) Ideal gas under gravity

Figure 11.6(a) shows an ideal gas contained in a cylindrical volume of base area A. It is assumed that the gas is under gravity of uniform acceleration g which is assumed to be constant. Let us consider a part of the gas enclosed between the two surfaces shown by dotted lines, the pressures at the upper and the lower surfaces are denoted respectively by P_{y+dY} and P_y where $P_y > P_{y+dy}$. Further, let us denote the difference $(P_y - P_{y+dy}) = dP$. Also let $\rho(y)$ denote the number density of gas molecules (i.e. the number of gas molecule in unit volume) in the volume enclosed by the surfaces shown by dotted lines and since dy is very small it may be taken to be constant within the enclosed volume.

Now
$$P_{y+dy} - P_y = -dP = \frac{\left(mass\,of\,molecules\,in\,volume\ A\,dy\right)x\,g}{area\,A}$$

Or
$$dP = -m\rho(y)g\,dy \qquad\qquad 11.38(a)$$

Here m denotes the mass of the gas molecule. It is, however, known from the equation of state of an ideal gas that

$$PV = Nk_BT \text{ or } P = \frac{N}{V}k_BT = \rho k_BT \text{ substituting this value of P in Eq. 11.38 (a) one gets,}$$

$$d\rho\,(y)\,k_BT = -m\rho\,(y)\,dy\,g$$

Or
$$\frac{d\rho(y)}{dy} = -\frac{mg}{k_BT}\rho(y) \qquad\qquad 11.38(b)$$

Equation 11.38(b) is a differential equation and has the solution

$$\rho(y) = \rho(y_0)e^{-\left(\frac{mg(y-y_0)}{k_BT}\right)} = \rho(y_0)e^{\left(\frac{mgy_0}{k_BT}\right)}e^{-\left(\frac{mgy}{k_BT}\right)} \qquad\qquad 11.38(c)$$

In Eq. 11.38(c) y_0 is an arbitrary reference height. If gas is a mixture of several gases with different masses, each component gas will have its own density distribution. Since the probability of finding a molecule at a certain height y is proportional to the number density $\rho(y)$, Eq. 11.38(c) also gives the probability distribution of gas molecules as a function of height y. This is also called "barometric distribution". Equation 11.38(c) tells that the probability of finding a particle at a height y is proportional

to $e^{-\left(\frac{mgy}{k_BT}\right)} = e^{-\left(\frac{E_P}{k_BT}\right)}$, where $E_p (=mgy)$ is the gravitational potential energy of the particle. Equation 11.38(c) is another manifestation of the Boltzmann distribution law.

11.4   The Equipartition Theorem

The equipartition theorem says that under certain conditions energy is evenly shared between all energetically accessible degrees of freedom of a system. This is not surprising and may be thought to be another way of saying that in an effort to maximize its entropy the system, under certain conditions, distributes the energy equally in all modes of motion available to it.

In general the total energy of a system is a function of many independent parameters. Each such parameter is called a degree of freedom, for example, velocity components v_x, v_y, v_z and frequency components ω_x, ω_y, ω_z etc may be called the degrees of freedom for a particles that have translational and vibrational motions. Similarly, the vertical height h of a body from the ground in gravitational field is also a degree of freedom since the gravitational potential energy of the body depends on h.

The theorem of equipartition of energy is specifically applicable to those modes of motion where the energy corresponding to that mode is proportional to the square of the parameter specifying the degree of freedom. Such modes are: the rectilinear motion, vibrational and rotational motions, etc.

since the kinetic energy E_x of rectilinear motion in x-direction is proportional to v_x^2 where v_x is the parameter that defines the degree of freedom. Similarly, the rotational energy $E_{\omega_x}^{rot}$ associated with the rotational frequency ω_x is proportional to ω_x^2. The same is true for other components of these motions. *The equipartition theorem states that in such cases where the energy is a quadric function of the parameter defining the degree of freedom, $\frac{1}{2}k_BT$ of energy is associated with each degree of freedom.* It may however, be noted that in the case of potential energy in gravitational field, $E_{pot}^{grav} \propto h$ *and not to* h^2. Hence the equipartition theorem is not applicable to the case of gravitational potential energy.

We will prove this theorem for quantum systems that follow either the classical or the Maxwell–Boltzmann distributions. For the classical and the Maxwell–Boltzmann distributions the average number $\overline{N_j}$ of particles with energy ϵ_j is given, respectively, as Eq. 10.42a and Eq. 10.48a in chapter-10,

$$\overline{N_j} = g_j e^{\frac{\mu}{k_BT}} e^{-\left(\frac{\epsilon_j}{k_BT}\right)} \quad \text{and} \quad \overline{N_j} = N g_j e^{\frac{\mu}{k_BT}} e^{-\left(\frac{\epsilon_j}{k_BT}\right)}$$

Both the above expressions may be represented by a single expression given below, where the constant A may have different values for the two distributions.

$$\overline{N_j} = A e^{-\left(\frac{\epsilon_j}{k_BT}\right)} \tag{11.39}$$

If it is now assumed that energy $\epsilon_{j(x)}$ is a function of some degree of freedom which is represented by x and that the energy and x may vary continuously say from 0 to ∞, then the average number of particles $\overline{dN_j^x}$ with energies in the small range dx of the degree of freedom will be given by,

$$\overline{dN_j^x} = A e^{-\left(\frac{\epsilon_j(x)}{k_BT}\right)} dx \tag{11.40}$$

The total number of particles N may be obtained by integrating Eq. 11.40,

$$N = \int_0^\infty A e^{-\left(\frac{\epsilon_j(x)}{k_BT}\right)} dx \tag{11.41}$$

Also the total energy ϵ_{total}, the sum of the energies of all particles, may be obtained by integrating the quantity obtained from the multiplication of $\epsilon_j(x)$ and $\overline{dN_j^x}$, i.e.,

$$\epsilon_{total} = \int_0^\infty A \epsilon_j(x) e^{-\left(\frac{\epsilon_j(x)}{k_BT}\right)} dx \tag{11.42}$$

Now the mean energy of the particles $\overline{\epsilon_j}(x)$ may be obtained by dividing the total energy ϵ_{total} by the total number of particles N,

Therefore,

$$\overline{\epsilon_j}(x) = \frac{\int_0^\infty \epsilon_j(x) e^{-\left(\frac{\epsilon_j(x)}{k_B T}\right)} dx}{\int_0^\infty e^{-\left(\frac{\epsilon_j(x)}{k_B T}\right)} dx} \qquad 11.43$$

If energy $\epsilon_j(x)$ is a quadratic function of the parameter x that defines the degree of freedom, i.e $\epsilon_j(x) = bx^2$, where b is a constant, then Eq. 11.43 becomes

$$\overline{\epsilon_j}(x) = \frac{\int_0^\infty bx^2 e^{-\left(\frac{bx^2}{k_B T}\right)} dx}{\int_0^\infty e^{-\left(\frac{bx^2}{k_B T}\right)} dx} \qquad 11.44$$

Now the definite integrals $\int_0^\infty bx^2 e^{-\left(\frac{bx^2}{k_B T}\right)} dx = \frac{b}{4}\sqrt{\frac{\pi k_B^3 T^3}{b^3}}$ and $\int_0^\infty e^{-\left(\frac{bx^2}{k_B T}\right)} dx = \frac{1}{2}\sqrt{\frac{\pi k_B T}{b}}$

Hence, Eq. 11.44 reduces to

$$\overline{\epsilon_j}(x) = \frac{1}{2} k_B T \qquad 11.45$$

Thus it may be observed that $\frac{1}{2} k_B T$ of energy is associated with each degree of freedom provided the energy is a quadratic function of the parameter that defines the degree of freedom. This proves the theorem of equipartition of energy. In case of a monatomic molecule there are three degrees of freedom (v_x, v_y, v_z) for the rectilinear motion and hence the total mean energy associated with each molecule at temperature T is $\frac{3}{2} k_B T$. Many motions, like the vibrational and rotational motions as discussed earlier fulfill the condition and each of these motions has $\frac{1}{2} k_B T$ energy associated with each degree of freedom at temperature T. It is worth mentioning that the proof of the theorem of equipartition given above is based on quantum statistics.

11.5 Heat Capacity of Polyatomic Gas Molecules

Monatomic gas molecule can have only three degrees of freedom on account of its translational motion. Polyatomic molecule, however, may have many degrees of freedom because atoms of the molecule may also have vibrational and rotational motions in addition to the translational motion of its center of mass. Further, a polyatomic molecule also has electronic excitations superimposed on rotational and vibrational excitations. As such a polyatomic molecule has an internal energy which is made up electronic, rotational and vibrational excitations. Study of the temperature dependence of the specific molar thermal capacity at constant volume of polyatomic gases suggests that the vibrational and rotational modes of motion in polyatomic molecules may be associated with quantized linear oscillators. We shall, therefore, first study the properties of an assembly of quantized linear oscillators.

11.5.1   An assembly of weakly interacting distinguishable non degenerate quantized linear oscillators

A particle of mass m executing simple harmonic motion (SHM) in a straight line about a fixed point constitutes a linear oscillator. The force F responsible for the SHM is proportional and opposite in direction to the displacement x from the fixed point, i.e.

$$F \propto -x \ or \ F = -Kx \qquad\qquad 11.46$$

Also,
$$F = m\frac{dx}{dt^2} = -Kx \qquad\qquad 11.47$$

The simple harmonic motion defined by Eq. 11.47 has the amplitude or maximum displacement from the fixed point which is often denoted by A and cyclic frequency $\omega = \sqrt{\frac{K}{m}}$. The linear frequency of the motion is denoted by v and is given by $v = \frac{\omega}{2\pi} = \frac{1}{2\pi}\sqrt{\frac{K}{m}}$. The total energy ϵ of the oscillator which is the sum of its kinetic and potential energies is given by

$$\epsilon = \frac{1}{2}KA^2 \qquad\qquad 11.48$$

It may be noted that the energy ϵ is totally kinetic at the fixed or equilibrium point and totally potential at the points of maximum displacement. Total energy is, therefore, proportional to the square of the amplitude A. Further, in classical physics the oscillator can have any and continuous values of total energy, amplitude or the frequency of vibrations. However, a quantum linear oscillator can have only discrete values for energy, amplitude and frequency. A quantum linear oscillator may have energy values given by the following set,

$$\epsilon_j = \left(n_j + \frac{1}{2}\right)hv \qquad\qquad 11.49$$

where n_j can have positive integer values including zero, i.e. 0, 1, 2, 3, ...; and h is Planck's constant. Eq. 11.49 shows that for $n_j = 0$, the lowest or the ground state energy ϵ_0 (also called the zero point energy) of a quantum linear oscillator is $\frac{1}{2}hv$ and not zero.

Let us now consider an assembly of identical but distinguishable quantum linear oscillators. The quantum oscillators of the assembly are assumed to interact with each other to the extent that the microstate of the assembly does not remain same indefinitely but go on changing with time scanning all possible microstates allowed by the total energy of the assembly. On the other hand the interaction between different oscillators is assumed to be so small that each oscillator may be treated as independent of each other. Since it has already been assumed in the beginning that oscillators are distinguishable, it is now possible to apply Maxwell–Boltzmann statistics to the assembly. Further, the oscillators are non-degenerate, there is only one state per level and $g_j = 1$ *for all values of j*. As such the partition function for the assembly may be written as (Eq. 10.50, chapter-10),

$$Z = \Sigma e^{-\left(\frac{\epsilon_j}{k_BT}\right)} \qquad\qquad 11.50$$

Substituting $\epsilon_j = \left(n_j + \frac{1}{2}\right)hv$ in Eq. 11.50 yields,

$$Z = \sum e^{-\left(\frac{\epsilon_j}{k_BT}\right)} = \sum_j e^{-\left[\frac{\left(n_j+\frac{1}{2}\right)hv}{k_BT}\right]} = e^{-\left(\frac{hv}{2k_BT}\right)}\left[\sum_j e^{-\left(\frac{n_jhv}{k_BT}\right)}\right]$$

$$= e^{-\left(\frac{hv}{2k_BT}\right)}\left[1 + e^{-\left(\frac{hv}{k_BT}\right)} + e^{-\left(\frac{2hv}{k_BT}\right)} + e^{-\left(\frac{3hv}{k_BT}\right)} + \ldots\ldots\right]$$

$$e^{-\left(\frac{hv}{2k_BT}\right)}\left[1 + \left\{e^{-\left(\frac{hv}{k_BT}\right)}\right\} + \left\{e^{-\left(\frac{hv}{k_BT}\right)}\right\}^2 + \left\{e^{-\left(\frac{hv}{k_BT}\right)}\right\}^3 + \ldots\ldots\right] \qquad 11.51$$

In Eq. 11.51 terms in the bracket on right hand side represent the sum of a series in geometric

progression (G.P) with common term $r = \left\{e^{-\left(\frac{hv}{k_BT}\right)}\right\}$. The sum of the G.P series is given by $\frac{1}{1-r}$.

Hence, $$Z = e^{-\left(\frac{hv}{2k_BT}\right)}\left[\frac{1}{1-e^{-\left(\frac{hv}{k_BT}\right)}}\right] = \frac{e^{-\left(\frac{hv}{2k_BT}\right)}}{1-e^{-\left(\frac{hv}{k_BT}\right)}} \qquad 11.52$$

The characteristic temperature Θ *of the assembly is defined by the relation,*

$$\Theta \equiv \frac{hv}{k_B} \qquad 11.53$$

The characteristic temperature is proportional to the frequency v of the oscillators in the assembly and may be calculated if the frequency is known.

The partition function of the assemble in terms of the characteristic temperature may be written as

$$Z = \frac{e^{-\left(\frac{\Theta}{2T}\right)}}{1-e^{-\left(\frac{\Theta}{T}\right)}} \qquad 11.54$$

One may calculate the average number of oscillators in the assemble that have energy, say ϵ_i, using the expression given by Eq. 10.52, chapter-10 with the substitution that $g_j = 1$,

$$\frac{\overline{N_i}}{N} = \frac{1}{Z}e^{-\left(\frac{\epsilon_i}{k_BT}\right)} = \frac{\left\{1-e^{-\left(\frac{\Theta}{T}\right)}\right\}}{e^{-\left(\frac{\Theta}{2T}\right)}}e^{-\left[\frac{\left(n_i+\frac{1}{2}\right)hv}{k_BT}\right]} = \frac{\left\{1-e^{-\left(\frac{\Theta}{T}\right)}\right\}}{e^{-\left(\frac{\Theta}{2T}\right)}}e^{-\left[\frac{\left(n_i+\frac{1}{2}\right)\Theta}{T}\right]}$$

Or $$\overline{N_i} = N\left\{1-e^{-\left(\frac{\Theta}{T}\right)}\right\}\left\{e^{-\left(\frac{n_i\Theta}{T}\right)}\right\} \qquad 11.55$$

Equation 11.55 tells that the average number of oscillators (occupation number) with energy ϵ_i decreases exponentially with n_i and the exponential decrease is faster if the temperature T is low. As a matter of fact it may be shown that at temperature $T \approx \Theta$, all most all the oscillators are within the first few energy levels. Sometimes confusion is created by the fact that large number of oscillators has same energy that is they are in the same energy level while it has been assumed that the system is non-degenerate and has only one state per level. However, there is no contradiction because the assembly of oscillators is assumed to follow Maxwell–Boltzmann distribution in which any number of particles (oscillators) may be accommodated in a given state.

11.5.1.1   *The internal energy of the assembly*

The internal energy U of the assembly of linear quantum oscillators may be calculated from its partition function using Eq. 10.57 of chapter-10,

$$U = NT^2 k_B \frac{\partial (\ln Z)}{\partial T}$$

Or

$$U = NT^2 k_B \frac{\partial}{\partial T} \left[\left(\ln \left[\frac{e^{-\left(\frac{\Theta}{2T}\right)}}{1 - e^{-\left(\frac{\Theta}{T}\right)}} \right] \right) \right]$$

$$U = Nk_B\Theta \left\{ \frac{1}{2} + \frac{1}{\left(e^{\frac{\Theta}{T}} - 1 \right)} \right\} \tag{11.56}$$

For a given system of oscillators, N the number of oscillators and the characteristic temperature Θ are fixed and, therefore, the total internal energy of the assembly of oscillators is a function only of the temperature T. Let us consider the following cases;

(a) Temperature T approaches absolute zero:

In this limit for a finite value of $\Theta \left(= \frac{h\nu}{k_B} \right)$ the denominator of the second term in the curly

bracket $\left(e^{\frac{\Theta}{T}} - 1 \right)$ approches ∞ and, therefore, the second term vanishes. The total internal

energy in this limit $U^{T \rightarrow 0}$ becomes

$$U^{T \rightarrow 0} = \frac{1}{2} Nk_B\Theta = \frac{1}{2} Nk_B \frac{h\nu}{k_B} \tag{11.57}$$

$$= N\left(\frac{1}{2}h\nu\right) = N \times \text{Ground state or zero point energy of the oscillator}$$

This shows that as the temperature approaches absolute zero all oscillators of the assembly stay in their ground states and each oscillator contributes energy equivalent to its zero point energy i.e. hν to the internal energy of the assembly.

(b) Temperature $T \gg \Theta$ (characteristic temperature of the system):

When $\dfrac{\Theta}{T} \ll 1$, $\left(e^{\frac{\Theta}{T}} - 1 \right) \approx \dfrac{\Theta}{T}$ and the internal energy of the system of oscillators becomes

$$U^{\left(\frac{\Theta}{T} \ll 1\right)} \approx Nk_B\Theta(\tfrac{1}{2} + \tfrac{T}{\Theta}) \approx Nk_B\Theta \dfrac{T}{\Theta} = Nk_BT = N\left[2x\left(\tfrac{1}{2}k_BT\right)\right] \qquad 11.58$$

Equation 11.58 shows that for higher temperatures larger than the characteristic temperature of the system, the total internal energy of the system become Nk_BT, indicating as if each oscillator is behaving like a particle having two degrees of freedom with $\tfrac{1}{2}k_BT$ of energy associated with each degree. Further the internal energy increases linearly with temperature T. The characteristic temperature plays an important role in characterizing the temperature dependence of the internal energy of the assembly.

11.5.1.2 *The heat capacity of the assembly at constant volume C_v*

The thermal capacity of the assembly may be calculated by taking the partial derivative of the internal energy with respect to temperature, i.e.;

$$C_v = \dfrac{\partial U}{\partial T} = \dfrac{\partial}{\partial T}\left[Nk_B\Theta\left\{\dfrac{1}{2} + \dfrac{1}{\left(e^{\frac{\Theta}{T}} - 1\right)}\right\}\right] = Nk_B\left(\dfrac{\Theta}{T}\right)^2 \left[\dfrac{e^{\frac{\Theta}{T}}}{\left(e^{\frac{\Theta}{T}} - 1\right)^2}\right] \qquad 11.59$$

As temperature T approaches $0\,K$, C_v approaches zero and when $T \gg \Theta$, C_v approaches the constant value Nk_B, as expected. The entropy of the system also approaches zero as temperature goes to absolute zero. Figure 11.7 shows the variation of the thermal capacity C_v of an assembly of N quantum linear oscillators with the ratio of the temperature T/Θ. At $0\,K$ temperature C_v is zero and as the ratio T/Θ approaches 1, C_v settles to a constant value Nk_B. Figure 11.7 shows the importance of the ratio T/Θ and how the thermal capacity varies with the ratio. Variation of C_V with $\dfrac{T}{\Theta}$ shows that it is only at temperatures $T \geq \Theta$ that significant number of oscillators moves to the state of higher excitation.

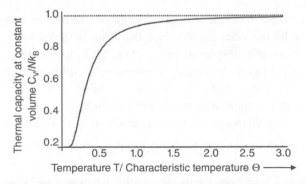

Fig. 11.7 Thermal capacity of quantum linear oscillator

11.6   Assembly of Quantum Oscillators and the Thermal Capacity of Polyatomic Gases

So long as the volume occupied by the gas molecules is negligible as compared to the volume of the container and in absence of any gravitational field the equation of state remains same both for monatomic and polyatomic gases. This is, however, not true for the internal energy and the heat capacity at constant volume of the two types of gases. It is because in a monatomic gas only the translational motion of the molecules contribute to the internal energy, while in polyatomic gases energy is contributed by the translational motion of the center of mass of the molecule as well as by the rotational and vibrational motions of the atoms of the molecule about the center of mass plus the electronic excitations. In the present discussion we do not consider the energy contributed by the electronic excitations. Further, the measured values of thermal capacities at constant volume for polyatomic gases and their temperature dependence indicate that both vibrational and rotational motions of atoms in these gases may be simulated by assemblies of quantum linear oscillators.

11.6.1   The diatomic molecule

Let us first consider the case of a diatomic gas. Translational motion of a diatomic molecule is essentially the motion of its center of mass. There are three degrees of freedom for the translational motion of the center of mass, represented by the three components of the velocity v_x, v_y and v_z and since the corresponding kinetic energy is proportional to the square of the velocity component, one may apply the theorem of equipartition of energy. Therefore, $3 \times \frac{1}{2} k_B T = 3/2 \, k_B T$ energy is contributed by the translational motion of each gas molecule to the internal energy of the assembly.

A diatomic molecule has a dumbbell like shape and may have rotational motions around three mutually perpendicular axes passing through the center of mass of the dumbbell. The energy E_a^{rot} associated with rotation about a particular axis, say a, is given by

$$E_a^{rot} = \frac{1}{2} I_a \omega_a^2$$

Here I_a is the moment of inertia of the dumbbell shaped molecule about the axis of rotation 'a' and $\omega_a \left(= 2\pi v_a^{rot} \right)$ the cyclic frequency. Since the moment of inertia about the axis passing through the centers of the two atoms of the molecule (i.e. axis of the dumbbell) is negligible, only the rotational motions around the other two axes will contribute energies to the molecule. As such there are only two degrees of freedom for the rotational motion of the atoms in the molecule and, therefore, from the theorem of equipartition of energy on average $2 \times \frac{1}{2} k_B T = k_B T$ of energy will be contributed by the rotational motion of the molecule when the temperature is sufficiently high so that significant number of molecules are in the excited state of rotational motion.

In order to understand the temperature dependence of the heat capacity at constant volume C_V for a diatomic gas, the rotational motion of the gas molecules may be assumed to be due to quantized linear oscillators and as such one may assign a characteristic temperature $\Theta^{rot} = \dfrac{h v^{rot}}{k_B}$ to the rotatory motion. For temperatures $T < \Theta^{rot}$ essentially all quantum rotors will be in their ground state and it will be only for temperatures $T \geq \Theta^{rot}$ that most of the rotors will populate the higher excitation state. The C_v verses $\dfrac{T}{\Theta^{rot}}$ graph will follow the general trend shown in Fig. 11.7.

Dumbbell shaped diatomic molecule is not a rigid body and the atoms of the molecule may also vibrate around the center of mass of the system. Vibratory motion may be along three mutually perpendicular axes passing through the center of mass, but vibrations along the axis of the molecule do not contribute and like the case of the rotatory motion, there are only two degrees of freedom associated with the vibrational motion also. Assuming vibrational motion due to an assembly of quantized oscillators, a characteristic temperature $\Theta^{vib} = \dfrac{hv^{vib}}{k_B}$ may be associated with the vibratory motion such that all vibrators are essentially in ground state at temperatures below Θ^{vib} and for temperatures $T \gg \Theta^{vib}$ all are in the excited state contributing on average $2 \times 1/2\, k_B T = k_B T$ energy per molecule to the internal energy and k_B per molecule to the thermal capacity at constant volume C_V.

Comparison with experiments:

The heat capacity at constant volume C_V and the molar specific heat capacity c_V of monatomic gases according to the theorem of equipartition of energy should be $(3/2)N\, k_B$ and $(1.5\ R)$ respectively at all temperatures. It is because in the case of a monatomic molecule the only type of motion is of translation with three degrees of freedom. The molar specific heat capacities in units of joule per mol per Kelvin and in terms of the gas constant R along with the ratio of the two specific heat capacities γ are listed in Table 11.1. It may be observed that there is reasonably good agreement between the values predicted by quantum thermodynamics and measured experimentally.

In the case of diatomic gases the measured values of the molar specific heat capacities at room temperature are around 2.5 R. If each degree of freedom contributes ½ R to the molar heat capacity, as is given by the quantum statistical theorem of equipartition of energy, then five degrees of freedom are taking part in the case of the diatomic molecules at room temperature (300 K). This means that apart from three degrees of freedom of translational motion additional two degrees of freedom of either rotational or vibrational motion are also participating at room temperature. The question whether the rotational degrees or the vibrational degrees of freedom are participating at room temperature will be decided by the value of the characteristic temperatures Θ^{rot} and Θ^{vib}. In case $\Theta^{rot} < 300$ K, then it will be rotational degrees of freedom and if $\Theta^{vib} < 300$ K then it will be vibrational degrees that are excited at room temperature. The experimental values of molar heat capacities also exclude the possibility of the simultaneous excitation of both rotational and vibrational modes at room temperature, because in that case the degrees of freedoms should be larger than five.

The characteristic temperature Θ^{rot} for the rotational motion of a molecule can be calculated and is inversely proportion to the moment of inertia of the molecule. Θ^{rot} has the largest value of 85.5 K for the lightest diatomic gas H_2 and has smaller values of 15.3 K, 27.5 K, and 2.77 K, for successively heavier molecules of HCl, OH, and CO respectively. On the other hand, the characteristic temperature for vibratory motion Θ^{vib} is inversely proportional to the mass of the molecule, being largest 6140 K for H_2, and 4300 K, 5360 K, and 3120 K, respectively for HCl, OH, and CO molecules. From these values of the characteristic temperatures it is clear that in case of all the above molecules only two degrees of freedom of rotational motion may contribute to the heat capacity at room temperature. Vibrational degrees could not be excited to any appreciable extent in any of these molecules because of the higher characteristic temperatures. Quantum thermodynamics *per se* not only explains the experimental data on the molar specific heat capacity of diatomic gases successfully but is also able to provide a rational reason why vibrational degrees of freedom are excited at higher temperatures. The classical thermodynamics failed in providing any reason for that.

Table 11.1 Molar specific heat capacity of some monatomic, diatomic and polyatomic gases at 300 K

Monatomic gases	c_V (J/mol.K)	c_V in units of R	$\gamma = \dfrac{C_P}{C_V}$
He	12.5	1.503 R	1.67
Ar	12.5	1.503 R	1.67
Ne	12.7	1.527 R	1.64
Kr	12.3	1.479 R	1.69
Diatomic gases			
H_2	20.4	2.453 R	1.41
N_2	20.8	2.501 R	1.40
O_2	21.1	2.537 R	1.40
CO	21.0	2.537 R	1.40
Cl_2	25.7	3.091 R	1.35
Polyatomic gases			
CO_2	28.5	3.427 R	1.30
SO_2	31.4	3.776 R	1.29
H_2O	27.0	3.247 R	1.31
CH_4	27.0	3.247 R	1.31

The real strength of the quantum approach comes from its success in explaining experimental data on the variation of molar heat capacity of a diatomic gas with temperature beyond the characteristic temperature for vibrational mode. However, it is only hydrogen that remains in gaseous state from very low ≈ 10 K to very high temperature $\approx 2500\ K$. A sketch of the variation of molar specific heat capacity c_V with temperature for hydrogen gas on a semi-log graph is shown in Fig. 11.8. As may be observed in the figure, c_V remains constant at the value 1.5 R (from point a to point b) for temperatures up to 50 K which is much lower than the characteristic temperature for rotational motion Θ^{rot}. During the part ab of the graph only the three degrees of freedom of translational motion contribute to the molar specific heat as oscillators corresponding to the rotational and vibrational motions are in their ground states. On further increasing the temperature to values $T \geq \Theta^{rot}$ (=85.5 K) c_V increases as some oscillators corresponding to the rotational mode shifts to excited states, which is shown by the part of the curve bc on the graph. When $T \gg \Theta^{rot}$, part cd of the graph, all rotor molecules go to the excited state contributing two additional degrees of freedom. The molar specific heat capacity attains the value 2.5 R from three degrees of freedom of translational and two degree of freedom of rotational motion of the molecule. Further increase of temperature pulls some oscillators to the excited level of the vibrational mode and at very high temperatures, it is expected that c_V will attain a value of 3.5 R when all vibrators are in excited states. The important property of the graph is that the shape of the curve from b to c and from d to e where the specific heat capacity rises from one steady value to the next is identical to the curve of Fig. 11.7 for quantum linear oscillator.

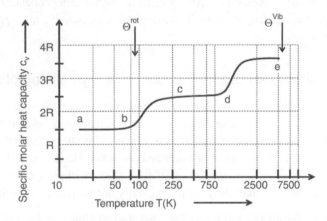

Fig. 11.8 Temperature dependence of specific molar heat capacity of hydrogen

11.6.2 Polyatomic molecules

Polyatomic molecules have large mass and moment of inertia. Therefore, their characteristic temperatures both for rotational and vibrational modes are quite low. As a result three degrees of freedom for translational motion along with two degrees each for rotational and vibrational modes contribute to the specific molar heat capacity even at room temperature. A total of seven degrees of freedom for polyatomic gases with ½ R of molar heat capacity associated with each degree gives a value of 3.5 R for c_V. This compares well with experimental values for polyatomic gases given in Table 11.1.

It may, however, be emphasized that in this simple treatment of diatomic and polyatomic molecules the electronic excitations have not been taken into account. Electronic excitations become important at higher temperatures. Also the moment of inertia and the mass of the molecule depends on the mass and size of each atom of the molecule and on their mutual separation. Detailed theories that take into account all these facts are quite successful in predicting thermal properties of polyatomic atoms.

11.7 Application of Quantum Statistics to Crystalline Solids

Molar specific heat capacity for crystalline solids may be obtained using the concepts of classical thermodynamics as well as that of quantum thermodynamics. Any theory for the molar specific heat capacity at constant volume for crystalline solids must explain the following two important experimental observations:

1. At and around room temperature molar heat capacity of most crystalline solids is ~ 3R, which is the well-known Dulong and Petit law.
2. At low temperatures the heat capacity decreases, becoming zero at T=0 K. The observed temperature dependence of heat capacity is of the form $\alpha T^3 + \gamma T$, where T^3 term arises from the lattice vibrations and the linear term γT from the conduction electrons.

Classical thermodynamics predicts a value of 3R for molar specific heat capacity of solids at all temperatures and is thus not successful in explaining the temperature dependence of the heat capacity. Einstein for the first time applied quantum thermodynamics for deriving an expression for the heat capacity of crystalline solids.

11.7.1   Einstein theory for the heat capacity of crystalline solids

Crystalline solids are characterized by a regular three dimensional arrangement of atoms. Though atoms of a solid cannot move freely like the atoms of the gas but at any temperature above absolute zero atoms may vibrate along three mutually perpendicular directions. Einstein assumed that the vibrational motion of atoms is like that of a quantized oscillator. He assumed that each atom of the solid is like a harmonic oscillator and that all oscillators vibrate with the same frequency. However, in his derivation he did not consider the zero point energy of the quantized oscillator and used the expression $\epsilon_i = n_i h v$ for the energy instead of the correct expression $\epsilon_i = \left(n_i + \dfrac{1}{2}\right) hv$. Since energy ϵ_i contributes to the internal energy U of the system while C_V is given by $\dfrac{\partial U}{\partial T}$, any constant term in the expression of ϵ_i will not affect the value of the heat capacity. Hence the results obtained by Einstein remain valid. In the following derivation we shall, however, use the correct expression for the energy of the quantized oscillator retaining the term ½ hv.

The internal energy U of a system having N linear quantum oscillators is given by Eq. 11.56 as

$$U = Nk_B\Theta \left\{ \frac{1}{2} + \frac{1}{\left(e^{\frac{\Theta}{T}} - 1\right)} \right\}$$

Einstein modified the above expression for the case of crystalline solids taking in to account the fact that in case of the solids each oscillator can vibrate in three directions independently and is, therefore, equivalent to three linear oscillators. If a solid has N atoms then it may be assumed that the atom has 3N linear oscillators. The internal energy of a solid with N atoms may then be given by

$$U^{sol} = 3Nk_B\Theta_{Ein} \left\{ \frac{1}{2} + \frac{1}{\left(e^{\frac{\Theta_{Ein}}{T}} - 1\right)} \right\} \tag{11.60}$$

Here Einstein temperature $\Theta_{Ein} \equiv \dfrac{h\nu}{k_B}$ is different for each solid and is related to the rigidity of the lattice.

The specific molar heat capacity at constant volume c_V of the solid may then be given by

$$c_v = \frac{\partial U^{sol}}{\partial T} = \frac{\partial}{\partial T}\left[3R\Theta_{Ein}\left\{\frac{1}{2} + \frac{1}{\left(e^{\frac{\Theta_{Ein}}{T}} - 1\right)}\right\}\right] = 3R\left(\frac{\Theta_{Ein}}{T}\right)^2\left[\frac{e^{\frac{\Theta_{Ein}}{T}}}{\left(e^{\frac{\Theta_{Ein}}{T}} - 1\right)^2}\right] \qquad 11.61$$

The c_v verses $\dfrac{T}{\Theta_{Ein}}$ graph is essentially similar to the graph of Fig. 11.7 and the best value for Einstein temperature Θ_{Ein} and hence the frequency ν for a given solid may be determined by getting the best fit between the experimental and the theoretical values of the molar specific heat capacities. However, it is observed that the same value of Θ_{Ein} does not reproduce the experimental data both at lower and higher temperatures. This is a big drawback of the Einstein theory.

When $T \ll \Theta_{Ein}$, $e^{\frac{\Theta_{Ein}}{T}} \gg 1$, 1 may be neglected and $\left[c_V\right]_{T \ll \Theta_{Ein}}$ becomes

$$\left[c_V\right]_{T \ll \Theta_{Ein}} = 3R\left(\frac{\Theta_{Ein}}{T}\right)^2 e^{-\left(\frac{\Theta_{Ein}}{T}\right)} \qquad 11.62$$

When $T \to 0$, the exponential term $e^{-\left(\frac{\Theta_{Ein}}{T}\right)}$ goes to zero more rapidly than the term $\left(\frac{\Theta_{Ein}}{T}\right)^2$ goes to infinity, therefore, c_v becomes zero at $T = 0$, as required. But because of the rapid decrease of the exponential term, the decrease in the theoretical value of c_v at low temperatures is faster than the decrease in the experimental values of the specific heat capacity. This is another discrepancy between the theory and the experiment that could not be explained by Einstein theory.

At $T \gg \Theta_{Ein}$, $\left(e^{\frac{\Theta_{Ein}}{T}} - 1\right) \approx \frac{\Theta_{Ein}}{T}$ and $e^{\frac{\Theta_{Ein}}{T}} \approx 1$. Eq. 11.61 then reduces to

$c_v \approx 3R$, Dulong and Petit value.

It may, therefore, be concluded that Einstein theory does not explain:

1. Why the same value of Einstein temperature Θ_{Ein} could not reproduce experimental data on c_V both at lower and higher temperatures.
2. Why at lower temperatures the decrease in the experimental values of c_V is slower than what is predicted by Einstein theory.

11.7.2 Debye theory for heat capacities of crystalline solids

Like Einstein, Debye also assumed that a crystalline solid with N atoms is like an assembly of 3N linear quantum oscillators, but unlike Einstein he assumed that all these oscillators do not vibrate with the same frequency. Debye assumed that these linear oscillators have a distribution of frequencies and may be divided into groups such that N_1 oscillators have frequencies in the range v_1 and $(v_1 + \Delta v_1)$; N_2 oscillators have frequencies in the range v_2 and $(v_2 + \Delta v_2)$; N_3 oscillators in the range v_3 and $(v_3 + \Delta v_3)$; and so on. However, there is an upper limit v_{max} of the frequency. Also, $N_1 + N_2 + N_3 + \ldots = 3N$. Debye further assumed that quantized frequencies of oscillators correspond to the frequencies of the elastic stationary waves that may be established in the solid by the coupled oscillations of its atoms. The upper limit to the frequency v_{max} comes from the maximum frequency of oscillations that may generate a standing wave in the solid. For vibrations of lower frequencies such that their wavelength λ is larger than the intra atomic distance 'a' of the crystal, the crystal lattice may be assumed to be continuous. For simplicity the crystal structure was taken to be isotropic so that it may be assumed that elastic waves travel with the same speed in all directions. Elastic stationary waves in the solid are produced by the quantized lattice vibrations in the crystal and, therefore, the energy associated with each mode of vibration is also quantized. The energy quantum of elastic lattice vibrations is called PHONON. The crystalline solid may, therefore, be treated as if it is filled with phonons of different frequencies corresponding to the stationary elastic waves. The frequency distribution of stationary waves in an isotropic continuous elastic solid may be calculated using the theory of elasticity that gives the number of stationary waves N_v (or the number of oscillators) having frequencies up to and including frequency v as,

$$N_v = \frac{4\pi V}{c^3} v^3 \qquad \qquad 11.63$$

In Eq. 11.63, 'c' is the velocity of the elastic wave in the given isotropic solid and should not be confused with the speed of light. The value of c depends on the rigidity of the crystal lattice. V is the volume of the solid.

Since the total number of oscillators (in a solid with N atoms) is 3N and the maximum frequency is v_{max}, one has

$$3N = \frac{4\pi V}{c^3} \left(v_{max}\right)^3 \qquad \qquad 11.64$$

And

$$\frac{4\pi V}{c^3} = \frac{3N}{\left(v_{max}\right)^3} \qquad \qquad 11.65$$

Eq. 11.63 may now be written as,

$$N_v = \frac{3N}{\left(v_{max}\right)^3} v^3 \qquad \qquad 11.66$$

The number ΔN_v of oscillators with frequencies in the range v and $v + \Delta v$ may be obtained by differentiating Eq. 11.66,

$$\Delta N_v = \frac{9N}{\left(v_{max}\right)^3} v^2 \, \Delta v \qquad \qquad 11.67$$

Or

$$\frac{\Delta N_v}{\Delta v} = \frac{9N}{\left(v_{max}\right)^3} v^2 \qquad \qquad 11.68$$

A graph of $\dfrac{\Delta N_v}{\Delta v}$ verses v is shown in Fig. 11.9, where shaded area up to $v = v_{max}$ is equal to 3N, the total number of oscillators.

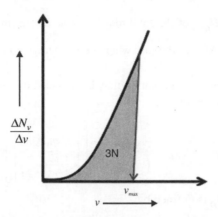

Fig. 11.9 Frequency distribution of stationary elastic waves

As has been stated earlier, it may be assumed that the crystalline solid contains groups of oscillators the number of oscillators in a group that has frequencies in the range v and $v + \Delta v$ is given by ΔN_v

(Eq. 11.67) and the energy of each oscillator of this group being $U = k_B \Theta \left\{ \dfrac{1}{2} + \dfrac{1}{\left(e^{\frac{\Theta}{T}} - 1 \right)} \right\}$ as given

by Eq. 11.56. Here $\Theta = \dfrac{hv}{k_B}$. Since a constant factor in the expression of energy does not contribute

to the heat capacity, the factor of ½ in the curly bracket for the energy U is dropped and then the energy of each oscillator of the group may be written as

$$U = k_B \Theta \left\{ \dfrac{1}{\left(e^{\frac{\Theta}{T}} - 1 \right)} \right\} = (hv) \left\{ \dfrac{1}{e^{\left(\frac{hv}{k_B T} \right)} - 1} \right\} \qquad 11.69$$

The total energy ΔU of all oscillators in this group may be obtained by multiplying the number of oscillators in the group ΔN_v with the energy of each oscillator U.

Therefore, $\Delta U = \left[\dfrac{9N}{\left(v_{max} \right)^3} v^2 \right] \left[(hv) \left\{ \dfrac{1}{e^{\left(\frac{hv}{k_B T} \right)} - 1} \right\} \right] \Delta v$

Or

$$\Delta U = \left[\frac{9Nh}{\left(v_{max}\right)^3} v^3 \right] \left[\left\{ \frac{1}{e^{\left(\frac{hv}{k_B T}\right)} - 1} \right\} \right] \Delta v \qquad 11.70$$

The total internal energy U_{total} of the solid may be found by summing the energies of all groups of oscillators, i.e. $U_{total} = \sum\limits_{v_{min}}^{v_{max}} \Delta U$. In the case when there are large numbers of oscillator groups with nearly continuous frequencies, the summation may be replaced by integration with the lower limit of integration starting with 0. Therefore,

$$U_{total} = \frac{9N}{\left(v_{max}\right)^3} \int\limits_0^{v_{max}} \left[\left\{ \frac{hv^3}{e^{\left(\frac{hv}{k_B T}\right)} - 1} \right\} \right] dv \qquad 11.71$$

The Debye temperature Θ_D is defined by

$$\Theta_D \equiv \frac{hv_{max}}{k_B} \qquad 11.72$$

To simplify Eq. 11.71, the following two substitutions may be made,

$$t = \frac{hv}{k_B T} \text{ so that } dv = \frac{k_B T}{h} dt \text{ and } t_m = \frac{hv_{max}}{k_B T} = \frac{\Theta_D}{T}$$

With these substitutions Eq. 11.71 becomes

$$U_{total} = 9NTk_B \frac{T^3}{\Theta_D^3} \int\limits_0^{t_m} \frac{t^3 dt}{\left(e^t - 1\right)} \qquad 11.73$$

Behavior at high temperatures:

At high temperatures i.e. for large T, t $\left(= \frac{hv}{k_B T} \right)$ has a small value and $\left(e^t - 1\right) \cong t$. Eq. 11.73 becomes

$$[U_{total}]_{high\ temp} = 9NTk_B \frac{T^3}{\Theta_D^3} \int\limits_0^{t_m} t^2 dt = 9NTk_B \frac{T^3}{\Theta_D^3} \left[\frac{t_m^3}{3} \right] = 3NTk_B$$

And the thermal capacity at constant volume at high temperature $[C_V]_{high\ temp}$ becomes

$$[C_V]_{high\ temp} = \frac{\partial}{\partial T} [U_{total}]_{high\ temp} = 3Nk_B \qquad 11.74$$

The corresponding molar specific heat capacity, $[c_V]_{high\ temp} = 3R$, which has the same value as given by Einstein theory and Dulong Petit law.

Behavior at low temperatures:

At low values of temperature the upper limit t_m of the integral has a large value and not much error is introduced if the upper limit is taken as infinity. Moreover, frequencies larger than v_{max} give negligible contribution to the integral. As such the total energy at low temperature $[U_{total}]_{low\ temp}$ may be written as,

$$[U_{total}]_{low\ temp} = 9NTk_B \frac{T^3}{\Theta_D^3} \int_0^\infty \frac{t^3 dt}{(e^t - 1)}$$

The definite integral $\int_0^\infty \frac{t^3 dt}{(e^t - 1)}$ has the value $\frac{\pi^4}{15}$, hence

$$[U_{total}]_{low\ temp} = 9NTk_B \frac{T^3}{\Theta_D^3} \frac{\pi^4}{15}$$

And the molar specific heat capacity at low temperatures

$$[c_V]_{low\ temp} = 233.308 \frac{R}{\Theta_D^3} T^3 \qquad\qquad 11.75$$

Since Debye temperature is a constant for a given solid, the molar specific heat capacity at lower temperatures varies as the cube of the temperature. This result is different from the Einstein theory in which at lower temperatures c_V decreases exponentially with temperature. Further, Debye theory also gives T^3 dependence of c_V, as required.

The total internal energy for any solid may be calculated from Eq. 11.73 by evaluating the integral $\int_0^{t_m} \frac{t^3 dt}{(e^t - 1)}$, and, therefore, is a function only of the ratio $(T/\Theta_D)^3$. Partial derivative of U with respect to temperature T then yields the heat capacity C_V and the specific molar heat capacity c_V. Both heat capacities are, therefore, functions of (T/Θ_D). Hence, from Debye theory, if a graph is plotted between the dimensionless ratios (c_V/R) and (T/Θ_D), data for all solids should fall on the same curve. That has been confirmed by experimental values of the heat capacities.

Application of quantum thermodynamic Debye theory successfully explains the T^3 dependence of the molar specific heat capacity of crystalline solids. Further, it also explains why the experimental rate of decrease of constant volume heat capacity with temperature is smaller than the exponential value predicted by Einstein's theory.

11.8 Contribution of Free Electrons of a Metal to the Heat Capacity

Free electrons in a metal behave like the molecules of an ideal gas. This sounds strange because electrons being charged particles are expected to interact strongly with each other and with the lattice of positively charged ions. However, these interactions are restricted to a large extent by the exclusion principle. Electrons are fermions that obey Pauli's exclusion principle, according to which no two electrons in a given assembly can have all their quantum numbers identical. This means that not more than one electron in the assembly of free electrons in a metal can be in a given state of energy. Since all most all energy states are already occupied in the assembly of free electrons and there are not many vacant energy states, the interactions between electrons and with lattice are severely restricted.

Further, free electrons of the metal are under electrostatic field that have screening effect and averages out as a whole. Thus free electrons of a metal may be treated as free molecules of an ideal gas.

From the point of view of the heat capacity, a metal may be considered to be a combination of a lattice of positive ions plus an ideal gas of free electrons. On heating, vibrations of the positive ion lattice produce stationary waves or phonons in the metal. The phonon contribution to the thermal capacity of the metal varies as the cube of the temperature (T^3) as given by Debye theory and confirmed by experiments.

According to the classical thermodynamics the ideal electron gas of N free electrons should contribute $\frac{3}{2} Nk_B$ to the heat capacity. However, experimental measurements of specific heat capacities of metals indicate that the contribution of free electrons is much smaller. In the following we will see how the application of quantum thermodynamics to the electron gas can explain this discrepancy.

Like the case of an ideal gas, the free electron gas also has large number of closely spaced energy levels. The concept of macro level (grouping of levels in a small energy interval) may also be applied to the electron gas. The degeneracy or the occupation number of the jth macro level for an ideal gas in terms of the velocity is derived in Eq. 11.7 and is given as $\Delta \mathcal{G}_j = \dfrac{4\pi m^3 V}{(2\pi)^3 \hbar^3} v_j^2 \Delta v_j$. The free electron gas differs from the ideal gas to the extent that two electrons, one spin down and one spin up, can be accommodated in each energy state. This effect may be included in the degeneracy of the micro level by multiplying the ideal gas degeneracy by a factor of two. Hence the degeneracy of the free electron macro level $\Delta \mathcal{G}_j^e$ may be given as,

$$\Delta \mathcal{G}_j^e = 2\,\Delta \mathcal{G}_j = \frac{8\pi m^3 V}{(2\pi)^3 \hbar^3} v_j^2 \Delta v_j \qquad\qquad 11.76$$

But $\qquad \epsilon_j = \dfrac{1}{2} m v_j^2$ and $\Delta \epsilon_j = m v_j\, \Delta v_j$ substituting these values in Eq. 11.76, one gets

$$\Delta \mathcal{G}_j^e = \left(\frac{\sqrt{2} V m^{3/2}}{\pi^2 \hbar^3} \right) \epsilon_j^{\frac{1}{2}} \Delta \epsilon_j = \mathcal{K}\, \epsilon_j^{\frac{1}{2}} \Delta \epsilon_j \qquad\qquad 11.77$$

Here, $\qquad\qquad \mathcal{K} = \left(\dfrac{\sqrt{2} V m^{3/2}}{\pi^2 \hbar^3} \right)$ is constant for a given system of volume V.

Electrons follow Fermi–Dirac statistics the distribution function for which is given by Eq. 10.41 of chapter-10. Accordingly the average number of particles (electrons) in the jth macro-level is given by,

$$\overline{N_j} = \frac{\Delta \mathcal{G}_j^e}{e^{\left(\frac{\epsilon_j - \mu}{k_B T} \right)} + 1} = \mathcal{K} \left[\frac{\epsilon_j^{\frac{1}{2}}}{e^{\left(\frac{\epsilon_j - \mu}{k_B T} \right)} + 1} \right] \Delta \epsilon_j \qquad\qquad 11.78$$

Since Eq. 11.78 involves $k_B T$ in the denominator of the exponential term, the average number of electrons in any macro level will depend on temperature T and for a fixed temperature will vary as the square root of the level energy. We evaluate the average number of electrons in the jth macro level at absolute zero temperature $\left[\overline{N_j} \right]_{T=0}$. The value will depend on the sign of the quantity $(\epsilon_j - \mu)$. If, the exponential term $e^{\left(\frac{\epsilon_j - \mu}{k_B T}\right)}$ at T = 0 will become $e^{\left(\frac{positive\ quantity}{0}\right)}$ which is infinite. Hence for those

macro levels which are at energies larger than μ the occupation number $\left[\overline{N_j}\right]_{T=0}^{\epsilon_j > \mu}$ will be zero.

However, if $(\epsilon_j < \mu)$; the exponential term at $T = 0$ will become $e^{-\infty}$ which is equal to zero. Therefore,

all levels below the energy μ will have $\left[\overline{N_j}\right]_{T=0}^{\epsilon_j < \mu} = \frac{\Delta \mathcal{G}_j^e}{0+1} = \Delta \mathcal{G}_j^e$. This means that at absolute zero

temperature all macro levels will be filled by the maximum number of electrons that the level may accommodate. The maximum number of electrons that a macro level may accommodate is equal to the degeneracy $\mathcal{G}_j^e$. Further all macro levels of energy larger than μ will have no electrons and will

be empty. The energy (μ) up to which all levels are completely filled at absolute zero temperature is called *Fermi Energy* and is denoted by ϵ_F. Thus the chemical potential μ is equal to Fermi energy and Eq. 11.79 in terms of Fermi energy may be written as

$$\overline{N_j} = \frac{\Delta \mathcal{G}_j^e}{e^{\left(\frac{\epsilon_j - \epsilon_F}{k_B T}\right)} + 1} = \mathcal{K}\left[\frac{\epsilon_j^{\frac{1}{2}}}{e^{\left(\frac{\epsilon_j - \epsilon_F}{k_B T}\right)} + 1}\right]\Delta\epsilon_j \qquad 11.79$$

The total number N of electrons may be obtained by summing the number of electrons in all macro levels and in the present case where macro-levels are themselves closely spaced and are large in number the summation may be replaced by integration, hence

$$N = \sum_j \overline{N_j} = \mathcal{K}\int_0^\infty \frac{\epsilon_j^{\frac{1}{2}}\, d\epsilon_j}{e^{\left(\frac{\epsilon_j - \epsilon_F}{k_B T}\right)} + 1} \qquad 11.80$$

As has already been said, for levels that have energies less than the Fermi energy ϵ_F at absolute

zero the denominator $[e^{\left(\frac{\epsilon_j - \epsilon_F}{k_B T}\right)} + 1]$ in the integral of Eq. 11.80 is unity and the total number of

electrons N is given by

$$N = \mathcal{K}\int_0^{\epsilon_F} \epsilon_j^{\frac{1}{2}}\, d\epsilon_j = \frac{2}{3}\mathcal{K}\epsilon_F^{3/2} \qquad 11.81$$

Putting the value of $\mathcal{K} = \left(\frac{\sqrt{2}V m^{3/2}}{\pi^2 \hbar^3}\right)$ in Eq. 11.81, the value of ϵ_F may be obtained as,

$$\epsilon_F = \frac{\hbar^2 \pi^{\frac{4}{3}}}{2m}\left(\frac{3N}{V}\right)^{2/3} \qquad 11.82$$

It is important to note that Fermi energy ϵ_F is independent of temperature T and depends only on

the ratio $\left(\frac{N}{V}\right)$, that is the number density of electrons in the metal.

As mentioned, at absolute zero all energy levels up to Fermi energy are completely filled and above Fermi energy all levels are empty. When temperature is raised above absolute zero some electrons from levels just below Fermi level get excited and move to the empty levels above Fermi energy. Electrons from levels deep below Fermi level do not get excited as all levels above them are completely filled and there is no empty state where they may go. This is the reason why only very few electrons (that are near the Fermi level) absorb heat energy and move to the empty levels when a metal is heated and most of the other electrons remain in the same state of energy (do not absorb energy). Now heat capacity at constant volume is the ratio of the heat absorbed ΔQ to rise in temperature ΔT, and for metals there will be two components of ΔQ, $\Delta Q_{lattice}$ and $\Delta Q_{electron}$. Since $\Delta Q_{electron}$ (due to free electrons) is very small compared to $\Delta Q_{lattice}$, electron gas contributes a very small amount to the total heat capacity of a metal. A major contribution to the heat capacity of metals comes from the quantized vibrations or the stationary elastic waves set up by the vibrations of the positive ion lattice, as shown in the preceding section on Debye theory.

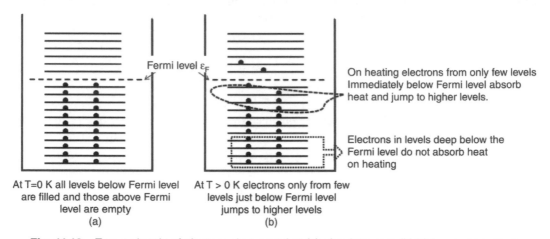

Fig. 11.10   Energy levels of electrons in a metal at (a) absolute zero (b) higher temperature

Let us compare the free electron gas in metals and a normal ideal gas under identical conditions. Molecules of a normal ideal gas obey classical statistics in which any number of particles may be accommodated in a given energy state and, therefore, energy levels are never completely filled. As a result when a normal ideal gas is heated each molecule of the gas may absorb energy and may shift to the next level of higher energy. On the other hand electrons obey Fermi–Dirac statistics and not more than two electrons, one with spin up and the other with spin down may be accommodated in a given energy level. On raising the temperature of a metal only few electrons near the Fermi level may absorb energy and may shift to the next unfilled levels above Fermi energy while most of the other electrons deep below the Fermi level cannot absorb energy and shift to higher levels as the levels above them are completely filled. Since each molecule of the ideal gas absorbs energy on heating, for the same rise of temperature ,the total amount of energy absorbed by the molecules of the normal ideal gas is much larger than the amount of energy absorbed by only a few electrons in case of the metal. Therefore, the molar specific heat capacity at constant volume for electron gas is much smaller than 3/2 R, the value for an ideal gas. Energy level diagram for an electron gas at absolute zero and at higher temperature are shown in Fig. 11.10.

Further, at absolute zero all macro levels up to Fermi energy are completely filled with maximum number of particles that may be accommodated in each macro level. The maximum number of particles that may be accommodated in any level is equal to the degeneracy of the level. The degeneracy $\Delta \mathcal{G}_j^e$ of the j-th macro level at energy ϵ_j is given by Eq. 11. 77 as

$$\Delta \mathcal{G}_j^e = \mathcal{K} \, \epsilon_j^{\frac{1}{2}} \, \Delta \epsilon_j$$

Thus the degeneracy of a level is proportional to the square root of the level energy. As such number of particles in macro levels increases as the square root of the energy of the macro level. This is shown in Fig. 11.11 where ordinate shows the number of particles per unit energy interval $\left(\dfrac{\Delta \mathcal{G}_j^e}{\Delta \epsilon_j} \right)$ and abscissa the energy ϵ_j. This graph is called the distribution function for free electrons

in metal or the state density function.

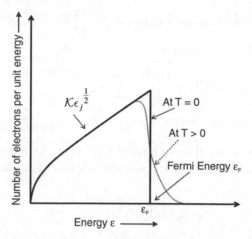

Fig. 11.11 Electron distribution graph for a metal

Fermi energy is the energy such that at absolute zero, all electron levels below Fermi energy are completely filled and above which all level are empty. Further it has been shown that the magnitude of Fermi energy does not depend on temperature. Therefore, when a metal is heated the Fermi energy does not change but few electrons from levels below the Fermi level jump to higher unoccupied levels across it. The electrons that have moved to higher levels leave behind 'holes' in the levels below and the number of holes is exactly equal to the number of excited electrons. The Fermi energy at any temperature $T > 0$ K, therefore, may also be defined as the energy such that the probability of finding an electron some E units of energy above the Fermi energy is equal to the probability of finding a hole E units of energy below it.

Next we derive expressions for the heat capacity at constant volume C_V and the molar specific heat capacity of the electron gas in metals. For that we calculate the total energy of all electrons of the system $U(T)$ at temperature T. $U(T)$ may be obtained by multiplying the number of electrons in the j-th macro level at temperature T with the energy ϵ_j of the level and integrating over all macro levels, i.e. from $\epsilon = 0$, to $\epsilon = \infty$.

$$U(T) = \int_0^\infty \overline{N_j} \epsilon_J = \mathcal{K} \int_0^\infty \left[\frac{\epsilon_j^{\frac{1}{2}}}{e^{\left(\frac{\epsilon_j - \epsilon_F}{k_B T}\right)} + 1} \right] \epsilon_j d\epsilon_J = \mathcal{K} \int_0^\infty \left[\frac{\epsilon_j^{3/2}}{e^{\left(\frac{\epsilon_j - \epsilon_F}{k_B T}\right)} + 1} \right] d\epsilon_J \qquad 11.83$$

The integral in Eq. 11.83 cannot be evaluated in a closed form. However, it can be written as an infinite series as given below,

$$U(T) = \mathcal{K} \left[\frac{2}{5} \epsilon_F^{5/2} + \frac{\pi^2}{6} (k_B T)^2 \epsilon_F^{1/2} - \frac{\pi^4}{40} (k_B T)^4 \epsilon_F^{-3/2} + ... \right] \qquad 11.84$$

The total internal energy of the electron gas may be obtained from Eq. 11.84 by putting the value of the temperature. The total internal energy of the electron gas at absolute zero is then,

$$U(0) = \frac{2}{5} \mathcal{K} \epsilon_F^{5/2} = \frac{3}{5} N \epsilon_F \qquad 11.85$$

But from Eq. 11.81, the total number of electrons $N = \frac{2}{3} \mathcal{K} \epsilon_F^{3/2}$ and Hence at absolute zero the average energy of electrons may be given as,

$$\bar{\epsilon}(0) = \frac{U(0)}{N} = \frac{3}{5} \epsilon_F \qquad 11.86$$

The heat capacity C_v at constant volume is given by $C_V = \frac{\partial}{\partial T} [U(T)]_V$ and may be obtained by partially differentiating the series given in Eq. 11.84. However, for temperatures that are not very high, it is sufficient to retain the first two terms of the series and under this assumption C_V is given by

$$C_V = \frac{1}{3} \mathcal{K} \pi^2 k_B^2 \epsilon_F^{1/2} T \qquad 11.87$$

Substituting the value of $\mathcal{K} \left(= \frac{3}{2} N \epsilon_F^{-\frac{3}{2}} \right)$ in above from Eq. 11.81 one gets,

$$C_V = \frac{1}{2} \pi^2 k_B^2 \epsilon_F^{-1} T N = \frac{1}{2} \pi^2 \left(\frac{k_B T}{\epsilon_F} \right) N k_B \qquad 11.88$$

But $N k_B$ = *Number of moles x gas constant* = $\mathbb{N} R$, and, therefore, specific molar heat capacity of electron gas is given by,

$$c_V = \frac{C_V}{\mathbb{N}} = \frac{1}{2} \pi^2 \left(\frac{k_B T}{\epsilon_F} \right) R \qquad 11.89$$

It may be observed that the specific molar heat capacity for electron gas depends linearly on temperature T and approaches zero at absolute zero, as required. On the other hand, from Debye theory the molar heat capacity due to lattice vibrations in solids vary as T^3. Therefore, at moderate temperatures electrons contribute very little to the heat capacity of metals. It is only at very low temperatures ($T < 1$) where T^3 is much smaller than T, that free electrons contribution to the heat capacity is significant.

The Fermi energy ϵ_F $\left(=\dfrac{\hbar^2 \pi^{\frac{4}{3}}}{2m}\left(\dfrac{3N}{V}\right)^{2/3}\right)$ for any metal may be calculated using Eq. 11.82 and has

a value ≈ 5.0 *eV*. Also, at around room temperature (300 K), $\left(\dfrac{k_B T}{\epsilon_F}\right) \approx 5 \times 10^{-3}$. Hence from Eq.

11.89, specific molar heat for metals at room temperature $\approx 2 \times 10^{-3}$ R, which is much smaller than 3/2 R, the value for an Ideal gas.

The entropy of the electron gas at S may be estimated making use of the fact that for reversible processes $[\Delta Q]_{reverible} = TdS = C_V\, dT$ and so

$$S = \int \frac{C_V\, dT}{T} = \int \frac{\frac{1}{2}\pi^2\left(\dfrac{k_B T}{\epsilon_F}\right)Nk_B}{T}\, dT = \frac{1}{2}\pi^2\left(\frac{k_B T}{\epsilon_F}\right)Nk_B \qquad 11.90$$

Also, from the relations $F = U - TS$ and $P = -\left(\dfrac{\partial F}{\partial V}\right)_T$

Now for electron gas at T = 0,

$$F(0) = U(0) = \frac{2}{5}N\epsilon_F \qquad 11.91$$

And electron gas pressure at absolute zero P(0) is,

$$P(0) = -\frac{2}{5}N\frac{\partial \epsilon_F}{\partial V} = -\frac{2}{5}N\left[-\frac{2}{3}\frac{\epsilon_F}{V}\right] = \frac{4}{15}\frac{N\epsilon_F}{V} \qquad 11.92$$

The equation of state at absolute zero of the electron gas is

$$P(0)V = 4/15\ N\epsilon_F \qquad 11.93$$

And the electron gas pressure $P(0) = \dfrac{4}{15}\dfrac{N\epsilon_F}{V}$ \qquad 11.94

The numerical value of P(0) $\approx 250 \times 10^3$ *atm*. It means that at absolute zero temperature the pressure of electron gas in a metal is very large. Free electrons at such a high pressure try to escape but are held back by the potential barrier which electrons find at the surface of the metal.

11.9 Energy Distribution of Blackbody Radiations: Application of Quantum Statistics

As discussed in chapter-8, blackbody radiation may be characterized as an assembly of electromagnetic radiations of all frequencies from zero to infinity with a well-defined strength of each frequency component. Though Wien, Rayleigh and Jeans gave empirical expressions for the energy distribution of blackbody radiations, former for high frequencies and the later for low frequencies, a unified formula that may reproduce experimental data at all frequencies could not be obtained on the basis of classical physics.

The empirical formula given by Wien and that correctly reproduced the energy density distribution of blackbody radiation only at higher frequencies may be written as

$$[u_v]_{Wien} = K_1 v^3 e^{-\left(\frac{k_2 v}{T}\right)} \Delta v$$

K_1 and K_2 in above expression are constants the values of which were obtained by fitting the experimental data. Though Wien's formula reproduced the experimental energy distribution at high frequencies but it totally failed at lower frequencies. On the other hand Rayleigh and Jean gave the following formula for the energy distribution of blackbody radiation that reproduced the experimental data at lower frequencies, however, at higher frequencies it led to infinite energy density which is unrealistic and is often called ultraviolet catastrophe. The constant K in Rayleigh–Jean's formula was evaluated by fitting experimental data.

$$[u_v]_{Rayleigh - Jean} = K v^2 T \Delta v$$

In order to develop an expression that reduces to Wien's formula at high frequencies and to Rayleigh–Jean's formula at lower frequency Planck made an ad hock assumption that an electromagnetic wave of frequency v is like a packet of energy hv, where h is a constant called Planck's constant, and gave the following empirical formula for the energy (or frequency) distribution of black body radiations.

$$\Delta U_v = \frac{a v^3 \, V}{e^{(bv/T)} - 1} \Delta v \qquad 11.95$$

In above expression ΔU_v is the energy of radiations in the frequency interval v and $v + \Delta v$ and V the volume of the cavity in which radiations are in equilibrium at temperature T. 'a' and' b' are two constants values for which depend on the system of units used.

As described in chapter-8, a metallic hollow sphere with inside walls blackened and a small opening may well serve as a blackbody. At a constant temperature T the volume of the spherical cavity is filled with blackbody radiations that are in equilibrium with the walls of the sphere. The blackbody radiation is also called cavity radiation. At a constant temperature T the cavity is filled with electromagnetic radiations of frequencies from zero to infinity with specific strength of each frequency component. The electromagnetic waves inside the cavity may be compared to the stationary elastic waves due to lattice vibrations that fill a crystalline solid. The quantized elastic waves are called phonon; similarly, quantized electromagnetic waves are called photons. An electromagnetic wave of frequency v is equivalent to a photon of energy hv, where h is Planck constant.

In quantum statistics a blackbody cavity is taken to be filled with a gas of photons of all frequencies. Since photons do not interact with each other, the photon gas is treated as an ideal gas. Further, photons are indistinguishable and obey Bose–Einstein statistics. In Bose–Einstein statistics any number of particles may be accommodated in a given energy state and, therefore, there is no restriction on the number of photons in a given energy state. In the state of equilibrium at a given fixed temperature, photons are continuously absorbed and re-emitted by the atoms of cavity walls hence the number of photons in the cavity keeps changing with time and fluctuates around a mean value. Since the total radiation energy of the cavity at a fixed temperature is fixed while the number of photons in the cavity fluctuates around a mean value and no photon leaves the cavity, the chemical potential for photon gas $\mu = 0$. Because of the large number of photons in the cavity energy levels of the photon assembly are very closely spaced and instead of considering each individual level one may lump together levels in a small energy interval into a macro-level.

The average number of particles in j-th level (of energy ϵ_j) for Bose–Einstein statistics is given by Eq. 10.34 of chapter-10 as,

$$N_j = \frac{g_j}{e^{-\left(\frac{\mu-\epsilon_j}{k_BT}\right)}-1}$$

For a macro-level of the photon gas that is formed by lumping all levels having photons of frequencies between v and $v + \Delta v$ (and energies from hv to $h(v + \Delta v)$), the above expression gets modified to,

$$\Delta N_v = \frac{\Delta\mathcal{G}_v}{e^{\left(\frac{hv}{k_BT}\right)}-1} \tag{11.96}$$

Here $\Delta\mathcal{G}_v$ is the total number of states in the macro-level, energy $\epsilon_j = hv$ and the chemical potential $\mu = 0$.

In order to calculate $\Delta\mathcal{G}_v$, the cavity may be considered to be filled with stationary electromagnetic waves, like the elastic stationary waves in the case of the crystalline solids. The theory of elastic waves gives the number $\mathcal{G}$ of stationary elastic waves up to and including frequency v in a container of volume V as,

$$\mathcal{G} = \frac{4\pi Vv^3}{3c^3} \tag{11.97}$$

Here c is the speed of the elastic wave in the isotropic crystal. While using Eq. 11.97 for electromagnetic waves one has to take into account the basic difference between the elastic and electromagnetic waves. Electromagnetic waves travel with the velocity of light hence c in the above expression will stand for the velocity of light. Also, electromagnetic waves are transverse waves and there may be waves in the cavity that are polarized in two mutually perpendicular directions. Therefore, the number of standing electromagnetic waves in volume V of the cavity will be twice as compared to the number of elastic waves. Hence for electromagnetic waves

$$\mathcal{G} = \frac{8\pi Vv^3}{3c^3} \tag{11.98}$$

And
$$\Delta\mathcal{G} = \frac{8\pi Vv^2}{c^3}\Delta v \tag{11.99}$$

Substituting the above value for the degeneracy of the macro-level in Eq. 11.96 one gets,

$$\Delta N_v = \frac{\Delta\mathcal{G}_v}{e^{\left(\frac{hv}{k_BT}\right)}-1} = \frac{\frac{8\pi Vv^2}{c^3}\Delta v}{e^{\left(\frac{hv}{k_BT}\right)}-1} = \frac{8\pi V}{c^3}\frac{v^2}{e^{\left(\frac{hv}{k_BT}\right)}-1}\Delta v \tag{11.100}$$

Eq. 11.100 gives the number of photons of frequency v and $v + \Delta v$ in the cavity of volume V and each of these photons has energy, hv. The total energy of all photons of frequency v is, therefore,

$$\Delta U_v = \Delta N_v \, x \, hv = \frac{8\pi hV}{c^3}\frac{v^3}{e^{\left(\frac{hv}{k_BT}\right)}-1}\Delta v \tag{11.101}$$

Eq. 11.101 derived using quantum statistics is identical to Eq. 11.95 which Planck obtained on the basis of the ad hock assumption of energy packets to reproduce the experimental energy distribution of blackbody radiation. A comparison of Eq. 11.101 and Eq. 11.95 gives 'a'= $\dfrac{8\pi h}{c^3}$ and 'b' = $\dfrac{h}{k_B}$,

which matched perfectly when values of h, c and the Boltzmann constant k_B were used.

The energy density Δu_v of photons of frequency v may be obtained by dividing Eq. 11.101 by the volume V of the cavity,

$$\Delta u_v = \frac{\Delta U_V}{V} = \frac{8\pi h}{c^3} \frac{v^3}{e^{\left(\frac{hv}{k_BT}\right)}-1} \Delta v \qquad 11.102$$

Let us now discuss the variation of the energy density Δu_v with frequency v at a fixed temperature T.

11.9.1   High frequency limit

For high frequencies such that $hv \gg k_BT$, the exponential term in the denominator of Eq. 11.102 becomes much larger than 1, which may be neglected. Hence,

$$\left[\Delta u_v\right]_{hv \gg k_BT} = \frac{8\pi h}{c^3} v^3 e^{-\left(\frac{hv}{k_BT}\right)} \Delta v \qquad 11.103$$

Thus under the high frequency limit the energy density distribution for blackbody radiation obtained using quantum statistics reduces to Wien's formula.

11.10.2   Low frequency limit

At lower frequencies where $hv \ll k_BT$, the term $\left(e^{\left(\frac{hv}{k_BT}\right)}-1\right)$ in the denominator of Eq. 11.103 is

nearly equal to $\left(\dfrac{hv}{k_BT}\right)$ and Eq. 11.102 reduces to,

$$\left[\Delta u_v\right]_{hv \gg k_BT} = \frac{8\pi k_BT}{c^3} v^2 \Delta v \qquad 11.104$$

Eq. 11.104 is the same as given by Rayleigh and Jean's empirically with value of the constant $K = \dfrac{8\pi k_B}{c^3}$.

Graphical representation of Eq. 11.102 ($\dfrac{u_v}{\Delta v}$ verse $\dfrac{hv}{k_BT}$) is given in Fig. 11.12. Since for lower

frequencies v^3 increases faster than the decrease in the exponential term $\left(\dfrac{1}{e^{\left(\frac{hv}{k_BT}\right)}-1}\right)$, the graph rises up to a maximum value after which the exponential decrease overrides the v^3 increase and the curve shows a maxima.

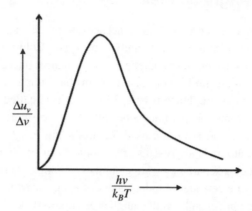

Fig. 11.12 Energy density distribution of blackbody radiation

Assuming a continuous distribution of frequencies from 0 to ∞, Eq. 11.102 may be integrated to obtain the energy density u in the cavity due to photons of all frequencies.

$$u = \int_0^\infty u_v dv = \frac{8\pi h}{c^3} \int_0^\infty \frac{v^3}{e^{\left(\frac{hv}{k_B T}\right)} - 1} dv \qquad 11.105$$

Evaluation of the definite integral in the limit zero to infinity in above equation may be facilitated by the following substitution;

$$\left(\frac{hv}{k_B T}\right) = t; \text{ so that } dv = \frac{k_B T}{h} dt \text{ and } v^3 = \left(\frac{k_B T}{h}\right)^3 t^3$$

And

$$u = \frac{8\pi h}{c^3} \left(\frac{k_B T}{h}\right)^3 \left(\frac{k_B T}{h}\right) \int_0^\infty \frac{t^3}{e^t - 1} dt$$

But

$$\int_0^\infty \frac{t^3}{e^t - 1} dt = \frac{\pi^4}{15}$$

Hence

$$u = \frac{8\pi^5 h}{15 c^3} \left(\frac{k_B T}{h}\right)^4 = \frac{8\pi^5 k_B^4}{15 c^3 h^3} T^4 = \sigma T^4 \qquad 11.106$$

where

$$\sigma = \frac{8\pi^5 k_B^4}{15 c^3 h^3} \qquad 11.107$$

Equation 11.106 represents Stefan–Boltzmann law and the value of σ obtained by substituting the values of c, h, k_B, and π in Eq. 11.107 is found to be in good agreement with the value obtained from experiments.

Assuming that the cavity (filled with blackbody radiation in thermal equilibrium) contains photon gas that obeys Bose–Einstein distribution, it is possible to derive the correct formula for energy density distribution of blackbody radiation and Stefan–Boltzmann law. This is another example of the success of quantum statistics.

11.10 Quantum Thermodynamics of a Paramagnetic Salt in External Magnetic Field

Each electron in an atom behaves like a tiny bar magnet because of the magnetic moment μ_e associated with the orbital motion of the electron. The magnitude of μ_e is 1 Bohr magneton denoted by μ_B. In atoms of normal substances the electron magnetic moments cancel out in pairs as a result the atom as a whole has no magnetic moment. However, some atoms have unpaired electrons and, therefore, a corresponding magnetic moment. This property of the atom is called paramagnetism and salts containing such atoms are termed as paramagnetic salts. Some crystalline salts of chromium and gadolinium are paramagnetic and are often used for producing low temperatures. In crystalline structure the paramagnetic atoms in above salts are in ionic form and in most of the cases they are far apart from each other, being separated by large number of nonmagnetic atoms and ions. As such the paramagnetic ions may be considered independent of each other in the crystal. The magnetic moment of a paramagnetic ion depends on the number of unpaired electrons which in the simplest case may be one but may be more in complex cases. For example, in the case of gadolinium, there are seven unpaired electrons. Though there may be weak magnetic fields present within the crystal while strong magnetic fields may be generated by aliening nuclear magnetic moments but at present we are interested only in the study of the simplest paramagnetic ion with only one unpaired electron, having the total angular momentum J = the spin S=1/2 , when it is subjected to an external magnetic field. Such a system is also called the ½ spin system. As already mentioned the magnetic moment of the simplest paramagnetic ion is 1 μ_B (Bohr magneton). When this ion is put in an external magnetic field, say B, the magnetic moment of the ion may align either in the direction of B, called the parallel or opposite to it, called the anti-parallel case. By convention the potential energy $\in$ of the interaction between the magnetic moment μ and the magnetic field B is given by the dot product $\in = -\vec{\mu} \cdot \vec{B}$. In the present case of the simplest paramagnetic ion with $\mu = 1\mu_B$ in the external magnetic field B, the potential energies for the parallel and the anti-parallel cases are respectively $\in_{\uparrow\uparrow} = -\mu_B B$ and $\in_{\uparrow\downarrow} = +\mu_B B$ as shown in Fig. 11.13.

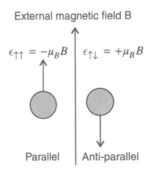

Fig. 11.13 Simplest paramagnetic ion in external magnetic field B

In quantum statistics a system consisting of paramagnetic ions of magnetic moments 1 μ_B in an external magnetic field B is like an assembly of particles that may have only two levels of energies $\in_{\uparrow\uparrow}$ and $\in_{\uparrow\downarrow}$. Further, the paramagnetic ions in a crystalline salt may be identified by their space parameters or coordinates, hence these ions are distinguishable. Maxwell–Boltzmann statistics may, therefore, be applied to this assembly of paramagnetic ions. In the present case there are only two levels each with only one state, either of parallel or anti-parallel particles, hence the levels are non-degenerate and g =1 for the two levels. However, any number of particles of a given kind may be accommodated in each state. The energy levels of spin 1/2 system at temperature $T > 0$ K are shown in Fig. 11.14.

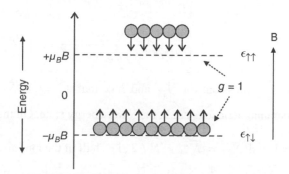

Fig. 11.14 Energy levels of a spin 1/2 system

11.10.1 The partition function

The partition function for a system obeying Maxwell–Boltzmann distribution is given by Eq. 10.50 of chapter-10 as,

$$Z = \sum_j g_j e^{-\left(\frac{\epsilon_j}{k_B T}\right)}$$

In the present case $j = 2$, $g_1 = g_2 = 1$ and $\epsilon_1 = \epsilon_{\uparrow\uparrow} = -\mu_B B$ and $= \epsilon_2 = \epsilon_{\uparrow\downarrow} = +\mu_B B$, hence

$$Z = e^{+\left(\mu_B B / k_B T\right)} + e^{-\left(\mu_B B / k_B T\right)} \qquad 11.108$$

Substituting $t = \left(\mu_B B / k_B T\right)$, Eq. 11.108 becomes,

$$Z = e^{+t} + e^{-t} \text{ But } e^{+t} + e^{-t} = 2 \, Cosh \, t,$$

Therefore, $\qquad\qquad Z = 2 \, Cosh \, t \qquad\qquad\qquad 11.109$

11.10.2 The magnetization

Also, the average number of particles in level j for Maxwell–Boltzmann distribution is given by Eq. 10.52, chapter-10 as,

$$\bar{N}_j = \frac{N}{Z} g_j e^{-\left(\frac{\epsilon_j}{k_B T}\right)}$$

For the present system we have only two non-degenerate levels of energies $\epsilon_{\uparrow\uparrow} = -\mu_B B$ and $\epsilon_{\uparrow\downarrow} = +\mu_B B$

$$\bar{N}_{\uparrow\uparrow} = \frac{N}{2\,Cosh\,t}e^{+\left(\frac{\mu_B B}{k_B T}\right)} \quad \text{and} \quad \bar{N}_{\uparrow\downarrow} = \frac{N}{2\,Cosh\,t}e^{-\left(\frac{\mu_B B}{k_B T}\right)} \qquad 11.110$$

Or

$$\bar{N}_{\uparrow\uparrow} = \frac{N}{\left[e^{+(\mu_B B/k_B T)} + e^{-(\mu_B B/k_B T)}\right]}e^{+\left(\frac{\mu_B B}{k_B T}\right)}$$

And

$$\bar{N}_{\uparrow\downarrow} = \frac{N}{\left[e^{+(\mu_B B/k_B T)} + e^{-(\mu_B B/k_B T)}\right]}e^{-\left(\frac{\mu_B B}{k_B T}\right)}$$

It is clear from the above expressions for $\bar{N}_{\uparrow\uparrow}$ and $\bar{N}_{\uparrow\downarrow}$ that at $T = 0$ K $\bar{N}_{\uparrow\uparrow} = N$ and $\bar{N}_{\uparrow\downarrow} = 0$

that means that all the paramagnetic ions are in the lower energy state. On the other hand at $T = \infty$,

$e^{+\left(\frac{\mu_B B}{k_B T}\right)} = e^{-\left(\frac{\mu_B B}{k_B T}\right)} = 1$, and $\bar{N}_{\uparrow\uparrow} = \bar{N}_{\uparrow\downarrow} = N/2$; i.e., half of the ions are in the lower state and

half in the higher energy state. At any intermediate temperature more ions will be in the lower energy state and a lesser number in higher energy state.

The resultant magnetic moment $\mathcal{M}$ of the assembly is due to the excess number ($\bar{N}_{\uparrow\uparrow} - \bar{N}_{\uparrow\downarrow}$) of

ions. Now each ion contributes a magnetic moment equal to μ_B, hence

$$\mathcal{M} = \mu_B\,(\bar{N}_{\uparrow\uparrow} - \bar{N}_{\uparrow\downarrow}) = \frac{N}{2\,Cosh\,t}\left[e^t - e^{-t}\right] = N\mu_B\frac{2\,Sinh\,t}{2\,Cosh\,t} = N\mu_B\,tanh\,t.$$

Or

$$\mathcal{M} = N\mu_B\,tanh\left(\frac{\mu_B B}{k_B T}\right) \qquad 11.111$$

Equation 11.111 gives a relation between magnetization $\mathcal{M}$ and the temperature T of the assembly and, therefore, the magnetic equation of state of the assembly. Further when the external magnetic

field B is large, i.e., $\mu_B B \gg k_B T$, $tanh\left(\frac{\mu_B B}{k_B T}\right) = 1$ and,

$$[\mathcal{M}]_{satu} = N\mu_B \qquad 11.112$$

Equation 11.112 gives the maximum value of magnetization (called the saturation value) that occurs when all the ionic dipoles are aligned in the direction of the applied field B.

In the case of the weak external field when $\mu_B B \ll k_B T$, $tanh\left(\frac{\mu_B B}{k_B T}\right) \approx \left(\frac{\mu_B B}{k_B T}\right)$ and from Eq.

11.111

$$\mathcal{M} = N\mu_B\left(\frac{\mu_B B}{k_B T}\right) = N\left(\frac{\mu_B^2}{k_B T}\right)\frac{B}{T} \qquad 11.113$$

Equation 11.113 gives the Curie law, derived on the basis of quantum statistics. According to the empirical law derived by Curie, at high temperatures and for low magnetic fields the magnetization $\mathcal{M}$ of paramagnetic salt is given by the relation,

$$\mathcal{M} = K_c\frac{B}{T} \qquad 11.114$$

Here K_c is a constant, called Curie constant, value for which may be obtained by comparing Eq. 11.113 and Eq. 11.114 as,

$$K_c = N\left(\frac{\mu_B^2}{k_B T}\right) \qquad 11.115$$

A satisfactory agreement is found between the values of the Curie constant determined experimentally and calculated using Eq. 11.115 which is based on quantum statistics.

The expression for magnetization $\mathcal{M}$ given by Eq. 11.113 holds for the case when the total angular momentum of the unpaired electron $J = S = 1/2$ and there are only two energy levels. However, if the total angular momentum is larger than ½, the system will have more than two energy levels and the particles will be distributed in all the possible levels. The same principle may be used to find the magnetization and the result is

$$\mathcal{M} = N g_J \mu_B \left[(J+1/2) Coth (J+1/2) C \right] - \left[\frac{Coth\dfrac{C}{2}}{2} \right] \qquad 11.116$$

where $\qquad C = \dfrac{g_J \mu_B B}{k_B T}$ and g_j is the Lande' g factor.

11.10.3 The interaction potential energy

For the spin half system the partition function Z is given by Eq. 11.109. Further, since the system follows Maxwell–Boltzmann statistics, the internal energy U of the system is related to the partition function through Eq. 10.57 of chapter-10. It may, however, be observed that in the present case of paramagnetic ions the internal energy U is the potential energy of the magnetic ions in the external magnetic field B.

$$U = N k_B T^2 \frac{\partial (\ln Z)}{\partial T} = N k_B T^2 \frac{\partial [\ln 2 Cosh t]}{\partial T} = N k_B T^2 \frac{\partial \left[\ln 2 Cosh \dfrac{\mu_B B}{k_B T} \right]}{\partial T}$$

Or $\qquad U = -N \mu_B B \tanh\left(\dfrac{\mu_B B}{k_B T}\right) \qquad 11.117$

But from Eq. 11.111 $\mathcal{M} = N\mu_B \tanh\left(\dfrac{\mu_B B}{k_B T}\right)$, therefore,

$$U = -\mathcal{M} B \qquad 11.118$$

Since the potential energy of interaction between a magnetic dipole and a magnetic field is taken as zero when the dipole is normal to the field, the negative sign in the above expression is due to the way the potential energy is defined.

The variation of the potential energy U with temperature T gives insight to the physics of the phenomenon. As shown in Fig. 11.15 (a) at 0 K $U = -N \mu_B B$. Leaving aside the negative sign, $N \mu_B B$

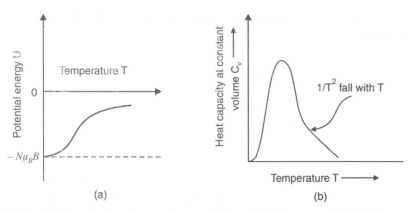

Fig. 11.15   (a) Temperature dependence of the potential energy U; (b) Temperature dependence of heat capacity at constant volume C_V

is the maximum magnitude of the potential energy when all the paramagnetic ions are aligned parallel to the applied magnetic field at zero temperature. On increasing the temperature some ions move to the higher energy level where ions are anti parallel to the field B. This reduces the magnitude of the potential energy. Ultimately at infinite temperature U approaches zero when roughly equal number of magnetic ions in the lower and the higher energy levels tend to contribute to the potential energy equal amounts but of opposite signs.

11.10.4   Heat capacity at constant volume

The heat capacity at constant volume C_V for the assembly may be obtained using the relation,

$$C_V = \left(\frac{\partial U}{\partial T}\right)_V = \left[\frac{\partial\left(-N\mu_B B \tanh\left(\frac{\mu_B B}{k_B T}\right)\right)}{\partial T}\right]_V$$

Or
$$C_V = Nk_B\left(\frac{\mu_B B}{k_B T}\right)^2 sech^2\left(\frac{\mu_B B}{k_B T}\right)$$ 11.119

Temperature dependence of C_V is shown Fig. 11.15 (b). At high temperatures for which $k_B T \gg \mu_B B$, the ratio $\frac{\mu_B B}{k_B T}$ tends to zero and $sech\left(\frac{\mu_B B}{k_B T}\right)$ approaches 1. Therefore, at higher temperatures heat capacity falls off as $1/T^2$ as shown in Fig. 11.15 (b).

$$\left[C_V\right]_{(k_B T \gg \mu_B B)} = Nk_B\left(\frac{\mu_B B}{k_B T}\right)^2 = \left(\frac{N\mu_B^2 B^2}{k_B}\right)\frac{1}{T^2}$$ 11.120

As shown in Fig. 11.15 (b) heat capacity shows a maximum which is rather unusual. Generally heat capacity is a smooth function of temperature and, therefore, this unusual peak in case of the ½ spin magnetic ions is called Schottky anomaly, which occurs around 1 K temperature. The reason for this lies in the fact that there are only two energy levels of the system and that high or low temperatures are relative to the energies of the magnetic levels.

When a paramagnetic crystalline solid is subjected to an external magnetic field and heated, a part of the heat energy absorbed goes in setting elastic waves due to lattice vibrations and another part in exciting magnetic ions from lower energy level to the higher one. As such both the lattice vibrations and magnetic excitation contribute to the heat capacity of the crystal. One may divide the total heat capacity at constant volume of the crystal into two components, one due to lattice vibrations and the second due to magnetic excitation. It can be shown that in a range of temperatures the contribution from magnetic heat capacity to the total heat capacity is order of magnitudes larger than the heat capacity due to the lattice vibration of the paramagnetic crystal, i.e. $[C_V]_{(magnet)} \gg [C_V]_{(lattice\ vib)}$.

11.10.5 The entropy

Once the partition function is known the entropy of the assembly may be calculated using Eq. 10.59 of chapter-10.

$$S(T) = N k_B \ln Z + N T k_B \frac{\partial (\ln Z)}{\partial T}$$

Putting $\quad Z = 2cosh\left(\frac{\mu_B B}{k_B T}\right) and \frac{\partial (\ln Z)}{\partial T} = \left[\frac{1}{Z}\frac{\partial Z}{\partial T}\right] = tanh\left(\frac{\mu_B B}{k_B T}\right)\left\{-\left(\frac{\mu_B B}{k_B}\right)\frac{1}{T^2}\right\}$, one gets

$$S(T) = N k_B \ln\left[2cosh\left(\frac{\mu_B B}{k_B T}\right)\right] - \frac{N\mu_B B}{T} tanh\left(\frac{\mu_B B}{k_B T}\right) \qquad 11.121$$

Temperature dependence of the entropy of a spin half system is shown in Fig. 11.16

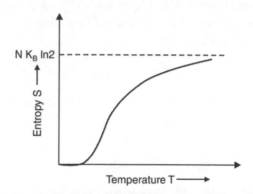

Fig. 11.16 Temperature dependence of the entropy of a spin half system

A simple way to understand the behavior of entropy S at low and at high temperatures is to look for the number of microstates Ω of the system and use the relation,

$$S = k_B \ln \Omega$$

At low temperature (when $T \to 0$) most of the paramagnetic ions will be in the ground state. There is only one way in which these ions can be arranged in the ground state, i.e. with their magnetic moments parallel to the field B. It means that there is only one possible microstate of the assembly at low temperatures, and, therefore, at low temperatures

$$S = k_B \ln \Omega = k_B \ln 1 = 0$$

The entropy will tend to zero as temperature goes to zero.

For temperatures such that $k_B T \gg \mu_B B$ some of the magnetic ions will move to the higher energy level. Every ion will have two possibilities: either in the lower level or in the upper level. For a single ion there will be 2 possible microstates and for N ions the number of microstates will be 2^N. Thus at higher temperatures $\Omega = 2^N$ and hence,

$$S = k_B \ln \Omega = k_B \ln 2^N = N k_B \ln 2$$

11.11 Population Inversion and Temperature beyond Infinity: The Concept of Negative Temperature

In a system that has two energy levels, like the assembly of half spin paramagnetic ions in external magnetic field, at low temperatures most of the particles live in the lower energy ground level and only very few particles are found in the level of higher energy. On increasing the temperature of the assembly, more and more particles shift to the higher energy level, but the number of particles in the lower level is always more than that in the level of higher energy. The usual situation, in which the lower level contains more particles as compared to the higher level, is called the normal population of the levels of the system. Now consider a situation when by some method more particles are excited to the level of higher energy as compared to the number of particles in the level of lower energy. This is called population inversion. If N_L and N_H are the average number of particles in the lower level of energy E_L and the higher level of energy E_H, respectively, at temperature T then the difference in the number of particles in the lower and the higher levels $(N_L - N_H)$ may be taken as a measure of the temperature T of the system, larger the magnitude of the difference lower the temperature. Also, the average number of particles is given by:

$$N_L = \frac{N}{Z} e^{-(E_L/k_B T)} \text{ and } N_H = \frac{N}{Z} e^{-(E_H/k_B T)} \qquad 11.122$$

where, N and Z are respectively the total number of particles and the partition function of the assembly. It follows from the above two equations that:

$$\frac{N_L}{N_H} = \frac{e^{-(E_L/k_B T)}}{e^{-(E_H/k_B T)}}$$

Taking natural log of the two sides of the equation one gets,

$$\ln(N_L - N_H) = \frac{(E_H - E_L)}{k_B T}$$

Or

$$T = \frac{(E_H - E_L)}{k_B \{\ln(N_L - N_H)\}} \qquad 11.123$$

Equation 11.123 may be used to define the quantum statistical temperature T of the system. The temperature T so defined remains positive so long as $E_H > E_L$ and $N_L > N_H$. The sign of the temperature T will become negative if $E_H > E_L$ but $N_L < N_H$. Thus in the state of population inversion the quantum statistical temperature of the system will become negative.

Earlier, for the case of spin half paramagnetic ions it was shown that at infinite temperature the average number of particles in both the lower and the upper levels are equal. Hence, in order to achieve population inversion one has to go to a temperature beyond infinity. It may thus be concluded that the negative temperature of quantum statistics is higher than the normal infinite temperature. This appears to be contradictory but it is because of the way how temperature is defined in quantum statistics.

Population inversion may be achieved in several different ways depending on the system. For example, in the case of paramagnetic ions if the direction of the magnetic field B is suddenly reversed so that the ions which were in the higher energy level (less in number) and had their magnetic moments anti parallel to the field now have them parallel to the new field direction and becomes particles of the lower energy level. Similarly, ionic magnets which were earlier in the lower energy level (larger in number) with their moments parallel to the field becomes particles in higher level of energy with their magnetic moments anti-parallel to the new field direction. Thus in this case population inversion may be achieved by simply reversing the direction of the externally applied magnetic field. In some cases, like the production of optical lasers, population inversion is achieved by pumping more particles to the level of higher energy from some nearby meta-stable state. However, the condition of population inversion does not last for long, generally with in a very short time $\approx 10^{-9}$ s, the system reverts back to the normal population. But in some special cases the state of population inversion stays for longer time up to a few minutes during which experiments may be done.

Solved Examples

1. Molecules of an ideal gas are confined to move only in a two dimensional plane. Derive expressions for (i) quantized energy of the molecule (ii) partition function (iii) equation of state and (iv) entropy of the assembly of molecules.

Solution:

(i) Quantized energy of the molecule

Each molecule of an ideal gas is considered to be a non-interacting indistinguishable particle. The quantized energy values of the molecule can be found by solving Schrödinger equation. Essentially, the energy Eigen values are given by the antinodes of stationary waves in the space where the particle is confined. Let us consider that the molecule or the particle is confined in a two dimensional space which is a square of length L. The wave length λ of the possible stationary waves will be given by the expression

$$\lambda = \frac{1}{n_j} 2L \qquad \text{S-11.1.1}$$

where n_j is an integer equal to the number of antinodes and can have values 1, 2, 3, ...

The wavelength λ is related to the momentum of the particle p by the relation

$$p = \frac{h}{\lambda} \qquad \text{S-11.1.2}$$

Here h is Planck's constant. Combining Eq. S-11.1.1 and Eq. S-11.1.2 one may get

$$P_j = n_j \frac{h}{2L}$$ S-11.1.3

Since particle can move only in a plane, i.e., along say x and y directions then;

$$p_x = n_x \frac{h}{2L} \text{ and } p_y = n_y \frac{h}{2L} \text{ and } p_j^2 = \left(n_x{}^2 + n_y{}^2 \right) \frac{h^2}{4L^2}$$

Or $$p_j^2 = n_j^2 \frac{h^2}{4L^2} \text{ where } n_j^2 = \left(n_x{}^2 + n_y{}^2 \right)$$ S-11.1.4

Since kinetic energy of the particle $\epsilon = \dfrac{p^2}{2m}$,

$$\epsilon_j = \frac{p_j^2}{2m} = n_j^2 \frac{h^2}{8mL^2} = n_j^2 \frac{h^2}{8mA}$$ S-11.1.5

where $A = L^2$ is the area of the two dimensional space in which the particle is free to move. Eq. S-11.1.5 gives the desired quantized energy levels of a free particle which is confined to move in a two dimensional space.

(ii) Partition function

Since the molecules of an ideal gas follow classical statistics for which the partition function is given as,

$$Z = \sum_j g_j e^{-\left(\frac{\epsilon_j}{k_B T} \right)}$$ S-11.1.6

Since the number of molecules (say N) is large, the energy levels are closely packed and one can use the concept of macro-level by lumping together all levels at energy ϵ_j within small energy interval $\Delta\epsilon_j$. We consider a two dimensional space made up of n_x and n_y, each point of intersection for $n_x = 1,2,3...$ and $n_y = 1,2,3...$ will give a possible value of n_j These intersection points are shown as dots in Fig.(S11.1). Since only non-zero positive integer values of n_x and n_y are allowed, number of points lying in $1/4^{th}$ parts of the circle of radius n_j gives the total number of particles up to and including energy ϵ_j which is denoted by $\mathcal{G}_j$.

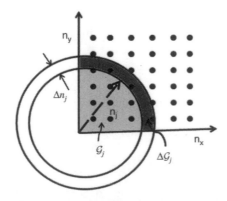

Fig. S-11.1   Degeneracy of the macro-level

The degeneracy of the macro-level may be obtained by differentiating $\mathcal{G}_j$. Now there is one point per unit area and hence, the number of points in ¼ part of the circle will be equal to the area of this part.

$$\mathcal{G}_j = \frac{1}{4}\pi n_j^2 \quad \text{and} \quad \Delta\mathcal{G}_j = \frac{1}{2}\pi n_j \, \Delta n_j \qquad\qquad \text{S-11.1.7}$$

Substituting the value of the degeneracy $\Delta\mathcal{G}_j$ from above in Eq. S-11.1.6 one gets,

$$Z = \sum_j \frac{1}{2}\pi n_j e^{-\left(\frac{\epsilon_j}{k_BT}\right)} \Delta n_j = \frac{1}{2}\pi \sum_j n_j e^{-\left(\frac{n_j^2 \frac{h^2}{8mA}}{k_BT}\right)} \Delta n_j$$

Substituting constant $b = \left(\dfrac{h^2}{8mAk_BT}\right)$ in the above equation and replacing summation by integration one obtains,

$$Z = \frac{1}{2}\pi \int_0^\infty n_j e^{-b n_j^2} \, dn_j$$

But the definite integral $\int_0^\infty n_j e^{-b n_j^2} \, dn_j = \dfrac{1}{2b} = \dfrac{4mAk_BT}{h^2}$

Hence, $$Z = \frac{2\pi mAk_BT}{h^2}$$

(iii) **Equation of state**

In the two dimensional case the role played by volume is played by the area A and the role played by pressure in 3-D case is played by force F in 2-D case.

Now $$\ln Z = \ln A + \ln\left(\frac{2\pi mk_BT}{h^2}\right) \quad \text{and} \quad \left[\frac{\partial(\ln Z)}{\partial A}\right]_{const.T} = \frac{1}{A}$$

Hence, $$\text{Force F} = Nk_B T \left[\frac{\partial(\ln Z)}{\partial A}\right]_{const.T} = N\,k_B\,T\,\frac{1}{A}$$

The equation of state is $FA = N\,k_B\,T$

(iv) **Entropy**

To calculate entropy we first calculate the internal energy U which for classical distribution is given by

$$U = Nk_BT^2\left(\frac{\partial(\ln Z)}{\partial T}\right)_{constant\,A} = Nk_BT^2 x\frac{1}{T} = Nk_BT$$

Hence $$\frac{U}{T} = Nk_B$$

And entropy S is,

$$S = \frac{U}{T} + Nk_B\{\ln Z - \ln N + 1\} = 2Nk_B + Nk_B\left[\ln\left(\frac{2\pi mk_B TA}{Nh^2}\right)\right]$$

Or

$$S = Nk_B\left[2 + \ln\left(\frac{2\pi mk_B TA}{Nh^2}\right)\right]$$

2. Taking helium at standard conditions as an ideal gas justify that classical distribution is appropriate to describe the gas. Following data is provided: mass of Helium molecule = 6.7×10^{-27} kg, 1.0 atm. Pressure = 1.01×10^5 Nm^{-2}, R = 8.31×10^3J kilomole^{-1} K^{-1}, Avogadro's number Av = 6.02×10^{26} kilomole^{-1}, k_B =R/Av , and h = 6.626×10^{-34} Js.

Solution:

Classical distribution is characterized by very small number of particles per state, i.e. a very small value of $\dfrac{N_j}{g_j}$ (<<1). In case of the ideal gas where one considers a macro-level, the above condition translates into $\dfrac{\Delta N_j}{\Delta g_j} \ll 1$. Here ΔN_j is the number of particles in the jth macro level and Δg_j is the degeneracy of the macro-level. If $\dfrac{N_j}{g_j}$ is very small, it will mean that there are only very few particles as compared to the degeneracy of the level which is one of the essential conditions for the validity of classical distribution. Now from Eq. 11.23 we have,

$$\Delta N_j = \frac{N}{Z}\Delta g_j\, e^{-\frac{\epsilon_j}{Tk_B}}$$

Or

$$\frac{\Delta N_j}{\Delta g_j} = \frac{N}{Z}\, e^{-\frac{\epsilon_j}{Tk_B}} \qquad \text{S-11.2.1}$$

Also, for an ideal gas the partition function is $Z = \left[V\left(\dfrac{mk_B T}{2\pi\hbar^2}\right)^{\!3/2}\right]$ in the above equation gives

$$\frac{\Delta N_j}{\Delta g_j} = \frac{N}{V}\left(\frac{mk_B T}{2\pi\hbar^2}\right)^{-(3/2)} e^{-\frac{\epsilon_j}{Tk_B}} \qquad \text{S-11.2.2}$$

Now we have to calculate the value of the expression on the right hand side of the above equation at standard conditions, i.e. at T= 273 K and 1 atm. Pressure. It is to be remembered that ϵ_j is the kinetic energy of the molecule at 273 K and from the principle of equipartition of energy it is of the order of 3/2 k_BT = 1.5 k_BT. Hence

$$e^{-\frac{\epsilon_j}{Tk_B}} = e^{-1.5} = 0.2231$$

For an ideal gas at 273 K and 1.0 atm. Pressure the volume of 1 kilomole of the gas v is

$v = RT/P$ and the number of molecules in one kilomole is equal to Avogadro number. So

$$\frac{N}{V} = \frac{Px\,Av}{RT} = \frac{1.01 \times 10^5 \times 6.02 \times 10^{26}}{8.31 \times 10^3 \times 273} = 2.68 \times 10^{25}$$

Also, on substituting the values of the given constants in the following expression, one gets

$$\left(\frac{mk_BT}{2\pi\hbar^2}\right)^{-\left(\frac{3}{2}\right)} = 1.45 \times 10^{-31}$$

Therefore,

$$\frac{\Delta N_j}{\Delta \mathcal{G}_j} = \frac{N}{V}\left(\frac{mk_BT}{2\pi\hbar^2}\right)^{-\left(\frac{3}{2}\right)} e^{-\frac{\epsilon_j}{Tk_B}} = 0.2231 \times 2.68 \times 10^{25} \times 1.45 \times 10^{-31}$$

Or

$$\frac{\Delta N_j}{\Delta \mathcal{G}_j} = 0.86 \times 10^{-6} \approx 1 \times 10^{-6}$$

Since $\frac{\Delta N_j}{\Delta \mathcal{G}_j} \approx 1 \times 10^{-6}$, which means that out of one billion possible states only one is filled

and hence the classical statistics is applicable to the assembly of helium gas at standard conditions.

3. Drive distribution function for the single component of velocity for the molecule of an ideal gas and show that it follows Gaussian distribution.

Solution:

Let us consider an ideal gas molecule which has a velocity v at some point in an assembly of ideal gas molecules. Let us consider an elementary volume of dimensions $(dv_x\, dv_y\, dv_z)$ around this molecule in the velocity space. If $\rho(v)$ is the volume density for velocity v, then the number dN_v of molecules that have velocities v in the elementary volume $(dx\, dy\, dz)$ will be given by

$$dN_v = \rho(v)\,(dv_x\, dv_y\, dv_z) \qquad \text{S-11.3.1}$$

Substituting the value of $\rho(v) = \frac{N}{\pi^{3/2}}\left(\frac{m}{2k_BT}\right)^{3/2} e^{-\left(\frac{mv^2}{2k_BT}\right)}$ from Eq. 11.38, one gets

$$dN_v = \frac{N}{\pi^{3/2}}\left(\frac{m}{2k_BT}\right)^{3/2} e^{-\left(\frac{mv^2}{2k_BT}\right)}(dv_x\, dv_y\, dv_z) \qquad \text{S-11.3.2}$$

The above equation may also be written in terms of the most probable velocity $v_m = \left(\frac{2k_BT}{m}\right)^{\frac{1}{2}}$

$$dN_v = \frac{N}{\pi^{3/2}}\left(\frac{1}{v_m}\right)^3 e^{-\left(\frac{v}{v_m}\right)^2}(dv_x\, dv_y\, dv_z) \qquad \text{S-11.3.3}$$

Or

$$dN_v = \frac{N}{\pi^{3/2}}\left(\frac{1}{v_m}\right)^3 e^{\left[-\left\{\frac{v_x^2}{v_m^2}+\frac{v_y^2}{v_m^2}+\frac{v_z^2}{v_m^2}\right\}\right]}\left(dv_x\,dv_y\,dv_z\right)$$

Or

$$dN_v = \frac{N}{\pi^{3/2}}\left(\frac{1}{v_m}\right)^3\left[\left\{e^{-\left(\frac{v_x^2}{v_m^2}\right)}dv_x\right\}\left\{e^{-\left(\frac{v_y^2}{v_m^2}\right)}dv_y\right\}\left\{e^{-\left(\frac{v_z^2}{v_m^2}\right)}dv_z\right\}\right] \qquad \text{S-11.3.4}$$

Suppose it is now required to find out the distribution function for only one component of the velocity, say v_z, then the above equation may be integrated for the dx and dy over the limits $-\infty$ to $+\infty$, leaving the dz term as such. Hence,

$$dN_{vz} = \frac{N}{\pi^{3/2}}\left(\frac{1}{v_m}\right)^3\left[\left\{\int_{-\infty}^{+\infty} e^{-\left(\frac{v_x^2}{v_m^2}\right)}dv_x\right\}\left\{\int_{-\infty}^{+\infty} e^{-\left(\frac{v_y^2}{v_m^2}\right)}dv_y\right\}\left\{e^{-\left(\frac{v_z^2}{v_m^2}\right)}dz\right\}\right] \qquad \text{S-11.3.5}$$

But definite integral $\int_{-\infty}^{+\infty} e^{-ax^2}dx = \left(\frac{\pi}{a}\right)^{\frac{1}{2}}$, therefore, each of the two integrals

$$\int_{-\infty}^{+\infty} e^{-\left(\frac{v_x^2}{v_m^2}\right)}dv_x = \int_{-\infty}^{+\infty} e^{-\left(\frac{v_y^2}{v_m^2}\right)}dv_y = v_m\sqrt{\pi}.$$

Therefore,

$$dN_{vz} = \frac{N}{\pi^{3/2}}\left(\frac{1}{v_m}\right)^3\left(v_m\sqrt{\pi}\right)^2\left\{e^{-\left(\frac{v_z^2}{v_m^2}\right)}dz\right\} = \frac{N}{v_m\sqrt{\pi}}\left\{e^{-\left(\frac{v_z^2}{v_m^2}\right)}dz\right\}$$

The desired Boltzmann distribution function for the z-component of velocity, irrespective of the values of the other two components v_x and v_y, is given as,

$$\frac{dN_{vz}}{dz} = \frac{N}{v_m\sqrt{\pi}}\left\{e^{-\left(\frac{v_z^2}{v_m^2}\right)}\right\} \qquad \text{S-11.3.6}$$

A general Gaussian function may be represented as

$$y(x) = A e^{-\left(\frac{(x-c)^2}{2\sigma^2}\right)} \qquad \text{S-11.3.7}$$

Here A is the height of the peak, c the distance of the peak from origin and σ the standard deviation of the curve. Comparing the one component velocity distribution curve given by Eq. (S11.3.6) with Gaussian curve, we find that c=0, the peak occurs at origin, height of the peak $A = \frac{N}{v_m\sqrt{\pi}}$ and the standard deviation $\sigma = \frac{1}{\sqrt{2}}v_m$.

Equation S-11.3.6 may also be written as

$$dN_{vz} = \frac{N}{v_m \sqrt{\pi}} \left\{ e^{-\left(\frac{v_z^2}{v_m^2}\right)} \right\} dv_z$$

And the number of molecules having positive z-components of velocities lower than and up to, say $(+V_0)$, may be obtained by integrating the above equation. The lower limit of integration becomes zero, since only positive z-components are to be counted. Further the integral assumes standard form by substituting $\frac{v_z}{v_m} = t$ *and* $dv_z = v_m dt$; that gives,

$$N_{(+v_z < V_o)} = \frac{N}{\sqrt{\pi}} \int_0^{\left(\frac{V_0}{v_m}\right)} e^{-t^2} dt = \frac{N}{2} \left[\frac{2}{\sqrt{\pi}} \int_0^{\left(\frac{V_0}{v_m}\right)} e^{-t^2} dt \right] \qquad \text{S-11.3.8}$$

The term in the big bracket $\left[\frac{2}{\sqrt{\pi}} \int_0^{\left(\frac{V_0}{v_m}\right)} e^{-t^2} dt \right]$ is often called the Error function and in the limit zero to infinity has a value 1. It follows from here that the number of molecules having positive z-component larger than V_0 will be given by:

$$N_{(+v_z > V_o)} = \frac{N}{2} \left[1 - \frac{2}{\sqrt{\pi}} \int_0^{\left(\frac{V_0}{v_m}\right)} e^{-t^2} dt \right] \qquad \text{S-11.3.9}$$

4. The flux and the number density of neutrons in the core of a reactor are respectively, 4×10^{16} neutrons m^{-2} s^{-1} and 6.34×10^{13} neutrons m^{-3}. Assuming neutrons to be an ideal gas obeying classical statistics calculate the temperature of the reactor core and the partial pressure due to neutrons in the core.

Solution:

For an ideal gas the molecular flux Φ is related to the number density *n and the average velocity* $\bar{v}$ by Eq.1.6(c) of Chapter-1 as,

$$\Phi = \frac{1}{4} n \bar{v} \qquad \text{S-11.4.1}$$

And $$\bar{v} = \sqrt{\frac{8 k_B T}{\pi m}} \text{ and } P = \frac{N k_B T}{V} = n k_B T$$

In this problem $\Phi = 4 \times 10^{16}$ neutron m^{-2} s^{-1}, n = 6.34×10^{13} m^{-3}, m (mass of neutron) = 1.67 $\times 10^{-27}$ kg, and k_B = 1.38×10^{-23} J K. Note that all quantities are in MKS system may be used directly in formulas.

So, $\dfrac{4\Phi}{n} = \sqrt{\dfrac{8k_BT}{\pi m}}$, or $T = \dfrac{16\pi m\Phi^2}{8n^2 k_B} = \dfrac{2\pi \times 1.67 \times 10^{-27} \times \left(4 \times 10^{16}\right)^2}{\left(6.34 \times 10^{13}\right)^2 \ \times 1.38\times10^{-23}} = 300.69\,K$

And partial pressure $P = 6.34 \times 10^{13} \times 1.38 \times 10^{-23} \times 300.69 = 2.63 \times 10^{-7}$ Nm^{-2}.

5. Calculate the percentage of ideal gas molecules that have (i) their velocities in the range of v_m and $v_m + 0.05v_m$ (ii) the x-component of velocity in the range v_m and $v_m + 0.05v_m$

Solution:

The number of molecules with velocity v_j is given by Eq. 11.29 as,

$$\Delta N_j^{v_j} = 4N(\pi)^{-\frac{1}{2}} \left(\dfrac{m}{2k_BT}\right)^{\frac{3}{2}} v_j^2\, e^{-\left(\dfrac{mv_j^2}{2k_BT}\right)} \Delta v_j \qquad \text{S-11.5.1}$$

And the number of molecules having the z-component of velocity is given by Eq. S-11.3.6 as,

$$dN_{vz} = \dfrac{N}{v_m\sqrt{\pi}} \left\{ e^{-\left(\dfrac{v_z^2}{v_m^2}\right)} dv_z \right\} \qquad \text{S-11.5.2}$$

Also, from Eq. 11.32 the most probable velocity of molecules is given by,

$$v_m = \sqrt{\dfrac{2k_BT}{m}} \qquad \text{S-11.5.3}$$

Equation S-11.5.1 in terms of the most probable speed v_m may be written as,

$$\Delta N_j^{v_j} = 4N(\pi)^{-\frac{1}{2}} \dfrac{1}{v_m^3} v_j^2\, e^{-\left(\dfrac{v_j^2}{v_m^2}\right)} \Delta v_j$$

The above equation may be written in integral form as,

$$\Delta N_j^{(v_m+0.05\,and\,v_m)} = 4N(\pi)^{-\frac{1}{2}} \dfrac{1}{v_m^3} \int\limits_{v_m}^{(v_m+0.05v_m)} v_j^2\, e^{-\left(\dfrac{v_j^2}{v_m^2}\right)} dv_j$$

$$\cong 4N(\pi)^{-\frac{1}{2}} \dfrac{1}{v_m^3} \left[e^{-1} v_m^2 \left((v_m + 0.05v_m) - v_m \right) \right]$$

$$= 4N(\pi)^{-\frac{1}{2}} \left[\dfrac{0.05}{e} \right] \qquad \text{S-11.5.4}$$

Hence, the percent of molecules with velocities between $(v_m + 0.05$ and $v_m)$

Is, $\dfrac{\Delta N_j^{(v_m+0.05\,and\,v_m)}}{N} \times 100 = 4(\pi)^{-\frac{1}{2}} \left[\dfrac{0.05}{e} \right] \times 100 = 4x\dfrac{1}{\sqrt{\pi}} \times \dfrac{0.05}{2.718} \times 100$

$$= 4.15\,\%$$

(ii) The number of molecules with Z-component of velocity between v_m and $v_m + 0.05v_m$ may be calculated by integrating Eq. S-11.5.2 within the given limits;

$$\int_{v_m}^{(v_m+0.05v_m)} dN_{vz} = \frac{N}{v_m\sqrt{\pi}} \left[\int_{v_m}^{(v_m+0.05v_m)} e^{-\left(\frac{v_z^2}{v_m^2}\right)} dv_z \right]$$

$$= \frac{N}{v_m\sqrt{\pi}} \left[v_m e^{-1} x 0.05 \right] = \frac{N}{\sqrt{\pi}} \left[e^{-1} x 0.05 \right]$$

Therefore, the percentage of particles with z-component of velocity between v_m and $(v_m + 0.05v_m)$ is,

$$\frac{N}{\sqrt{\pi}} \left[e^{-1} x 0.05 \right] x \frac{100}{N} = 1.04 \%$$

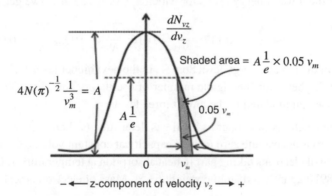

Fig. S11.5.1

Figure S-11.5.1 shows the number of particles per unit velocity $\dfrac{dN_{vz}}{dv_z}$ distribution for a single component of molecular velocity v_z, given by Eq. S – 11.5.2. The distribution curve has Gaussian shape and at $v_z = v_m$ falls to $1/e$ of its peak value A. The number of particles with their velocities between v_m and $(v_m + 0.05v_m)$ is equal to the shaded area in the figure. Since v_m and $(v_m + 0.05v_m)$ are very close the shaded area is equal to $\dfrac{A}{e} \times 0.05v_m$.

6. The frequency of an assembly of quantized linear oscillators is 5×10^{13} *cycl*/sec. Determine the characteristic temperature of the assembly and show that for temperature $T = \theta/2$ most of the oscillators are accommodated in the few levels of lowest energy.

Solution:

The characteristic temperature θ of an assembly of quantized linear oscillators is given by Eq. 11.53 as,

$$\theta = \frac{h\nu}{k_B} = \frac{6.626x10^{-34} \, x5x10^{13}}{1.3807x10^{-23}} = 2399 \, K \qquad \text{S-11.6.1}$$

The average number of oscillators in i^{th} level is given by Eq. 11.55 as,

$$\overline{N_i} = N\left\{1 - e^{-\left(\frac{\Theta}{T}\right)}\right\}\left\{e^{-\left(\frac{n_i\Theta}{T}\right)}\right\}$$

S-11.6.2

Substituting $T = \theta/2$, one gets;

$$\frac{\overline{N_i}}{N} = \left\{1 - e^{-\left(\frac{\Theta}{\theta/2}\right)}\right\}\left\{e^{-\left(\frac{n_i\Theta}{\theta/2}\right)}\right\} = \left(1 - e^{-2}\right)\left(e^{-2n_i}\right)$$

Or
$$\frac{\overline{N_i}}{N} = \left(1 - e^{-2}\right)\left(e^{-2i}\right) = 0.864\left(e^{-2n_i}\right)$$

Therefore, for the lowest energy levels for which $n_i = 0, 1, 2, 3, 4 \dots$ we get

$$\frac{\overline{N_0}}{N} = 0.864; \frac{\overline{N_1}}{N} = 0.117; \frac{\overline{N_2}}{N} = 0.016; \frac{\overline{N_3}}{N} = 0.002; \frac{\overline{N_1}}{N} = 0.00002$$

It means that in the lowest energy level 86.4 %, in the next higher level 11.7 % in the third level 1.6 % in other higher levels negligible number of oscillators are found. That is nearly 99.7% oscillators are accommodated in the lowest three levels.

7. Assuming free electrons in copper as ideal gas determine (a) Fermi energy, (b) mean kinetic energy of electrons at absolute zero (c) heat capacity at constant volume C_v at temperature T K and the fraction of electrons taking part in heat absorption at temperature T. Copper is divalent, its density is 8920 kg m^{-3}; and atomic weight 63.51. Mass of electron is 9.11×10^{-31} kg.

Solution:

(a) The Fermi energy of the electron gas is given by Eq. 11.82 as,

$$\epsilon_F = \frac{\hbar^2 \pi^{\frac{4}{3}}}{2m}\left(\frac{3N}{V}\right)^{2/3}$$

S-11.7.1

To calculate ϵ_F one has to calculate N/V, number of electron per cubic meter. It is given that the density of Copper is 8920 kg m^{-3} which means that the mass of 1.0 m^3 of Copper is 8920 kg. Further, the atomic weight of Copper is 63.51 hence from Avogadro's hypothesis in 63.51 kg of Copper there are 6.03×10^{26} atoms of Copper. It is given that copper is divalent so each Copper atom on ionization gives two electrons. As such, the number of free electrons in 1.0 m^3 of copper N is

$$\frac{N}{V} = \frac{2 \times 6.02 \times 10^{26} \times 8920}{63.51} = 16.91 \times 10^{28}$$

S-11.7.2

Putting this value of N/V in Eq. S-11.7.1 and taking mass of electron m = 9.11×10^{-31} one gets,

$$\epsilon_F = 18.036 \times 10^{-19} \, J$$

And since $\quad$ $1.6 \times 10^{-19} \, J = 1 \; eV$;

$$\epsilon_F = 11.27 \; eV \qquad\qquad\qquad \text{S-11.7.3}$$

(b) The mean kinetic energy of electrons at absolute zero is given by Eq. 11.86,

$$\overline{\epsilon}(0) = \frac{U(0)}{N} = \frac{3}{5}\epsilon_F = 10.82 \times 10^{-19} \, J \text{ or } 6.75 \, eV \qquad \text{S-11.7.4}$$

It may be remarked that the mean kinetic energy of an ideal gas molecule at room temperature of about 300 K is only 0.025 eV. On the other hand the average kinetic energy of an electron in metals at absolute zero temperature is as high as 6.75 eV.

We now calculate the absolute temperature T at which an ideal gas molecule will have the mean kinetic energy 6.75 eV (or $10.82 \times 10^{-19} \, J$). From the theorem of equipartition of energy a molecule of an ideal gas at temperature T has the mean kinetic energy $3/2 k_B \, T$. So,

$$3/2 k_B \, T = 10.82 \times 10^{-19} \, J$$

Or $\qquad\qquad\qquad T = \dfrac{2 \times 10.82 \times 10^{-19}}{3 \times 1.3807 \times 10^{-23}} = 5.2 \times 10^4 \, K$

Thus, the average kinetic energy of free electrons in copper at 0 K is so large that an ideal gas must be heated to a temperature of $5.2 \times 10^4 \, K$ so that the gas molecules have the same kinetic energy.

(c) The heat capacity at temp T K due to N-free electrons in a metal is given by Eq. 11.88 as,

$$C_V = \frac{1}{2}\pi^2 \left(\frac{k_B T}{\epsilon_F} \right) N k_B \qquad\qquad\qquad \text{S-11.7.5}$$

Let us compare this with the heat capacity C_V^{gas} of N_{eff} molecules of an ideal gas at temperature T, which is given by $C_V^{gas} = \dfrac{3}{2} N_{eff} k_B T$.

$$\frac{1}{2}\pi^2 \left(\frac{k_B T}{\epsilon_F} \right) N k_B = \frac{3}{2} N_{eff} k_B T$$

Or $\qquad\qquad\qquad \dfrac{N_{eff}}{N} = \dfrac{\pi^2 k_B}{3\epsilon_F} = \dfrac{\pi^2 \times 1.3807 \times 10^{-23}}{3 \times 18.036 \times 10^{-19}} = 25 \times 10^{-6} \qquad \text{S-11.7.6}$

Equation S-11.7.6 shows that out of some billion electrons only about 25 electrons participate in absorbing heat.

8. N particles obeying classical distribution populate two energy levels of same degeneracy at energies ϵ_0 and ϵ_1 at temperature T. (i) derive an expression for the average energy $\overline{\epsilon}$ per particle and evaluate $\overline{\epsilon}$ in the limits $T \rightarrow 0$ and $T \rightarrow \infty$ (ii) derive an expression for the specific molar heat capacity and evaluate it in the limits of $T \rightarrow 0$ and $T \rightarrow \infty$.

Solution:

(*i*) The average number of particles in a level at energy ϵ for classical distribution is given by:

$\overline{N_\epsilon} = \dfrac{N}{Z} g_\epsilon \, e^{-\left(\frac{\epsilon}{k_B T}\right)}$. If N_0 and N_1 are respectively the average number of particles in levels

of energies ϵ_0 and ϵ_1, then,

$$N_0 = \frac{N}{Z} g \, e^{-\left(\frac{\epsilon_0}{k_B T}\right)} \quad and \; N_1 = \frac{N}{Z} g \, e^{-\left(\frac{\epsilon_1}{k_B T}\right)} \qquad \text{S-11.8.1}$$

The total energy of all particles E and the total number of particles N are respectively given by,

$$E = N_0 \epsilon_0 + N_1 \epsilon_1 = \frac{N}{Z} g \left[\epsilon_0 \, e^{-\left(\frac{\epsilon_0}{k_B T}\right)} + \epsilon_1 e^{-\left(\frac{\epsilon_1}{k_B T}\right)} \right]$$

$$N = N_0 + N_1 = \frac{N}{Z} g \left[e^{-\left(\frac{\epsilon_0}{k_B T}\right)} + e^{-\left(\frac{\epsilon_1}{k_B T}\right)} \right]$$

The average energy per particle =

$$\frac{E}{N} = \bar{\epsilon} = \frac{\left[\epsilon_0 \, e^{-\left(\frac{\epsilon_0}{k_B T}\right)} + \epsilon_1 e^{-\left(\frac{\epsilon_1}{k_B T}\right)} \right]}{\left[e^{-\left(\frac{\epsilon_0}{k_B T}\right)} + e^{-\left(\frac{\epsilon_1}{k_B T}\right)} \right]} = \frac{\left\{ \epsilon_0 + \epsilon_1 e^{-\left\{\frac{(\epsilon_1 - \epsilon_0)}{k_B T}\right\}} \right\}}{\left\{ 1 + e^{-\left\{\frac{(\epsilon_1 - \epsilon_0)}{k_B T}\right\}} \right\}}$$

Let us assume that $\epsilon_1 > \epsilon_0$ and we denote $\epsilon_1 - \epsilon_0 = \Delta\epsilon$ and $\dfrac{1}{k_B T} = \beta$; so that

$$\bar{\epsilon} = \frac{\left\{ \epsilon_0 + \epsilon_1 e^{-\beta\Delta\epsilon} \right\}}{\left\{ 1 + e^{-\beta\Delta\epsilon} \right\}} \qquad \text{S-11.8.2}$$

In the limit when $T \to 0$; $\beta \to \infty$

$$\bar{\epsilon} \approx \left\{ \epsilon_0 + \epsilon_1 e^{-\beta\Delta\epsilon} \right\}\left\{ 1 - e^{-\beta\Delta\epsilon} \right\} = \epsilon_0 + \epsilon_1 e^{-\beta\Delta\epsilon} - \epsilon_0 e^{-\beta\Delta\epsilon} - \epsilon_1 e^{-2\beta\Delta\epsilon}$$

Since $\quad e^{-2\beta\Delta\epsilon} = negligible$,

$$\bar{\epsilon} \approx \epsilon_0 + (\epsilon_1 - \epsilon_0) e^{-\beta\Delta\epsilon} = \epsilon_0 + \Delta\epsilon \, e^{-\beta\Delta\epsilon} \qquad \text{S-11.8.3}$$

In the limit $T \to \infty$ or $\beta \to 0$

$$\bar{\epsilon} = \frac{\left\{ \epsilon_0 + \epsilon_1 e^{-\beta\Delta\epsilon} \right\}}{\left\{ 1 + e^{-\beta\Delta\epsilon} \right\}} \approx \frac{\left\{ \epsilon_0 + \epsilon_1 e^{-\beta\Delta\epsilon} \right\}}{\left\{ 1 + 1 \right\}} = \frac{1}{2}\left[\epsilon_0 + \epsilon_1 (1 + \beta\Delta\epsilon) \right] = \frac{1}{2}\left[\epsilon_0 + \epsilon_1 \right] \qquad \text{S-11.8.4}$$

(ii) Heat capacity per particle may be given by

$$c = \frac{\partial \bar{\epsilon}}{\partial T} = \frac{\partial \bar{\epsilon}}{\partial \beta}\frac{\partial \bar{\beta}}{\partial T} = k_B \left(\frac{\Delta\epsilon}{k_B T}\right)^2 \left\{ \frac{e^{-\left(\frac{\Delta\epsilon}{k_B T}\right)}}{\left(1 + e^{-\left(\frac{\Delta\epsilon}{k_B T}\right)}\right)^2} \right\} \qquad \text{S-11.8.5}$$

To get the molar specific heat capacity we multiply the above expression for c by the number of particles in one mole that is by the Avogadro's number A_V.

$$\text{Molar heat capacity } C = A_V \times c = A_v k_B \left(\frac{\Delta\epsilon}{k_B T}\right)^2 \left\{ \frac{e^{-\left(\frac{\Delta\epsilon}{k_B T}\right)}}{\left(1 + e^{-\left(\frac{\Delta\epsilon}{k_B T}\right)}\right)^2} \right\}$$

Or   putting $A_v k_B = R$, $C = R \left(\frac{\Delta\epsilon}{k_B T}\right)^2 \left\{ \dfrac{e^{-\left(\frac{\Delta\epsilon}{k_B T}\right)}}{\left(1 + e^{-\left(\frac{\Delta\epsilon}{k_B T}\right)}\right)^2} \right\}$ S-11.8.5

In the limit $T \to 0$ $\qquad C \approx R\left(\frac{\Delta\epsilon}{k_B T}\right)^2 e^{-\left(\frac{\Delta\epsilon}{k_B T}\right)}$

In the limit $T \to \infty$; $\qquad C \approx \frac{1}{4} R\left(\frac{\Delta\epsilon}{k_B T}\right)^2$

9. A system containing particles that obey Maxwell–Boltzmann statistics is in contact with a heat reservoir at temperature T. Calculate the temperature of reservoir if particles of the system are distributed in single particle levels according to the following scheme

Level energy in mille-electron volt (10^{-3} eV)	Population
4.3	63 %
12.9	23 %
21.5	8.5 %
30.1	3.1 %

Solution:

According to Maxwell–Boltzmann distribution the ratio of average number of particles N_1 in level of energy E_1 and N_2 in level of energy E_2 is given by,

$$\frac{N_1}{N_2} = e^{\frac{(E_2-E_1)}{k_BT}} \quad \text{or} \quad \frac{(E_2-E_1)}{k_BT} = \ln\frac{N_1}{N_2}$$

Therefore,

$$T = \frac{(E_2-E_1)}{k_B\left(\ln\frac{N_1}{N_2}\right)} \qquad \text{S-11.9.1}$$

Taking $E_1 = 4.3 \times 10^{-3}$ eV, $E_2 = 12.9 \times 10^{-3}$ eV; $N_1 = 63$, $N_2 = 23$ and $k_B = 8.617 \times 10^{-5}$ eV

One gets,
$$T = \frac{8.6\times10^{-3}}{8.617 \times 10^{-5} \times \ln\frac{63}{23}} = 99.4\,K$$

Similarly for the other pairs $E_1 = 21.5\times10^{-3}$ eV, $E_2 = 30.1\times10^{-3}$ eV, $N_1 = 8.5$, $N_2 = 3.1$, one gets

$$T = \frac{8.6 \times 10^{-3}}{8.617 \times 10^{-5} \times \ln\frac{8.5}{3.1}} = 99.7\,K$$

Also by taking $E_1 = 30.1\times10^{-3}$, $N_1 = 3.1$, $E_2 = 4.3\times10^{-3}$, $N_2 = 63$

$$T = \frac{25.8 \times 10^{-3}}{8.617 \times 10^{-5} \times \ln\frac{63}{3.1}} = 99.4$$

The mean value of the temperature may be taken as 99.5 K

10. (i) Assuming that sun is made up entirely of ionized hydrogen and has a mass 2×10^{30} kg calculate the number of electrons in the sun. (ii) If sun turns into a white dwarf of radius 2×10^7 m, calculate the Fermi energy of electrons in the white dwarf.

Solution:

(i) Since each hydrogen atom on ionization gives one electron the number of electrons is equal to the number of hydrogen atoms in the sun. The mass of one hydrogen atom is 1.67×10^{-27} kg, hence the number of electrons N_e in sun is:

$$N_e = \frac{2\times10^{30}}{1.67\times10^{-27}} = 1.197\times10^{57}$$

(ii) If sun converts into a white dwarf the number of electrons will remain to be N_e, but the volume of the white dwarf V will be

$$V = \frac{4}{3}\pi\left(2\times10^7\right)^3 = 33.51\times10^{21}$$

The Fermi energy of electrons is given by Eq. 11.82 as,

$$\epsilon_F = \frac{\hbar^2 \pi^{\frac{4}{3}}}{2m} \left(\frac{3N}{V}\right)^{2/3} \text{ where m is the mass of electron} = 9.1 \times 10^{-31} kg.$$

Or $\quad \epsilon_F = \dfrac{\left(1.0545 \times 10^{-34}\right)^2 \times \pi^{\frac{4}{3}}}{2 \times 9.1 \times 10^{-31}} \left(\dfrac{3 \times 1.197 \times 10^{57}}{33.51 \times 10^{21}}\right)^{2/3} = 6.37 \times 10^{-15} J = 3.98 \times 10^4 eV$

11. Compute the minimum number density of electrons in a gas of free electrons at absolute zero so that electrons may initiate the following reaction

$$proton + electron \rightarrow neutron - 0.8 \; MeV$$

Solution:

Since the reaction has a negative Q value, the minimum energy of electron must be 0.8 MeV. At absolute zero, the maximum energy of electrons in an electron gas is equal to Fermi energy ϵ_F. Therefore, to initiate the reaction the Fermi energy of electrons must be at least 0.8 MeV, hence

$$\epsilon_F = \frac{\hbar^2 \pi^{\frac{4}{3}}}{2m} \left(\frac{3N}{V}\right)^{2/3} = 0.8 \; MeV$$

Or

$$\left(\frac{3N}{V}\right)^{2/3} = \frac{2 \times 9.1 \times 10^{-31} \times 0.8 \times 10^6 \times 1.6 \times 10^{-19}}{\left(1.0545 \times 10^{-34}\right)^2 \pi^{\frac{4}{3}}} = 4.55 \times 10^{24}$$

Or $\qquad \dfrac{N}{V} = 3.24 \times 10^{36} \; m^{-3}$

The minimum number density $\dfrac{N}{V}$ of electrons to initiate the reaction should be $3.24 \times 10^{36} \; m^{-3}$

12. Write expression for the energy density of blackbody radiations in terms of the wavelength and hence drive Wien's displacement formula. Also calculate the wavelength of maximum emission for Earth assuming it as a blackbody at a temperature of 300 K.

Solution:

The energy density (energy per unit volume of the cavity) of blackbody radiation in the frequency interval v and $v + \Delta v$ is given by Eq. 11.102 as,

$$\Delta u_v = \frac{\Delta U_v}{V} = \frac{8\pi h}{c^3} \frac{v^3}{e^{\left(\frac{hv}{k_B T}\right)} - 1} \Delta v \qquad\qquad \text{S-11.12.1}$$

Using the relation $v = \frac{c}{\lambda} \; and \; \Delta v = -\frac{c}{\lambda^2} \Delta \lambda$ and remembering that negative sign in the

expression for Δv indicates that when frequency increases wavelength decreases, has no significance for energy density, which may be written as,

$$\Delta u_\lambda = \frac{8\pi c h}{\lambda^5} \frac{1}{e^{\left(\frac{hc}{k_B T \lambda}\right)} - 1} \Delta\lambda \qquad \text{S-11.12.2}$$

Above equation gives the energy density distribution of blackbody radiation in terms of the wavelength.

To get the value of the wavelength λ_m that corresponds to the maximum emission one has to partially differentiate Eq. S-11.12.2 with respect to T and equate it to zero,

$$\frac{\partial \Delta u_\lambda}{\partial \lambda} = 0 \text{ that gives } \frac{\partial}{\partial \lambda}\left[\frac{1}{\lambda^5} \frac{1}{\left(e^{\frac{\alpha}{\lambda T}} - 1\right)} \right] = 0 \qquad \text{S-11.12.3}$$

Here

$$\alpha = \frac{hc}{k_B} \qquad \text{S-11.12.4}$$

Equation S-11.12.3 on differentiation gives

$$\left[\frac{-5}{\lambda^6} \frac{1}{\left(e^{\frac{\alpha}{\lambda T}} - 1\right)} + \frac{1}{\lambda^5} \frac{\left(-e^{\frac{\alpha}{\lambda T}}\right)}{\left(e^{\frac{\alpha}{\lambda T}} - 1\right)^2} \frac{a}{T\lambda^2} \right] = 0$$

Or

$$\lambda_m T = \frac{a}{5\left[1 - e^{-\frac{a}{\lambda_m T}}\right]} \qquad \text{S-11.12.5}$$

Equation S-11.12.5 may be solved numerically to give

$$\lambda_m T = 2.898 \times 10^{-3} \ m \ K \qquad \text{S-11.12.6}$$

Treating earth to be a blackbody at temperature 300 K one gets,

$$\lambda_m = \frac{2.898 \times 10^{-3}}{300} \ m = 9.66 \times 10^{-6} \ m$$

13. A photon gas in equilibrium at temperature T is enclosed in a cavity of volume V. (i) what is the value of chemical potential μ of the photon gas? (ii) What is the number of photons in the cavity and how this average number depends on the temperature? (iii) If the energy density $\frac{E}{V}$ is written as $\frac{E}{V} = \int \rho(\omega) d\omega$, where ω is the angular frequency, determine $\rho(\omega)$. Investigate temperature dependence of U/V.

Solution:

(i) In a photon gas at equilibrium the number of photons does not remain constant and fluctuate around a mean value that is given by the condition $\left(\dfrac{\partial F}{\partial N}\right)_{T,V} = 0$. However from Eq. 7.63 of chapter-7, $\left(\dfrac{\partial F}{\partial N}\right)_{T,V} = \mu$, therefore, chemical potential for a photon gas is zero.

(ii) The angular frequency and the cyclic frequency are related by the relation

$$\omega = 2\pi v \qquad\qquad \text{S-11.13.1}$$

Therefore, $$\Delta\omega = 2\pi\Delta v \text{ and } v^2 = \frac{\omega^2}{4\pi^2} \qquad\qquad \text{S-11.13.2}$$

And the degeneracy of the macro-level with frequency v and $v + \Delta v$ is given by Eq. 11.99

$$\Delta\mathcal{G} = \frac{8\pi V v^2}{c^3}\Delta v$$

Substituting the values of v^2 and Δv in terms of ω from Eq.(S11.13.2) in the above equation one gets,

$$\Delta\mathcal{G}_\omega = \frac{V\omega^2}{\pi^2 c^3}\Delta\omega \qquad\qquad \text{S-11.13.3}$$

Also, the number of photons with angular frequency ω and $\omega + \Delta\omega$ is given by (see Eq. 11.100),

$$\Delta N_\omega = \frac{\Delta\mathcal{G}_\omega}{e^{\left(\frac{h\omega}{2\pi k_B T}\right)} - 1} = \frac{V\omega^2}{\pi^2 c^3}\frac{1}{e^{\left(\frac{h\omega}{2\pi k_B T}\right)} - 1}\Delta\omega \qquad\qquad \text{S-11.13.4}$$

And the total number of photons may be obtained by assuming ω to be continuous and integrating the above equation in the limit ω going from zero to infinity.

$$N = \frac{V}{\pi^2 c^3}\int_0^\infty \frac{\omega^2}{e^{\left(\frac{h\omega}{2\pi k_B T}\right)} - 1}\, d\omega \qquad\qquad \text{S-11.13.5}$$

One may now change the variable in the above Eq. from ω to x where

$$x = \frac{h\omega}{2\pi k_B T} \text{ so that } d\omega = \frac{2\pi k_B T}{h}dx \text{ and } \omega^2 = \frac{(2\pi k_B T)^2}{h^2}x^2$$

With these substitution Eq. S-11.13.5 becomes,

$$N = \frac{8\pi V k_B^3 T^3}{h^3 c^3}\int_0^\infty \frac{x^2}{e^x - 1}\, d\omega \qquad\qquad \text{S-11.13.6}$$

In Eq. S-11.13.6 the integral is a definite integral and has a fixed value that does not depend on x. As such the number of photons N in the cavity depends on the cube of the temperature.

(iii) Since each photon of cyclic frequency ω has the energy $\left(\dfrac{h\omega}{2\pi}\right)$, the total energy E of N photons may be obtained by multiplying N with $\left(\dfrac{h\omega}{2\pi}\right)$

$$E = \left(\frac{h\omega}{2\pi}\right) \frac{V}{\pi^2 c^3} \int_0^\infty \frac{\omega^2}{e^{\left(\frac{h\omega}{2\pi k_B T}\right)} - 1} \, d\omega$$

And

$$\frac{E}{V} = \int_0^\infty \frac{h\omega^3}{2\mathring{A}^3 c^3 \left[e^{\left(\frac{h\omega}{2\pi k_B T}\right)} - 1 \right]} d\omega \qquad\qquad \text{S-11.13.7}$$

Therefore,

$$\rho(\omega) = \frac{h\omega^3}{2\mathring{A}^3 c^3 \left[e^{\left(\frac{h\omega}{2\pi k_B T}\right)} - 1 \right]}$$

To investigate the temperature dependence of $\dfrac{E}{V}$ we substitute $\dfrac{h\omega}{2\pi k_B T} = y$ in Eq. S-11.13.7 to get,

$$\frac{E}{V} = \frac{8\pi k_B^4 T^4}{h^3 c^3} \int_0^\infty \frac{y^3}{2\mathring{A}^3 c^3 \left[e^y - 1 \right]} d\omega \qquad\qquad \text{S-11.13.8}$$

In the above equation integral is a definite integral and has a fixed value hence, $\dfrac{E}{V}$ varies as the fourth power of temperature.

14. Calculate the Fermi momentum, the Fermi temperature and Fermi velocity of electrons in Aluminum metal using following data; density of Al =2700 kg m^{-3}; 1 kilomole of Al weighs 26.98 kg, each atom releases 3 electrons.

Solution:

Fermi momentum $P_F \equiv \sqrt{2m\epsilon_F}$, Fermi temperature $\theta_F \equiv \dfrac{\epsilon_F}{k_B}$ and Fermi velocity $v_F \equiv \left(\dfrac{2\epsilon_F}{m}\right)^{1/2}$

Number of electrons in $1 m^3$ of Al $= \dfrac{3 \times 2700 \times 6.02 \times 10^{26}}{26.98} = 18.07 \times 10^{28}$

$$\epsilon_F = \frac{\hbar^2 \pi^{\frac{4}{3}}}{2m} \left(\frac{3N}{V}\right)^{2/3} = \frac{(1.0545)^2 \, \hbar^{4/3}}{2 \times 9.11 \times 10^{-31}} \left(3 \times 18.07 \times 10^{28}\right)^{2/3} = 18.56 \times 10^{-19} \, J$$

$$P_F \equiv \sqrt{2m\epsilon_F} = \left(2 \times 9.11 \times 10^{-31} \times 18.56 \times 10^{-19}\right)^{1/2} = 18.39 \times 10^{-25} \, kg \, m \, s^{-1}$$

$$\theta_F \equiv \frac{\epsilon_F}{k_B} = \frac{18.56 \times 10^{-19}}{1.3807 \times 10^{-23}} = 13.44 \times 10^4 \, K$$

$$v_F \equiv \left(\frac{2\epsilon_F}{m}\right)^{1/2} = \left(\frac{2 \times 18.56 \times 10^{-19}}{9.11 \times 10^{-31}}\right)^{0.5} = 2.02 \times 10^6 \, ms^{-1}$$

15. An ideal gas in equilibrium at temperature T K is contained in a right circular cylinder of height L and base area A. Molecules of the gas of mass m are acted upon by the gravitational field of Earth of constant acceleration g. Calculate the constant volume heat capacity per molecule c_v of the gas and obtain its value for the limiting cases of $T \to 0$ and $T \to \infty$

Solution:

The total energy E_{total} of each gas molecule is

$$E_{total} = \frac{3}{2} k_B T + mg\bar{y} , \qquad\qquad \text{S-11.15.1}$$

Here $\bar{y}$ is the average height of the molecule from the base of the cylinder. To determine the average height we make use of Boltzmann distribution which tells that the probability distribution of heights is proportional to $e^{-\left(\frac{mgy}{k_B T}\right)}$. Hence average height is;

$$\bar{y} = \frac{\int_0^L y e^{-\left(\frac{mgy}{k_B T}\right)} dy}{\int_0^L e^{-\left(\frac{mgy}{k_B T}\right)} dy} = \frac{k_B T}{mg} \left[1 - \frac{mgL}{\left(e^{\left(\frac{mgL}{k_B T}\right)} - 1\right)} \right]$$

Or

$$mg\bar{y} = k_B T \left[1 - \frac{mgL}{\left(e^{\left(\frac{mgL}{k_B T}\right)} - 1\right)} \right]$$

Hence from Eq. S-11.15.1

$$E_{total} = \frac{5}{2} k_B T - \frac{mgL}{\left(e^{\left(\frac{mgL}{k_B T}\right)} - 1\right)}$$

Now constant volume heat capacity per molecule may be obtained by differentiating the above expression w.r.t T,

$$c_v = \frac{5}{2}k_BT - \frac{k_B(mgL)^2}{(k_B)^2} \frac{\left(e^{\left(\frac{mgL}{k_BT}\right)}\right)}{\left(e^{\left(\frac{mgL}{k_BT}\right)} - 1\right)^2} \qquad \text{S-11.15.2}$$

When $\quad T \to 0$, $\lim_{T \to 0} E_{total} \approx \frac{5}{2}k_BT$ as $\left(e^{\left(\frac{mgL}{k_BT}\right)} - 1\right)$ approaches ∞

Hence $c_{vT \to 0} = \frac{\partial}{\partial T}\left(\lim_{T \to 0} E_{total}\right) = \frac{\partial}{\partial T}\left(\frac{5}{2}k_BT\right) = \frac{5}{2}k_B$

When $T \to \infty$ $\lim_{T \to 0} E_{total} \approx \frac{3}{2}k_BT$ as $\left(e^{\left(\frac{mgL}{k_BT}\right)} - 1\right)$ approaches $\frac{mgL}{k_BT}$ and $\frac{mgL}{\left(e^{\left(\frac{mgL}{k_BT}\right)} - 1\right)} \approx k_BT$

Hence $\quad c_{vT \to \infty} = \frac{\partial}{\partial T}\left(\lim_{T \to \infty} E_{total}\right) = \frac{\partial}{\partial T}\left(\frac{3}{2}k_BT\right) = \frac{3}{2}k_B$

16. Calculate the constant volume heat capacity per atom for a monatomic gas the atoms of which have only two energy levels, the ground level with degeneracy 2 and the other level with energy ϵ and degeneracy 5. Assume that the gas obeys Maxwell–Boltzmann distribution and is in equilibrium at temperature T.

Solution:

We first calculate the average energy of an atom of the gas. Let N_0 and N_1 denote the number of atoms in ground and the excited levels respectively. Let us also assume that the energy scale starts from the ground state i.e. ground level energy is zero. Since the gas obeys Boltzmann distribution the number of particle in energy level E will be proportional to ge^{-E/k_BT}, hence;

$$N_0 = A2e^{-\left(\frac{0}{k_BT}\right)} \quad and \; N_1 = A5e^{-\left(\frac{\epsilon}{k_BT}\right)}, \text{ here A is a constant.}$$

Therefore, total number of atoms $\quad N = N_0 + N_1 = A\left[2 + 5e^{-\left(\frac{\epsilon}{k_BT}\right)}\right]$

The total energy of these atoms $\quad E = N_0 \times 0 + N_1\epsilon = 0 + A5\epsilon e^{-\left(\frac{\epsilon}{k_BT}\right)}$

Hence the average energy per atom $\bar{E} = \dfrac{E}{N} = \dfrac{A5\epsilon \ e^{-\left(\frac{\epsilon}{k_B T}\right)}}{A\left[2 + 5e^{-\left(\frac{\epsilon}{k_B T}\right)}\right]} = \dfrac{5\epsilon \ e^{-\left(\frac{\epsilon}{k_B T}\right)}}{\left[2 + 5e^{-\left(\frac{\epsilon}{k_B T}\right)}\right]}$

Apart from $\bar{E}$ each molecule also has an average energy equal to $(3/2\ k_B\ T)$ due to the thermal motion with three degrees of freedom. Hence the total average energy of each atom becomes,

$$U_{atom} = \frac{3}{2}k_B T + \frac{5\epsilon \ e^{-\left(\frac{\epsilon}{k_B T}\right)}}{\left[2 + 5e^{-\left(\frac{\epsilon}{k_B T}\right)}\right]}$$

The constant volume heat capacity $c_v = \dfrac{\partial}{\partial T}\left(U_{atom}\right) = \dfrac{3}{2}k_B + \dfrac{\partial}{\partial T}\dfrac{5\ \epsilon\ e^{-\left(\frac{\epsilon}{k_B T}\right)}}{\left[2 + 5e^{-\left(\frac{\epsilon}{k_B T}\right)}\right]}$

Or $\qquad c_v = \dfrac{3}{2}k_B + \dfrac{\partial}{\partial T}\dfrac{5\epsilon}{\left[2e^{\left(\frac{\epsilon}{k_B T}\right)} + 5\right]} = \dfrac{3}{2}k_B + \dfrac{10\epsilon^2 \ e^{\left(\frac{\epsilon}{k_B T}\right)}}{k_B T^2 \left(2e^{\left(\frac{\epsilon}{k_B T}\right)} + 5\right)^2}$

17. A system contains large number N of spin half paramagnetic ions in equilibrium at temperature *T*. The system is subjected to an external magnetic field B in z-direction. Determine (i) M_z, the z-component of total spin in z-direction, (ii) total number of states Ω accessible to the system as a function of M_z. (iii) The value of M_z for which Ω is a maximum.(iv) Draw a rough sketch for the variation of entropy of the system as a function of temperature and indicate the region of negative temperature.

Solution:

(i) Let $N_\uparrow$ and $N_\downarrow$ respectively denote the ions aligned parallel (in +z direction) and opposite (in the –z direction) to the applied magnetic field B. Generally, $N_\uparrow > N_\downarrow$. The z-component of total spin $M_z = 1/2(N_\uparrow - N_\downarrow)$.

(ii) But $(N_\uparrow + N_\downarrow) = N$ and $(N_\uparrow - N_\downarrow) = 2\ M_z$. Hence, $N_\uparrow = \dfrac{1}{2}N + M_z$ and $N_\downarrow = \dfrac{1}{2}N - M_z$

The total number of states assessable to the system is,

$$\Omega = \frac{(N)!}{(N_\uparrow)!(N_\downarrow)!} = \frac{(N)!}{\left(\frac{1}{2}N + M_z\right)!\left(\frac{1}{2}N - M_z\right)!}$$

Using Sterling's formula one can write In (N)! = N In N

Therefore, In Ω = N In N – $N_\uparrow$ In $N_\uparrow$ – $N_\downarrow$ In $N_\downarrow$ = N In N – $N_\uparrow$ In $N_\uparrow$ – $(N - N)$ In $(N - N_\uparrow)$

(iii) To obtain the maximum value of In Ω (or Ω) we differentiate the above expression w.r.t. $N_\uparrow$ and equate it to zero.

$$\frac{\partial \ln \Omega}{\partial N_\uparrow} = -\text{In } N_\uparrow + \text{In } (N - N_\uparrow) = 0$$

That gives $$N_\uparrow = \frac{N}{2} = N_\downarrow$$

It may be observed that the maximum number of assessable states occur when equal number of ions are in the parallel and anti-parallel states when M_z =0. This is possible at infinite temperature.

(iv) Since entropy $S = k_B$ In Ω and In Ω is maximum when the number of ions parallel and anti parallel states are equal, that occurs at infinite temperature, entropy of the system increases with temperature, attains the maximum value at infinite temperature. On further increasing temperature (which is the region of negative temperature) population inversion occurs and the entropy decreases again reaching a value zero at highly negative temperature when all ions are in the higher energy state.

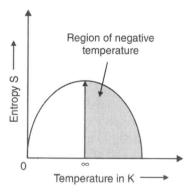

Fig. S-11.17

Problems

1. The partition function for photon gas has the form In $Z = aT^3 V$, show that the (i) internal energy $U = 3aVT^4 = 3PV$ (ii) entropy $S = 4aVk_BT^4$

2. The maximum frequency of coupled oscillators for diamond is 87×10^{13}. Calculate the Debye temperature and the heat capacity at constant volume C_v for diamond at 200 K and compare it to the experimental value of 2.12 x 10^3 J kilomole^{-1} K^{-1}.

3. The energy E of a particle is a function of some parameter z such that $E = az^2$ and z may have values from $-\infty$ to $+\infty$. Calculate the average energy of the particle if it obeys Boltzmann statistics and explain how the obtained value for the average energy may be explained by the principle of equipartition of energy.

4. A system of two atoms each having only three levels at energies 0, E and 2E, is in contact with a heat reservoir at temperature T. Write down the partition functions for the system when (i) atoms are distinguishable (ii) atoms are indistinguishable.

5. In a photon gas that is in equilibrium at temperature T show that (i) the number of photons is proportional to T^3 and (ii) the heat capacity C_V also varies as T^3.

6. Spherical colloidal particles of radius 0.25×10^{-6} and density 1100 kg m^{-3} are suspended in water of density 1000 kg m^{-3}. Taking into account the effects of buoyancy calculate the vertical distance over which the number density of particles falls to 1/e of its original value.

   [Hint: *buoyancy effectively reduces the mass m of the particle to a value $m_{eff} = m(1100 - 1000)$.*]

   [The vertocal height in which number density falls by $\dfrac{1}{e}$ *is equal to* $\dfrac{k_B T}{m_{eff}\, g}$]

7. A cavity of initial volume V is filled with black body radiations in equilibrium at temperature T for which the maximum emission of energy density per unit frequency interval $\dfrac{\Delta u}{\Delta v}$ occurs at a frequency v_i. The cavity is then adiabatically and reversibly expended to a volume 2V. Determine the final frequency corresponding to maximum energy emission.

   [Hint: for blackbody radiation it is known that U= 3 P V; v_m = *constant x T* and

   $dU = TdS - PdV$; *So* $dU = 3PdV + 3VdP$ *and for adiabatic change dS = 0, Hence*

   $dU = -PdV$ *Or* $- PdV = 3PdV + 3VdP$ *that gives* $V^4 P^3 =$ *constant. But* $P \propto T^4$

   Therefore, $T^3 V =$ *constant. Now* $T_i^3 V_i = T_f^3 V_f$ *But final volume* $V_f = 2V_i$

   So $T_f = T_i \left(\dfrac{V_i}{V_f}\right)^{1/3} = T_i \left(\dfrac{1}{2}\right)^{1/3}$. Now use $v_m =$ *constant x T*]

8. The universe is filled with blackbody radiation characteristic of temperature 3K. This radiation is the reminiscent of the photon cloud that was created at the time of big bang and has since cooled down adiabatically. If in the next say 10^{15} years the universe adiabatically expands to N times of its present volume, what will be the characteristic temperature of the blackbody radiation?

9. Vibrational levels of a diatomic gas are given by

$$\varepsilon_v = (n + 1/2)\hbar\omega_0 \text{ where n=0, 1, 2, 3,.........}$$

   Derive partition function and hence the vibrational Helmholtz free energy per mole f and molar constant volume heat capacity c_V of the gas.

10. A system of large number of particles N = $(N_1 + N_2)$ has only two levels at energies E_1 and E_2 $(E_2 > E_1)$. The system in contact with a reservoir at temperature T and is in equilibrium with N_1 particles in level of energy E_1 and N_2 particles in level with energy E_2. When a particle from level E_2 goes to level E_1 a quantum of energy is delivered to the reservoir. Calculate the change in entropy of the system and the reservoir in this de-excitation process.

11. The energy levels of a three dimensional rigid rotator of moment of inertia I are given by

$$\varepsilon_{J,M} = \frac{J(J+1)}{2I}\hbar^2$$

Here J = 0, 1, 2, 3, ... and M may vary from – J to + J In steps of unity.

Obtain the partition function for the system in the limit of high temperature when summation over J may be replaced by an integral.

[Hint: The partition function $Z = \sum\limits_{J=0}^{\infty} (2J + 1)e^{-\left(\frac{J(J+1)\hbar^2}{2Ik_BT}\right)}$

In the limit of high temperature such that $k_BT \gg \dfrac{\hbar^2}{2I}$, $\dfrac{J(J+1)\hbar^2}{2Ik_BT}$ varies slowly with J and the

summation may be replaced by the integral. Putting $J(J+1) = t$, one gets

$$Z = \int\limits_{0}^{\infty} e^{-\left(\frac{\hbar^2}{2Ik_BT}t\right)} dt]$$

Short Answer Questions

1. Free electrons in a metal essentially do not interact. Explain.

2. Only very few free electrons take part in heat absorption when a metal is heated. Why?

3. Write a short note on Boltzmann distribution and obtain the distribution function for vertical height of ideal gas molecules under gravity.

4. A blackbody cavity of volume V in equilibrium at temperature T is adiabatically and reversible compressed to half of the original volume. What will be the temperature of the new cavity?

5. Obtain an expression for the degeneracy of a macro-level of a system that obeys classical statistics.

6. Without derivation write the expression for the entropy of an ideal gas obtained using quantum thermodynamics, and compare it with the value obtained using classical thermodynamics. Hence indicate the advantage of using quantum thermodynamics.

7. Show that the principle of equipartition of energy is valid only if the energy is a quadratic function of the parameter defining the degrees of freedom.

8. Show that the equation of state of a polyatomic gas will be same as that for a monatomic gas if the molecules of the polyatomic gas do not interact with each other and their energy does not depend on the coordinates of the center of mass.

9. Draw a rough sketch for the variation of the entropy of a system of half spin paramagnetic ions subjected to an external magnetic field with the temperature of the system and hence indicate the region of negative temperature.

10. For an ideal gas under gravity, prove that the number density of molecules at height H is proportional to $\dfrac{1}{e^H}$.

11. What is meant by population inversion and what is its significance in case of the half spin paramagnetic crystals?

12. The characteristic temperatures for rotational and vibrational motions of a diatomic gas are respectively, 2.5 K and 2800 K. What will be the molar heat capacity of the gas at 5000 K?

13. What is Debye temperature and how is it related to the molar heat capacity at constant volume for a crystalline substance at very low temperatures?

14. Draw rough sketches for (i) the heat capacity at constant volume for a system of harmonic oscillators (ii) heat capacity at constant magnetic field for a paramagnetic salt and explain why the two curves are different in shape.

15. Draw rough sketches for (i) distribution function for free electrons in metals at T = 0 K, and (ii) occupation number verses level energy at T = 0 K, for free electrons in metals and explain these curves.

16. Show that in case of an ideal gas obeying classical statistics the number of particles ΔN_i in a level 'i' is always much small than the degeneracy ΔG_i of the level.

Long Answer Questions

1. Explain Bose Statistics, Fermi statistics and Boltzmann statistics indicating the points of difference between them. How these statistics are related to the indistinguishability of identical particles?

2. Assuming that monatomic ideal gas obeys classical statistics derive expressions for constant volume heat capacity and the entropy of the gas. Is it possible to explain the increase of entropy when two dissimilar ideal gases at same temperature and pressure are mixed (Gibbs paradox)?

3. Write a detailed note on Boltzmann distribution law and show that classical statistics is appropriate for a monatomic ideal gas.

4. Define density of velocity ρ_v in velocity space, and derive an expression for it. A graph of ρ_v against v shows that ρ_v is a maximum at $v = 0$, on the other hand it is known that hardly any gas molecule is at rest. How the two facts can be reconciled?

5. What is the main point of difference between Einstein and Debye theories for the constant volume heat capacity of crystalline solids? Derive an expression for the maximum frequency of Debye oscillators in an isotropic crystalline solid.

6. Which statistics is followed by an assembly of photons? Derive expression for the degeneracy of a macro-level of a photon gas. Write partition function and obtain expression for the energy density distribution of the photon gas which is in equilibrium at temperature T.

7. Obtain partition function for a paramagnetic salt that is subjected to an external magnetic field, assume that there is no interaction between the paramagnetic ions and that electronic interactions are switched off.

8. What is meant by population inversion? How the concept of population inversion leads to the negative temperature in case of a spin half paramagnetic ion?

9. Explain why free electrons in a metal may be treated as a non-interacting electron gas. Define Fermi energy and show that it does not depend on the temperature. Drive expression for the degeneracy of a macro-level of the electron gas and show that it varies as the square root of energy of the level.

10. Show that the constant volume heat capacity of electron gas at not too high temperature is $\dfrac{N\pi^2 k_B{}^2}{2\epsilon_F} T$. Why only very few electrons participate in heat absorption?

Multiple Choice Questions

Note: Some questions may have more than one correct alternatives all correct alternatives must be marked for complete answer in such cases.

1. The volume dependent term in Sackur–Tetrode equation for molar specific entropy of an ideal gas is
   (a) R lnV
   (b) V lnR
   (c) R/lnV
   (d) V/lnR

2. At a constant temperature T the average number density of ideal gas molecules with speed v and $v + \Delta v$ is proportional to;

   (a) v^2
   (b) $e^{-\left(\frac{mv^2}{2k_B T}\right)}$
   (c) $v^2\, e^{\left(\frac{mv^2}{2k_B T}\right)}$
   (d) $v^2\, e^{-\left(\frac{mv^2}{2k_B T}\right)}$

3. If $\bar{v}$, v_m and v_{rms} respectively denote the average, most probable and root mean square speeds of the molecules of a gas at temperature T, then
   (a) $\bar{v} > v_m > v_{rms}$
   (b) $\bar{v} < v_m < v_{rms}$
   (c) $\bar{v} > v_m < v_{rms}$
   (d) $\bar{v} < v_{rms} > v_m$

4. The partition function Z (at temperature T) for a system of quantized linear oscillators of frequency v and characteristic temperature θ is

   (a) $Z = \dfrac{\exp\left(\dfrac{\theta}{2T}\right)}{1 - \exp\left(-\dfrac{\theta}{T}\right)}$
   (b) $Z = \dfrac{\exp\left(-\dfrac{\theta}{2T}\right)}{1 - \exp\left(-\dfrac{\theta}{T}\right)}$

   (c) $Z = \dfrac{\exp\left(-\dfrac{\theta}{2T}\right)}{1 - \exp\left(\dfrac{\theta}{T}\right)}$
   (d) $Z = \dfrac{\exp\left(-\dfrac{\theta}{T}\right)}{1 - \exp\left(-\dfrac{\theta}{2T}\right)}$

5. The ground state energy of a quantized linear oscillator of frequency v is
   (a) $\dfrac{3}{2} hv$
   (b) hv
   (c) $\dfrac{1}{2} hv$
   (d) zero

6. According to Debye theory the total number N_v of quantum linear oscillators that have frequencies up to and including v is given by;
   (a) $N_v = \dfrac{3N}{v_m^3} v^3$
   (b) $N_v = \dfrac{3N}{v^3} v_m^3$

(c) $N_v = \dfrac{v_m^3}{3N} v^3$

(d) $\dfrac{v^3}{3N\,v_m^3}$

Here N is the number of atoms in the crystal and v_m is the maximum frequency

7. Atoms of a crystalline solid have only two levels at energies +E and –E, the mean energy per atom at temperature T is

(a) $E \sinh\left(\dfrac{E}{k_B T}\right)$

(b) $E \cosh\left(\dfrac{E}{k_B T}\right)$

(c) $E \tanh\left(\dfrac{E}{k_B T}\right)$

(d) $E \coth\left(\dfrac{E}{k_B T}\right)$

8. A one dimensional quantum linear oscillator is in thermal equilibrium with a constant temperature heat source of temperature T. The average energy of the oscillator is;

(a) $\dfrac{3}{2} hv \coth\left(\dfrac{hv}{2K_B T}\right)$

(b) $\dfrac{1}{2} hv \coth\left(\dfrac{hv}{2K_B T}\right)$

(c) $\dfrac{3}{2} hv \tanh\left(\dfrac{hv}{2K_B T}\right)$

(d) $\dfrac{1}{2} hv \tanh\left(\dfrac{hv}{2K_B T}\right)$

9. An ideal gas that obeys classical distribution is held in a container of height L in a gravitational field of constant acceleration g. The average height $\bar{H}$ of gas molecules of mass m is

(a) $\dfrac{k_B T}{mg}\left[1 + \dfrac{L}{e^{mgL} - 1}\right]$

(b) $\dfrac{k_B T}{mg}\left[1 - \dfrac{L}{e^{-mgL} - 1}\right]$

(c) $\dfrac{mg}{k_B T}\left[1 - \dfrac{L}{e^{mgL} - 1}\right]$

(d) $\dfrac{k_B T}{mg}\left[1 - \dfrac{L}{e^{mgL} - 1}\right]$

10. A crystalline paramagnetic salt subjected to an external magnetic field is in the state of population inversion. Its absolute temperature is
   (a) Negative
   (b) beyond $+\infty$
   (c) below absolute zero
   (d) below $-\infty$

11. The maximum or cutoff frequency v_m in Debye theory is inversely proportional to
   (a) Number density of atoms in the crystal   (b) volume of the crystal
   (c) inter atomic distance in the crystal   (d) electron density in the crystal

12. The Fermi energy of a metal of volume V and total number of free electrons N in thermal equilibrium at temperature T depends on

   (a) T

   (b) $\left(\dfrac{N}{V}\right)^{2/3}$

   (c) $\left(\dfrac{N}{V}\right)^{3/2}$

   (d) $\left(\dfrac{V}{N}\right)^{2/3}$

13. At absolute zero temperature the chemical potential of the free electron gas in a metal μ is
   (a) $\mu = 0$
   (b) $\mu = \epsilon_F$
   (c) $\mu > \epsilon_F$
   (d) $\mu < \epsilon_F$

14. A piece of metal has N free electrons in volume V. Assuming free electrons as a non-degenerate gas that obeys Fermi –Dirac distribution, the total number of energy levels of the system up to Fermi energy and at absolute zero temperature will be
   (a) Infinite   (b) N^2
   (c) N   (d) N/2

15. In Fermi–Dirac distribution the degeneracy $\Delta \mathcal{G}\epsilon$ of a macro-level at energy ϵ and $\epsilon + \Delta\epsilon$ is proportional to
   (a) $\epsilon^{1/2}$ (b) ϵ
   (c) $\epsilon^{-1/2}$ (d) ϵ^{-1}

Answers to Problems and Multiple Choice Questions

Answer to problems

2. Debye temp. 1858 K, 1.93×10^3 J kilomole $^{-1}$ K^{-1}

3. Average energy $\frac{1}{2} k_B T$, since energy depends only on one parameter it has only one degree of freedom.

4. (i) $Z = \left(1 + e^{-\left(\frac{E}{k_B T}\right)} + e^{-\left(\frac{2E}{k_B T}\right)}\right)^2$

   (ii) $\frac{1}{2}\left(1 + e^{-\left(\frac{E}{k_B T}\right)} + e^{-\left(\frac{2E}{k_B T}\right)}\right)^2$

6. 6.3×10^{-5} m

7. $\dfrac{v_i}{2^{1/3}}$

8. $\dfrac{3}{N^{1/3}} K$

9. $Z_v = \dfrac{e^{-\left(\hbar\omega_0/2k_B T\right)}}{1 - e^{-\left(\hbar\omega_0/k_B T\right)}}$, $F(per\, mole) = \frac{1}{2} A_v \hbar\omega_0 + A_v k_B T \ln\left[1 - e^{-\left(\hbar\omega_0/k_B T\right)}\right]$,

   Molar heat capacity at constant volume $c_v = R \{e^{-\left(\hbar\omega_0/k_B T\right)}\}^2 \dfrac{e^{\left(\hbar\omega_0/k_B T\right)}}{\left[e^{\left(\hbar\omega_0/k_B T\right)} - 1\right]^2}$

10. $\Delta S_{system} = k_B \ln\left(\dfrac{N_2}{N_1 + 1}\right)$; $\Delta S_{reservoir} = \dfrac{E_2 - E_1}{T}$

11. $Z = \dfrac{2 I k_B T}{h^2}$

Answers to multiple choice questions

1. (a)	2. (d)	3. (c), (d)	4. (c)
5. (c)	6. (a)	7. (c)	8. (b)
9. (d)	10. (a), (b)	11. (c)	12. (b)
13. (c)	14. (d)	15. (a)	

Revision

1. A monatomic ideal gas follows classical statistics of indistinguishable particles with much smaller number of particles in each level than the degeneracy of the level. Molecules are considered to be free and non-interacting and their energy levels are given by the expression

$$\epsilon_j = n_j^2 \frac{\hbar^2 (2\pi)^2}{8m} V^{-\frac{2}{3}}$$

2. Since the number of levels is very large, one uses the concept of macro-level which is made by lumping together all levels in a small energy interval $\Delta\epsilon$ around energy ϵ. The degeneracy of the macro-level is given by $\Delta\mathcal{G}_j = \frac{1}{2}\pi n_j^2 \, dn_j$ and the partition function for a monatomic ideal

gas is $Z = \left[V \left(\frac{mk_BT}{2\pi\hbar^2} \right)^{3/2} \right]$; Using the tools of quantum thermodynamics for classical

distribution, it can be shown that the internal energy $U = NT^2 k_B \frac{\partial(\ln Z)}{\partial T} = \frac{3}{2}nRT$;

$$C_v = \left(\frac{\partial U}{\partial T} \right)_V = \frac{3}{2}Nk_B = \frac{3}{2}nR; \; c_v = \frac{C_v}{n} = \frac{3}{2}R; \; S(T) = \frac{U}{T} +$$

$$N k_B (\ln Z - \ln N + 1) = Nk_B \left[\frac{5}{2} + \ln\{ \frac{V}{N} \left(\frac{mk_BT}{2\pi\hbar^2} \right)^{\frac{3}{2}} \} \right]$$

$$= \frac{5}{2}nR + nR \ln V + nR \ln\{ \frac{1}{N} \left(\frac{mk_BT}{2\pi\hbar^2} \right)^{\frac{3}{2}} \}$$

and $s(T) = \frac{5}{2}R + R \ln V + \frac{3}{2}R \ln T + R \left[\ln\{ \frac{1}{N} \left(\frac{mk_B}{2\pi\hbar^2} \right)^{\frac{3}{2}} \} \right]$ Sackur–Tetrode Equation.

An important property of the quantum derivation of entropy is that it gives the absolute value of entropy without any undetermined constant. On the other hand in classical thermodynamics entropy is always relative and it is only the difference in entropies of the two states of the system that can be calculated.

3. Speed distribution of ideal gas molecules may be obtained as

$$\frac{\Delta N^v}{\Delta v} = 4N(\pi)^{-\frac{1}{2}} \left(\frac{m}{2k_BT} \right)^{\frac{3}{2}} v^2 e^{-\left(\frac{mv^2}{2k_BT} \right)}$$

The peak of the speed or velocity distribution curve shifts to the higher velocity side on increasing the temperature of the gas but the height of the peak decreases such that the total area under the distribution curve remains constant and is equal to the total number of gas molecules N.

4. The most probable, the average and the rms speeds of gas molecules are given by

$$v_m = \left(\frac{2k_BT}{m} \right)^{\frac{1}{2}} = 1.414 \sqrt{\frac{k_B}{m}} \; ; \; \bar{v} = \frac{2}{\sqrt{\pi}} \left(\frac{m}{2k_BT} \right)^{-\frac{1}{2}} = \sqrt{\frac{8}{\pi} \frac{k_BT}{m}} = 1.1281 \, v_m;$$

$$v_{rms} = \sqrt{\frac{3k_BT}{m}} = \sqrt{3/2}\, v_m = 1.225\, v_m$$

These are the same values that were obtained using the Kinetic theory.

5. Maxwell–Boltzmann velocity distribution function: Maxwell–Boltzmann velocity distribution function tells how many molecules of the gas with speeds between v_j and $(v_j + dv_j)$ may be found in a unit volume anywhere in the gas. In other words it tells about the density distribution of gas molecules with speeds between v_j and $(v_j + dv_j)$

$$\rho_{vj} \equiv \frac{dN_j^{v_j}}{\Delta V_j} = \frac{N}{\pi^{3/2}}\left(\frac{m}{2k_BT}\right)^{3/2} e^{-\left(\frac{mv_j^2}{2k_BT}\right)}$$

6. Boltzmann distribution law: The law is applicable to systems that obey either Maxwell–Boltzmann distribution or classical distribution and states that the frequency of occurrence of the system in the state of energy E is proportional to $e^{-\left(\frac{E}{k_BT}\right)}$ where symbols have their usual meanings. Using this law one may determine the number density of gas molecules at a certain height in a gravitational field.

7. The law of equipartition of energy for systems in which energy is a quadratic function of the parameter that specify the degrees of freedom may be derived using the formulations of quantum thermodynamic s.

8. The energy values of a quantized linear oscillator is given by $\epsilon_j = \left(n_j + \frac{1}{2}\right)h\nu$; its zero point

energy $= \frac{1}{2}h\nu$;

9. The partition function for an assembly of quantized linear oscillators is given by $Z = \dfrac{e^{-\left(\frac{h\nu}{2k_BT}\right)}}{1 - e^{-\left(\frac{h\nu}{k_BT}\right)}}$

$= \dfrac{e^{-\left(\frac{\Theta}{2T}\right)}}{1 - e^{-\left(\frac{\Theta}{T}\right)}}$ where the characteristic temperature $\Theta = \dfrac{h\nu}{k_B}$; the internal energy of the assembly

$$U = Nk_B\Theta\left\{\frac{1}{2} + \frac{1}{\left(e^{\frac{\Theta}{T}} - 1\right)}\right\}$$; At absolute zero all oscillators are in their ground state and

contribute energy equal to their zero point energy to the total energy,

$U^{T\to 0} = N\left(\frac{1}{2}h\nu\right)$; At high temperatures T>> θ, $U^{\left(\frac{\Theta}{T}\ll 1\right)} = Nk_BT$: that means as if each oscillator

is behaving as if it has two degrees of freedoms. The constant volume heat capacity of the assembly of linear quantum oscillators is given by

$$C_v = \frac{\partial U}{\partial T} = N k_B \left(\frac{\Theta}{T}\right)^2 \left[\frac{e^{\frac{\Theta}{T}}}{\left(e^{\frac{\Theta}{T}} - 1\right)^2} \right]$$. As temperature T approaches 0K, C_v approaches zero and

when $T \gg \Theta$, C_v approaches the constant value $N k_B$, as expected.

10. Measurement of constant volume heat capacity of diatomic molecules indicates as if these molecules have 5 degrees of freedom at moderate temperatures and 7 degrees of freedom at higher temperatures. Assuming that rotational and vibrational motions in the molecule may also be represented by quantized linear oscillators with characteristic temperatures Θ^{rot} and Θ^{vib} respectively, it may be shown that only two degrees of freedom will contribute for each of the motion respectively at temperatures greater than Θ^{rot} and Θ^{vib}.

11. In the case of polyatomic gases, the moment of inertia of the molecule is large and, therefore, the characteristic temperatures for rotational and vibrational motions are quite low. As a result even at moderate temperatures both the rotational and vibrational mode are excited along with the translational mode, hence the molar constant volume heat capacity has the value $\approx$ 7/2 R.

12. Classical theory predicts constant volume molar heat capacity for solids a value 3R at all temperatures which is Dulong and Petit law. Einstein for the first time applied quantum thermodynamics and assumed that in a solid there are three linear oscillators, one each for each direction for each molecule of the solid and these oscillators vibrate with same frequency.

13. The specific molar heat capacity at constant volume c_v of the solid may then be given by

$$c_v = \frac{\partial U^{sol}}{\partial T} = \frac{\partial}{\partial T}\left[3R\Theta_{\mathrm{Ein}} \left\{ \frac{1}{2} + \frac{1}{\left(e^{\frac{\Theta_{\mathrm{Ein}}}{T}} - 1\right)} \right\} \right] = 3R \left(\frac{\Theta_{\mathrm{Ein}}}{T}\right)^2 \left[\frac{e^{\frac{\Theta_{\mathrm{Ein}}}{T}}}{\left(e^{\frac{\Theta_{\mathrm{Ein}}}{T}} - 1\right)^2} \right]$$

Though this formula gives the value of molar C_v = 3R at moderate temperatures but a big drawback of Einstein theory is that the same value of Θ_{Ein} does not reproduce the experimental data both at lower and higher temperature ends. Further, at lower temperatures the decrease of the theoretical C_v is faster than the decrease in the experimental values.

14. Debye theory: Debye modified Einstein's approach by assuming that the 3N oscillators in a crystalline solid having N atoms do not all vibrate with the same frequency. But instant they vibrate as coupled oscillators and set stationary elastic waves in the crystal. These stationary elastic waves set in the crystal by the lattice vibrations are termed as phonons. Thus it may be assumed that the solid is filled with phonons of different frequencies up to a maximum value denoted by v_{max}. The total internal energy of the solid is given by $U_{total} = 9NTk_B \dfrac{T^3}{\Theta_D^3} \displaystyle\int_0^{t_m} \dfrac{t^3 dt}{\left(e^t - 1\right)}$.

Here $\Theta_D = \dfrac{h v_{max}}{k_B}$ is called Debye temperature and has different value for different materials.

The molar constant volume heat capacity at high temperatures $\approx$ 3R, and at low temperatures it varies as T^3. This behavior is consistent with experimental observations.

15. Contribution of free electrons in metals to the constant volume heat capacity: Free electrons in metals may be treated as non-interacting, free particles like the molecules of an ideal gas. This non-interaction between free electrons originates from Pauli's Exclusion principle since most of the energy states of electrons are already occupied and are not available for scattered electrons if they interact. At absolute zero all electron states up to Fermi level are completely filled while those above it are all vacant. At higher temperatures some electrons from levels just below Fermi energy are excited to the vacant levels above it. As such, though there are very large number of free electrons in a piece of metal but only very few electrons that are in levels just near to Fermi level may take part in absorption of heat. Hence the contribution of free electrons in metals to the constant volume heat capacity is very small as compared to that of the molecules of the ideal gas. The specific molar heat capacity of electron gas is given by,

$$c_V = \frac{1}{2}\pi^2 \left(\frac{k_B T}{\epsilon_F}\right) R$$

It may be observed that c_V varies linearly with temperature T.

16. Application of quantum statistics to blackbody radiations: In quantum statistics a blackbody cavity is taken to be filled with a gas of photons of all frequencies. Since photons do not interact with each other, the photon gas is treated as an ideal gas. Further, photons are indistinguishable and obey Bose–Einstein statistics. In Bose–Einstein statistics any number of particles may be accommodated in a given energy state and, therefore, there is no restriction on the number of photons in a given energy state. In the state of equilibrium at a given fixed temperature, photons are continuously absorbed and re-emitted by the atoms of cavity walls hence the number of photons in the cavity keeps changing with time and fluctuates around a mean value. Since the total radiation energy of the cavity at a fixed temperature is fixed while the number of photons in the cavity fluctuates around a mean value and no photon leaves the cavity, the chemical potential for photon gas $\mu = 0$. The energy density of blackbody radiations of frequency v is given by,

$$\Delta u_v = \frac{8\pi h}{c^3} \frac{v^3}{e^{\left(\frac{hv}{k_B T}\right)} - 1} \Delta v$$

At high frequency this reduces to $\left[\Delta u_v\right]_{(hv \gg k_B T)} = \frac{8\pi h}{c^3} v^3 e^{-\left(\frac{hv}{k_B T}\right)} \Delta v$, which is Wien's formula.

And at lower frequencies where $hv \ll k_B T$, the term $\left(e^{\left(\frac{hv}{k_B T}\right)} - 1\right)$ in the denominator of above

equation is nearly equal to $\left(\frac{hv}{k_B T}\right)$ and Eq. 11.102 reduces to,

$$\left[\Delta u_v\right]_{(hv \ll k_B T)} = \frac{8\pi k_B T}{c^3} v^2 \Delta v$$

The total energy density may be found by integrating over all frequencies and is given by

$$u = \frac{8\pi^5 k_B^4}{15c^3 h^3} T^4 = \sigma T^4$$

17. Quantum statistics of a paramagnetic salt in external magnetic field: In quantum statistics a system consisting of paramagnetic ions of magnetic moments 1 μ_B in an external magnetic field B is like an assembly of particles that may have only two levels of energies $\epsilon_{\uparrow\uparrow}$ *and* $\epsilon_{\uparrow\downarrow}$. Further,

the paramagnetic ions in a crystalline salt may be identified by their space parameters or coordinates, hence these ions are distinguishable. Maxwell–Boltzmann statistics may, therefore, be applied to this assembly of paramagnetic ions. In the present case there are only two levels each with only one state, either of parallel or anti-parallel particles, hence the levels are non-degenerate and g =1 for the two levels. However, any number of particles of a given kind may be accommodated in each state. The partition function for spin half paramagnetic salt in an external magnetic field is given as,

$$Z = 2\ Cosh\ t, \text{ where } t = \left(\mu_B B\ /\ k_B T \right),$$

It is simple to show that at T=0,all the paramagnetic ions are in the lower energy state and on

the other hand at T = ∞, $e^{+\left(\frac{\mu_B B}{k_B T} \right)} = e^{-\left(\frac{\mu_B B}{k_B T} \right)} = 1$, and $\bar{N}_{\uparrow\uparrow} = \bar{N}_{\uparrow\downarrow} = N\ /\ 2$; i.e. half of the

ions are in the lower state and half in the higher energy state. At any intermediate temperature more ions will be in the lower energy state and a lesser number in higher energy state.
The resultant magnetic moment $\mathcal{M}$ of the assembly is due to the excess number ($\bar{N}_{\uparrow\uparrow} - \bar{N}_{\uparrow\downarrow}$)

of ions. Now each ion contributes a magnetic moment equal to μ_B, hence:

$$\mathcal{M} = N \mu_B \tanh\left(\frac{\mu_B B}{k_B T} \right)$$

Above equation is the magnetic equation of estate of the assembly. Further when the external

magnetic field B is large, i.e. $\mu_B B \gg k_B T$, $\tanh\left(\frac{\mu_B B}{k_B T} \right) = 1$ and,

$$[\mathcal{M}]_{satu} = N \mu_B$$

Above equation gives the maximum value of magnetization, (called the saturation value) that occurs when all the ionic dipoles are aligned in the direction of the applied field B. In the case

of the weak external field when $\mu_B B \gg k_B T$, $\tanh\left(\frac{\mu_B B}{k_B T} \right) \approx \left(\frac{\mu_B B}{k_B T} \right)$ and,

$$\mathcal{M} = N \mu_B \left(\frac{\mu_B B}{k_B T} \right) = N \left(\frac{\mu_B^2}{k_B T} \right) \frac{B}{T} = K_c \frac{B}{T}$$

Thus for weak external fields the expression for magnetization reduces to Curie law with Curie

constant $K_c = N \left(\frac{\mu_B^2}{k_B T} \right)$.

The internal energy U in the present case is the potential energy of the interaction of the dipoles and may be calculated from the partition function as,

$$U = -N\mu_B B \tanh\left(\frac{\mu_B B}{k_B T}\right) = -\mathcal{M}B$$

When a paramagnetic crystalline solid is subjected to an external magnetic field and heated, a part of the heat energy absorbed goes in setting elastic waves due to lattice vibrations and another part in exciting magnetic ions from lower energy level to the higher one. As such both the lattice vibrations and magnetic excitation contribute to the heat capacity of the crystal. One may divide the total heat capacity at constant volume of the crystal into two components, one due to lattice vibrations and the second due to magnetic excitation. It can be shown that in a range of temperatures the contribution from magnetic heat capacity to the total heat capacity is order of magnitudes larger than the heat capacity due to the lattice vibration of the paramagnetic crystal, i.e. $\left[C_V\right]_{(magnet)} \gg \left[C_V\right]_{(lattice\ vib)}$.

The entropy of the system that may be calculated from the partition function is given by

$$S(T) = N k_B \ln\left[2\cosh\left(\frac{\mu_B B}{k_B T}\right)\right] - \frac{N\mu_B B}{T}\tanh\left(\frac{\mu_B B}{k_B T}\right)$$

At absolute zero temperature (T=0 K) all the paramagnetic ions are in their ground state and therefore there is only one possible microstate of the system and hence,

$$S = k_B \ln \Omega = k_B \ln 1 = 0$$

At higher temperatures the number of microstates is 2^N, and hence,

$$S = k_B \ln \Omega = k_B \ln 2^N = Nk_B \ln 2$$

18. Population inversion and temperature beyond infinity: The concept of negative temperature
    In the case of a system that has only two possible energy states, the lower state at energy E_L and the higher at energy E_H, the statistical temperature may be defined as,

$$T = \frac{\left(E_H - E_L\right)}{k_B \left\{\ln\left(N_L - N_H\right)\right\}}$$

The temperature T so defined remains positive so long as $E_H > E_L$ and $N_L > N_H$. The sign of the temperature T will become negative if $E_H > E_L$ but $N_L < N_H$. Thus in the state of population inversion the quantum statistical temperature of the system will become negative.

Earlier, for the case of spin half paramagnetic ions it was shown that at infinite temperature the average number of particles in both the lower and the upper levels are equal. Hence, in order to achieve population inversion one has to go to a temperature beyond infinity. It may thus be concluded that the negative temperature of quantum statistics is higher than the normal infinite temperature. This appears to be contradictory but it is because of the way how temperature is defined in quantum statistics.

Population inversion may be achieved in several different ways depending on the system. For example, in the case of paramagnetic ions if the direction of the magnetic field B is suddenly reversed so that the ions which were in the higher energy level (less in number) and had their magnetic moments anti parallel to the field now have them parallel to the new field direction and becomes particles of the lower energy level. Similarly, ionic magnets which were earlier in the lower energy level (larger in number) with their moments parallel to the field becomes particles in higher level of energy with their magnetic moments anti-parallel to the new field direction. Thus in this case population inversion may be achieved by simply reversing the direction of the externally applied magnetic field. In some cases, like the production of optical lasers, population inversion is achieved by pumping more particles to the level of higher energy from some nearby meta-stable state. However, the condition of population inversion does not last for long, generally with in a very short time $\approx 10^{-9}$ s, the system reverts back to the normal population. But in some special cases the state of population inversion stays for longer time up to a few minutes during which experiments may be done.

Introduction to the Thermodynamics of Irreversible Processes

12.0   Introduction

It may be recalled that both classical and quantum thermodynamics consider systems in equilibrium, although theoretically the state of equilibrium takes an infinite time to reach. As such, the classical and quantum thermodynamics describe idealized systems that, in general, do not exist. The concept of reversibility is also very intimately related to the concept of equilibrium as any reversible process passes through a succession of equilibrium states. The fact is that most of the real life phenomenon are non-equilibrium processes and hence irreversible. The branch of science that deals with the flow of energy and mass or both of them between non-equilibrium systems via irreversible paths is called the thermodynamics of irreversible processes. Now there are two possibilities: the non- equilibrium systems involved in the irreversible processes are only marginally away from their equilibrium states or they are far away from their equilibrium states. In the former case the thermodynamics is called the linear thermodynamics of irreversible processes while in the latter case it is termed as the non-linear thermodynamics of irreversible processes.

Before the modern tools of handling irreversible processes were developed, the method for treating a general problem of thermodynamics was to separate out the truly reversible and irreversible components of the problem and work out the thermodynamics of the truly reversible component ignoring the irreversible component. In some cases both the truly reversible and the irreversible components were analyzed in the frame work of equilibrium thermodynamics. This methodology, though basically wrong, yielded correct results some times, but in some other cases it led to results that were not consistent with observations. However, need for a separate frame work for the treatment of irreversible processes was felt even earlier, and it was L. Onsager (27 Nov 1903–5 Oct, 1976, Norwegian born American physical chemist and theoretical physicist), who in the year 1931 developed mathematical representation of irreversible processes. Onsager was later awarded Nobel Prize of 1968 for this work.

12.1   Entropy Generation in Irreversible Processes

One essential difference between a reversible and an irreversible process is that in a reversible process the entropy of the isolated system remains unaltered while in case of an irreversible process it always increases. Let us consider a part of the universe that contains a system and its environment as an isolated system as shown in Fig. 12.1 (a). The system may interact with its environment and, therefore, heat (energy) and matter may be interchanged between the system and its environment, if the system boundary is transparent to both the heat and the matter. On the other hand if the system boundary is diathermic and does not allow the flow of matter, only the flow of heat may take place between the system and its environment. Next let us assume that some irreversible process occurs within the system as a result of which the entropy of the system plus environment changes by the amount dS. If the changes in the entropies of the environment and the system are respectively denoted by $(dS)_{env}$ and $(dS)_{sys}$ then,

$$dS = (dS)_{env} + (dS)_{sys} \qquad\qquad 12.1$$

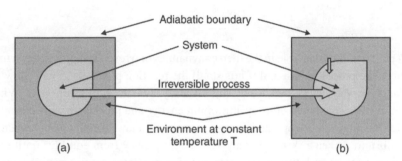

Fig. 12.1   Initial and final states of system that undergoes irreversible process

It may now be noted that dS will always be positive, i.e., $dS > 0$ as the process responsible for this is an irreversible process. Now $(dS)_{env}$ may be positive or negative depending on the nature of the boundary between the system and the environment and the type of the irreversible process. For example, if the boundary between the system and its environment is diathermic and as a result of the irreversible process the temperature of the system decreases, then the system may draw some heat, say dQ, from the environment which is at a constant temperature T, then:

$$(dS)_{env} = -\frac{dQ}{T} \qquad\qquad 12.2$$

And

$$(dS)_{sys} = \left\{\frac{dQ}{T} + (dS)_{sys}^{int}\right\} \qquad\qquad 12.3$$

In Eq. 12.3 the first term $\frac{dQ}{T}$ in the curly bracket on the right hand side gives the increase in the entropy of the system due to the flow of heat from the environment and is equal to $-(dS)_{env}$, while the second term denoted by $(dS)_{sys}^{int}$ is the increase in entropy of the system due to internal changes in the system, like the change in temperature of the system and the change in the number of particles in the system if the boundary between the system and its environment allows the transfer of matter. On substituting the values of $(dS)_{env}$ and $(dS)_{sys}$, Eq. 12.1 reduces to

$$dS = (dS)_{env} + (dS)_{sys} = (dS)_{sys}^{int} \qquad\qquad 12.4$$

And since dS is positive, $\left(dS\right)_{sys}^{int}$ must also be positive. It may further be noted that instead of absorbing heat from the environment if some heat dQ flows out from the system as a result of the irreversible process then $(dS)_{env}$ will be positive but the sign of the first term in the curly bracket on the right will be negative. Therefore, Eq. 12.4 will still hold in the latter case also. This shows that whenever a system undergoes some irreversible process some entropy is always generated within the system, this internally generated entropy, that is denoted by $\left(dS\right)_{sys}^{int}$ is in addition to the change in the entropy of the system due to its interactions with other systems. The rate of increase of internal entropy, denoted by Greek letter sigma σ, and is given by the relation

$$\sigma = \frac{\left(dS\right)_{sys}^{int}}{dt} \qquad\qquad 12.5$$

Sigma (σ) plays an important role in the thermodynamics of irreversible processes. It may be noted that if σ is large the process is fast and if it is small the reaction proceeds slowly. Slow irreversible processes take place when the system is only slightly away from its equilibrium state. When a system is in equilibrium it has maximum entropy, minimum energy and no change in its entropy or energy may take place. Changes in entropy and energy of a system may take place only when the system is not in equilibrium. When a system is only very slightly away from equilibrium it may undergo extremely slow processes such that at each step the system attains a quasi-equilibrium state. Such extremely slow processes passing through quasi-equilibrium states constitute reversible processes and no change in the entropy of the system takes place in such reversible processes. In an irreversible process, on the other hand, energy is continuously dissipated and entropy is continuously created.

12.2 Matter and/or Energy Flow in Irreversible Processes: Flux and Affinity

An irreversible process is always accompanied by the flow of matter or energy or both. For example, if temperature gradient exists in a fluid, heat (energy) flows from the region of higher temperature to the region of lower temperature. Similarly, if a system has density gradient, matter flows from the region of higher density to the region of lower density. It may be noted that both these processes of heat flow and of matter flow are irreversible processes. Many more examples of flow of some physical quantity in irreversible processes may be given. If the system is made up of charged particles, these energy and matter flows will also constitute electric currents. Thus, every irreversible process is associated with the flow of some physical quantity which in the language of thermodynamics is called 'flux' or 'current' and is denoted by J_i. Since there may be many different types of fluxes in a given system undergoing some irreversible process, they may be distinguished using the subscript í. Every flux may be assigned a source that is responsible for the flux. The source of a flux is termed as 'affinity', 'potential' or 'driving force' and is denoted by X_i. In the example of heat flow, the flux is the heat flow and the corresponding affinity is the temperature gradient while in the case of matter flow, the affinity or potential is the density gradient. An irreversible process may have several different types of fluxes and affinities.

In an irreversible process it often happens that a particular flux is generated by two or more different sources or affinities. For example, if a system is made up of moveable charged particles, like free electrons in a metal, flow of electrons or electric current may be constituted both by the temperature gradient and the electric potential gradient. This dependence of a particular flux on different affinities may be represented by the following mathematical relation,

$$J_i = J_i\,(X_1, X_2, X_3, \ldots) \qquad\qquad 12.6$$

Here J_i represents the ith type of flux and X_1, X_2, X_3, etc., are different affinities each of which is partially responsible for the flux J_i. Such fluxes that partially depend on many affinities are called coupled fluxes.

12.3 Linear Irreversible Process

In the case of irreversible processes that occur in a system which is not far away from its equilibrium state, affinities or the flux driving forces X_1, X_2, X_3, are not too large. In such cases $J_i = J_i\,(X_1, X_2, X_3, \ldots)$ may be expanded using Taylor's expansion and one may retain only the first order (or linear) terms neglecting the higher order (or non-linear) terms in the expansion. Eq. 12.6 may then be written as:

$$J_i\,(X_1, X_2, X_3, \ldots) = \left(\frac{\partial J_i}{\partial X_1}\right)_0 X_1 + \left(\frac{\partial J_i}{\partial X_2}\right)_0 X_2 + \left(\frac{\partial J_i}{\partial X_3}\right)_0 X_3 \ldots.$$

Or
$$J_i\,(X_1, X_2, X_3, \ldots) = \sum_{j=1}^{n} \left(\frac{\partial J_i}{\partial X_j}\right)_0 X_j \qquad\qquad 12.7$$

Here $\left(\dfrac{\partial J_i}{\partial X_j}\right)_0$ is the rate of change of J_i with X_j when system approaches the equilibrium state.

Eq. 12.7 may be written as,

$$J_i = \sum_{j=1}^{n} L_{ij} X_j \qquad\qquad 12.8$$

where
$$L_{ij} = \left(\frac{\partial J_i}{\partial X_j}\right)_0$$

A relation of the type represented by Eq. 12.8 is called a 'Linear phenomenological relation' which represents an irreversible process that takes place in a system that is only so far away from the equilibrium state such that only the linear terms of the Taylor's expansion remain important.

Irreversible processes that may be described by the linear terms of Taylor's expansion, as given by Eq. 12.8 are called linear irreversible processes. However, when the system is far away from the corresponding equilibrium state, the affinities and fluxes are sufficiently large and higher order terms in Taylor's expansion cannot be neglected. Irreversible process taking place in such systems are called non-linear irreversible processes. Onsager developed thermodynamics for linear irreversible processes, which is outlined here.

Onsager used these linear phenomenological relations to describe irreversible processes containing n-fluxes and n-affinities in terms of the coefficients L_{ij},

$$J_1 = L_{11}X_1 + L_{12}X_2 + L_{13}X_3 + \ldots L_{1n}X_n$$

$$J_2 = L_{21}X_1 + L_{22}X_2 + L_{23}X_3 + \ldots L_{2n}X_n$$

$$J_3 = L_{31}X_1 + L_{32}X_2 + L_{33}X_3 + \ldots L_{3n}X_n \qquad 12.9$$

$$\ldots\ldots\ldots\ldots\ldots\ldots\ldots\ldots\ldots\ldots\ldots\ldots\ldots\ldots\ldots\ldots\ldots\ldots$$

$$\ldots\ldots\ldots\ldots\ldots\ldots\ldots\ldots\ldots\ldots\ldots\ldots\ldots\ldots\ldots\ldots\ldots$$

$$J_n = L_{n1}X_1 + L_{n2}X_2 + L_{n3}X_3 + \ldots L_{nn}X_n \quad .$$

Here L_{ij} are called phenomenological coefficients and have the properties of generalized mobility or conductance.

Conversely, the flux deriving forces or affinities, X_i may also be written in terms of the fluxes J_i as follows,

$$X_i = \sum_{j=1}^{n} R_{ij}J_j \ , \ i = 1, 2, 3, \ldots n \qquad 12.10$$

The coefficients R_{ij} in Eq. 12.10 have the properties of generalized resistance or friction. It may be noted that coefficients like $L_{11}, L_{22}, L_{33} \ldots L_{nn}$ and $R_{11}, R_{22}, R_{33} \ldots R_{nn}$ represent interactions between the same type of fluxes and affinities and are, therefore, called 'Straight coefficients'. On the other hand, coefficients like L_{ij} and R_{ij} ; $i \neq j$ represent interactions between different fluxes and affinities and are called 'Cross coefficients'. In general both L_{ij} and R_{ij} form matrices. Knowledge of phenomenological coefficients provides complete description of the fluxes and their driving forces and, therefore, the thermodynamics of the irreversible process.

A given irreversible process may be characterized either by L_{ij} or by R_{ij} and, therefore, they are related to each other. In the simple case of two fluxes J_1, J_2 and two affinities X_1, X_2 one may write:

$$J_1 = L_{11} X_1 + L_{12} X_2 \text{ and } X_1 = R_{11} J_1 + R_{12} J_2 \qquad 12.11$$

$$J_2 = L_{21} X_1 + L_{22} X_2 \text{ and } X_2 = R_{21} J_1 + R_{22} J_2 \qquad 12.12$$

Solving Eqs. 12.11 and 12.12 one gets,

$$L_{11} = R_{22}\Big[\big|L_{11}L_{22} - L_{12}L_{21}\big|\Big] \ ; \ L_{22} = R_{11}\Big[\big|L_{11}L_{22} - L_{12}L_{21}\big|\Big] \qquad 12.13$$

And $$L_{12} = -R_{12}\Big[\big|L_{11}L_{22} - L_{12}L_{21}\big|\Big] \ ; \ L_{21} = -R_{21}\Big[\big|L_{11}L_{22} - L_{12}L_{21}\big|\Big] \qquad 12.14$$

It may be observed that in the simple case of just two fluxes there are four linear coefficients, L_{11}, L_{22}, L_{12} and L_{21} or R_{11}, R_{22}, R_{12} and R_{21} that must be experimentally determined for understanding the thermodynamics of the irreversible process. In general when there are large number of fluxes and affinities, the number of coefficients required to be known is much larger.

12.4 Onsager's Theorem

Onsager's theorem reduces the number of linear coefficients under some conditions. The theorem says that if the currents and affinities are chosen properly the cross coefficients are symmetrical, i.e.,

$$L_{21} = L_{12} \, and \, R_{21} = R_{12} \quad or \quad L_{ij} = L_{ji} \, and \, R_{ij} = R_{ji} \qquad 12.15$$

Equation 12.15 represents Onsager's reciprocal relation of linear coefficients.

The choice of proper affinities and currents that satisfy Onsager's reciprocal relations are those which satisfy the following relation of the rate of generation of internal entropy

$$\sigma = \frac{(dS)_{sys}^{int}}{dt} = \sum_{i=1}^{n} J_i X_i \qquad 12.16$$

Onsager used the following Gibb's equation for calculating the rate of internal energy generation,

$$T\left(ds\right)_{sys}^{int} = du + P\, dv - \sum_i \mu_i c_i \qquad 12.17$$

Here μ_i and c_i are respectively, the chemical potential and the concentration of the ith component of the system. Equation 12.17, known as Gibb's equation, is strictly valid only for systems that are in equilibrium. Use of this equation which is true only for reversible processes by Onsager in the case of irreversible process where the system is not in equilibrium, invited lot of criticism and discussion. However, using perturbation theory, it was shown that Gibb's equation remains valid for systems that are displaced from equilibrium by a small amount. Further, it was also shown that the limit of displacement from equilibrium for the validity of Gibb's equation is more than the limit of displacement of the system (from the equilibrium state) for which first order Taylor's expansion is valid. As such it is justified to use Gibb's equation in the case of linear irreversible processes.

Having selected the proper fluxes (currents) and affinities (current driving forces), the next thing is to experimentally determine the phenomenological coefficients L_{ij} or R_{ij}. In general, these coefficients are likely to vary with time as the fluxes grow with time in the initial phase of the process. No measurement can be made in this phase. After some time, however, the system attains a steady state, when net flux at a particular point of the system becomes zero, i.e., as much flux reaches the point at each instant as leaves it. This is a time invariant state. A time invariant or steady state is different from an equilibrium state. In equilibrium state energy is minimum, entropy maximum and there is no change in the entropy of the state with time, at constant volume. There are no fluxes and affinities in an equilibrium state. On the other hand in a steady state or time invariant state, entropy is generated at constant rate which may be either maximum or minimum. As a matter of fact time invariant state or steady state is more general state and equilibrium state is a special case of steady state.

12.5 Prigogine's Theorem of Minimum Entropy Generation in Steady State

Prigogine's theorem states that the rate of production of entropy in a non-equilibrium steady state of a system is a minimum. The theorem can be easily proved for a simple case of two currents and two affinities. According to Onsager criteria imbedded in Eq. 12.16, the rate of entropy production in a linear irreversible process having two fluxes and two affinities is given by:

$$\sigma = \frac{(dS)_{sys}^{int}}{dt} = \sum_{i=1}^{n} J_i X_i = J_1 X_1 + J_2 X_2 \qquad 12.18$$

And
$$J_1 = L_{11}X_1 + L_{12}X_2 \quad and \quad J_2 = L_{21}X_1 + L_{22}X_2 \qquad 12.19$$

Substituting the values of J_1 and J_2 from Eq. 12.19 in Eq. 12.18, one gets;

$$\sigma = L_{11}X_1^2 + L_{12}X_2X_1 + L_{21}X_1X_2 + L_{22}X_2^2$$

But from Onsager's reciprocal relation $L_{12} = L_{21}$, and the above relation reduces to

$$\sigma = L_{11}X_1^2 + 2L_{21}X_1X_2 + L_{22}X_2^2 \qquad 12.20$$

Now suppose that affinity X_1 is kept constant and Eq. 12.20 is differentiated with respect to X_2 in order to get the condition for the minimum or maximum of:

$$\left(\frac{d\sigma}{dX_2}\right)_{X_1} = \left(\frac{d\dfrac{d(dS)^{int}_{sys}}{dt}}{dX_2}\right)_{X_1} = 0 = 2\,L_{21}X_1 + 2L_{22}X_2 = 2\left(L_{21}X_1 + L_{22}X_2\right) = 2J_2 \qquad 12.21$$

Equation 12.21 tells that in the steady state the conjugate flux, i.e., J_2 must vanish and that in that time invariant state the rate of internal entropy production $\left(\dfrac{d\sigma}{dX_2}\right)_{X_1}$ will either be minimum or a maximum. However, it is already known that $\sigma = \dfrac{(dS)^{int}_{sys}}{dt}$ is always positive in the case of a linear irreversible process. Hence, the internal entropy production rate is a minimum in a steady state.

12.6   Order of a Steady State

In a general case a linear irreversible process may have n-fluxes and n-affinities, out of which some m-fluxes are kept constant so that the remaining (n-m) fluxes adjust themselves to obtain steady state. The steady state so obtained is called a steady state of mth order. It follows from this definition that the equilibrium state is a steady state of $zero^{th}$ order.

12.7   Matrix Representation of Coupled Linear Phenomenological Relations

It is often convenient to represent interdependent relations using matrix algebra, for example, the flux and affinity relations given by Eq. 12.11 and Eq. 12.12 may also be written in the compact matrix form as,

$$\begin{bmatrix} J_1 \\ J_2 \end{bmatrix} = \begin{bmatrix} L_{11} & L_{12} \\ L_{21} & L_{22} \end{bmatrix} \begin{bmatrix} X_1 \\ X_2 \end{bmatrix} \quad and \quad \begin{bmatrix} X_1 \\ X_2 \end{bmatrix} = \begin{bmatrix} R_{11} & R_{12} \\ R_{21} & R_{22} \end{bmatrix} \begin{bmatrix} J_1 \\ J_2 \end{bmatrix} \qquad 12.22$$

These matrices on expansion give back the relations represented by Eqs. 12.11 and 12.12. In the case of a linear irreversible process that have many coupled fluxes and forces (affinities) the matrix representation will be

$$
\begin{bmatrix} J_1 \\ J_2 \\ J_3 \\ J_4 \\ \cdot \\ \cdot \\ \cdot \\ J_n \\ \cdot \end{bmatrix} = \begin{bmatrix} L_{11}\, L_{12}\, L_{13}\, \ldots L_{1n} \\ L_{21}\, L_{22}\, L_{23}\, \ldots L_{2n} \\ L_{31}\, L_{32}\, L_{33}\, \ldots L_{3n} \\ L_{41}\, L_{42}\, L_{43}\, \ldots L_{4n} \\ \cdot \\ , \\ , \\ L_{n1}\, L_{n2}\, L_{n3}\, \ldots L_{nn} \end{bmatrix} \begin{bmatrix} X_1 \\ X_2 \\ X_3 \\ X_4 \\ \cdot \\ \cdot \\ \cdot \\ X_n \\ \cdot \end{bmatrix}
$$

12.23

Equation 12.23 on expansion gives,

$$
J_1 = L_{11}X_1 + L_{12}X_2 + L_{13}X_3 + \ldots L_{1n}X_n
$$
$$
J_2 = L_{21}X_1 + L_{22}X_2 + L_{23}X_3 + \ldots L_{2n}X_n
$$
$$
J_3 = L_{31}X_1 + L_{32}X_2 + L_{33}X_3 + \ldots L_{3n}X_n
$$

12.24

$$
\ldots\ldots\ldots\ldots\ldots\ldots\ldots\ldots\ldots\ldots\ldots\ldots\ldots\ldots\ldots
$$
$$
\ldots\ldots\ldots\ldots\ldots\ldots\ldots\ldots\ldots\ldots\ldots\ldots\ldots
$$
$$
J_n = L_{n1}X_1 + L_{n2}X_2 + L_{n3}X_3 + \ldots L_{nn}X_n
$$

12.8   Application of Onsager's Method for Linear Irreversible Processes to Thermoelectricity

Thermoelectricity, the generation of electric potential difference between the two ends of a metal wire when the ends are kept at different temperatures, which was discovered in the beginning of the nineteenth century, provides an example of linear irreversible process in which two fluxes, the flux of charged particles (electrons) and the flux of thermal energy (heat) are coupled with each other. It was John Seebeck who in the year 1821 for the first time observed the coupling of two potentials (affinities), the electrochemical potential and the temperature difference. Later, in 1834 Jean Peltier demonstrated that the heat flux and the flux of electric current could also be coupled. In 1931, these coupled thermodynamic forces and fluxes were adequately described by Lars Onsager using the method of thermodynamics of linear irreversible processes developed by him.

Free electrons in a metal may be treated as a perfect gas, but of charged fermions, i.e., electrons. The equivalent of partial pressure in an electron gas is the *electrochemical potential*, μ_e given by,

$$
\mu_e = qV + \mu_c
$$

12.25

where q is the charge of the carrier particle (electron), V electric potential and μ_c the chemical potential, i.e., the energy needed to give one carrier particle (electron) to the system.

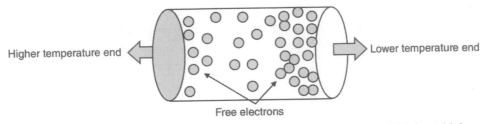

Fig. 12.2   Distribution of free electrons in a conductor one end of which is at higher temperature

As shown in Fig. 12.2, the electron density at the higher temperature side of a conductor is low and the average speed of electrons is higher while at the lower temperature end average electron speed is lower and density larger. Accumulation of electrons at the lower temperature end develops an electric potential difference V between the two ends, with the higher temperature end at higher potential. The electric field thus generated forces free electrons to move from lower temperature end towards the higher temperature end. In the steady state the electron flux from higher temperature end towards the lower temperature end due to the higher speeds at the end is on average balanced by the flux in the opposite direction due to the electric field. Free electrons moving from the higher temperature end carries heat energy towards the other end. It may be noted that the temperature difference is the force (affinity) that drives the gradient of electron (carrier) density. Not only that, temperature difference also creates an electric potential difference that also contributes to the motion of free electrons in the reverse direction. Thus both the electron flux and the heat flux are coupled with each other so also the temperature difference and the electrochemical potential difference, the two affinities in the system are coupled. Thus thermoelectric material is an excellent system in which irreversible process due to coupled fluxes and affinities takes place when a temperature gradient is established in the system. Further, the system eventually attains a steady state. The thermodynamic state variables of the system are temperature T and electrochemical potential μ_e. The system is essentially a non-equilibrium system in steady state, but as assumed by Onsager and Prigogine the Gibb's equation for the system holds good in the steady state and, therefore,

$$G = U - TS - N\mu_e$$

Or
$$dG = dU - TdS - SdT - Nd\mu_e - \mu_e dN \qquad 12.26$$

If δQ and δW are respectively the change in the heat and the work done by the system, then

$$\delta Q = TdS \text{ and } \delta W = \mu_e dN \qquad 12.27$$

Here U is the internal energy and N the number of carriers (free electrons) in the system. Also, the change in the internal energy

$$dU = \delta Q + \delta W = TdS + \mu_e dN \qquad 12.28$$

Substituting the value of dU from Eq. 12.28 in Eq. 12.26 one gets,

$$dG = -SdT - Nd\mu_e \qquad 12.29$$

And
$$S = -\left(\frac{\partial G}{\partial T}\right)_{constant\ \mu_e} \quad and \quad N = -\left(\frac{\partial G}{\partial \mu_e}\right)_{constant\ T} \qquad 12.30$$

If $\overrightarrow{J_N}, \overrightarrow{J_S}, \overrightarrow{J_Q}$ and $\overrightarrow{J_e}$ respectively represent particle flux, entropy flux, heat flux and electrical flux (electric current density) then using Eq. 12.27:

$$\overrightarrow{J_Q} = T\overrightarrow{J_S} \quad and \quad \overrightarrow{J_e} = q\overrightarrow{J_N} \qquad 12.31$$

Here q is the electric charge of the carrier particles (which in this case are free electrons in the metal each with charge -e).

The driving forces or affinities for $\overrightarrow{J_N}$ (or $\overrightarrow{J_e}$) and $\overrightarrow{J_S}$ (or $\overrightarrow{J_Q}$) are respectively the difference in electrochemical potential $\Delta\mu_e$ and the difference in the temperature ΔT at the two ends of the conductor. If $\vec{E}$ denotes the intensity of the electric field generated because of $\Delta\mu_e$ then the force (or affinity) F_e experienced by a carrier particle of charge q is,

$$\overrightarrow{F_e} = q\vec{E} \qquad 12.32$$

Further, the force F_T or affinity due to the temperature difference ΔT may be written as:

$$\overrightarrow{F_T} = -\overrightarrow{\nabla T} \qquad 12.33$$

Using Onsager method one may write the coupled equations for the particle flux and entropy flux in the following determinant form,

$$\begin{bmatrix} \overrightarrow{J_N} \\ \overrightarrow{J_S} \end{bmatrix} = \begin{bmatrix} L_{11} & L_{12} \\ L_{21} & L_{22} \end{bmatrix} \begin{bmatrix} \overrightarrow{F_e} \\ \overrightarrow{F_T} \end{bmatrix} \qquad 12.34$$

The same may also be written a

$$\begin{bmatrix} \dfrac{\overrightarrow{J_e}}{q} \\ \dfrac{\overrightarrow{J_Q}}{T} \end{bmatrix} = \begin{bmatrix} L_{11} & L_{12} \\ L_{21} & L_{22} \end{bmatrix} \begin{bmatrix} q\vec{E} \\ -\overrightarrow{\nabla T} \end{bmatrix} \qquad 12.35$$

Or
$$\overrightarrow{J_e} = q^2 L_{11}\vec{E} + qL_{12}\left(-\overrightarrow{\nabla T}\right) \qquad 12.36$$

and
$$\overrightarrow{J_Q} = qTL_{21}\vec{E} + TL_{22}\left(-\overrightarrow{\nabla T}\right) \qquad 12.37$$

In steady state particle flux or electric flux $\overrightarrow{J_e} = 0$; that means:

$$q^2 L_{11}\vec{E} = qL_{12}\left(\overrightarrow{\nabla T}\right) \qquad 12.38$$

The ratio of the electric field to the gradient of temperature when there is steady state and no particle or electric flux i.e., $\left(\dfrac{E}{\nabla T}\right)_{no\ particle\ or\ electric\ flux}$ is called the *Seebeck coefficient* which is denoted by α. It follows from Eq. 12.38 that,

$$\alpha = \left(\frac{E}{\nabla T}\right)_{no\ particle\ or\ electric\ flux} = \frac{L_{12}}{qL_{11}} \qquad 12.39$$

The terms $qL_{12}\left(-\overrightarrow{\nabla T}\right)$ and $TL_{22}\left(-\overrightarrow{\nabla T}\right)$ respectively in Eqs.12.36 and 12.27 arise because of the externally imposed temperature difference between the two ends of a conductor. However, if there is no such externally imposed temperature difference on the conductor but an electric flux $\overrightarrow{J_e}$ is made to flow through the conductor by applying an external electrostatic potential difference or electric field E between the ends of the conductor, a temperature difference will be created between the ends of the conductor. This generation of temperature difference between the two ends of a conductor when an electric current is passed through it, is called *Peltier effect*. The ratio of the electric flux to the heat flux under the condition of $\left(-\overrightarrow{\nabla T}\right) = 0$ is called Peltier coefficient which is represented by Π.

So,

$$\Pi = \left(\frac{J_Q}{J_e}\right)_{\left(\overrightarrow{\nabla T}=0\right)} = T\frac{L_{12}}{qL_{11}} \qquad 12.40$$

It is easy to see that

$$\Pi = T\frac{L_{12}}{qL_{11}} = T\pm \qquad 12.41$$

Experimental verification of relation 12.41 proves the correctness of the method of linear irreversible process developed by Onsager.

12.9   The Network Thermodynamics

Classical and quantum thermodynamics deal with initial and final equilibrium states without spelling out how starting from the initial equilibrium state the system reached the final equilibrium state. Onsager's theory for linear irreversible processes successfully explained many simple physical, chemical and biological processes but it is applicable only to linear irreversible processes and does not take into account non-linear irreversible processes and also situations where both reversible and irreversible processes occur in the system together. In biological systems it often happen that reversible and non-linear processes proceed together. Thus both classical thermodynamic theories and Onsager's approach of linear irreversible processes are, therefore, not adequate to treat complex biological processes. A new approach based on the analogy of non-equilibrium systems and electrical networks, that overcomes all the shortcomings of the earlier classical and Onsager's dynamic theory has been developed, particularly for biological systems, to describe reversible, linear irreversible and non-linear irreversible processes occurring simultaneously in a system. Since it uses elements of the electrical network theory, it is called the network thermodynamics. Network thermodynamics developed out of the synthesis of concepts based on general theory of linear passive systems, thermodynamics, circuit theory, graph theory and differential geometry.

Any dynamic system, including an irreversible process, is accompanied with the flow of some sort of fluxes that may be compared with the flow of currents in electrical networks. As such, in principle network theory may be applied to any dynamic process. Further, if the circuit elements are non-dissipative, pure capacitances and inductances, then there will be no loss of energy which may be compared to a reversible process. In electrical circuits the lumped parameter approximation to Maxwell's equations gives rise to electrical network theory. In a similar way a lumped parameter approximation to the field equations and continuum mechanics gives rise to the network thermodynamics. Network thermodynamics makes use of the experience of electrical engineers of solving large and complex non-linear systems by decomposing them into smaller subsystems interconnected with each other. The simpler subsystems may then be separately solved and their solutions are suitable coupled to get the solution of the parent complex system. Coupling of the solutions of subsystems depends on how they are connected in the complex system. This brings in the topology of the problem, i.e., how the subsystems are interconnected. Topology of the system is very important. This may be understood from the fact that all electrical systems are made up of resistances, capacitances, inductances, diodes, etc., but different ways of interconnection between these circuit elements make amplifiers, radios, and counters, etc., which are totally different from each other in their working. Thus topology or interconnection between elements and subsystems decides the performance of the network.

12.9.1   Postulates of network thermodynamics

The structure of the network thermodynamics is based on the following three postulates:

1. **Local phase equilibrium**: Although in the case of irreversible process the system is not in equilibrium, but in network thermodynamics it is assumed that at the local level the system may be described in terms of the thermodynamic variables like temperature, pressure and chemical potential.
2. **Discrete space**: In general space may be treated either continuum or discrete. In mathematical terms the continuum space may be described by partial differential equations, vector calculus and point set topology, while the discrete space by simple differential equations, and algebraic topology. One advantage of discrete space is that one can make finite size models. Network thermodynamics assumes discrete structure for the space.
3. **Conceptual separation of reversible and irreversible processes:** Although in every volume element of the complex system both reversible and irreversible processes may be going on simultaneously, but conceptually one decouples them into independent reversible and irreversible components.

12.9.2   Terminology of network thermodynamics

In network thermodynamics a system is regarded as a black box and any process of interaction in the system that may change the energy of the system is called a *port*. A port is represented by a line. Each port has two variables attached to it and, therefore, a port is treated as a two terminal network. A mechanical port (like a piston) has pressure P and specific volume v as two variables associated with it, a membrane that may allow one particular kind of particles to pass through may be treated as a port with the two associated variables N, the number of particles and μ the chemical potential. A diathermic boundary is represented by a thermal port with temperature T and specific entropy s as the two associated variables or terminals. An adiabatic boundary does not allow any interaction of the system with the surroundings and, therefore, is not a port. A typical system enclosed by three ports and adiabatic boundary is shown in Fig. 12.3.

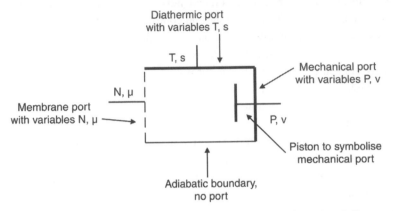

Fig. 12.3   Representation of system interactions through ports in network thermodynamics

12.9.3   Equilibrium network thermodynamics

In classical thermodynamics the internal energy U of a system is treated as a real valued function that depends on several thermodynamic variables like, the volume V, temperature T, entropy S, the number of particles N_i of type i, etc. In network thermodynamics the internal energy U is assumed to be a function of generalized displacements q_i. The partial derivative $\dfrac{\partial U}{\partial q_i} \equiv e_i$ defines the potential conjugate to displacement q_i. Conjugate potentials e_i are also called efforts. The generalized displacements q_i and the conjugate potentials e_i are the new primitive variables used in network thermodynamics. Thus,

In classical equilibrium thermodynamics $U = U\,(V,\,S,\,T,\,N_i,\,\mu_i,\,...)$ 12.42

However, in equilibrium network thermodynamics, $U = U\,(q_i,\,q_2,\,q_3\,\cdots\,q_n)$ 12.43

And $\dfrac{\partial U}{\partial q_1} \equiv e_1,\, \dfrac{\partial U}{\partial q_2} \equiv e_2,\, \dfrac{\partial U}{\partial q_3} \equiv e_3,\, \cdots \dfrac{\partial U}{\partial q_n} \equiv e_n$ 12.44

In network thermodynamics, each port (that indicates a particular type of interaction of the system) is characterized by a relationship between the port variables, i.e., the displacement q_i and the effort e_i. These relationships are called 'constitutive relations' or the 'equations of state'.

Now, from Eq. 12.43, it is easy to obtain the following Gibb's equation

$$dU = \sum_{i=1}^{n} \frac{\partial U}{\partial q_i}\, dq_i \equiv TdS - pdV + \sum_i \mu_i dN_i$$ 12.45

One important feature in-built in Eq. 12.45 is that it provides a correct equilibrium theory starting from a dynamical theory.

Network approach may be applied only to those systems for which the rate of energy transmission; that may be energy loss (dissipation) or energy storage; is finite and may be written as the product of an 'effort' (or force variable) e, and a 'flow' variable f, i.e.,

$$\text{energy rate or power } P = e.f \qquad\qquad 12.46$$

These effort and flow variables have different names in different processes, for example in case of electrical network, effort is electric potential difference and flow is current; in mechanical flow it is force and velocity, in chemical reactions affinity and reaction rate and in diffusion process, the difference in chemical potential and mass flow.

12.9.4 The generalized displacement and the generalized momentum

In order to retain the dynamic nature of the network thermodynamics, two additional state variables may be defined by integrating over time the effort e and the flow f, in the following way,

$$\text{The generalized momentum } P(t) \equiv P(0) + \int_0^t e(t)\,dt \qquad\qquad 12.47$$

$$\text{And} \quad \text{The generalized displacement } q(t) \equiv q(0) + \int_0^t f(t)\,dt \qquad\qquad 12.48$$

Retaining the analogy between the electrical networks and the network thermodynamics, generalized momentum P(t) or effort e(t) obeys continuity law equivalent to Kirchhoff's voltage law (KVL) regarding the continuity of the function in a closed loop, while the generalized displacement q(t) or flow f(t) obeys the law of conservation equivalent of Kirchhoff's current law (KCL).

12.9.5 Three types of (energy rate or power) interactions

As has already been mentioned, in network thermodynamics each energy interaction is characterized by a port. A port may store the energy or it may dissipate it. In electrical networks energy may be stored either in a pure capacitor or in a pure inductor while the energy is dissipated by a pure resistor. In network thermodynamics pure capacitor, pure inductor and pure resistive elements are defined through the integration relation

$$P = e.f = \sum_i e_i f_i = \int e\,f \qquad\qquad 12.49$$

Equation 12.49 may be evaluated in the following three different ways, each giving rise to a different type of energy interaction:

(a) Capacitative or displacement energy store: When integration of the energy rate is done in the following way it defines a Capacitative energy store as,

$$\in_{cap}(t) \equiv \int_0^t e.f\,dt = \int_{q(0)}^{q(t)} e.dq \qquad\qquad 12.50$$

In order to carry out the above integration it is required to have constitutive relation between the generalized displacement q and the effort e which has the form,

$$q = \psi_c(e) \qquad\qquad 12.51$$

Constitutive relations, like Eq. 12.51, in phenomenological theories are obtained either experimentally or from some other theory like the statistical mechanics. It follows from Eq. 12.51 that

$$\frac{dq}{dt} = \frac{d\psi_c}{de}\frac{de}{dt} \qquad 12.52$$

The reversible flow on the capacitor f_{rev} and the incremental capacitance C is then defined as:

$$f_{rev} \equiv \frac{dq}{dt} \text{ and } C \equiv \frac{d\psi_c}{de} \qquad 12.53$$

This leads to,

$$f_{rev} \equiv C\frac{de}{dt} \qquad 12.54$$

(b) Inductive or kinetic energy store: On the other hand, if integration is carried out between the initial and final values of the generalized momentum, the resulting interaction gives rise to inductive or kinetic energy store:

$$\in_{ind}(t) \equiv \int_0^t f.edt = \int_{P(0)}^{P(t)} f.dP \qquad 12.55$$

Equation 12.55 gives the energy stored as a result of the relative motion of mass or charge. Evaluation of the above integral requires a constitutive relation between generalized momentum P and the flow f. This constitutive relation may be written as,

$$P = \psi_L(f) \qquad 12.56$$

This type of energy storage device is called an inductor and is denoted by L.

(c) Resistive energy dissipation: Time integral of the multiplication of electric potential difference and the current in electrical networks give the energy dissipated by a resistive element. In a similar way, energy dissipation via an equivalent resistive element in network thermodynamics is given by the following integral:

$$\in_{res}(t) \equiv \int_0^t e.fdt \qquad 12.57$$

To compute the energy dissipation in a resistance like device a constitutive relationship between effort e and flow f is required. This relation may be of the form,

$$e = \psi_R(f) \qquad 12.58$$

The incremental resistance R may then be defined as,

$$R \equiv \frac{\partial \psi_R}{\partial f} \qquad 12.59$$

12.9.6   The bond graph notation

Graphical representation of dynamic process is a well-known technique; however, the use of graphs for complex flows becomes unwieldy. In network thermodynamics which is usually applied to biological systems, there are many concurrent flows of different types of energies. Bond graph notation was developed by H. Paynter to deal with such complex flows of non-equilibrium thermodynamics. The properties of the bond graph may be summarized as follows:

1. Bond graphs treat all energy flows on equal footing and, therefore, they may be used as notations for energy conservation and coupling.
2. Two kinds of junctions are used in bond graph notations to couple the elements of the system. They are:

    (a) *The parallel junction or the '0' junction*: This junction is denoted by the notation

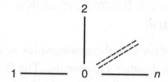

Fig. 12.4 Notation for '0'-junction

For a '0'-junction, applying Kirchhoff's current law one obtains:

$$e_1 = e_2 = e_3 \ldots\ldots\ldots = e_n \ or \ \sum_i f_i = 0 \qquad\qquad 12.60$$

(b) *The series junction or '1'-junction:* This junction is denoted by the notation given in Fig. 12.5. For a '1'-junction:

$$f_1 = f_2 = f_3 \ldots\ldots\ldots = f_n \ or \ \sum_i e_i = 0 \qquad\qquad 12.61$$

Equation 12.61 represents the application of Kirchhoff's Voltage Law to the junction.

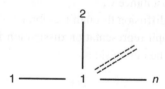

Fig. 12.5 Notation for '1'-junction

(c) The bonds connecting the network elements to a junction are denoted by half arrows ($\rightharpoonup$) that also indicate the direction of energy flow. Each bond is also associated with an effort variable and a flow variable. For example, $\begin{bmatrix} e_1 \\ \rightharpoonup C \\ f_1 \end{bmatrix}$ represents energy flow into the capacitive element while, $\begin{bmatrix} e_2 \\ \leftharpoonup R \\ f_2 \end{bmatrix}$ represents the energy flow out from a resistive element.

(3) *The transducer device (TD):* An additional ideal two port device called transducer is also often used in bond notation of network thermodynamics. The compact notation for the transducer is:

$$
\begin{bmatrix}
e_1 \ (r) \ e_2 \\
\rightarrow TD \rightarrow \\
f_1 \quad f_2
\end{bmatrix}
, \text{ where (r), called the transfer ratio or transducer modulus provides the}
$$

relationships between the input and output variables, i.e.,

$$
e_1 = \frac{e_2}{r} \ and \ f_1 = rf_2 \qquad\qquad 12.62
$$

12.9.7 Application of network thermodynamics to the diffusion of a fluid through a membrane

As an example of the application of network thermodynamics we consider a very simple case of the diffusion of fluid of only one kind through a membrane. The fluid that may pass through a given membrane is called a permeant. For simplicity we are taking a single permeant in part-I of the container that diffuses through the membrane in to part II. Let us conceptually divide the membrane into small volumes and consider one of these small volumes (Fig. 12.6a).

As the diffusion process beguines the fluid or the permeant first stick to the membrane, this is called the saturation of the membrane. Once the membrane is fully saturated, then the fluid diffuses on to the other side giving rise to the steady flow of the permeant. Similarly, when the steady flow ends the permeant attached to the membrane goes out. Thus saturation of the membrane and leaving of the attached permeant from the membrane at the end of the flow may be well compared to the charging and discharging of the capacitor. It may be noted that the membrane and the two compartments -I and II have the property of holding the permeant, and in this respect they behave like capacitors. The equivalent capacitance of the three components may be lumped together and may be represented by a single capacitance C.

It may further be realized that diffusion through membrane is a dissipative process, i.e., energy is lost in the process. In bond graph representation dissipation is included by assuming resistive elements both in the input and on the output sides.

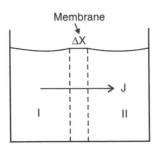

Fig. 12.6   (a) Diffusion of fluid through a membrane

For simplicity, it may be assumed that: (a) all constitutive relations are linear, and (b) the membrane divides the chamber in two symmetrical parts I and II. The network equivalent of the diffusion process is shown in Fig. 12.6(b). Figure 12.6(c) shows the bond graph representation of the diffusion process. Each branch of the bond graph has definite value of the potential or force e_n, and the flow

f_n, $n = 1$ to 7, with $e_1 = E_{in}$ and $e_7 = E_{out}$. It may be noted that across the resistive element (R) there is drop in the chemical potential but no accumulation of the permeant. Hence the resistive element is connected through a parallel or '1'-junction.

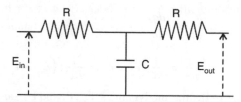

Fig. 12.6   (b) Network equivalent of the diffusion process

On the other hand, the mass flow splits when it passes through the capacitor and therefore, the capacitor is coupled by a series or '0'-junction.

$$E_{in} \xrightarrow[(e_1, f_1)]{} 1 \xrightarrow[(e_3, f_3)]{\substack{R \\ \uparrow \\ (e_2, f_2)}} 0 \xrightarrow[(e_5, f_5)]{\substack{C \\ \uparrow \\ (e_4, f_4)}} 1 \xrightarrow[(e_7, f_7)]{\substack{R \\ \uparrow \\ (e_6, f_6)}} E_{out}$$

Fig. 12.6   (c) Bond graph representation of diffusion through the membrane

The constitutive relation between the membrane capacitance C, the force e_4 and the generalized displacement q_4 may be written as:

$$Ce_4 = q_4 \qquad\qquad 12.63$$

So that

$$C\frac{de_4}{dt} = \frac{dq_4}{dt} \equiv f_4 \qquad\qquad 12.64$$

Also, from the definition of a '0'-junction it follows that:

$$f_3 = f_4 + f_5 \ or \ f_4 = f_3 - f_5 \qquad\qquad 12.65$$

Similarly, from the definition of a '1'-junction:

$$f_1 = f_2 = f_3 \ and \ f_5 = f_6 = f_7 \qquad\qquad 12.67$$

And,

$$e_1 = e_2 + e_3 \ and \ e_5 = e_6 + e_7$$

Or

$$e_2 = E_{in} - e_3 \ and \ e_6 = e_5 - E_{out} \qquad\qquad 12.68$$

where we have made use of the fact that $e_1 = E_{in}$ and $e_7 = E_{out}$

Further, it follows from the dissipative flow through resistive elements that:

$$f_2 = \frac{e_2}{R} \ and \ f_6 = \frac{e_6}{R} \qquad\qquad 12.69$$

Now from Eqs. 12.65, 12.67, and 12.69 it follows that:

$$f_4 = f_3 - f_5 = f_2 - f_6 = \frac{e_2}{R} - \frac{e_6}{R}$$ 12.70

But from Eqs. 12.64 and 12.70

$$C\frac{de_4}{dt} = \frac{dq_4}{dt} \equiv f_4 = \frac{e_2}{R} - \frac{e_6}{R}$$

Or $$\frac{de_4}{dt} = \frac{1}{RC}\left[e_2 - e_6\right] = \frac{1}{RC}\left[\left(E_{in} - e_3\right) - \left(e_5 - E_{out}\right)\right]$$ 12.71

Now from the point of view of the thermodynamics, E_{in}, e_4 and E_{out} are respectively the chemical potential of the part-I of the container, chemical potential at the membrane, which we now denote by $\mu_m (= e_4 = e_3 = e_5)$ and chemical potential of part-II of the container.

It is convenient to define the mean chemical potential $\langle \mu \rangle$ of the container as:

$$\langle \mu \rangle = \frac{E_{in} + E_{out}}{2}$$ 12.72

Using the above relations, one gets:

$$\frac{d\mu_m}{dt} = \frac{2}{RC}\left[\langle \mu \rangle - \mu_m\right] \equiv \frac{\left(\langle \mu \rangle - \mu_m\right)}{\tau_m}$$ 12.73

$\tau_m = \frac{RC}{2}$ in Eq. 12.73 is the relaxation time of the membrane process and as expected has the form of the RC time constant of electrical network theory. The magnitude of the resistive element R of the system may be experimentally determined at steady state and the magnitude of the Capacitative element, C, at equilibrium, only equilibrium and steady state measurements are required to completely characterize the dynamic properties of the system in network thermodynamics. Further, it may be shown that:

$$\tau_m = \frac{(\Delta x)^2}{2D_m}$$

Here Δx is the thickness of the membrane and D_m is the diffusion coefficient of the permeant within the membrane.

Short Answer Questions

1. Without deriving, write expressions for the relaxation time for a single permeant through a linear symmetrical membrane in terms of the chemical potentials and in terms of the thickness and the diffusion coefficient of the permeant through the membrane.

2. Define the properties of (i) a '0'-junction, and (ii) a '1'-junction used in bond representation of network thermodynamics.

3. Briefly discuss how energy interactions of a system are described in network thermodynamics.

4. Define the transducer element used in network thermodynamics.

5. State the postulates of the network thermodynamics.

6. Discuss the statement 'entropy is internally generated in every irreversible process'.

7. Write a brief note on linear and non-linear irreversible processes.

8. Distinguish between a reversible and an irreversible process in steady state.

9. State and explain Onsager's theorem.

10. Write a note on Prigogine's theorem of minimum entropy generation in steady state.

Long Answer Questions

1. Give a brief description of Onsager's methodology for treating linear irreversible processes and apply it to obtain Seebeck coefficient of thermoelectricity.

2. State and Explain: (a) Onsager's theorem and (b) Prigogine's theorem of minimum entropy generation in steady state, with reference to linear irreversible processes.

3. Write a detailed note on the need and postulates of the network thermodynamics.

4. Explain the following terms as used in network thermodynamics;

   (a) Port (b) generalized displacement and momentum (c) displacement and inductive energy stores (d) resistive energy dissipation and (e) the transducer and its
      properties.

5. Write a note on bond notation representation in network thermodynamics.

6. Give an account of the application of network thermodynamics to the problem of diffusion of a single fluid through a linear membrane.

7. Take the example of thermoelectricity to explain how linear irreversible process may be analyzed.

Multiple Choice Questions

Note: Some of the following questions may have more than one correct alternative. All correct alternatives must be selected for the complete answer of the question in such cases.

1. In bond representation of network thermodynamics the relation $f_1 = f_2 = f_3 \ldots\ldots = f_n$, holds for:
   (a) Series junction
   (b) Parallel junction
   (c) '0'-junction
   (d) '1'-junction

2. In network thermodynamics a port represents:
   (a) Energy interaction with the system
   (b) Inflow of flux into the system
   (c) Out flow of flux from the system
   (d) A single terminal element

3. The dynamic variables of a diathermic port in network thermodynamics are:
   (a) P, v
   (b) T, s
   (c) P, s
   (d) v, s

4. The rate of generation of entropy is a minimum in the steady state of a
   (a) non-linear irreversible process
   (b) non-linear reversible process
   (c) linear irreversible process
   (d) linear reversible process.

5. Tick the correct alternative that holds for equilibrium network thermodynamics,

(a) $\dfrac{\partial U}{\partial e_i} = q_i$

(b) $\dfrac{\partial U}{\partial q_i} = e_i$

(c) $\dfrac{\partial q_i}{\partial U} = e_i$

(d) $\dfrac{\partial e_i}{\partial U} = q_i$

Answers to Multiple Choice Questions

1. (a), (d)   2. (a)   3. (b)   4. (c)
5. (b)

Revision

1. Both classical and quantum thermodynamics deal essentially with systems that are in equilibrium, and ,therefore, a more appropriate name for this subject may be thermo-statics rather than thermodynamics. However, as a convention the study of the equilibrium systems is still called thermodynamics.

2. Actual processes in Nature are neither reversible nor in such reactions the system ever pass through a sequence of quasi-equilibrium states. All natural processes are irreversible processes in which there is dissipation of energy. The study of such irreversible processes is true thermodynamic as such processes are accompanied with the flow of some physical quantity.

3. Whenever some irreversible process takes place in an isolated system and its environment the entropy of the system increases. This means that entropy is generated in irreversible processes. The rate of generation of entropy is denoted by $\sigma = \dfrac{(dS)_{sys}^{int}}{dt}$. Sigma plays important role in irreversible processes. If σ is large the process is fast and if it is small the process proceeds slowly. In an irreversible process energy is continuously dissipated and entropy is continuously created.

4. An irreversible process is always accompanied by the flow of matter or energy or both. Thus, every irreversible process is associated with the flow of some physical quantity which in the language of thermodynamics is called 'flux" or "current' and is denoted by J_i. Since there may be many different types of fluxes in a given system undergoing some irreversible process, they may be distinguished using the subscript í. Every flux may be assigned a source that is responsible for the flux. The source of a flux is termed as 'affinity', 'potential' or 'driving force' and is denoted by X_i. In irreversible processes it often happens that a particular flux is generated by two or more different sources or affinities. This dependence of a particular flux on different affinities may be represented by the following mathematical relation,

$$J_i = J_i(X_1, X_2, X_3, \dots)$$

In the case of irreversible processes that occur in a system which is not far away from its equilibrium state, affinities or the flux driving forces X_1, X_2, X_3, are not too large. In such cases $J_i = J_i(X_1, X_2, X_3, \dots)$ may be expanded using Taylor's expansion and one may retain only the first order (or linear) terms neglecting the higher order (or non-linear) terms in the expansion. One may then write the above equation as:

$$J_i(X_1, X_2, X_3, \dots) = \left(\frac{\partial J_i}{\partial X_1}\right)_0 X_1 + \left(\frac{\partial J_i}{\partial X_2}\right)_0 X_2 + \left(\frac{\partial J_i}{\partial X_3}\right)_0 X_3 \dots$$

Or
$$J_i(X_1, X_2, X_3, \dots) = \sum_{j=1}^{n} \left(\frac{\partial J_i}{\partial X_j}\right)_0 X_j \qquad 12.7$$

Here $\left(\dfrac{\partial J_i}{\partial X_j}\right)_0$ is the rate of change of J_i with X_j when system approaches the equilibrium state

which may be written as,

$$J_i = \sum_{j=1}^{n} L_{ij} X_j \; ; \text{ where } L_{ij} = \left(\frac{\partial J_i}{\partial X_j}\right)_0$$

Irreversible processes that may be described by the linear terms of Taylor's expansion, as given by the above relations, are called linear irreversible processes. However, when the system is far away from the corresponding equilibrium state, the affinities and fluxes are sufficiently large and higher order terms in Taylor's expansion cannot be neglected. Irreversible process taking place in such systems are called non-linear irreversible processes. Onsager developed thermodynamics for linear irreversible processes. Onsager used these linear phenomenological relations to describe irreversible processes containing n-fluxes and n-affinities in terms of the coefficients L_{ij},

$$J_n = L_{n1} X_1 + L_{n2} X_2 + L_{n3} X_3 + \dots L_{nn} X_n$$

The flux deriving forces or affinities, X_i may also be written in terms of the fluxes J_i as follows,

$$X_i = \sum_{j=1}^{n} R_{ij} J_j, \qquad i = 1, 2, 3, \dots\dots n$$

Coefficients like $L_{11}, L_{22}, L_{33} \dots L_{nn}$ and $R_{11}, R_{22}, R_{33} \dots R_{nn}$ represent interactions between the same type of fluxes and affinities and are, therefore, called 'Straight coefficients'. On the other hand, coefficients like L_{ij} and R_{ij} ; $i \neq j$ represent interactions between different fluxes and affinities and are called 'Cross coefficients'.

5. Onsager's theorem: The theorem says that if the currents and affinities are chosen properly the cross coefficients are symmetrical, i.e.,

$$L_{21} = L_{12} \text{ and } R_{21} = R_{12} \text{ or } L_{ij} = L_{ji} \text{ and } R_{ij} = R_{ji}$$

The above equation represents Onsager's reciprocal relation of linear coefficients.

The choice of proper affinities and currents that satisfy Onsager's reciprocal relations are those which satisfy the following relation of the rate of generation of internal entropy;

$$\sigma = \frac{(dS)_{sys}^{int}}{dt} = \sum_{i=1}^{n} J_i X_i$$

Onsager used the following Gibb's equation for calculating the rate of internal energy generation,

$$T(ds)_{sys}^{int} = du + P\,dv - \sum_i \mu_i c_i$$

Here μ_i and c_i are respectively, the chemical potential and the concentration of the ith component of the system.

6. A time invariant or steady state is different from an equilibrium state. In equilibrium state energy is minimum, entropy maximum and there is no change in the entropy of the state with time, at constant volume. There are no fluxes and affinities in an equilibrium state. On the other hand in a steady state or time invariant state, entropy is generated at constant rate which may be either maximum or minimum. As a matter of fact a steady state is more general state and equilibrium state is a special case of steady state.

7. Prigogine's theorem of minimum entropy generation in steady state: Prigogine's theorem states that the rate of production of entropy in a non-equilibrium steady state of a system is a minimum.

8. Order of a steady state: In a general case a linear irreversible process may have n-fluxes and n-affinities, out of which some m-fluxes are kept constant so that the remaining (n–m) fluxes adjust themselves to obtain steady state. The steady state so obtained is called a steady state of mth order. It follows from this definition that the equilibrium state is a steady state of zeroth order.

9. Application of Onsager's method to thermoelectricity: Free electrons in a metal may be treated as a perfect gas, but of charged fermions, i.e., electrons. The equivalent of partial pressure in an electron gas is the *electrochemical potential* μ_e given by,

$$\mu_e = qV + \mu_c$$

where q is the charge of the carrier particle (electron), V electric potential and μ_c the chemical potential, i.e., the energy needed to give one carrier particle (electron) to the system.

The system is essentially a non-equilibrium system in steady state, but as assumed by Onsager and Prigogine the Gibb's equation for the system holds good in the steady state and, therefore,

$$dG = dU - T\,dS - S\,dT - N\,d\mu_e - \mu_e\,dN$$

If δQ and δW are respectively the change in the heat and the work done by the system, then

$$\delta Q = T\,dS \quad and \quad \delta W = \mu_e\,dN$$

Here U is the internal energy and N the number of carriers (free electrons) in the system. Also, the change in the internal energy

$$dU = \delta Q + \delta W = T\,dS + \mu_e\,dN$$

Substituting the value of dU from above one gets,

$$dG = -S\,dT - N\,d\mu_e$$

And

$$S = -\left(\frac{\partial G}{\partial T}\right)_{constant\ \mu_e} \quad and \quad N = -\left(\frac{\partial G}{\partial \mu_e}\right)_{constant\ T}$$

If $\overrightarrow{J_N}, \overrightarrow{J_S}, \overrightarrow{J_Q}$ and $\overrightarrow{J_e}$ respectively represent particle flux, entropy flux, heat flux and electrical flux (electric current density) then using Eq. 12.27:

$$\overrightarrow{J_Q} = T\overrightarrow{J_S} \text{ and } \overrightarrow{J_e} = q\overrightarrow{J_N} \qquad\qquad 12.31$$

Here q is the electric charge of the carrier particles (which in this case are free electrons in the metal each with charge -e).

The driving forces or affinities for $\overrightarrow{J_N}$ (or $\overrightarrow{J_e}$) and $\overrightarrow{J_S}$ (or $\overrightarrow{J_Q}$) are respectively the difference in electrochemical potential $\Delta\mu_e$ and the difference in the temperature ΔT at the two ends of the conductor. If $\vec{E}$ denotes the intensity of the electric field generated because of $\Delta\mu_e$ then the force (or affinity) F_e experienced by a carrier particle of charge q is,

$$\overrightarrow{F_e} = q\vec{E}$$

Further, the force F_T or affinity due to the temperature difference ΔT may be written as:

$$\overrightarrow{F_T} = -\overrightarrow{\nabla T}$$

Using Onsager method one may write the coupled equations for the particle flux and entropy flux in the following determinant form,

$$\begin{bmatrix} \overrightarrow{J_N} \\ \overrightarrow{J_S} \end{bmatrix} = \begin{bmatrix} L_{11} & L_{12} \\ L_{21} & L_{22} \end{bmatrix} \begin{bmatrix} \overrightarrow{F_e} \\ \overrightarrow{F_T} \end{bmatrix}$$

The same may also be written a

$$\begin{bmatrix} \dfrac{\overrightarrow{J_e}}{q} \\ \dfrac{\overrightarrow{J_Q}}{T} \end{bmatrix} = \begin{bmatrix} L_{11} & L_{12} \\ L_{21} & L_{22} \end{bmatrix} \begin{bmatrix} q\vec{E} \\ -\overrightarrow{\nabla T} \end{bmatrix}$$

Or

$$\overrightarrow{J_e} = q^2 L_{11}\vec{E} + qL_{12}\left(-\overrightarrow{\nabla T}\right)$$

and

$$\overrightarrow{J_Q} = qTL_{21}\vec{E} + TL_{22}\left(-\overrightarrow{\nabla T}\right)$$

In steady state particle flux or electric flux $\overrightarrow{J_e} = 0$; that means;

$$q^2 L_{11}\vec{E} = qL_{12}\left(\overrightarrow{\nabla T}\right)$$

The ratio of the electric field to the gradient of temperature when there is steady state and no particle or electric flux i.e., $\left(\dfrac{E}{\nabla T}\right)_{no\ particle\ or\ electric\ flux}$ is called *Seebeck coefficient*, denoted by α. It follows that,

$$\alpha = \left(\dfrac{E}{\nabla T}\right)_{no\ particle\ or\ electric\ flux} = \dfrac{L_{12}}{qL_{11}}$$

The terms $qL_{12}\left(-\overrightarrow{\nabla T}\right)$ and $TL_{22}\left(-\overrightarrow{\nabla T}\right)$ respectively arise because of the externally imposed temperature difference between the two ends of a conductor. However, if there is no such externally imposed temperature difference on the conductor but an electric flux $\overrightarrow{J_e}$ is made to flow through the conductor by applying an external electrostatic potential difference or electric field E between the ends of the conductor, a temperature difference will be created between the ends of the conductor. This generation of temperature difference between the two ends of a conductor when an electric current is passed through it, is called *Peltier effect*. The ratio of the electric flux to the heat flux under the condition of $\left(-\overrightarrow{\nabla T}\right) = 0$ is called Peltier coefficient which is represented by Π.

So,
$$\Pi = \left(\frac{J_Q}{J_e}\right)_{(\overrightarrow{\nabla T}=0)} = T\frac{L_{12}}{qL_{11}}$$

It is easy to see that
$$\Pi = T\frac{L_{12}}{qL_{11}} = T\pm$$

Experimental verification of relation 12.41 proves the correctness of the method of linear irreversible process developed by Onsager.

10. Network thermodynamics: Both classical thermodynamic theories and Onsager's approach of linear irreversible processes are not adequate to treat complex biological processes where reversible, linear irreversible and non-linear irreversible processes proceed simultaneously. A new approach based on the analogy of non-equilibrium systems and electrical networks, has been developed, particularly for biological systems, to describe reversible, linear irreversible and non-linear irreversible processes occurring simultaneously in a system. Since it uses elements of the electrical network theory, it is called the network thermodynamics. Network thermodynamics developed out of the synthesis of concepts based on general theory of linear passive systems, thermodynamics, circuit theory, graph theory and differential geometry. A lumped parameter approximation to the field equations and continuum mechanics gives rise to the network thermodynamics.

11. In network thermodynamics large and complex non-linear systems are decomposing into smaller subsystems interconnected with each other. The simpler subsystems are then separately solved and their solutions are suitable coupled to get the solution of the parent complex system. Coupling of the solutions of subsystems depends on how they are connected in the complex system. This brings in consideration the topology of the problem, i.e., how the subsystems are interconnected. Topology of the system is very important.

    Postulates of network thermodynamics: (a) *Local phase equilibrium*: Although in the case of irreversible process the system is not in equilibrium, but in network thermodynamics it is assumed that at the local level the system may be described in terms of the thermodynamic variables like temperature, pressure and chemical potential. (b) *Discrete space*: Network thermodynamics assumes discrete structure for the space which may be described by simple differentials and algebraic topology. (c) *Conceptual separation of reversible and irreversible processes*: Although in every volume element of the complex system both reversible and irreversible processes may be going on simultaneously, but conceptually one decouples them into independent reversible and irreversible components.

12. Terminology of network thermodynamics: In network thermodynamics a system is regarded as a black box and any process of interaction in the system that may change the energy of the system is called a port. A port is represented by a line. Each port has two variables attached to

it and, therefore, a port is treated as a two terminal network. Pressure P and volume V are the two variables associated with a mechanical port, temperature T and entropy S are the variables of a diathermic boundary port and number of particles N and the chemical potential μ that of a membrane port. An adiabatic boundary is not a port as neither energy nor matter can flow across it.

13. Equilibrium network thermodynamics: In equilibrium network thermodynamics the internal energy U is assumed to be a function of generalized displacements q_i. The partial derivative $\frac{\partial U}{\partial q_i} \equiv e_i$ defines the potential conjugate to displacement q_i. Conjugate potentials e_i are also called efforts. Thus in equilibrium network thermodynamics q_i and e_i are primitive variables

$$U = U\,(q_i, q_2, q_3 \cdots q_n)$$

And

$$\frac{\partial U}{\partial q_1} \equiv e_1, \frac{\partial U}{\partial q_2} \equiv e_2, \frac{\partial U}{\partial q_3} \equiv e_3, \cdots \frac{\partial U}{\partial q_n} \equiv e_n$$

Gibb's equation in case of network thermodynamics may be written as,

$$dU = \sum_{i=1}^{n} \frac{\partial U}{\partial q_i} dq_i \equiv TdS - pdV + \sum_i \mu_i dN_i$$

14. Network approach may be applied only to those systems for which the rate of energy transmission; that may be energy loss (dissipation) or energy storage; is finite and may be written as the product of an 'effort' (or force variable) e, and a 'flow' variable f, i.e.,

energy rate or power P = e.f

These effort and flow variables have different names in different processes, for example in case of electrical network, effort is electric potential difference and flow is current; in mechanical flow it is force and velocity, in chemical reactions affinity and reaction rate and in diffusion process, the difference in chemical potential and mass flow.

15. The generalized momentum and the generalized displacement: Two additional variables, called generalized momentum $P(t) \equiv P(0) + \int_0^t e(t)dt$ and generalized displacement

$q(t) \equiv q(0) + \int_0^t f(t)dt$ are also defined to retain the dynamic nature of the problem.

16. Retaining the analogy between the electrical networks and the network thermodynamics, generalized momentum P(t) or effort e(t) obeys continuity law equivalent to Kirchhoff's voltage law (KVL) regarding the continuity of the function in a closed loop, while the generalized displacement q(t) or flow f(t) obeys the law of conservation equivalent of Kirchhoff's current law (KCL).

17. Three types of (energy rate or power) interactions: In network thermodynamics pure capacitor, pure inductor and pure resistive elements (responsible for reversible processes) are defined through the integration relation
$P = e.f = \sum_i e_i f_i = \int e\,f$ which may be evaluated in the following three different ways, each giving rise to a different type of energy interaction:

(a) Capacitative or displacement energy store: When integration of the energy rate is done in the following way It defines a capacitive energy store as, $\in_{cap}(t) \equiv \int_0^t e.fdt = \int_{q(0)}^{q(t)} e.dq$ in this case constitutive relation between the generalized displacement q and the effort e has the form, $q = \psi_c(e)$ and $\dfrac{dq}{dt} = \dfrac{d\psi_c}{de}\dfrac{de}{dt}$

The reversible flow on the capacitor f_{rev} and the incremental capacitance C is then defined as;

$$f_{rev} \equiv \frac{dq}{dt} \ and \ C \equiv \frac{d\psi_c}{de}$$

(b) Inductive or kinetic energy store: On the other hand if integration is carried out between the initial and final values of the generalized momentum, the resulting interaction gives rise to inductive or kinetic energy store;

$$\in_{ind}(t) \equiv \int_0^t f.edt = \int_{P(0)}^{P(t)} f.dP$$

That gives the energy stored as a result of the relative motion of mass or charge. Evaluation of the above integral requires a constitutive relation between generalized momentum P and the flow f. This constitutive relation may be written as,

$$P = \psi_L(f)$$

This type of energy storage device is called an inductor and is denoted by L.

(c) Resistive energy dissipation: Time integral of the multiplication of electric potential difference and the current in electrical networks give the energy dissipated by a resistive element. In a similar way, energy dissipation via an equivalent resistive element in network thermodynamics is given by the following integral;

$$\in_{res}(t) \equiv \int_0^t e.fdt$$

To compute the energy dissipation in a resistance like device a constitutive relationship between effort e and flow f is required. This relationship may be of the form,

$$e = \psi_R(f)$$

The incremental resistance R may then be defined as,

$$R \equiv \frac{\partial \psi_R}{\partial f}$$

18. The bond graph notation: Bond graphs treat all energy flows on equal footing and, therefore, they may be used as notations for energy conservation and coupling. Two kinds of junctions are used in bond graph notations to couple the elements of the system. They are:

(a) *The parallel junction or the '0' junction:* For a '0'-junction, applying Kirchhoff's Current Law one obtains;

$$e_1 = e_2 = e_3 \dots = e_n \ or \ \sum_i f_i = 0$$

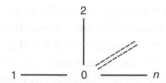

Fig. 12.4   Notation for '0'-junction

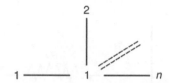

Fig. 12.5   Notation for '1'-junction

(b) *The series junction or '1'-junction:* This junction is denoted by the notation given in Fig. 12.5. For a '1'-junction

$$f_1 = f_2 = f_3 \ldots = f_n \text{ or } \sum_i e_i = 0$$

(c) The bonds connecting the network elements to a junction are denoted by half arrows ($\rightarrow$) that also indicate the direction of energy flow. Each bond is also associated with an effort variable and a flow variable. For example, $\begin{bmatrix} e_1 \\ \rightarrow C \\ f_1 \end{bmatrix}$ represents energy flow into the capacitive element while, $\begin{bmatrix} e_2 \\ \leftarrow R \\ f_2 \end{bmatrix}$ represents the energy flow out from a resistive element.

(d) *The transducer device (TD):* An additional ideal two port device called transducer is also often used in bond notation of network thermodynamics. The compact notation for the transducer is: $\begin{bmatrix} e_1\ (r)\ e_2 \\ \rightarrow TD \rightarrow \\ f_1 \qquad f_2 \end{bmatrix}$, where (r), called the transfer ratio or transducer modulus provides the relationships between the input and output variables, i.e.,

$$e_1 = \frac{e_2}{r} \text{ and } f_1 = rf_2$$

19. Application of network thermodynamics to the diffusion of a fluid through a membrane:
For simplicity we are taking a single permeant in part-I of the container that diffuses through the membrane in to part II. Let us conceptually divide the membrane into small volumes and consider one of these small volumes. Thus saturation of the membrane and leaving of the attached permeant from the membrane at the end of the flow may be well compared to the charging and discharging of the capacitor. The membrane and the two compartments -I and II

have the property of holding the permeant, and in this respect they behave like capacitors. The equivalent capacitance of the three components may be lumped together and may be represented by a single capacitance C. It may further be realized that diffusion through membrane is a dissipative process, i.e., energy is lost in the process. In bond graph representation dissipation is included by assuming resistive elements both in the input and on the output sides. Figures 12.6 (a), (b) and (c) respectively shows the diffusion of the fluid through a membrane, network equivalent of the diffusion process and the bond graph representation of it.

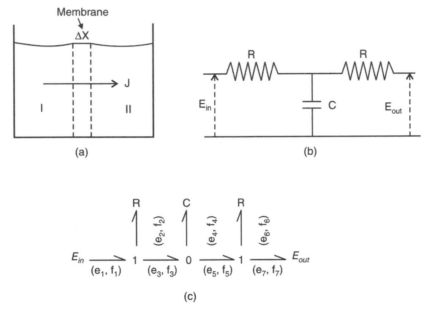

Fig. 12.6   (a) Diffusion of a fluid through a membrane; (b) Network equivalent of the diffusion process; (c) Bond graph representation of diffusion through the membrane

The constitutive relation between the membrane capacitance C, the force e_4 and the the generalized displacement q_4 may be written as:

$$Ce_4 = q_4 \qquad \qquad 12.63$$

So that

$$C\frac{de_4}{dt} = \frac{dq_4}{dt} \equiv f_4 \qquad \qquad 12.64$$

Also, from the definition of a '0'-junction it follows that:

$$f_3 = f_4 + f_5 \ or \ f_4 = f_3 - f_5 \qquad \qquad 12.65$$

Similarly, from the definition of a '1'-junction:

$$f_1 = f_2 = f_3 \ and \ f_5 = f_6 = f_7 \qquad \qquad 12.67$$

And,

$$e_1 = e_2 + e_3 \ and \ e_5 = e_6 + e_7$$

Or
$$e_1 = E_{in} - e_3 \quad and \quad e_6 = e_5 - E_{out} \qquad \text{12.68}$$

Where we have made use of the fact that $e_1 = E_{in} \quad and \, e_7 = E_{out}$

Further, it follows from the dissipative flow through resistive elements that

$$f_2 = \frac{e_2}{R} \quad and \quad f_6 = \frac{e_6}{R} \qquad \text{12.69}$$

Now from Eqs.12.65, 12.67 and 12.69 it follows that

$$f_4 = f_3 - f_5 = f_2 - f_6 = \frac{e_2}{R} - \frac{e_6}{R} \qquad \text{12.70}$$

But from Eqs.12.64 and 12.70

$$C\frac{de_4}{dt} \equiv \frac{dq_4}{dt} \equiv f_4 = \frac{e_2}{R} - \frac{e_6}{R}$$

Or
$$\frac{de_4}{dt} = \frac{1}{RC}\left[e_2 - e_6\right] = \frac{1}{RC}\left[\left(E_{in} - e_3\right) - \left(e_5 - E_{out}\right)\right] \qquad \text{12.71}$$

Now from the point of view of the thermodynamics, E_{in}, e_4 and E_{out} are respectively the chemical potential of the part-I of the container, chemical potential at the membrane, which we now denote by $\mu_m \left(= e_4 = e_3 = e_5\right)$ and chemical potential of part-II of the container.

It is convenient to define the mean chemical potential $<\mu>$ of the container as:

$$<\mu> = \frac{E_{in} + E_{out}}{2} \qquad \text{12.72}$$

Using the above relations, one gets

$$\frac{d\mu_m}{dt} = \frac{2}{RC}\left[<\mu> -\mu_m\right] \equiv \frac{\left(<\mu> -\mu_m\right)}{\tau_m} \qquad \text{12.73}$$

$\tau_m = \dfrac{RC}{2}$ in Eq. 12.73 is the relaxation time of the membrane process and as expected has the

form of the RC time constant of electrical network theory. The magnitude of the resistive element R of the system may be experimentally determined at steady state and the magnitude of the capacitive element, C, at equilibrium, only equilibrium and steady state measurements are required to completely characterize the dynamic properties of the system in network thermodynamics. Further, it may be shown that

$$\tau_m = \frac{\left(\Delta x\right)^2}{2D_m}$$

Here Δx is the thickness of the membrane and D_m is the diffusion coefficient of the permeant within the membrane.

Index